Equations, Data, and
Rules of Thumb

About the Author

Arthur A. Bell, Jr., PE, is a registered professional engineer and certified mechanical, plumbing, fire, and energy conservation code official in the Commonwealth of Pennsylvania with more than 31 years of experience in the design of HVAC systems. In addition, he has been involved in the design of plumbing systems, fire protection systems, and construction field engineering-mechanical systems. Mr. Bell is also the author of *HVAC Design Portfolio*, published by McGraw-Hill.

W. Larsen Angel, PE, is a registered professional engineer in the State of Maryland with more than 25 years of experience in the design of HVAC systems. In addition, he has been involved in the design of plumbing systems and electrical systems. Mr. Angel is also the author of *HVAC Design Sourcebook*, published by McGraw-Hill.

HVAC
Equations, Data, and Rules of Thumb

Arthur A. Bell, Jr., PE

W. Larsen Angel, PE

Third Edition

Mc Graw Hill Education

New York   Chicago   San Francisco   Athens
London   Madrid   Mexico City   Milan
New Delhi   Singapore   Sydney   Toronto

Library of Congress Cataloging-in-Publication Data

Bell, Arthur A.
    HVAC equations, data, and rules of thumb / Arthur A. Bell, Jr., W. Larsen Angel—
3rd ed.
        p.   cm.
    Includes bibliographical references and index.
    ISBN 978-0-07-148242-4 (alk. paper)
        1. Heating—Mathematics—Handbooks, manuals, etc.   2. Ventilation—
Mathematics—Handbooks, manuals, etc.   3. Air conditioning—
Mathematics—Handbooks, manuals, etc.   4. Engineering mathematics—
Formulae—Handbooks, manuals, etc.   I. Title.
TH7225.B45   2008
697.000291—dc22   2007025494

1 2 3 4 5 6 7 8 9 0   DOC/DOC   10 9 8 7 6 5

ISBN 978-0-07-182959-5
MHID 0-07-182959-8

Sponsoring Editor: Lauren Poplawski
Production Supervisor: Pamela A. Pelton
Editing Supervisor: Donna M. Martone
Project Manager: Tanya Punj, Cenveo® Publisher Services
Copy Editor: Cenveo Publisher Services
Proofreader: Cenveo Publisher Services
Art Director, Cover: Jeff Weeks
Composition: Cenveo Publisher Services

Figures 41.13 and 41.14 are courtesy of Cummins-Wagner, Inc.

Printed and bound by RR Donnelley.

McGraw-Hill books are available at special quantity discounts to use as premiums and sales promotions, or for use in corporate training programs. For more information, please write to the Director of Special Sales, McGraw-Hill Professional, Two Penn Plaza, New York, NY 10121-2298. Or contact your local bookstore.

This book is printed on acid-free paper.

To Lisa, my lovely wife of twenty-five years, and to my two wonderful sons. *Soli Deo gloria*

Contents

Additional bonus content for select chapters can be found at www.mheducation.com/ HVACequations.

*This chapter is available online.

Introduction

1.01 Background

A. The heating, ventilation, and air conditioning (HVAC) equations, data, rules of thumb, and other information contained within this reference manual were assembled to aid the beginning engineer and designer in the design of HVAC systems. In addition, the experienced engineer or designer may find this manual useful as a quick design reference guide, field manual, and teaching tool.

B. The following pages compile information from various reference sources listed in Part 53 of this manual, from college HVAC class notes, from continuing education design seminars and classes, from engineers, and from personal experience. This document was put together as an encyclopedic type reference in contract specification outline format where information could be looked up quickly, in lieu of searching through volumes of textbooks, reference books and manuals, periodicals, trade articles, and product catalogs.

1.02 Rules of Thumb

A. Rules of thumb listed herein should be used considering the following:

1. Building loads are based on building gross square footage.
2. Building loads generally include ventilation and make-up air requirements.
3. These rules of thumb may be used to estimate system loads during the preliminary design stages of a project.
4. Building loads for construction documents should be calculated using the *ASHRAE Handbook of Fundamentals* or similar computational procedure in lieu of using these rules of thumb for final designs. When calculating heating and cooling loads, actual occupancy, lighting, and equipment information should be obtained from the owner, architect, electrical engineer, other design team members, or from technical publications such as ASHRAE.

B. Many of the rules of thumb listed within this reference manual were developed many years ago. I have received many questions when conducting seminars regarding these rules of thumb. The most often asked question is "Are the cooling and heating load rules of thumb still accurate with the mandate of energy codes and tighter and improved building envelope construction?" The answer to this question is yes. The reason the cooling rules of thumb are still accurate is that the internal loads have increased substantially and cooling loads have switched from building-envelope-dependent, to lighting-dependent, and now to people-and-equipment-dependent (more people and equipment placed in the same area). The reason the heating load rules of thumb are still reasonably accurate is that the ventilation air (outdoor air load dictated by code) has increased.

1.03 Codes and Standards

A. Code items contained herein were included more for comparison purposes than for use during design. All code items (i.e., ICC, ASHRAE, NFPA) are subject to change, and federal, state, and local codes should be consulted for applicable regulations and requirements.

B. The following codes were used unless otherwise noted.

1. 2015 International Code Council Series of Codes (ICC):
   a. 2015 International Building Code (herein referred to as 2015 IBC).
   b. 2015 International Mechanical Code (herein referred to as 2015 IMC).
   c. 2015 International Energy Conservation Code (herein referred to as 2015 IECC).
   d. 2015 International Plumbing Code (herein referred to as 2015 IPC).
   e. 2015 International Fire Code (herein referred to as 2015 IFC).
   f. 2015 International Fuel Gas Code (herein referred to as 2015 IFGC).
   g. 2015 International Residential Code (herein referred to as 2015 IRC).
   h. 2015 International Existing Building Code.
   i. 2015 International Performance Code for Buildings and Facilities.
   j. 2015 International Private Sewage Disposal Code.
   k. 2015 International Property Maintenance Code.
   l. 2015 International Zoning Code.
   m. 2015 International Wildland-Urban Interface Code.
2. American Society of Heating, Refrigerating, and Air Conditioning Engineers (ASHRAE):
   a. ASHRAE Standard 15—2013 (herein referred to as ASHRAE 15-2013).
   b. ASHRAE Standard 55—2013 (herein referred to as ASHRAE 55-2013).
   c. ASHRAE Standard 62.1—2013 (herein referred to as ASHRAE 62.1-2013).
   d. ASHRAE Standard 62.2—2013 (herein referred to as ASHRAE 62.2-2013).
   e. ASHRAE Standard 90.1—2013 (herein referred to as ASHRAE 90.1-2013).
   f. ASHRAE Standard 90.2—2007 (herein referred to as ASHRAE 90.2-2007).
3. National Fire Protection Association Codes (NFPA):
   a. NFPA 90A—2015 Installation of Air-Conditioning and Ventilating Systems.
   b. NFPA 96—2014 Ventilation Control and Fire Protection of Commercial Cooking Operations.

Definitions

2.01 General

A. *Furnish.* Except as otherwise defined in greater detail, the term *furnish* is used to mean "supply and deliver to the project site, ready for unloading, unpacking, assembly, installation, and similar operations" as applicable to each instance.

B. *Install.* Except as otherwise defined in greater detail, the term *install* is used to describe operations at the project site including actual "unloading, unpacking, assembly, erection, placing, anchoring, connecting, applying, working to dimension, finishing, curing, protecting, testing to demonstrate satisfactory operation, cleaning, and similar operations" as applicable in each instance.

C. *Provide.* Except as otherwise defined in greater detail, the term *provide* means to furnish and install, complete and ready for intended use and successfully tested to demonstrate satisfactory operation as applicable in each instance.

D. *Remove.* Except as otherwise defined in greater detail, the term *remove* means to disassemble, dismantle, and/or cut into pieces in order to remove the equipment from the site and to properly dispose of the removed equipment and pay for all associated costs incurred.

E. *Replace.* Except as otherwise defined in greater detail, the term *replace* means to remove the existing equipment and to provide new equipment of the same size, capacity, electrical characteristics, function, etc., as the existing equipment.

F. *Relocate.* Except as otherwise defined in greater detail, the term *relocate* means to carefully remove without damaging item and to install where shown on the contract documents and/or as directed by the design professional and/or owner.

G. *Shall. Shall* indicates action that is mandatory on the part of the contractor.

H. *Will. Will* indicates action that is probable on the part of the contractor.

I. *Should. Should* indicates action that is probable on the part of the contractor.

J. *May. May* indicates action that is permissible on the part of the contractor.

K. *Indicated.* The term *indicated* is a cross-reference to graphic representations, details, notes, or schedules on the drawings; to other paragraphs or schedules in the specifications; and to similar means of recording requirements in the contract documents. Where terms such as *shown, noted, scheduled,* and *specified* are used in lieu of *indicated,* it is for the purpose of helping the reader locate the cross-reference, and no limitation is intended except as specifically noted.

L. *Shown.* The term *shown* is a cross-reference to graphic representations, details, notes, or schedules on the contract drawings and to similar means of recording requirements in the contract documents.

M. *Detailed.* The term *detailed* is a cross-reference to graphic representations, details, notes, or schedules on the contract drawings and to similar means of recording requirements in the contract documents.

N. *Specified.* The term *specified* is a cross-reference to paragraphs or schedules in the specifications and to similar means of recording requirements in the contract documents. The specifications include the general provisions, special provisions, and the technical specifications for the project.

O. *Including, Such as.* The terms *including* and *such as* shall always be taken in the most inclusive sense, namely "including, but not limited to" and "such as, but not limited to."

P. *Supply, Procurement.* The terms *supply* and *procurement* shall mean to purchase, procure, acquire, and deliver complete with related accessories.

Q. *At No Additional Cost.* The phrase "at no additional cost" shall mean at no additional cost to the owner and at no additional cost to the design professional or construction manager.

R. *Approved, Accepted.* Where used in conjunction with the design professional's response to submittals, requests, applications, inquiries, reports, and claims by the contractor, the meaning of the terms *approved* and *accepted* shall be held to the limitations of the design professional's responsibilities to fulfill requirements of the contract documents. The terms *approved* and *accepted* shall also mean to permit the use of material, equipment, or methods conditional upon compliance with the contract documents.

S. *Approved Equal, Approved Equivalent.* The terms *approved equal* and *approved equivalent* shall mean possessing the same performance qualities and characteristics and fulfilling the same utilitarian function and approved by the design professional.

T. *Directed, Requested, Required, etc.* Where not otherwise explained, terms such as *directed, requested, required, authorized, selected, approved, accepted, designated, prescribed, ordered,* and *permitted* mean "directed by the design professional," "requested by the design professional," "required by the design professional," and similar phrases. However, no such implied meaning will be interpreted to expand the design professional's responsibility into the contractor's area of construction supervision.

U. *Review.* The term *review* shall mean limited observation or checking to ascertain general conformance with the design concept of the work and with information given in the contract documents. Such action does not constitute a waiver or alteration of the contract document requirements.

V. *Suitable, Reasonable, Proper, Correct, and Necessary.* Such terms shall mean as suitable, reasonable, proper, correct, or necessary for the purpose intended as required by the contract documents, subject to the judgment of the design professional or the construction manager.

W. *Option.* The term *option* shall mean a choice from the specified products, manufacturers, or procedures which shall be made by the contractor. The choice is not "whether" the work is to be performed, but "which" product, "which" manufacturer, or "which" procedure is to be used. The product or procedure chosen by the contractor shall be provided at no increase or additional cost to the owner, design professional, or construction manager, and with no lessening of the contractor's responsibility for its performance.

X. *Similar.* The term *similar* shall mean generally the same but not necessarily identical; details shall be worked out in relation to other parts of the work.

Y. *Submit.* The term *submit* shall mean, unless otherwise defined in greater detail, transmit to the design professional for approval, information, and record.

Z. *Project Site, Work Site.* The term *project site* shall be defined as the space available to the contractor for performance of the work, either exclusively or in conjunction with others performing other work as part of the project or another project. The extent of the project site is shown on the drawings or specified and may or may not be identical with the land upon which the project is to be built. The project site boundaries may include public streets, highways, roads, interstates, etc., public easements, and property under ownership of someone other than the client and are not available for performance of work.

AA. *Testing Laboratories.* The term *testing laboratories* shall be defined as an independent entity engaged to perform specific inspections or tests of the work, either at the project site or elsewhere, and to report and, if required, interpret the results of those inspections or tests.

BB. *Herein.* The term *herein* shall mean the contents of a particular section where this term appears.

CC. *Singular Number.* In all cases where a device or part of equipment or system is herein referred to in the singular number (such as fan, pump, cooling system, heating system, etc.), it is intended that such reference shall apply to as many such items as are required by the contract documents and to complete the installation.

DD. *No Exception Taken.* The term *no exception taken* shall mean the same as *approved.*

EE. *Approved as Noted, Make Corrections Noted, or Revise—No Resubmittal Required.* The terms *approved as noted, make corrections noted,* and *revise—no resubmittal required* shall mean the submittal essentially complies with the contract documents except for a few minor discrepancies that have been annotated directly on the submittal that will have to be corrected on the submittal and the work correctly installed in the field by the contractor.

FF. *Revise and Resubmit.* The term *revise and resubmit* shall mean the contractor shall revise the submittal to conform with the contract documents by correcting moderate errors, omissions, and/or deviations from the contract documents and resubmit it for review prior to approval and before any material and/or equipment can be fabricated, purchased, or installed by the contractor.

GG. *Disapproved/Resubmit.* The term *disapproved/resubmit* shall mean the contractor shall revise the submittal to conform with the contract documents by correcting serious errors, omissions, and/or deviations from the contract documents and resubmit it for review prior to approval and before any material and/or equipment can be fabricated, purchased, or installed by the contractor.

HH. *Disapproved or Rejected.* The terms *disapproved* and *rejected* shall mean the contractor shall discard and replace the submittal because the submittal did not comply with the contract documents in a major way.

II. *Submit Specified Item.* The term *submit specified item* shall mean the contractor shall discard and replace the submittal with a submittal containing the specified items because the submittal contained improper manufacturer, model number, material, etc.

JJ. *Acceptance.* The formal acceptance by the owner or design professional of the work, as evidenced by the issuance of the acceptance certificate.

KK. *Contract Item, Pay Item, Contract Fixed Price Item.* A specifically described item of work that is priced in the contract documents.

LL. *Contract Time, Time of Completion.* The number of calendar days (not working days) set forth in the contract documents for completion of the work.

MM. *Failure.* Any detected inability of material or equipment, or any portion thereof, to function or perform in accordance with the contract documents.

NN. *Substantial Completion. Substantial completion* shall be defined as the sufficient completion and accomplishment by the contractor of all work or designated portions thereof essential to fulfillment of the purpose of the contract, so the owner can occupy or utilize the work or designated portions thereof for the use for which it is intended.

OO. *Final Completion, Final Acceptance. Final completion* or *final acceptance* shall be defined as completion and accomplishment by the contractor of all work including contractual administrative demobilization work, all punch list items, and all other contract requirements essential to fulfillment of the purpose of the contract, so the owner can occupy or utilize the work for the use for which it is intended.

PP. *Pre-Final Inspection or Observation.* The term *pre-final inspection or observation* shall be held to the limitations of the design professional's responsibilities to fulfill the requirements of the contract documents and shall not relieve the contractor from contract obligations. The term *pre-final inspection* shall also mean all inspections conducted prior to the final inspection by the owner, the design professional, or both, verifying that all the work, with the exception of required contractual administrative demobilization work, inconsequential punch list items, and guarantees, has been satisfactorily completed in accordance with the contract documents.

QQ. *Final Inspection or Observation.* The term *final inspection or observation* shall be held to the limitations of the design professional's responsibilities to fulfill the requirements of the contract documents and shall not relieve the contractor from contract obligations. The term *final inspection* shall also mean the inspection conducted by the owner, the design professional, or both, verifying that all the work, with the exception of required contractual administrative demobilization work, inconsequential punch list items, and guarantees, has been satisfactorily completed in accordance with the contract documents.

RR. *Reliability.* The probability that a system will perform its intended functions without failure and within design parameters under specified operating conditions for which it is designed and for a specified period of time.

SS. *Testing.* The term *testing* may be described as the inspection, investigation, analysis, and diagnosis of all systems and components to ensure that the systems are operable, meet the requirements of the contract documents, and are ready for operation. Included are such items as:

1. Verification that the system is filled with water and is not air bound.
2. Verification that expansion tanks of the proper size are connected at the correct locations and that they are not waterlogged.
3. Verification that all system components are in proper working order and properly installed. Check for proper flow directions.
4. Checking of all voltages for each motor in the system.
5. Checking that all motors rotate in the correct direction and at the correct speed.
6. Checking all motors for possible overload (excess amperage draw) on initial start-up.
7. Checking of each pump for proper alignment.
8. Checking all systems for leaks, etc.
9. Checking all systems and components to ensure they meet the contract document requirements as far as capacity, system operation, control function, and other items required by the contract documents.

TT. *Adjusting.* The term *adjusting* may be described as the final setting of balancing devices such as dampers and valves, establishing and setting minimum variable frequency controller speed, in addition to automatic control devices, such as thermostats and pressure/temperature controllers to achieve maximum system performance and efficiency during normal operation. Adjusting also includes final adjustments for pumps by regulation of motor speed, partial close-down of pump discharge valve or impeller trim (preferred over the partial close-down of pump discharge valve).

UU. *Balancing.* The term *balancing* is the methodical regulation of system fluid flow-rates (air and water) through the use of workable and industry-accepted procedures as specified to achieve the desired or specified flow rates (CFM or GPM) in each segment (main, branch, or subcircuit) of the system.

VV. *Commissioning.* The term *commissioning* is the methodical procedures and methods for documenting and verifying the performance of the building envelope, HVAC, plumbing, fire protection, electrical, life safety, and telecom/data systems so that the systems operate in conformity with the design intent. Commissioning will include testing; adjusting; balancing; documentation of occupancy requirements and design assumptions; documentation of design intent for use by contractors, owners, and operators; functional performance testing and documentation necessary for evaluating all systems for acceptance; and adjusting the building systems to meet actual occupancy needs within the capability of the systems. The purpose of commissioning of building systems is to achieve the end result of a fully functional, fine-tuned, and operational building.

WW. *Functional Performance Testing.* The term *functional performance testing* shall mean the full range of checks and tests carried out to determine if all components, subsystems, systems, and interfaces between systems function in accordance with the contract documents. In this context, *function* includes all modes and sequences of control operation, all interlocks and conditional control responses, and all specified responses to abnormal emergency conditions.

XX. *Confined Spaces.* Confined spaces (according to OSHA regulations) are spaces which must have these three characteristics:

1. The space must be large enough and configured to permit personnel to enter and work.
2. The space is not designed for continuous human occupancy.
3. The space has limited or restricted means of entry and exit.
4. Two categories of confined spaces exist:
   a. *Non-Permit Required Confined Spaces (NRCS).* Spaces that contain no physical hazards that could cause death or serious physical harm, and cannot possibly contain any atmospheric hazards.
   b. *Permit Required Confined Spaces (PRCS).* Spaces that contain or may contain a hazardous atmosphere (atmospheric hazards—oxygen deficiency or enrichment 19.5 percent acceptable minimum and 23.5 percent acceptable maximum; flammable contaminants; and toxic contaminants—product, process, or reactivity); a liquid or finely divided solid material such as grain, pulverized coal, etc., that could surround or engulf a person; or some other recognized serious safety or health hazard such as temperature extremes or mechanical or electrical hazards (boilers, open transformers, tanks, vaults, sewers, manholes, pits, machinery enclosures, vats, silos, storage bins, rail tank cars, and process or reactor vessels).

YY. Hazardous Location Classifications

1. Hazardous locations are those areas where a potential for explosion and fire exist because of flammable gases, vapors, or dust in the atmosphere, or because of the presence of easily ignitable fibers or flyings in accordance with the National Electric Code (*NEC—NFPA 70*).
2. *Class I Locations.* Class I locations are those locations in which flammable gases or vapors are, or may be, present in the air in quantities sufficient to produce explosive or ignitable mixtures.
   a. *Class I, Division 1 Locations.* These are Class I locations where the hazardous atmosphere is expected to be present during normal operations. It may be present continuously, intermittently, periodically, or during normal repair or maintenance operations. Division 1 locations are also those locations where a breakdown in the operation of processing equipment results in the release of hazardous vapors while providing a source of ignition with the simultaneous failure of electrical equipment.
   b. *Class I, Division 2 Locations.* These are Class I locations in which volatile flammable liquids or gases are handled, processed, or used, but in which they can escape only in the case of accidental rupture or breakdown of the containers or systems. The hazardous conditions will occur only under abnormal conditions.

3. *Class II Locations.* Class II locations are those locations that are hazardous because of the presence of combustible dust.
   a. *Class II, Division 1 Locations.* These are Class II locations where combustible dust may be in suspension in the air under normal conditions in sufficient quantities to produce explosive or ignitable mixtures. This may occur continuously, intermittently, or periodically. Division 1 locations also exist where failure or malfunction of machinery or equipment might cause a hazardous location to exist while providing a source of ignition with the simultaneous failure of electrical equipment. Included also are locations in which combustible dust of an electrically conductive nature may be present.
   b. *Class II, Division 2 Locations.* These are Class II locations in which combustible dust will not normally be in suspension in the air, and normal operations will not put the dust in suspension, but where accumulation of the dust may interfere with the safe dissipation of heat from electrical equipment or where accumulations near electrical equipment may be ignited by arcs, sparks, or burning material from the equipment.
4. *Class III Locations.* Class III locations are those locations that are hazardous because of the presence of easily ignitable fibers or flyings, but in which the fibers or flyings are not likely to be in suspension in the air in quantities sufficient to produce ignitable mixtures.
   a. *Class III, Division 1 Locations.* These are locations in which easily ignitable fibers or materials producing combustible flyings are handled, manufactured, or used.
   b. *Class III, Division 2 Locations.* These locations are where easily ignitable fibers are stored or handled.

2.02 Systems

A. *Mechanical Systems.* The term *mechanical systems* shall mean for the purposes of these contract documents all heating, ventilating, and air conditioning systems and all piping systems as specified and as shown on the mechanical drawings and all services and appurtenances incidental thereto.

B. *Plumbing Systems.* The term *plumbing systems* shall mean for the purposes of these contract documents all plumbing fixtures, plumbing systems, piping systems, medical vacuum, medical compressed air, medical gas, laboratory vacuum, laboratory compressed air, and all laboratory gas systems as specified and as shown on the plumbing drawings and all services and appurtenances incidental thereto.

C. *Fire Suppression Systems.* The term *fire suppression systems* shall mean for the purposes of these contract documents all fire protection piping systems, standpipe, wet-pipe, dry-pipe, preaction, foam suppression, and all fire protection systems as specified and as shown on the fire protection drawings and all services and appurtenances incidental thereto.

D. *Ductwork.* The term *ductwork* shall include ducts, fittings, flanges, dampers, insulation, hangers, supports, access doors, housings, and all other appurtenances comprising a complete and operable system.

1. *Supply Air Ductwork.* The term *supply air ductwork* shall mean for the purposes of these contract documents all ductwork carrying air from a fan or air handling unit to the room, space, or area to which it is introduced. The air may be conditioned or unconditioned. Supply air ductwork extends from the fan or air handling unit to all the diffusers, registers, and grilles.
2. *Return Air Ductwork.* The term *return air ductwork* shall mean for the purposes of these contract documents all ductwork carrying air from a room, space, or area to a fan or air handling unit. Return air ductwork extends from the registers, grilles, or other return openings to the return fan (if used) and the air handling unit.

3. *Exhaust Air Ductwork.* The term *exhaust air ductwork* shall mean for the purposes of these contract documents all ductwork carrying air from a room, space, area, or equipment to a fan and then discharged to the outdoors. Exhaust air ductwork extends from the registers, grilles, equipment, or other exhaust openings to the fan, and from the fan to the outdoor discharge point.

4. *Relief Air Ductwork.* The term *relief air ductwork* shall mean for the purposes of these contract documents all ductwork carrying air from a room, space, or area without the use of a fan or with the use of a return fan to be discharged to the outdoors. Relief air ductwork extends from the registers, grilles, or other relief openings to the outdoor discharge point, or from the return fan discharge to the outdoor discharge point.

5. *Outside Air Ductwork.* The term *outside air ductwork* shall mean for the purposes of these contract documents all ductwork carrying unconditioned air from the outside to a fan or air handling unit. Outdoor air ductwork extends from the intake point or louver to the fan, air handling unit, or connection to the return air ductwork.

6. *Mixed Air Ductwork.* The term *mixed air ductwork* shall mean for the purposes of these Contract Documents all ductwork carrying a mixture of return air and outdoor air. Mixed air ductwork extends from the point of connection of the return air and outdoor air ductwork to the fan or air handling unit.

7. *Supply Air Plenum.* The term *supply air plenum* shall mean for the purposes of these contract documents all ductwork in which the discharges of multiple fans or air handling units connect forming a common supply header, or all ductwork or ceiling construction forming a common supply box where supply air ductwork discharges into the box at limited locations for air distribution to supply diffusers which are directly connected to the plenum.

8. *Return Air Plenum.* The term *return air plenum* shall mean for the purposes of these contract documents all ductwork in which the suctions of multiple return fans or the discharges of multiple return fans connect forming a common suction or discharge return header or the space above the architectural ceiling and below the floor or roof structure used as return air ductwork.

9. *Exhaust Air Plenum.* The term *exhaust air plenum* shall mean for the purposes of these contract documents all ductwork in which the suctions of multiple exhaust fans or the discharges of multiple exhaust fans connect forming a common suction or discharge exhaust header or the ductwork formed around single or multiple exhaust air discharge openings or louvers to create a connection point for exhaust air ductwork.

10. *Relief Air Plenum.* The term *relief air plenum* shall mean for the purposes of these contract documents all ductwork in which multiple relief air ductwork connections are made forming a common relief air header.

11. *Outdoor Air Plenum.* The term *outdoor air plenum* shall mean for the purposes of these contract documents all ductwork in which the suctions of multiple fans or air handling units connect to form a common outside air header or the ductwork formed around single or multiple outside air openings or louvers to create a connection point for outside air ductwork.

12. *Mixed Air Plenum.* The term *mixed air plenum* shall mean for the purposes of these contract documents all ductwork in which multiple return air and multiple outdoor air ductwork connections are made forming a common mixed air header.

13. *Vents, Flues, Stacks, and Breeching.* The terms *vents, flues, stacks*, and *breeching* shall mean for the purposes of these contract documents ductwork conveying the products of combustion to atmosphere for safe discharge.

E. **Piping.** The term **piping** shall include pipe, fittings, valves, flanges, unions, traps, drains, strainers, insulation, hangers, supports, and all other appurtenances comprising a complete and operable system.

F. **Wiring.** The term **wiring** shall include wire, conduit, raceways, bus duct, fittings, junction and outlet boxes, switches, cutouts, receptacles, and all other appurtenances comprising a complete and operable system.

G. *Product.* The term *product* shall include materials, equipment, and systems for a complete and operable system.

H. *Motor Controllers.* The term *motor controllers* shall be manual or magnetic starters (with or without switches), variable frequency controllers, individual push buttons, or hand-off-automatic (HOA) switches controlling the operation of motors.

I. *Control Devices.* The term *control devices* shall be automatic sensing and switching devices such as thermostats, float, and electro-pneumatic switches controlling the operations of mechanical and electrical equipment.

J. *Work, Project.* The terms *work* and *project* shall mean labor, operations, materials, supervision, services, machinery, equipment, tools, supplies, and facilities to be performed or provided including work normally done at the location of the project to accomplish the requirements of the contract including all alterations, amendments, or extensions to the contract made by change order.

K. *Extra Work.* The term *extra work* shall be any item of work not provided for in the awarded contract as previously modified by change order (change bulletin) or supplemental agreement, but which is either requested by the owner or found by the design professional to be necessary to complete the work within the intended scope of the Contract as previously modified.

L. *Concealed.* The term *concealed* shall mean hidden from normal sight; includes work in crawl spaces, above ceilings, in walls, in chases, and in building shafts.

M. *Exposed.* The term *exposed* shall mean not concealed.

N. *Below Ground.* The term *below ground* shall mean installed underground, buried in the earth, or buried below the ground floor slab.

O. *Above Ground.* The term *above ground* shall mean not installed underground, not buried in the earth, and not buried below the ground floor slab.

P. *Conditioned.* The term *conditioned* shall mean for the purposes of these contract documents rooms, spaces, or areas that are provided with mechanical heating and cooling.

Q. *Unconditioned and Nonconditioned.* The terms *unconditioned* and *nonconditioned* shall mean for the purposes of these contract documents rooms, spaces, or areas that are not provided with mechanical heating or cooling.

R. *Heated.* The term *heated* shall mean for the purposes of these contract documents rooms, spaces, or areas that are provided with mechanical heating only.

S. *Air Conditioned.* The term *air conditioned* shall mean for the purposes of these contract documents rooms, spaces, or areas that are provided with mechanical cooling only.

T. *Unheated.* The term *unheated* shall mean for the purposes of these contract documents rooms, spaces, or areas that are not provided with mechanical heating.

U. *Ventilated Spaces.* The term *ventilated spaces* shall mean for the purposes of these contract documents rooms, spaces, or areas supplied with outdoor air on a continuous or intermittent basis. The outdoor air may be conditioned, heated, unconditioned, or unheated.

V. *Indoor.* The term *indoor* shall mean for the purposes of these contract documents items or devices contained within the confines of a building, structure, or facility and items or devices that are not exposed to weather. The term *indoor* shall generally reference ductwork, piping, or equipment location (indoor ductwork, indoor piping, indoor equipment).

W. *Outdoor.* The term *outdoor* shall mean for the purposes of these contract documents items or devices not contained within the confines of a building, structure, or facility and items or devices that are exposed to weather. The term *outdoor* shall generally reference ductwork, piping, or equipment (outdoor ductwork, outdoor piping, outdoor equipment).

X. *Hot.* The term *hot* shall mean for the purposes of these contract documents the temperature of conveyed solids, liquids, or gases that are above the surrounding ambient temperature or above 100°F (hot supply air ductwork, heating water piping).

Y. *Cold.* The term *cold* shall mean for the purposes of these contract documents the temperature of conveyed solids, liquids, or gases that are below the surrounding ambient temperature or below 60°F (cold supply air ductwork, chilled water piping).

Z. *Warm.* The term *warm* shall mean for the purposes of these contract documents the temperature of conveyed solids, liquids, or gases that are at the surrounding ambient temperature or between 60°F and 100°F (condenser water piping).

AA. *Hot/Cold.* The term *hot/cold* shall mean for the purposes of these contract documents the temperature of conveyed solids, liquids, or gases that can be either hot or cold depending on the season of the year (heating and air conditioning supply air ductwork, dual temperature piping systems).

BB. *Removable.* The term *removable* shall mean detachable from the structure or system without physical alteration or disassembly of the materials or equipment or disturbance to other construction.

CC. *Temporary Work.* Work provided by the contractor for use during the performance of the work, but which is to be removed prior to final acceptance.

DD. *Normally Closed (NC).* The term *normally closed* shall mean the valve, damper, or other control device shall remain in, or go to, the closed position when the control air pressure, the control power, or the control signal is removed. The position the device will assume when the control signal is removed.

EE. *Normally Open (NO).* The term *normally open* shall mean that the valve, damper, or other control device shall remain in, or go to, the open position when the control air pressure, the control power, or the control signal is removed. The position the device will assume when the control signal is removed.

FF. *Traffic Level or Personnel Level.* The term *traffic level or personnel level* shall mean for the purposes of these contract documents all areas, including process areas, equipment rooms, boiler rooms, chiller rooms, fan rooms, air handling unit rooms, and other areas where insulation may be damaged by normal activity and local personnel traffic. The area extends vertically from the walking surface to 8'0" above walking surface and extends horizontally 5'0" beyond the edge of the walking surface. The walking surface shall include floors, walkways, platforms, catwalks, ladders, and stairs.

2.03 Contract Documents

A. *Contract Drawings.* The terms *contract drawings* and *drawings* shall mean all drawings or reproductions of drawings pertaining to the construction or plans, sections, elevations, profiles, and details of the work contemplated and its appurtenances.

B. *Contract Specifications.* The terms *contract specifications* and *specifications* shall mean the description, provisions, and other requirements pertaining to the method and manner of performing the work and to the quantities and qualities of materials to be furnished under the contract. The specifications shall include the general provisions, the special provisions, and the technical specifications.

C. *Contract Documents.* The term *contract documents* shall include contract drawings, contract specifications, addenda, amendments, shop drawings, coordination draw-ings, general provisions, special provisions, the executed agreement and other items required for, or pertaining to, the contract, including the executed contract.

D. *Addenda.* Addenda are issued as changes, amendments, or clarifications to the original or previously issued contract documents. Addenda are issued in written and/or draw-ing form prior to acceptance or signing of the construction contract.

E. *Amendments (Change Orders, Change Bulletins).* Amendments (change orders, change bulletins) are issued changes or amendments to the contract documents. Amendments are issued in written and/or drawing form after acceptance or signing of the contract.

F. *Submittals or Shop Drawings.* The term *submittals* or *shop drawings* shall include draw-ings, coordination drawings, diagrams, schedules, performance characteristics, charts, brochures, catalog cuts, calculations, certified drawings, and other materials prepared by the contractor, subcontractor, manufacturer, or distributor that illustrate some por-tion of the work in accordance with the requirements of the contract documents used by the contractor to order, fabricate, and install the general construction, mechanical, plumbing, fire protection, and electrical equipment and systems in a building.

   The corrections or comments annotated on a shop drawing during the design professional's review do not relieve the contractor from full compliance with the contract documents regarding the work. The design professional's check is only a review of the shop drawing's general compliance with the information shown in the contract documents. The contractor remains responsible for continuing the correlation of all material and component quantities and dimensions, coor-dination of the contractor's work with that of other trades, selection of suitable fabrication and installation techniques, and performance of work in a safe and satisfactory manner.

G. *Product Data.* Illustrations, standard schedules, performance charts, instructions, brochures, diagrams, and other information furnished by the contractor to illustrate a material, product, or system for some portion of the work.

H. *Samples.* Physical examples that illustrate material, equipment, or workmanship and establish standards to which the work will be judged.

I. *Coordination Drawings.* The terms *coordination drawings* and *composite drawings* are drawings created by the respective contractors showing work of all contractors super-imposed on the sepia or mylar of the basic shop drawing of one of the contractors to coordinate and verify that all work in a congested area will fit in an acceptable manner.

J. *Contract.* A set of documents issued by the owner for the work, which may include the contract documents, the advertisement, form of proposal, free competitive bid-ding affidavit, affidavit as to taxes, certification of bidder, buy America requirements, disadvantaged business enterprise forms, bid bond, agreement, waiver of right to file mechanics lien, performance bond, payment bond, maintenance bond(s), certification regarding lobbying, disclosure form to report lobbying, and other forms that form part of the contract as required by the owner and the contract documents.

K. *Payment Bond.* The approved form of security furnished by the contractor and its surety as a guarantee to pay promptly, or cause to be paid promptly, in full, such items as may be due for all material furnished, labor supplied or performed, rental of equipment used, and services rendered in connection with the work.

L. *Maintenance Bond.* The approved form of security furnished by the contractor and its surety as a guarantee on the part of the contractor to remedy, without cost to the owner, any defects in the work that may develop during a period of twelve (12) months from the date of substantial completion.

M. *Performance Bond.* The approved form of security furnished by the contractor and its surety as a guarantee on the part of the contractor to execute the work.

N. *Working Drawings.* Drawings and calculations prepared by the contractor, subcontractor, supplier, distributor, etc., that illustrate work required for the construction of, but which will not become an integral part of, the work. These shall include, but are not limited to, drawings showing contractor's plans for temporary work such as decking, temporary bulkheads, support of excavation, support of utilities, groundwater control systems, forming and false-work, erection plans, and underpinning.

O. *Construction Drawings or Coordination Drawings.* Detailed drawings prepared by the contractor, subcontractor, supplier, distributor, etc., that illustrate in exact and intricate detail, work required for the construction contract. These drawings often show hanger locations, vibration isolators, ductwork and pipe fittings, sections, dimensions of ducts and pipes, and other items required to construct the work.

P. *Project Record Documents.* A copy of all contract drawings, shop drawings, working drawings, addenda, change orders, contract documents, and other data maintained by the contractor during the work. The contractor's recording, on a set of prints, of accurate information and sketches regarding the exact detail and location of the work as actually installed, recording such information as the exact location of all underground utilities, contract changes, and contract deviations. The contractor's information is then transferred to the original contract documents by the design professional for the owner's permanent record unless otherwise directed or specified.

Q. *Proposal Guarantee.* Cashier's check, certified check, or bid bond accompanying the proposal submitted by the bidder as a guarantee that the bidder will enter into a contract with the owner for the performance of the work indicated and file acceptable bonds and insurance if the contract is awarded to it.

R. *Project Schedule.* The schedule for the work as prepared and maintained by the contractor in accordance with the contract documents.

S. *Certificate of Substantial Completion.* Certificate issued by the owner or design professional certifying that a substantial portion of the work has been completed in accordance with the contract documents with the exception of contractual administrative demobilization work, inconsequential punch list items, and guarantees. The certificate of substantial completion shall establish the date of substantial completion, shall state the responsibilities of the owner and the contractor for security, maintenance, heat, utilities, damage to the work, and insurance, and shall fix the time within which the contractor shall complete the items listed therein. Warranties required by the contract documents shall commence on the date of substantial completion of the work or a designated portion thereof unless otherwise provided in the certificate of substantial completion or the contract documents.

T. *Certificate of Final Completion (Final Acceptance).* Certificate issued by the owner or design professional certifying that all of the work has been completed in accordance with the contract documents to the best of the owner's or design professional's knowledge, information, and belief, and on the basis of that person's observations and inspections including contractual administrative demobilization work and all punch list items. The certificate of final completion shall establish the date of owner acceptance. Warranties required by the contract documents shall commence on the date of final completion of the work unless otherwise provided in the certificate of substantial completion or the contract documents.

U. *Acceptance Certificate.* Certificate to be issued by the owner or design professional certifying that all the work has been completed in accordance with the contract documents.

V. *Award.* The acceptance by the owner of the bid from the responsible bidder (sometimes the lowest responsible bidder) as evidenced by the written notice to award to the bidder tendering said bid.

W. *Bid (Proposal).* The proposal of the bidder for the work, submitted on the prescribed bid form, properly signed, dated, and guaranteed, including alternates, the unit price schedule, bonds, and other bidding requirements as applicable.

X. *Certificate of Compliance.* Certificate issued by the supplier certifying that the material or equipment furnished is in compliance with the contract documents.

Y. *Agreement.* The instrument executed by the owner and the contractor in conformance with the contract documents for the performance of the work.

Z. *Field Order.* A notice issued to the contractor by the design professional specifying an action required of the contractor.

AA. *Request for Information or Request for Interpretation (RFI).* A notice issued by the contractor to the design professional or owner requesting a clarification of the contract documents.

BB. *Notice to Proceed.* A written notice from the owner to the contractor or design professional directing the contractor or design professional to proceed with the work.

CC. *Advertisement, Invitation to Bid.* The public or private announcement, as required by law or the owner, inviting bids for the work to be performed, material to be furnished, or both.

2.04 Contractors/Manufacturers/Authorities

A. *Contractor.* The term *contractor* shall mean the individual, firm, partnership, corporation, joint venture, or any combination thereof or their duly authorized representatives who have executed a contract with the client for the proposed work.

B. *Subcontractor or Trade Contractor.* The terms *subcontractor* or *trade contractor* shall mean all the lower-tier contractors, material suppliers, and distributors that have executed a contract with the contractor for the proposed work.

C. *Furnisher or Supplier.* The terms *furnisher* or *supplier* shall be defined as the "entity" (individual, partnership, firm, corporation, joint venture, or any combination thereof) engaged by the contractor, its subcontractor, or sub-subcontractor, to furnish a particular unit of material or equipment to the project site. It shall be a requirement that the furnisher or supplier be experienced in the manufacture of the material or equipment they are to furnish.

D. *Installer.* The term *installer* shall be defined as the "entity" (individual, partnership, firm, corporation, joint venture, or any combination thereof) engaged by the contractor, its subcontractor, or sub-subcontractor to install a particular unit of work at the project site, including installation, erection, application, and similar required operations. It shall be a requirement that the installer be experienced in the operations they are engaged to perform.

E. *Provider.* The term *provider* shall be defined as the "entity" (individual, partnership, firm, corporation, joint venture, or any combination thereof) engaged by the contractor, its subcontractor, or sub-subcontractor to provide a particular unit of material or equipment at the project site. It shall be a requirement that the provider be experienced in the operations they are engaged to perform.

F. *Bidder.* An individual, firm, partnership, corporation, joint venture, or any combination thereof submitting a bid for the work as a single business entity and acting directly or through a duly authorized representative.

G. *Authority Having Jurisdiction.* The term *authority having jurisdiction* shall mean federal, state, and/or local authorities or agencies thereof having jurisdiction over work to which reference is made and authorities responsible for "approving" equipment, installation, and/or procedures.

H. *Surety.* The corporate body that is bound with, and for, the contractor for the satisfactory performance of the work by the contractor, and the prompt payment in full for materials, labor, equipment, rentals, and services, as provided in the bonds.

I. *Acceptable Manufacturers.* The term *acceptable manufacturers* shall mean the specified list of manufacturers considered acceptable to bid the project for a specific piece of equipment. Only the equipment specified has been checked for spatial compatibility. If the contractor elects to use an optional manufacturer from the acceptable manufacturers list in the specifications, it shall be the contractor's responsibility to determine and ensure the spatial compatibility of the manufacturer's equipment selected.

Equations

3.01 Airside System Equations and Derivations

A. Equations

$$H_S = 1.08 \times CFM \times \Delta T$$

$$H_L = 0.68 \times CFM \times \Delta W_{GR.}$$

$$H_L = 4840 \times CFM \times \Delta W_{LB.}$$

$$H_T = 4.5 \times CFM \times \Delta h$$

$$H_T = H_S + H_L$$

$$SHR = \frac{H_S}{H_T} = \frac{H_S}{H_S + H_L}$$

H_S = sensible heat (Btu/h)
H_L = latent heat (Btu/h)
H_T = total heat (Btu/h)
ΔT = temperature difference (°F)
$\Delta W_{GR.}$ = humidity ratio difference (Gr.H_2O/lbs.DA)
$\Delta W_{LB.}$ = humidity ratio difference (lbs.H_2O/lbs.DA)
Δh = enthalpy difference (Btu/lbs.DA)
CFM   = air flow rate (cubic feet per minute)
SHR   = sensible heat ratio
m   = mass flow (lbs.DA/h)
c_a = specific heat of air (0.24 Btu/lbs.DA °F)
DA   = dry air

B. Derivations

1. Standard air conditions:
   a. Temperature: 60°F
   b. Pressure: 14.7 psia (sea level)
   c. Specific volume: 13.33 ft.^{3}/lbs.DA
   d. Density: 0.075 lbs./ft.^{3}
   e. L_V = Latent heat of water @60°F: 1060 Btu/lbs.
2. Sensible heat equation:
   $H_S = m \times c_a \times \Delta T$
   c_P = 0.24 (Btu/~~lbs.DA~~. °F) × 0.075 ~~lbs.DA~~/ft.^{3} × 60 min./h
      = 1.08 Btu min./h ft.^{3} °F
   H_S = 1.08 (Btu min./h ft.^{3} °F) × CFM (ft.^{3}/min.) × ΔT (°F)
   H_S = 1.08 × CFM × ΔT
3. Latent heat equation:
   $H_L = m \times L_V \times \Delta W_{GR}$
   L_V = 1060 Btu/~~lbs.H₂O~~ × 0.075 lbs.DA/ft.^{3} × 60 min./h × 1.0 ~~lbs.H₂O~~/7,000 Gr.H_2O
      = 0.68 Btu lbs.DA min./hft.^{3} Gr.H_2O
   H_L = 0.68 (Btu lbs.DA min./hft.^{3} Gr.H_2O) × CFM (ft.^{3}/min.) × ΔW_{GR} (Gr.H_2O/lbs.DA)
   H_L = 0.68 × CFM × ΔW_{GR}
4. Total heat equation:
   H_T = m × Δh
   Factor = 0.075 lbs.DA/ft.^{3} × 60 min./h = 4.5 lbs.DA min./hft.^{3}
   H_T = 4.5 (lbs.DA min./hft.^{3}) × CFM (ft.^{3}/min.) × Δh (Btu/lbs.DA)
   H_T = 4.5 × CFM × Δh

3.02 Waterside System Equations and Derivations

A. Equations

$$H = 500 \times GPM \times \Delta T$$

$$GPM_{EVAP.} = \frac{TONS \times 24}{\Delta T}$$

$$GPM_{COND.} = \frac{TONS \times 30}{\Delta T}$$

H	= total heat (Btu/h)
GPM	= water flow rate (gallons per minute)
ΔT	= temperature difference (°F)
TONS	= air conditioning load (tons)
$GPM_{EVAP.}$	= evaporator water flow rate (gallons per minute)
$GPM_{COND.}$	= condenser water flow rate (gallons per minute)
c_w	= specific heat of water (1.0 Btu/lbs.H_2O)

B. Derivations

1. Standard water conditions:
   a. Temperature: 60°F
   b. Pressure: 14.7 psia (sea level)
   c. Density: 62.4 lbs./ft.^{3}
2. Water equation
   $H = m \times c_w \times \Delta T$
   c_w = 1.0 Btu/~~Lb H_2O~~ °F × 62.4 ~~lbs.H_2O~~/ft^3 × 1.0 ft^3/7.48052 gal. × 60 min./h
       = 500 Btu min./h °F gal.
   H = 500 Btu min./h °F gal. × GPM (gal./min.) × ΔT (°F)
   H = 500 × GPM × ΔT
3. Evaporator equation:
   GPM_{EVAP} = H/(500 × ΔT)
   Factor   = 12,000 ~~Btu/h~~/1.0 tons ÷ 500 ~~Btu~~ min./~~h~~ °F gal.
           = 24°F gal./tons min.
   GPM_{EVAP} = tons (tons) × 24 (°F gal./tons min.)/ΔT (°F)
   GPM_{EVAP} = tons × 24/ΔT
4. Condenser equation:
   GPM_{COND} = 1.25 × GPM_{EVAP} = 1.25 × tons × 24/ΔT
   GPM_{COND} = tons × 30/ΔT

3.03 Air Change Rate Equations

$$\frac{AC}{HR} = \frac{CFM \times 60}{VOLUME}$$

$$CFM = \frac{\frac{AC}{HR} \times VOLUME}{60}$$

AC/H	= air change rate per hour
CFM	= air flow rate (cubic feet per minute)
VOLUME	= space volume (cubic feet)

3.04 English/Metric Airside System Equations Comparison

A. Sensible Heat Equations

$$H_S = 1.08 \; \frac{Btu \; min.}{Hr \; ft^3 \, °F} \times CFM \times \Delta T$$

$$H_{SM} = 72.42 \; \frac{kJ \; min.}{hr. \; m^3 \, °C} \times CMM \times \Delta T_M$$

B. Latent Heat Equations

$$H_L = 0.68 \; \frac{Btu \; min. \; Lb \; DA}{hr. \; ft^3 \; Gr \; H_2O} \times CFM \times \Delta W$$

$$H_{LM} = 177{,}734.8 \; \frac{kJ \; min. \; kg \; DA}{hr. \; m^3 kg \; H_2O} \times CMM \times \Delta W_M$$

C. Total Heat Equations

$$H_T = 4.5 \; \frac{lb \; min.}{hr. \; ft.^3} \times CFM \times \Delta h$$

$$H_{TM} = 72.09 \; \frac{kg \; min.}{hr. \; m^3} \times CMM \times \Delta h_M$$

$$H_T = H_S + H_L$$

$$H_{TM} = H_{SM} + H_{LM}$$

H_S = sensible heat (Btu/h)
H_{SM} = sensible heat (kJ/h)
H_L = latent heat (Btu/h)
H_{LM} = latent heat (kJ/h)
H_T = total heat (Btu/h)
H_{TM} = total heat (kJ/h)
ΔT = temperature difference (°F)
ΔT_M = temperature difference (°C)
ΔW = humidity ratio difference (Gr.H_2O/lbs.DA)
ΔW_M = humidity ratio difference (kg.H_2O/kg.DA)
Δh = enthalpy difference (Btu/lbs.DA)
Δh_M = enthalpy difference (kJ/lbs.DA)
CFM = air flow rate (cubic feet per minute)
CMM = air flow rate (cubic meters per minute)

3.05 English/Metric Waterside System Equation Comparison

$$H = 500 \; \frac{Btu \; min.}{hr. \; gal. \; °F} \times GPM \times \Delta T$$

$$H_M = 250.8 \; \frac{kJ \; min.}{hr. \; Liters \; °C} \times LPM \times \Delta T_M$$

H   = total heat (Btu/h)
H_M = total heat (kJ/h)
ΔT = temperature difference (°F)
ΔT_M = temperature difference (°C)
GPM = water flow rate (gallons per minute)
LPM = water flow rate (liters per minute)

3.06 English/Metric Air Change Rate Equation Comparison

$$\frac{AC}{HR} = \frac{CFM \times 60\ \frac{min.}{h}}{VOLUME}$$

$$\frac{AC}{HR_M} = \frac{CMM \times 60\ \frac{min.}{h}}{VOLUME_M}$$

AC/H   = air change rate per hour – English
AC/H_M = air change rate per hour – Metric
AC/H   = AC/H_M
VOLUME   = space volume (cubic feet)
$VOLUME_M$ = space volume (cubic meters)
CFM   = air flow rate (cubic feet per minute)
CMM   = air flow rate (cubic meters per minute)

3.07 English/Metric Temperature and Other Conversions

°F = 1.8°C + 32

$$°C = \frac{°F - 32}{1.8}$$

°F   = degrees Fahrenheit
°C   = degrees Celsius
kJ/h   = Btu/h × 1.055
CMM   = CFM × 0.02832
LPM   = GPM × 3.785
kJ/kg   = Btu/lbs. × 2.326
meters   = ft. × 0.3048
sq. meters   = sq. ft. × 0.0929
cu. meters   = cu. ft. × 0.02832
kg   = lbs. × 0.4536
1.0 GPM   = 500 lbs. steam/h
1.0 lbs. stm./h = 0.002 GPM
1.0 lbs. H_2O/h = 1.0 lbs. steam/h
kg/cu. meter   = lbs./cu. ft. × 16.017   (Density)
cu. meters/kg   = cu. ft./lbs. × 0.0624   (Specific Volume)
kg H_2O/kg DA = Gr.H_2O/lbs.DA/7,000 = lbs.H_2O/lbs.DA

3.08 Steam and Condensate Equations

A. General

$$LBS.STM./HR = \frac{BTU/HR}{H_{FG}} = \frac{BTU/HR}{960}$$

$$LBS.STM.COND./HR = \frac{EDR}{4}$$

$$EDR = \frac{BTU/HR}{240}$$

$$LBS.STM.COND./HR = \frac{GPM \times 500 \times SP.GR. \times C_w \times \Delta T}{H_{FG}}$$

$$LBS.STM.COND./HR = \frac{CFM \times 60 \times D \times C_a \times \Delta T}{H_{FG}}$$

B. Approximating Condensate Loads

$$LBS.STM.COND./HR = \frac{GPM(WATER) \times \Delta T}{2}$$

$$LBS.STM.COND./HR = \frac{GPM(FUEL\ OIL) \times \Delta T}{4}$$

$$LBS.STM.COND./HR = \frac{CFM(AIR) \times \Delta T}{900}$$

stm.   = steam
GPM   = quantity of liquid (gallons per minute)
CFM   = quantity of gas or air (cubic feet per minute)
SP.GR. = specific gravity
D   = density (lbs./cubic feet)
C_a = specific heat of air (0.24 Btu/lbs.)
Cw   = specific heat of water (1.00 Btu/lbs.)
H_{FG} = latent heat of steam (Btu/lbs.) at steam design pressure (ASHRAE Fundamentals)
ΔT = final temperature minus initial temperature
EDR   = equivalent direct radiation

3.09 Building Envelope Heating Equation and R-Values/U-Values

$$H = U \times A \times \Delta T$$

$$R = \frac{1}{C} = \frac{1}{K} \times Thickness \ (in.)$$

$$R = \frac{1}{\Sigma R}$$

H   = heat flow (Btu/h)
ΔT = temperature difference (°F)
A   = area (sq.ft.)
U   = U-Value (Btu./h sq.ft. °F): See Part 35 for definitions.

R = R-Value (h sq.ft. °F/Btu.): See Part 35 for definitions.
C = conductance (Btu./h sq.ft. °F): See Part 35 for definitions.
K = conductivity (Btu. in./h sq.ft. °F): See Part 35 for definitions.
ΣR = sum of the individual R-Values

3.10 Fan Laws

$$\frac{CFM_2}{CFM_1} = \frac{RPM_2}{RPM_1}$$

$$\frac{SP_2}{SP_1} = \left[\frac{CFM_2}{CFM_1}\right]^2 = \left[\frac{RPM_2}{RPM_1}\right]^2$$

$$\frac{BHP_2}{BHP_1} = \left[\frac{CFM_2}{CFM_1}\right]^3 = \left[\frac{RPM_2}{RPM_1}\right]^3 = \left[\frac{SP_2}{SP_1}\right]^{1.5}$$

$$BHP = \frac{CFM \times SP \times SP.GR.}{6356 \times FAN_{EFF.}}$$

$$MHP = \frac{BHP}{M/D_{EFF.}}$$

CFM	= cubic feet/minute	Air Density	= constant
RPM	= revolutions/minute	SP.GR.(Air)	= 1.0
SP	= static pressure, in. W.G.	FAN_{EFF}	= 65–85%
BHP	= brake horsepower	M/D_{EFF}	= 80–95%
Fan Size	= constant	M/D	= motor/drive

3.11 Pump Laws

$$\frac{GPM_2}{GPM_1} = \frac{RPM_2}{RPM_1}$$

$$\frac{HD_2}{HD_1} = \left[\frac{GPM_2}{GPM_1}\right]^2 = \left[\frac{RPM_2}{RPM_1}\right]^2$$

$$\frac{BHP_2}{BHP_1} = \left[\frac{GPM_2}{GPM_1}\right]^3 = \left[\frac{RPM_2}{RPM_1}\right]^3 = \left[\frac{HD_2}{HD_1}\right]^{1.5}$$

$$BHP = \frac{GPM \times HD \times SP.GR.}{3960 \times PUMP_{EFF.}}$$

$$MHP = \frac{BHP}{M/D_{EFF.}}$$

$$VH = \frac{V^2}{2g}$$

$$HD = \frac{P \times 2.31}{SP.GR.}$$

GPM	= gallons/minute
RPM	= revolutions/minute
HD	= head in ft. H_2O
BHP	= brake horsepower
Pump Size	= constant

Water Density = constant
SP.GR.   = specific gravity of liquid with respect to water
SP.GR.(Water) = 1.0
$PUMP_{EFF}$ = 60–80%
M/D_{EFF} = 85–95%
M/D   = motor/drive
P   = pressure (psi)
VH   = velocity head in ft. H_2O
V   = velocity (ft./sec.)
g   = acceleration due to gravity (32.16 ft./sec.²)

3.12 Pump Net Positive Suction Head (NPSH) Calculations

$$NPSH_{AVAIL} > NPSH_{REQ'D}$$

$$NPSH_{AVAIL} = H_A \pm H_S - H_F - H_{VP}$$

$NPSH_{AVAIL}$ = net positive suction available at pump (feet)
$NPSH_{REQ'D}$ = net positive suction required at pump (feet)
H_A = pressure at liquid surface (feet – 34 feet for water at atmospheric pressure)
H_S = height of liquid surface above (+) or below (−) pump (feet)
H_F = friction loss between pump and source (feet)
H_{VP} = absolute pressure of water vapor at liquid temperature (feet – ASHRAE Fundamentals)

Note: Calculations may also be performed in psig, provided that all values are in psig.

3.13 Mixed Air Temperature

$$T_{MA} = \left(T_{ROOM} \times \frac{CFM_{RA}}{CFM_{SA}} \right) + \left(T_{OA} \times \frac{CFM_{OA}}{CFM_{SA}} \right)$$

$$T_{MA} = \left(T_{RA} \times \frac{CFM_{RA}}{CFM_{SA}} \right) + \left(T_{OA} \times \frac{CFM_{OA}}{CFM_{SA}} \right)$$

CFM_{SA} = supply air CFM
CFM_{RA} = return air CFM
CFM_{OA} = outside air CFM
T_{MA} = mixed air temperature (°F)
T_{ROOM} = room design temperature (°F)
T_{RA} = return air temperature (°F)
T_{OA} = outside air temperature (°F)

3.14 Psychrometric Equations

$$W = 0.622 \times \frac{P_W}{P - P_W}$$

$$RH \cong \frac{W_{ACTUAL}}{W_{SAT}} \times 100\%$$

$$RH = \frac{P_W}{P_{SAT}} \times 100\%$$

$$H_S = m \times c_P \times \Delta T$$

$$H_L = L_v \times m \times \Delta W$$

$$H_T = m \times \Delta h$$

$$W = \frac{(2501 - 2.381\,T_{WB})(W_{SAT\;WB}) - (T_{DB} - T_{WB})}{(2501 + 1.805\,T_{DB} - 4.186\,T_{WB})}$$

$$W = \frac{(1093 - 0.556\,T_{WB})(W_{SAT\;WB}) - (0.240)(T_{DB} - T_{WB})}{(1093 + 0.444\,T_{DB} - T_{WB})}$$

W = specific humidity, lbs.H_2O/lbs.DA or Gr.H_2O/lbs.DA
W_{ACTUAL} = actual specific humidity, lbs.H_2O/lbs.DA or Gr.H_2O/lbs.DA
W_{SAT} = saturation specific humidity at the dry bulb temperature
$W_{SAT\;WB}$ = saturation specific humidity at the wet bulb temperature
P_W = partial pressure of water vapor, lbs./sq.ft.
P = total absolute pressure of air/water vapor mixture , lbs./sq.ft.
P_{SAT} = saturation partial pressure of water vapor at the dry bulb temperature, lbs./sq.ft.
RH = relative humidity, %
H_S = sensible heat, Btu/h
H_L = latent heat, Btu/h
H_T = total heat, Btu/h
m = mass flow rate, lbs.DA/h or lbs.H_2O/h
c_P = specific heat, Air—0.24 Btu/lbs.DA, Water—1.0 Btu/lbs.H_2O
T_{DB} = dry bulb temperature, °F
T_{WB} = wet bulb temperature, °F
ΔT = temperature difference, °F
ΔW = specific humidity difference, lbs.H_2O/lbs.DA or Gr.H_2O/lbs.DA
Δh = enthalpy difference, Btu/lbs.DA
L_V = latent heat of vaporization, Btu/lbs.H_2O

3.15 Ductwork Equations

$$TP = SP + VP$$

$$VP = \left[\frac{V}{4005}\right]^2 = \frac{(V)^2}{(4005)^2}$$

$$V = \frac{Q}{A} = \frac{Q \times 144}{W \times H}$$

$$D_{EQ} = \frac{1.3 \times (A \times B)^{0.625}}{(A + B)^{0.25}}$$

TP = total pressure
SP = static pressure, friction losses
VP = velocity pressure, dynamic losses
V = velocity, ft./min.
Q = air flow rate through duct, CFM
A = area of duct, sq.ft.
W = width of duct, in.

H   = height of duct, in.
D_{EQ} = equivalent round duct size for rectangular duct, in.
A   = one dimension of rectangular duct, in.
B   = adjacent side of rectangular duct, in.

3.16 Equations for Flat Oval Ductwork

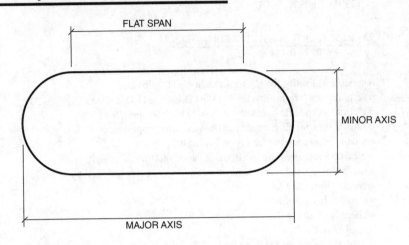

$$FS = MAJOR - MINOR$$

$$A = \frac{(FS \times MINOR) + \frac{(\pi \times MINOR^2)}{4}}{144}$$

$$P = \frac{(\pi \times MINOR) + (2 \times FS)}{12}$$

$$D_{EQ} = \frac{1.55 \times (A)^{0.625}}{(P)^{0.25}}$$

FS   = flat span dimension in inches
MAJOR = major axis dimension in inches (larger dimension)
MINOR = minor axis dimension in inches (smaller dimension)
A   = cross-sectional area in square feet
P   = perimeter or surface area in square feet per lineal feet
D_{EQ} = equivalent round duct diameter

3.17 Steel Pipe Equations

$$A   = 0.785 \times ID^2$$

$$W_P = 10.6802 \times T \times (OD - T)$$

$$W_W = 0.3405 \times ID^2$$

$$OSA = 0.2618 \times OD$$

$$ISA = 0.2618 \times ID$$

$$A_M = 0.785 \times (OD^2 - ID^2)$$

A $\quad$ = cross sectional area (sq.in.)
W_P $\quad$ = weight of pipe per foot (lbs.)
W_W $\quad$ = weight of water per foot (lbs.)
T $\quad$ = pipe wall thickness (in.)
ID $\quad$ = inside diameter (in.)
OD $\quad$ = outside diameter (in.)
OSA $\quad$ = outside surface area per foot (sq.ft.)
ISA $\quad$ = inside surface area per foot (sq.ft.)
A_M $\quad$ = area of the metal (sq.in.)

3.18 Steam and Steam Condensate Pipe Sizing Equations

A. Steam Pipe Sizing Equations

$$\Delta P = \frac{(0.01306) \times W^2 \times \left(1 + \frac{3.6}{ID}\right)}{3600 \times D \times ID^5}$$

$$W = 60 \times \sqrt{\frac{\Delta P \times D \times ID^5}{0.01306 \times \left(1 + \frac{3.6}{ID}\right)}}$$

$$W = 0.41667 \times V \times A_{INCHES} \times D = 60 \times V \times A_{FEET} \times D$$

$$V = \frac{2.4 \times W}{A_{INCHES} \times D} = \frac{W}{60 \times A_{FEET} \times D}$$

ΔP $\quad$ = pressure drop per 100 ft. of pipe, psig/100 ft.
W $\quad$ = steam flow rate, lbs./h
ID $\quad$ = actual inside diameter of pipe, in.
D $\quad$ = average density of steam at system pressure, lbs./cu.ft.
V $\quad$ = velocity of steam in pipe, ft./min.
A_{INCHES} = actual cross-sectional area of pipe, sq.in.
A_{FEET} = actual cross-sectional area of pipe, sq.ft.

B. Steam Condensate Pipe Sizing Equations

$$FS = \frac{H_{S_{SS}} - H_{S_{CR}}}{H_{L_{CR}}} \times 100$$

$$W_{CR} = \frac{FS}{100} \times W$$

FS $\quad$ = flash steam, percentage %
$H_{S_{SS}}$ = sensible heat at steam supply pressure, Btu/lbs.
$H_{S_{CR}}$ = sensible heat at condensate return pressure, Btu/lbs.
$H_{L_{CR}}$ = latent heat at condensate return pressure, Btu/lbs.
W $\quad$ = steam flow rate, lbs./h
W_{CR} = condensate flow based on percentage of flash steam created during condensing process, lbs./h. Use this flow rate in the preceding steam equations to determine the condensate return pipe size.

3.19 Air Conditioning Condensate

$$GPM_{AC\ COND} = \frac{CFM \times \Delta W_{LB.}}{SpV \times 8.33}$$

$$GPM_{AC\ COND} = \frac{CFM \times \Delta W_{GR.}}{SpV \times 8.33 \times 7000}$$

$GPM_{AC\ COND}$ = air conditioning condensate flow (gal./min.)
CFM = air flow rate (cu.ft./min.)
SpV = specific volume of air (cu.ft./lbs.DA) = 13.33 ft³/lbs. DA at 60°F
$\Delta W_{LB.}$ = specific humidity (lbs.H$_2$O/lbs.DA)
$\Delta W_{GR.}$ = specific humidity (Gr.H$_2$O/lbs.DA)

3.20 Humidification

$$GRAINS_{REQ'D} = \left(\frac{W_{GR.}}{SpV}\right)_{ROOM\ AIR} - \left(\frac{W_{GR.}}{SpV}\right)_{SUPPLY\ AIR}$$

$$POUNDS_{REQ'D} = \left(\frac{W_{LB.}}{SpV}\right)_{ROOM\ AIR} - \left(\frac{W_{LB.}}{SpV}\right)_{SUPPLY\ AIR}$$

$$LBS.STM./HR = \frac{CFM \times GRAINS_{REQ'D} \times 60}{7000} = CFM \times POUNDS_{REQ'D} \times 60$$

$GRAINS_{REQ'D}$ = grains of moisture required (Gr.H$_2$O/cu.ft.)
$POUNDS_{REQ'D}$ = pounds of moisture required (lbs.H$_2$O/cu.ft.)
CFM = air flow rate (cu.ft./min.)
SpV = specific volume of air (cu.ft./lbs.DA)
$W_{GR.}$ = specific humidity (Gr.H$_2$O/lbs.DA)
$W_{LB.}$ = specific humidity (lbs.H$_2$O/lbs.DA)

3.21 Humidifier Sensible Heat Gain

$$H_S = (0.244 \times Q \times \Delta T) + (L \times 380)$$

H_S = sensible heat gain (Btu/h)
Q = steam flow (lbs. steam/h)
ΔT = steam temperature – supply air temperature (°F)
L = length of humidifier manifold (ft.)

3.22 Expansion Tanks

$$CLOSED\ V_T = V_S \times \frac{\left[\left(\frac{v_2}{v_1}\right) - 1\right] - 3\alpha\Delta T}{\left[\frac{P_A}{P_1} - \frac{P_A}{P_2}\right]}$$

$$OPEN \ V_T = 2 \times \left\{ \left[V_S \times \left[\left(\frac{v_2}{v_1} \right) - 1 \right] \right] - 3\alpha\Delta T \right\}$$

$$DIAPHRAGM \ V_T = V_S \times \frac{\left[\left(\frac{v_2}{v_1} \right) - 1 \right] - 3\alpha\Delta T}{1 - \left(\frac{P_1}{P_2} \right)}$$

V_T = volume of expansion tank (gallons)
V_S = volume of water in piping system (gallons)
$\Delta T = T_2 - T_1$ (°F)
T_1 = lower system temperature (°F)

    Heating Water T_1 = 45–50°F temperature at fill condition
    Chilled Water T_1 = supply water temperature
    Dual Temperature T_1 = chilled water supply temperature

T_2 = higher system temperature (°F)

    Heating Water T_2 = supply water temperature
    Chilled Water T_2 = 95°F ambient temperature (design weather data)
    Dual Temperature T_2 = heating water supply temperature

P_A = atmospheric pressure (14.7 psia)
P_1 = system fill pressure/minimum system pressure (psia)
P_2 = system operating pressure/maximum operating pressure (psia)
v_1 = SpV of H_2O at T_1 (cu.ft./lbs.H_2O) ASHRAE Fundamentals
v_2 = SpV of H_2O at T_2 (cu.ft./lbs.H_2O) ASHRAE Fundamentals
α = linear coefficient of expansion

    α_{STEE} = 6.5×10^{-6}
    α_{COPPER} = 9.5×10^{-6}

System Volume Estimate:
    12 gal./ton
    35 gal./BHP
System Fill Pressure/Minimum System Pressure Estimate:
    Height of System + 5 to 10 psi OR 5–10 psi, whichever is greater.
System Operating Pressure/Maximum Operating Pressure Estimate:
    150 lbs. Systems   45–125 psi
    250 lbs. Systems   125–225 psi

3.23 Air Balance Equations

SA   = Supply Air
RA   = Return Air
OA   = Outside Air
EA   = Exhaust Air
RFA   = Relief Air
SA   = RA + OA = RA + EA + RFA

If minimum OA (ventilation air) is greater than EA, then

OA = EA + RFA

If EA is greater than minimum OA (ventilation air), then

OA = EA   RFA = 0

For Economizer Cycle:

$$OA = SA = EA + RFA \qquad RA = 0$$

3.24 Efficiencies

$$COP = \frac{BTU\ OUTPUT}{BTU\ INPUT} = \frac{EER}{3.413}$$

$$EER = \frac{BTU\ OUTPUT}{WATTS\ INPUT} = COP \times 3.413$$

$$KW/TON = \frac{12,000\ BTU/HR\ TON}{COP \times 3,517\ BUT/HR\ KW}$$

Turndown Ratio = Maximum Firing Rate: Minimum Firing Rate (e.g., 5:1, 10:1, 25:1)

$$OVERALL\ THERMAL\ EFF. = \frac{GROSS\ BTU\ OUTPUT}{GROSS\ BTU\ INPUT} \times 100\%$$

$$COMBUSTION\ EFF. = \frac{BTU\ INPUT - BTU\ STACK\ LOSS}{BTU\ INPUT} \times 100\%$$

Overall Thermal Efficiency Range   75–90%

Combustion Efficiency Range   85–95%

3.25 Cooling Towers and Heat Exchangers

$$APPROACH_{CT'S} = LWT - AWB$$

$$APPROACH_{HE'S} = EWT_{HS} - LWT_{CS}$$

$$RANGE = EWT - LWT$$

EWT = entering water temperature (°F)
LWT = leaving water temperature (°F)
AWB = ambient wet bulb temperature (Design WB – °F)
HS   = hot side
CS   = cold side

3.26 Cooling Tower/Evaporative Cooler Blowdown Equations

$$C = \frac{(E+D+B)}{(D+B)}$$

$$B = \frac{E-[(C-1) \times D]}{(C-1)}$$

$$E = GPM_{COND.} \times R \times 0.0008$$

$$D = GPM_{COND.} \times 0.0002$$

$$R = EWT - LWT$$

B = blowdown, GPM
C = cycles of concentration
D = drift, GPM
E = evaporation, GPM
EWT = entering water temperature, °F
LWT = leaving water temperature, °F
R = range, °F

3.27 Electricity

A. General

$$KVA = KW + KVAR$$

$$PF = KW/KVA$$

B. Single-Phase Power

$$KW_{1\phi} = \frac{V \times A \times PF}{1000}$$

$$KVA_{1\phi} = \frac{V \times A}{1000}$$

$$BHP_{1\phi} = \frac{V \times A \times PF \times DEVICE_{EFF.}}{746}$$

$$MHP_{1\phi} = \frac{BHP_{1\phi}}{M/D_{EFF.}}$$

C. Three-Phase Power

$$KW_{3\phi} = \frac{\sqrt{3} \times V \times A \times PF}{1000}$$

$$KVA_{3\phi} = \frac{\sqrt{3} \times V \times A}{1000}$$

$$BHP_{3\phi} = \frac{\sqrt{3} \times V \times A \times PF \times DEVICE_{EFF.}}{746}$$

$$MHP_{3\phi} = \frac{BHP_{3\phi}}{M/D_{EFF.}}$$

KVA = total power (kilovolt amps)
KW = real power, electrical energy (kilowatts)
KVAR = reactive power or "imaginary" power (kilovolt amps reactive)
V = voltage (volts)
A = current (amps)
PF = power factor (0.75–0.95)
BHP = brake horsepower
MHP = motor horsepower
EFF = efficiency
M/D = motor drive

3.28 Moisture Condensation on Glass

$$T_{GLASS} = T_{ROOM} - \left[\frac{R_{IA}}{R_{GLASS}} \times (T_{ROOM} - T_{OA}) \right]$$

$$T_{GLASS} = T_{ROOM} - \left[\frac{U_{GLASS}}{U_{IA}} \times (T_{ROOM} - T_{OA}) \right]$$

If $T_{GLASS} < DP_{ROOM}$ condensation occurs

T = temperature (°F)
R = R-Value (h sq.ft. °F/Btu)
U = U-Value (Btu./h sq.ft. °F)
IA = inside airfilm
OA = design outside air temperature
DP = dewpoint

3.29 Calculating Heating Loads for Loading Docks, Heavily Used Vestibules and Similar Spaces

A. Find volume of space to be heated (cu.ft.).

B. Determine acceptable warm-up time for space (min.).

C. Divide volume by time (CFM).

D. Determine inside and outside design temperatures—assume inside space temperature has dropped to the outside design temperature because doors have been open for an extended period of time.

E. Use sensible heat equation to determine heating requirement using CFM and inside and outside design temperatures determined earlier in this Part.

3.30 Ventilation of Mechanical Rooms with Refrigeration Equipment

A. For a more detailed description of ventilation requirements for mechanical rooms with refrigeration equipment, see ASHRAE Standard 15 and Part 8.

B. Completely Enclosed Equipment Rooms

$$CFM = 100 \times G^{0.5}$$

CFM = exhaust air flow rate required (cu.ft./minute)
G   = mass of refrigerant of largest system (pounds)

C. Partially Enclosed Equipment Rooms

$$FA = G^{0.5}$$

FA = ventilation free opening Area (sq.ft.)
G = mass of refrigerant of largest system (pounds)

3.31 Pipe Expansion Equations

A. L-Bends

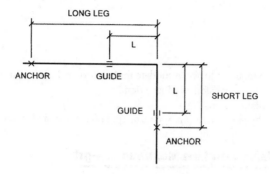

$$L = 6.225 \times \sqrt{\Delta D}$$

$F = 500 \text{ LB./PIPE DIA.} \times \text{PIPE DIA.}$

L = length of leg required to accommodate thermal expansion or contraction, feet
Δ = thermal expansion or contraction of long leg, inches
D = pipe outside diameter, inches
F = force exerted by pipe expansion or contraction on anchors and supports, lbs.
    See Tables in Part 18 for solved equations.

B. Z-Bends

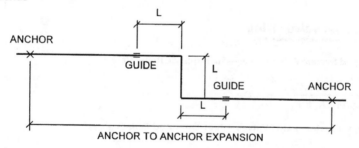

$$L = 4 \times \sqrt{\Delta D}$$

$F = 200 - 500 \text{ LB./PIPE DIA.} \times \text{PIPE DIA.}$

L = length of offset leg required to accommodate thermal expansion or contraction, feet
Δ = anchor to anchor expansion or contraction, inches
D = pipe outside diameter, inches
F = force exerted by pipe expansion or contraction on anchors and supports, lbs. See
    Tables in Part 18 for solved equations.

C. U-Bends or Expansion Loops

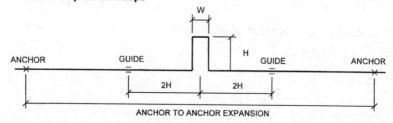

$$L = 6.225 \times \sqrt{\Delta D}$$

$$F = 200 \text{ LB./PIPE DIA.} \times \text{PIPE DIA.}$$

$$L = 2H + W$$

$$H = 2W$$

$$L = 5W$$

L = length of loop required to accommodate thermal expansion or contraction, ft.

D = anchor to anchor expansion or contraction, in.

D = pipe outside diameter, in.

F = force exerted by pipe expansion or contraction on anchors and supports, lbs.

3.32 Relief Valve Vent Line Maximum Length

$$L = \frac{9 \times P_1^2 \times D^5}{C^2} = \frac{9 \times P_2^2 \times D^5}{16 \times C^2}$$

$$P_1 = 0.25 \times [(PRESSURE\ SETTING \times 1.1) + 14.7]$$

$$P_2 = [(PRESSURE\ SETTING \times 1.1) + 14.7]$$

L = maximum length of relief vent line in feet

D = inside diameter of pipe in inches

C = minimum discharge of air in lbs./min.

3.33 Relief Valve Sizing

A. **Liquid System Relief Valves—Spring-Style Relief Valves**

$$A = \frac{GPM \times \sqrt{G}}{28.14 \times K_B \times K_V \times \sqrt{\Delta P}}$$

B. **Liquid System Relief Valves—Pilot-Operated Relief Valves**

$$A = \frac{GPM \times \sqrt{G}}{36.81 \times K_V \times \sqrt{\Delta P}}$$

C. **Steam System Relief Valves**

$$A = \frac{W}{51.5 \times K \times P \times K_{SH} \times K_N \times K_B}$$

D. **Gas and Vapor System Relief Valves—lbs./h**

$$A = \frac{W \times \sqrt{TZ}}{C \times K \times P \times K_B \times \sqrt{M}}$$

E. **Gas and Vapor System Relief Valves—SCFM**

$$A = \frac{SCFM \times \sqrt{TGZ}}{1.175 \times C \times K \times P \times K_B}$$

F. Relief Valve Equation Definitions

1. A   = Minimum required effective relief valve discharge area (sq.in.)
2. GPM = Required relieving capacity at flow conditions (gal./min.)
3. W   = Required relieving capacity at flow conditions (lbs./h)
4. SCFM = Required relieving capacity at flow conditions (standard cu.ft./min.)
5. G   = Specific gravity of liquid, gas, or vapor at flow conditions
             Water = 1.0 for most HVAC applications
             Air = 1.0
6. C   = Coefficient determined from expression of ratio of specific heats
             C = 315 if value is unknown
7. K   = Effective coefficient of discharge
             K = 0.975
8. K_B = Capacity correction factor due to back pressure
             K_B = 1.0 for atmospheric discharge systems
9. K_V = Flow correction factor due to viscosity
             K_V = 0.9 to 1.0 for most HVAC applications with water
10. K_N = Capacity correction factor for dry saturated steam at set pressures above
             1500 psia and up to 3200 psia
             K_N = 1.0 for most HVAC applications
11. K_{SH} = Capacity correction factor due to the degree of superheat
             K_{SH} = 1.0 for saturated steam
12. Z   = Compressibility factor
             Z = 1.0 if value is unknown
13. P   = Relieving pressure (psia)
             P = Set pressure (psig) + over pressure (10% psig) + atmospheric pressure
             (14.7 psia)
14. ΔP = Differential pressure (psig)
             ΔP = Set pressure (psig) + over pressure (10% psig) – back pressure (psig)
15. T   = Absolute temperature (°R = °F + 460)
16. M   = Molecular weight of the gas or vapor

G. Relief Valve Sizing Notes

1. When multiple relief valves are used, one valve shall be set at or below the maximum allowable working pressure, and the remaining valves may be set up to 5 percent over the maximum allowable working pressure.
2. When sizing multiple relief valves, the total area required is calculated on an over pressure of 16 percent or 4 psi, whichever is greater.
3. For superheated steam, the following correction factor values may be used:
   a. Superheat up to 400°F: 0.97   (range 0.979–0.998)
   b. Superheat up to 450°F: 0.95   (range 0.957–0.977)
   c. Superheat up to 500°F: 0.93   (range 0.930–0.968)
   d. Superheat up to 550°F: 0.90   (range 0.905–0.974)
   e. Superheat up to 600°F: 0.88   (range 0.882–0.993)
   f. Superheat up to 650°F: 0.86   (range 0.861–0.988)
   g. Superheat up to 700°F: 0.84   (range 0.841–0.963)
   h. Superheat up to 750°F: 0.82   (range 0.823–0.903)
   i. Superheat up to 800°F: 0.80   (range 0.805–0.863)
   j. Superheat up to 850°F: 0.78   (range 0.786–0.836)
   k. Superheat up to 900°F: 0.75   (range 0.753–0.813)
   l. Superheat up to 950°F: 0.72   (range 0.726–0.792)
   m. Superheat up to 1000°F: 0.70   (range 0.704–0.774)

4. Gas and vapor properties:

Gas and vapor properties

Gas or Vapor	Molecular Weight	Ratio of Specific Heats	Coefficient C	Specific Gravity
Acetylene	26.04	1.25	342	0.899
Air	28.97	1.40	356	1.000
Ammonia (R-717)	17.03	1.30	347	0.588
Argon	39.94	1.66	377	1.379
Benzene	78.11	1.12	329	2.696
N-Butane	58.12	1.18	335	2.006
Iso-Butane	58.12	1.19	336	2.006
Carbon Dioxide	44.01	1.29	346	1.519
Carbon Disulphide	76.13	1.21	338	2.628
Carbon Monoxide	28.01	1.40	356	0.967
Chlorine	70.90	1.35	352	2.447
Cyclohexane	84.16	1.08	325	2.905
Ethane	30.07	1.19	336	1.038
Ethyl Alcohol	46.07	1.13	330	1.590
Ethyl Chloride	64.52	1.19	336	2.227
Ethylene	28.03	1.24	341	0.968
Helium	4.02	1.66	377	0.139
N-Heptane	100.20	1.05	321	3.459
Hexane	86.17	1.06	322	2.974
Hydrochloric Acid	36.47	1.41	357	1.259
Hydrogen	2.02	1.41	357	0.070
Hydrogen Chloride	36.47	1.41	357	1.259
Hydrogen Sulphide	34.08	1.32	349	1.176
Methane	16.04	1.31	348	0.554
Methyl Alcohol	32.04	1.20	337	1.106
Methyl Butane	72.15	1.08	325	2.491
Methyl Chloride	50.49	1.20	337	1.743
Natural Gas	19.00	1.27	344	0.656
Nitric Oxide	30.00	1.40	356	1.036
Nitrogen	28.02	1.40	356	0.967
Nitrous Oxide	44.02	1.31	348	1.520
N-Octane	114.22	1.05	321	3.943
Oxygen	32.00	1.40	356	1.105
N-Pentane	72.15	1.08	325	2.491
Iso-Pentane	72.15	1.08	325	2.491
Propane	44.09	1.13	330	1.522
R-11	137.37	1.14	331	4.742
R-12	120.92	1.14	331	4.174
R-22	86.48	1.18	335	2.985
R-114	170.93	1.09	326	5.900
R-123	152.93	1.10	327	5.279
R-134a	102.03	1.20	337	3.522
Sulfur Dioxide	64.04	1.27	344	2.211
Toluene	92.13	1.09	326	3.180

3.34 Motor Drive Formulas

$$D_{FP} \times RPM_{FP} = D_{MP} \times RPM_{MP}$$

$$BL = [(D_{FP} + D_{MP}) \times 1.5708] + (2 \times L)$$

D_{FP} = fan pulley diameter
D_{MP} = motor pulley diameter
RPM_{FP} = fan pulley RPM
RPM_{MP} = motor pulley RPM
BL = belt length
L = center to center distance of fan and motor pulleys

3.35 Domestic Water Heater Sizing

$$H_{OUTPUT} = GPH \times 8.34 \; LBS./GAL. \times \Delta T \times 1.0$$

$$H_{INPUT} = \frac{GPH \times 8.34 \; LBS./GAL. \times \Delta T}{\% EFFICIENCY}$$

$$GPH = \frac{H_{INPUT} \times \% EFFICIENCY}{\Delta T \times 8.34 \; LBS./GAL.} = \frac{KW \times 3413 \; BTU / KW}{\Delta T \times 8.34 \; LBS./GAL.}$$

$$\Delta T = \frac{H_{INPUT} \times \% EFFICIENCY}{GPH \times 8.34 \; LBS./GAL.} = \frac{KW \times 3413 \; BTU / KW}{GPH \times 8.34 \; LBS./GAL.}$$

$$KW = \frac{GPH \times 8.34 \; LBS./GAL. \times \Delta T \times 1.0}{3413 \; BTU / KW}$$

$$\% COLD \; WATER = \frac{T_{HOT} - T_{MIX}}{T_{HOT} - T_{COLD}}$$

$$\% HOT \; WATER = \frac{T_{MIX} - T_{COLD}}{T_{HOT} - T_{COLD}}$$

H_{OUTPUT} = heating capacity – output
H_{INPUT} = heating capacity – input
GPH   = recovery rate – gallons per hour
ΔT = temperature rise – °F
kW   = kilowatts
T_{COLD} = temperature – cold water – °F
T_{HOT} = temperature – hot water – °F
T_{MIX} = temperature – mixed water – °F

3.36 Domestic Hot Water Recirculation Pump/Supply Sizing

A. Determine the approximate total length of all hot water supply and return piping.

B. Multiply this total length by 30 Btu/ft. for insulated pipe and 60 Btu/ft. for uninsulated pipe to obtain the approximate heat loss.

C. Divide the total heat loss by 10,000 to obtain the total pump capacity in GPM.

D. Select a circulating pump to provide the total required GPM and obtain the head created at this flow.

E. Multiply the head by 100 and divide by the total length of the longest run of the hot water return piping to determine the allowable friction loss per 100 feet of pipe.

F. Determine the required GPM in each circulating loop and size the hot water return pipe based on this GPM and the allowable friction loss as determined earlier.

3.37 Swimming Pools

A. Sizing Outdoor Pool Heater

1. Determine pool capacity in gallons – obtain from architect if available.
   Length × Width × Depth × 7.5 gal./cu.ft. (If depth is not known, assume an average depth of 5.5 feet.)
2. Determine heat pick-up time in hours from owner.
3. Determine pool water temperature in °F from the owner. If owner does not specify temperature, assume 80°F.
4. Determine the average air temperature on the coldest month in which the pool will be used.
5. Determine the average wind velocity in miles per hour. For pools less than 900 square feet and where the pool is sheltered by nearby buildings, fences, shrubs, etc., from the prevailing wind, an average wind velocity of less than 3.5 mph may be assumed. The surface heat loss factor of 5.5 Btuh/sq.ft. °F in the following equation assumes a wind velocity of 3.5 mph. If a wind velocity of less than 3.5 mph is used, multiply the equation by 0.75; for 5.0 mph, multiply the equation by 1.25; and for 10 mph, multiply the equation by 2.0.
6. Pool heater equations:

$$H_{POOLHEATER} = H_{HEAT-UP} + H_{SURFACE\ LOSS}$$

$$H_{HEAT-UP} = \frac{GAL. \times 8.34\ LBS./GAL. \times \Delta T_{WATER} \times 1.0\ BTU/LBS.°F}{HEAT\ PICK-UP\ TIME}$$

$$H_{SURFACELOSS} = 5.5\ BTU/HRSQ.FT.°F \times \Delta T_{WATER/ARI} \times POOL\ AREA$$

$$\Delta T_{WATER} = T_{FINAL} - T_{INITIAL}$$

$$T_{FINAL} = POOL\ WATER\ TEMPERATURE$$

$$T_{INITIAL} = 50°F$$

$$\Delta T_{WATER/AIR} = T_{FINAL} - T_{AVERAGEAIR}$$

H = heating capacity (Btu/h)
ΔT = temperature difference (°F)

PART **4**

Conversion Factors

4.01 Length

1 mile $= 1760$ yds. $= 5280$ ft. $= 63,360$ in. $= 1.609$ km
1 ft. $= 0.3048$ m $= 30.48$ cm $= 304.8$ mm
1 in. $= 2.54$ cm $= 25.4$ mm
1 cm $= 0.3937$ in.
1 m $= 39.37$ in. $= 3.2808$ ft. $= 1.094$ yds.
1 km $= 3281$ ft. $= 0.6214$ miles $= 1094$ yds.
1 fathom $= 6$ feet $= 1.828804$ meters
1 furlong $= 660$ feet

4.02 Weight

1 gal.H_2O $= 8.33$ lbs.H_2O
1 lb. $= 16$ oz. $= 7000$ grains $= 0.4536$ kg
1 ton $= 2000$ lbs. $= 907$ kg
1 kg $= 2.205$ lbs.
1 lb.steam $= 1$ lb.H_2O

4.03 Area

1 sq.ft. $= 144$ sq.in.
1 acre $= 43,560$ sq.ft. $= 4840$ sq.yds. $= 0.4047$ hectares
1 sq.mile $= 640$ acres
1 sq.yd. $= 9$ sq.ft. $= 1296$ sq.in.
1 hectare $= 2.417$ acres
1 sq.m $= 1,550$ sq.in. $= 10.7639$ sq.ft. $= 1.1968$ sq.yds.

4.04 Volume

1 cu.yd. $= 27$ cu.ft. $= 46,656$ cu.in. $= 1616$ pints $= 807.9$ quarts $= 764.6$ liters
1 cu.ft. $= 1,728$ cu.in.
1 liter $= 0.2642$ gallons $= 1.057$ quarts $= 2.113$ pints
1 gallon $= 4$ quarts $= 8$ pints $= 3.785$ liters
1 cu.m $= 61,023$ cu.in. $= 35.3134$ cu.ft. $= 1.3093$ cu.yds.
1 barrel oil $= 42$ gallons oil
1 barrel beer $= 31.5$ gallons beer
1 barrel wine $= 31.0$ gallons wine
1 bushel $= 1.2445$ cu.ft. $= 32$ quarts (dry) $= 64$ pints (dry) $= 4$ pecks
1 hogshead $= 63$ gallons $= 8.42184$ cu.ft.

4.05 Velocity

1 mph $= 5280$ ft./h $= 88$ ft./min. $= 1.467$ ft./sec. $= 0.8684$ knot
1 knot $= 1.1515$ mph $= 1.8532$ km/hr. $= 1.0$ nautical mile/h
1 league $= 3.0$ miles (approx.)

4.06 Speed of Sound in Air

1128.5 ft./sec. $= 769.4$ mph

4.07 Pressure

14.7 psi = 33.95 ft. H_2O = 29.92 in. Hg = 407.2 in. W.G. = 2116.8 lbs./sq.ft.
1 psi   = 2.307 ft. H_2O = 2.036 in. Hg = 16 oz = 27.7 in. WC
1 ft. H_2O = 0.4335 psi = 62.43 lbs./sq.ft.
1 oz   = 1.73 in. WC

4.08 Density

A. Water

62.43 lbs./cu.ft.   = 8.33 lbs./gal. = 0.1337 cu.ft./gal.
1 cu.ft.   = 7.48052 gallons = 29.92 quarts = 62.43 lbs.H_2O

B. Standard Air @ 60°F, 14.7 psi

13.329 cu.ft./lbs. = 0.0750 lbs./cu.ft.
1 lb./cu.ft.   = 177.72 cu.ft./lb.
1 cu.ft./lb.   = 0.00563 lbs./cu.ft.
1 kg/cu.m   = 16.017 lbs./cu.ft.
1 cu.m/kg   = 0.0624 cu.ft./lb

4.09 Energy

1 hp	= 0.746 kW = 746 watts = 2,545 Btuh. 1.0 kva
1 kW	= 1,000 watts = 3413 Btuh = 1.341 hp
1 watt	= 3.413 Btuh
1 ton AC	= 12,000 Btuh cooling = 15,000 Btuh heat rejection
1 Btuh	= 1 Btu/h
1 bhp	= 34,500 Btuh (33,472 Btuh) = 34.5 lbs.stm/h = 34.5 lbs.H_2O/h = 0.069 gpm = 4.14 gph = 140 edr (sq.ft. of equivalent direct radiation)
1 therm	= 100,000 Btuh
1 mbh	= 1,000 Btuh
1 lb.stm/h	= 0.002 gpm
1 gpm	= 500 lbs.stm./h
1 edr (equivalent direct radiation)	= 0.000496 gpm = 0.25 lbs.stm.cond./h
1,000 edr	= 0.496 gpm
1 edr hot water	= 150 Btu/h
1 edr steam	= 240 Btu/h
1 edr	= 240 Btu/h (up to 1,000 ft. above sea level)
1 edr	= 230 Btu/h (1,000 ft.–3,000 ft. above sea level)
1 edr	= 223 Btu/h (3,000 ft. –5,000 ft. above sea level)
1 edr	= 216 Btu/h (5,000 ft.–7,000 ft. above sea level)
1 edr	= 209 Btu/h (7,000 ft.–10,000 ft. above sea level)

4.10 Flow

1 mgd (million gal./day) = 1.547 cu.ft./sec. = 694.4 gpm
1 cu.ft./min.   = 62.43 lbs.H_2O/min. = 448.8 gph

4.11 HVAC Metric Conversions

kJ/h	= Btu/h × 1.055
cmm	= cfm × 0.02832
lpm	= gpm × 3.785
kJ/kg	= Btu/lb. × 2.326
meters	= ft. × 0.3048
sq. meters	= sq. ft. × 0.0929
cu. meters	= cu. ft. × 0.02832
kg	= lbs. × 0.4536
1.0 gpm	= 500 lbs.steam/h
1.0 lb.stm./h	= 0.002 gpm
1.0 lb.H_2O/h	= 1.0 lb.steam/h
kg/cu.m	= lbs./cu.ft. × 16.017 (density)
cu.m/kg	= cu.ft./lb. × 0.0624 (specific volume)

kg H_2O/kg DA = Gr H_2O/lb. DA/7000 = lb. H_2O/lb. DA

4.12 Fuel Conversion Factors

A. Electric Baseboard to Hydronic Baseboard

1. KWH × 1.19   = KWH for electric boiler
2. KWH × 0.033 = gal. for oil-fired boiler
3. KWH × 0.046 = therms for gas-fired boiler

B. Electric Furnace to Hydronic Baseboard

1. KWH × 1.0   = KWH for electric boiler
2. KWH × 0.028 = gal. for oil-fired boiler
3. KWH × 0.038 = therms for gas-fired boiler

C. Ceiling Cable to Hydronic Baseboard

1. KWH × 1.06   = KWH for electric boiler
2. KWH × 0.03   = gal. for oil-fired boiler
3. KWH × 0.041 = therms for gas-fired boiler

D. Heat Pump to Hydronic Baseboard

1. KWH × 1.88   = KWH for electric boiler
2. KWH × 0.052 = gal. for oil-fired boiler
3. KWH × 0.073 = therms for gas-fired boiler

E. Electric Baseboard to Warm Air Furnace

1. KWH × 1.19   = KWH for electric furnace
2. KWH × 0.039 = gal. for oil-fired furnace
3. KWH × 0.054 = therms for gas-fired furnace

F. Electric Furnace to Fuel-Fired Furnace

1. KWH × 0.032 = gal. for oil-fired furnace
2. KWH × 0.045 = therms for gas-fired furnace

G. Ceiling Cable to Warm Air Furnace

1. KWH × 1.06   = KWH for electric furnace
2. KWH × 0.034 = gal. for oil-fired furnace
3. KWH × 0.048 = therms for gas-fired furnace

H. Heat Pump to Warm Air Furnace

1. KWH × 1.88 = KWH for electric furnace
2. KWH × 0.061 = gal. for oil-fired furnace
3. KWH × 0.085 = therms for gas-fired furnace

I.   Warm Air Systems to Hydronic Baseboard System

1. gal. oil for W.A. × 0.857   = gal. for hydronics
2. therms gas for W.A. × 0.857 = therms for hydronics
3. gal. oil for W.A. × 1.2   = therms for hydronics
4. therms gas for W.A. × 0.612 = gal. for hydronics

Cooling Load Rules of Thumb

5.01 Offices, Commercial

A. General

1. Total Heat	300–400 sq.ft./ton	(Range 230–520)
2. Total Heat	30–40 Btuh/sq.ft.	(Range 23–52)
3. Room Sens. Heat	25–28 Btuh/sq.ft.	(Range 19–37)
4. SHR	0.75–0.93	
5. Perimeter Spaces	1.0–3.0 cfm/sq.ft.	
6. Interior Spaces	0.5–1.5 cfm/sq.ft.	
7. Building Block cfm	1.0–1.5 cfm/sq.ft.	
8. Air Change Rate	4–10 AC/h	

B. Large, Perimeter

1. Total Heat	225–275 sq.ft./ton
2. Total Heat	43–53 Btuh/sq.ft.

C. Large, Interior

1. Total Heat	300–350 sq.ft./ton
2. Total Heat	34–40 Btuh/sq.ft.

D. Small

1. Total Heat	325–375 sq.ft./ton
2. Total Heat	32–37 Btuh/sq.ft.

5.02 Banks, Court Houses, Municipal Buildings, Town Halls

A. Total Heat	200–250 sq.ft./ton	(Range 160–340)
B. Total Heat	48–60 Btuh/sq.ft.	(Range 35–75)
C. Room Sens. Heat	28–38 Btuh/sq.ft.	(Range 21–48)
D. SHR	0.75–0.90	
E. Air Change Rate	4–10 AC/h	

5.03 Police Stations, Fire Stations, Post Offices

A. Total Heat	250–350 sq.ft./ton	(Range 200–400)
B. Total Heat	34–48 Btuh/sq.ft.	(Range 30–60)
C. Room Sens. Heat	25–35 Btuh/sq.ft.	(Range 20–40)
D. SHR	0.75–0.90	
E. Air Change Rate	4–10 AC/h	

5.04 Precision Manufacturing

A. Total Heat	50–300 sq.ft./ton
B. Total Heat	40–240 Btuh/sq.ft.
C. Room Sens. Heat	32–228 Btuh/sq.ft.

D. SHR	0.80–0.95
E. Air Change Rate	10–50 AC/h

5.05 Computer Rooms

A. Total Heat	50–150 sq.ft./ton
B. Total Heat	80–240 Btuh/sq.ft.
C. Room Sens. Heat	64–228 Btuh/sq.ft.
D. SHR	0.80–0.95
E. Air Flow	2.0–4.0 cfm/sq.ft.
F. Air Change Rate	15–20 AC/h

5.06 Restaurants

A. Total Heat	100–250 sq.ft./ton	(Range 75–300)
B. Total Heat	48–120 Btuh/sq.ft.	(Range 40–155)
C. Room Sens. Heat	21–62 Btuh/sq.ft.	(Range 20–80)
D. SHR	0.65–0.80	
E. Air Flow	1.5–4.0 cfm/sq.ft.	
F. Air Change Rate	8–12 AC/h	

5.07 Kitchens (Depends Primarily on Kitchen Equipment)

A. Total Heat	150–350 sq.ft./ton	(at 85°F space)
B. Total Heat	34–80 Btuh/sq.ft.	(at 85°F space)
C. Room Sens. Heat	20–56 Btuh/sq.ft.	(at 85°F space)
D. SHR	0.60–0.70	
E. Air Flow	1.5–2.5 cfm/sq.ft.	
F. Air Change Rate	12–15 AC/h	

5.08 Cocktail Lounges, Bars, Taverns, Clubhouses, Nightclubs

A. Total Heat	150–200 sq.ft./ton	(Range 75–300)
B. Total Heat	60–80 Btuh/sq.ft.	(Range 40–155)
C. Room Sens. Heat	27–40 Btuh/sq.ft.	(Range 20–80)
D. SHR	0.65–0.80	

E. Spaces   1.5–4.0 cfm/sq.ft.

F. Air Change Rate   15–20 AC/h   **Cocktail Lounges, Bars, Taverns, Clubhouses**

G. Air Change Rate   20–30 AC/h   **Night Clubs**

5.09 Hospital Patient Rooms, Nursing Home Patient Rooms

A. Total Heat   250–300 sq.ft./ton   (Range 200–400)

B. Total Heat   40–48 Btuh/sq.ft.   (Range 30–60)

C. Room Sens. Heat   32–46 Btuh/sq.ft.   (Range 25–50)

D. SHR   0.75–0.85

5.10 Buildings w/100 percent OA Systems (e.g., Laboratories, Hospitals)

A. Total Heat   100–300 sq.ft./ton

B. Total Heat   40–120 Btuh/sq.ft.

5.11 Medical/Dental Centers, Clinics, and Offices

A. Total Heat   250–300 sq.ft./ton   (Range 200–400)

B. Total Heat   40–48 Btuh/sq.ft.   (Range 30–60)

C. Room Sens. Heat   32–46 Btuh/sq.ft.   (Range 25–50)

D. SHR   0.75–0.85

E. Air Change Rate   8–12 AC/h

5.12 Residential

A. Total Heat   500–700 sq.ft./ton

B. Total Heat   17–24 Btuh/sq.ft.

C. Room Sens. Heat   12–20 Btuh/sq.ft.

D. SHR   0.80–0.95

5.13 Apartments (Eff., One-Room, Two-Room)

A. Total Heat   350–450 sq.ft./ton   (Range 300–500)

B. Total Heat   27–34 Btuh/sq.ft.   (Range 24–40)

C. Room Sens. Heat   22–30 Btuh/sq.ft.   (Range 20–35)

D. SHR   0.80–0.95

5.14 Motel and Hotel Public Spaces

A. Total Heat   250–300 sq.ft./ton   (Range 160–375)

B. Total Heat   40–48 Btuh/sq.ft.   (Range 32–74)

C. Room Sens. Heat   32–46 Btuh/sq.ft.   (Range 25–60)

D. SHR   0.75–0.90

5.15 Motel and Hotel Guest Rooms, Dormitories

A. Total Heat   400–500 sq.ft./ton   (Range 300–600)

B. Total Heat   24–30 Btuh/sq.ft.   (Range 20–40)

C. Room Sens. Heat   20–25 Btuh/sq.ft.   (Range 15–35)

D. SHR   0.80–0.95

5.16 School Classrooms

A. Total Heat   225–275 sq.ft./ton   (Range 150–350)

B. Total Heat   43–53 Btuh/sq.ft.   (Range 35–80)

C. Room Sens. Heat   25–42 Btuh/sq.ft.   (Range 20–65)

D. SHR   0.65–0.80

E. Air Change Rate   4–12 AC/h

5.17 Dining Halls, Lunch Rooms, Cafeterias, Luncheonettes

A. Total Heat   100–250 sq.ft./ton   (Range 75–300)

B. Total Heat   48–120 Btuh/sq.ft.   (Range 40–155)

C. Room Sens. Heat   21–62 Btuh/sq.ft.   (Range 20–80)

D. SHR   0.65–0.80

E. Spaces   1.5–4.0 cfm/sq.ft.

F. Air Change Rate   12–15 AC/h

5.18 Libraries, Museums

A. Total Heat   250–350 sq.ft./ton   (Range 160–400)

B. Total Heat   34–48 Btuh/sq.ft.   (Range 30–75)

C. Room Sens. Heat   22–32 Btuh/sq.ft.   (Range 20–50)

D. SHR   0.80–0.90

E. Air Change Rate   8–12 AC/h

5.19 Retail, Department Stores

A. Total Heat   200–300 sq.ft./ton   (Range 200–500)

B. Total Heat   40–60 Btuh/sq.ft.   (Range 24–60)

C. Room Sens. Heat   32–43 Btuh/sq.ft.   (Range 16–43)

D. SHR   0.65–0.90

E. Air Change Rate   6–10 AC/h

5.20 Drug, Shoe, Dress, Jewelry, Beauty, Barber, and Other Shops

A. Total Heat   175–225 sq.ft./ton   (Range 100–350)

B. Total Heat   53–69 Btuh/sq.ft.   (Range 35–115)

C. Room Sens. Heat   23–54 Btuh/sq.ft.   (Range 15–90)

D. SHR   0.65–0.90

E. Air Change Rate   6–10 AC/h

5.21 Supermarkets

A. Total Heat   250–350 sq.ft./ton   (Range 150–400)

B. Total Heat   34–48 Btuh/sq.ft.   (Range 30–80)

C. Room Sens. Heat   25–40 Btuh/sq.ft.   (Range 22–67)

D. SHR   0.65–0.85

E. Air Change Rate   4–10 AC/h

5.22 Malls, Shopping Centers

A. Total Heat   150–350 sq.ft./ton   (Range 150–400)

B. Total Heat   34–80 Btuh/sq.ft.   (Range 30–80)

C. Room Sens. Heat   25–67 Btuh/sq.ft.   (Range 22–67)

D.	SHR	0.65–0.85
E.	Air Change Rate	6–10 AC/h

5.23 Jails

A.	Total Heat	350–450 sq.ft./ton	(Range 300–500)
B.	Total Heat	27–34 Btuh/sq.ft.	(Range 24–40)
C.	Room Sens. Heat	22–30 Btuh/sq.ft.	(Range 20–35)
D.	SHR	0.80–0.95	

5.24 Auditoriums, Theaters

A.	Total Heat	0.05–0.07 tons/seat
B.	Total Heat	600–840 Btuh/seat
C.	Room Sens. Heat	325–385 Btuh/seat
D.	SHR	0.65–0.75
E.	Air Flow	15–30 cfm/seat
F.	Air Change Rate	8–15 AC/h

5.25 Churches

A.	Total Heat	0.04–0.06 tons/seat
B.	Total Heat	480–720 Btuh/seat
C.	Room Sens. Heat	260–330 Btuh/seat
D.	SHR	0.65–0.75
E.	Air Flow	15–30 cfm/seat
F.	Air Change Rate	8–15 AC/h

5.26 Bowling Alleys

A.	Total Heat	1.5–2.5 tons/alley
B.	Total Heat	18,000–30,000 Btuh/alley
C.	Air Change Rate	10–15 AC/h

5.27 All Spaces

A.	Total Heat	300–500 cfm/ton@20°F ΔT
B.	Total Heat	400 cfm/ton ± 20%@20°F ΔT

   C. **Perimeter Spaces**   **1.0–3.0 cfm/sq.ft.**

   D. **Interior Spaces**   **0.5–1.5 cfm/sq.ft.**

   E. **Building Block cfm**   **1.0–1.5 cfm/sq.ft.**

   F. **Air Change Rate**   **4 AC/h minimum**

   G. **Total heat includes ventilation. Room sensible heat does not include ventilation.**

5.28 Cooling Load Calculation Procedure

A. Obtain Building Characteristics

1. Construction materials.
2. Construction material properties: U-values, R-values, shading coefficients, solar heat gain coefficients.
3. Size.
4. Color.
5. Shape.
6. Location.
7. Orientation, N, S, E, W, NE, SE, SW, NW, etc.
8. External/internal shading.
9. Occupancy type and time of day.

B. Select Outdoor Design Weather Conditions

1. Temperature.
2. Wind direction and speed.
3. Conditions in selecting outdoor design weather conditions:
   a. Type of structure, heavy, medium, or light.
   b. Is structure insulated? If the structure is heated or cooled, the structure must be insulated by code.
   c. Is structure exposed to high winds?
   d. Infiltration or ventilation load.
   e. Amount of glass.
   f. Time of building occupancy.
   g. Type of building occupancy.
   h. Length of reduced indoor temperature.
   i. What is daily temperature range, minimum/maximum?
   j. Are there significant variations from ASHRAE weather data?
   k. What type of heating devices will be used?
   l. Expected cost of fuel.
4. See Part 15 for code restrictions on the selection of outdoor design conditions.

C. Select the indoor design temperature to be maintained in each space. See Part 15 for code restrictions on the selection of indoor design conditions.

D. Estimate temperatures in unconditioned spaces.

E. Select and/or compute U-values for walls, roof, windows, doors, partitions, etc.

F. Determine the area of walls, windows, floors, doors, partitions, etc.

G. Compute the conduction heat gains for all walls, windows, floors, doors, partitions, skylights, etc.

H. Compute the solar heat gains for all walls, windows, floors, doors, partitions, skylights, etc.

I. Infiltration heat gains are generally ignored unless space temperature and humidity tolerance are critical.

J. Compute the ventilation heat gain required.

K. Compute the internal heat gains from lights, people, and equipment.

L. Compute the sum of all heat gains indicated in items G, H, I, J, and K earlier in this list.

M. Include morning cool-down for buildings with intermittent use and night setup. See Part 15 for code restrictions on the excess HVAC system capacity permitted for morning cool-down.

N. Consider equipment and materials that will be brought into the building above the inside design temperature.

O. Cooling load calculations should be conducted using industry-accepted methods to determine the actual cooling load requirements.

P. Cooling load calculations are often performed using computer simulation programs. These programs greatly simplify the calculation process; however, the basic procedures and input information required are the same.

5.29 Cooling Load Peak Time Estimate (for Calculating Cooling Loads by Hand)

MONTH OF PEAK ROOM COOLING LOAD FOR VARIOUS EXPOSURES

Window Characteristics			Probable Month of Peak Room Cooling Load							
% Glass	Shade Coef.	Overhang	N	S	E	W	NE	SE	SW	NW
25	0.4	0	July	Sept.	July	July	July	Sept.	Sept.	July
25	0.4	1:2	July	Oct.	July	Aug.	July	Sept.	Sept.	July
25	0.4	1:1	July	Oct.	July	July	July	Sept.	Oct.	July
25	0.6	0	July	Sept.	July	July	July	Sept.	Sept.	July
25	0.6	1:2	July	Oct.	July	Aug.	July	Sept.	Sept.	July
25	0.6	1:1	July	Dec.	July	Sept.	July	Sept.	Oct.	July
50	0.4	0	July	Sept.	July	July	July	Sept.	Sept.	July
50	0.4	1:2	July	Oct.	July	Aug.	July	Sept.	Sept.	July
50	0.4	1:1	July	Dec.	July	Sept.	July	Sept.	Oct,	July
50	0.6	0	July	Oct.	July	July	July	Sept.	Sept.	July
50	0.6	1:2	July	Dec.	July	Aug.	July	Sept.	Oct.	July
50	0.6	1:1	July	Dec.	July	Sept.	July	Sept.	Dec.	July

Notes:
1 Percent glass is the percent of gross wall area for the particular exposure.
2 The shading coefficient refers to the overall shading coefficient. A shading coefficient of 0.4 is approximately equal to double-pane glass with the heat-absorbing plate out and the regular plate in, combined with medium-color Venetian blinds.
3 Although the room peak for south, southeast, and southwest exposures is September or later, the system peak will likely be in July.
4 The value for the overhang is the ratio of the depth of the overhang to the height of the window with the overhang at the same elevation as the top of the window.
5 The roof will peak in June or July.

Heating Load Rules of Thumb

6.01 All Buildings and Spaces

A. 20–60 Btuh/sq.ft.

B. 25–40 Btuh/sq.ft. Average

6.02 Buildings w/100 Percent OA Systems (i.e., Laboratories, Hospitals)

A. 40–120 Btuh/sq.ft.

6.03 Buildings w/Ample Insulation, Few Windows

A. AC tons × 12,000 Btuh/ton × 1.2

6.04 Buildings w/Limited Insulation, Many Windows

A. AC tons × 12,000 Btuh/ton × 1.5

6.05 Walls Below Grade (Heat Loss at Outside Air Design Condition)

A. –30°F – 6.0 Btuh/sq.ft.

B. –25°F – 5.5 Btuh/sq.ft.

C. –20°F – 5.0 Btuh/sq.ft.

D. –15°F – 4.5 Btuh/sq.ft.

E. –10°F – 4.0 Btuh/sq.ft.

F. –5°F – 3.5 Btuh/sq.ft.

G. 0°F – 3.0 Btuh/sq.ft.

H. 5°F – 2.5 Btuh/sq.ft.

I. 10°F – 2.0 Btuh/sq.ft.

J. 15°F – 1.9 Btuh/sq.ft.

K. 20°F – 1.8 Btuh/sq.ft.

L. 25°F – 1.7 Btuh/sq.ft.

M. 30°F – 1.5 Btuh/sq.ft.

6.06 Floors Below Grade (Heat Loss at Outside Air Design Condition)

A. –30°F – 3.0 Btuh/sq.ft.

B. –25°F – 2.8 Btuh/sq.ft.

C. –20°F – 2.5 Btuh/sq.ft.

D. –15°F – 2.3 Btuh/sq.ft.

E. –10°F – 2.0 Btuh/sq.ft.

F. –5°F – 1.8 Btuh/sq.ft.

G. 0°F – 1.5 Btuh/sq.ft.

H. 5°F – 1.3 Btuh/sq.ft.

I. 10°F – 1.0 Btuh/sq.ft.

J. 15°F – 0.9 Btuh/sq.ft.

K. 20°F – 0.8 Btuh/sq.ft.

L. 25°F – 1.7 Btuh/sq.ft.

M. 30°F – 0.5 Btuh/sq.ft.

6.07 Heating System Selection Guidelines

A. If heat loss exceeds 450 Btu/h per lineal feet of wall, heat should be provided from under the window or from the base of the wall to prevent downdrafts.

B. If heat loss is between 250 and 450 Btu/h per lineal feet of wall, heat should be provided from under the window or from the base of the wall, or it may be provided from overhead diffusers, located adjacent to the perimeter wall, discharging air directly downward, blanketing the exposed wall and window areas.

C. If heat loss is less than 250 Btu/h per lineal feet of wall, heat should be provided from under the window or from the base of the wall, or it may be provided from overhead diffusers, located adjacent to or slightly away from the perimeter wall, discharging air directed at, or both directed at and directed away from, the exposed wall and window areas.

6.08 Heating Load Calculation Procedure

A. **Obtain Building Characteristics**
1. Construction materials.
2. Construction material properties: U-values, R-values, shading coefficients, solar heat gain coefficients.
3. Size.
4. Color.
5. Shape.
6. Location.

7. Orientation, N, S, E, W, NE, SE, SW, NW, etc.
8. External/internal shading.
9. Occupancy type and time of day.

B. Select Outdoor Design Weather Conditions

1. Temperature.
2. Wind direction and speed.
3. Conditions in selecting outdoor design weather conditions:
   a. Type of structure: heavy, medium, or light.
   b. Is structure insulated? If the structure is heated or cooled, it must be insulated according to code.
   c. Is structure exposed to high wind?
   d. Infiltration or ventilation load.
   e. Amount of glass.
   f. Time of building occupancy.
   g. Type of building occupancy.
   h. Length of reduced indoor temperature.
   i. What is daily temperature range, minimum/maximum?
   j. Are there significant variations from ASHRAE weather data?
   k. What type of heating devices will be used?
   l. Expected cost of fuel.
4. See Part 15 for code restrictions on selection of outdoor design conditions.

C. Select indoor design temperature to be maintained in each space. See Part 15 for code restrictions on selection of indoor design conditions.

D. Estimate temperatures in unheated spaces.

E. Select and/or compute U-values for walls, roof, windows, doors, partitions, etc.

F. Determine area of walls, windows, floors, doors, partitions, etc.

G. Compute heat transmission losses for all walls, windows, floors, doors, partitions, etc.

H. Compute heat losses from basement and/or grade level slab floors.

I. Compute infiltration heat losses.

J. Compute ventilation heat loss required.

K. Compute sum of all heat losses indicated in items G, H, I, and J shown earlier.

L. For a building with sizable and steady internal heat release, a credit may be taken, but only a portion of the total. Use extreme caution!!! For most buildings, credit for heat gain should not be taken.

M. Include morning warm-up for buildings with intermittent use and night set-back. See Part 15 for code restrictions on excess HVAC system capacity permitted for morning warm-up.

N. Consider equipment and materials that will be brought into the building below the inside design temperature.

O. Heating load calculations should be conducted using industry accepted methods to determine actual heating load requirements.

P. Heating load calculations are often performed using computer simulation programs. These programs greatly simplify the calculation process; however, the basic procedures and input information required are the same.

Infiltration Rules of Thumb

7.01 General

A. **Below Grade or Interior Spaces—No infiltration losses or gains are taken for rooms located below grade or interior spaces.**

B. **Buildings that are not humidified have no latent infiltration heating load.**

C. **Winter sensible infiltration loads will generally be 1/2 to 3 times the conduction heat losses (average 1.0 to 2.0 times).**

7.02 Heating Infiltration (15-mph wind)

A. **Air Change Rate Method**

1. Range 0 to 10 AC/h
2. Commercial buildings:
   a. 1.0 AC/h one exterior wall
   b. 1.5 AC/h two exterior walls
   c. 2.0 AC/h three or four exterior walls
3. Vestibules 3.0 AC/h

B. **CFM/sq.ft. of Wall Method**

1. Range   0 to 1.0 CFM/sq.ft.
2. Tight buildings   0.1 CFM/sq.ft.
3. Average buildings 0.3 CFM/sq.ft.
4. Leaky buildings   0.6 CFM/sq.ft.

C. **Crack Method**

1. Range   0.12 to 2.8 CFM/ft. of crack
2. Average 1.0 CFM/ft. of crack

7.03 Cooling Infiltration (7.5-mph wind)

A. **Cooling load infiltration is generally ignored unless close tolerances in temperature and humidity control are required.**

B. **Cooling infiltration values are generally taken as 1/2 of the values listed earlier for heating infiltration.**

Ventilation Rules of Thumb

8.01 2015 IMC and ASHRAE Standard 62.1-2013

MINIMUM VENTILATION RATES

Occupancy Classification	Occupant Density People/1,000 SF[a]	CFM per Person	CFM per SF[a]	Exhaust Airflow Rate CFM/SF[a]
Correctional Facilities				
Booking/waiting	50	7.5	0.06	–
Cell—with plumbing fixtures[b]	25	5	0.12	1.0
Cell—without plumbing fixtures	25	5	0.12	–
Day room	30	5	0.06	–
Dining halls (see food and beverage service)	–	–	–	–
Guard Stations	15	5	0.06	–
Dry Cleaners, Laundries				
Coin-operated dry cleaner[c]	20	15	–	–
Coin-operated laundries	20	7.5	0.06[d], 0.12[e]	20
Commercial dry cleaner[c]	30	30	–	–
Commercial laundry[c]	10	25	–	–
Storage, pick-up[c]	30	7.5	0.12	–
Education				
Art classroom[b]	20	10	0.18	0.7
Auditoriums	150	5	0.06	–
Classrooms (ages 5 to 8)	25	10	0.12	–
Classrooms (ages 9 plus)	35	10	0.12	–
Computer lab	25	10	0.12	–
Corridors (see public spaces)	–	–	–	–
Daycare (through age 4)	25	10	0.18	–
Daycare sickroom[g]	25	10	0.18	–
Lecture classroom	65	7.5	0.06	–
Lecture hall (fixed seats)	150	7.5	0.06	–
Locker/dressing rooms[b]	–	–	–	0.25
Media center	25	10	0.12	–
Multiuse assembly	100	7.5	0.06	–
Music/theater/dance	35	10	0.06	–
Science laboratories[b]	25	10	0.18	1.0
Smoking lounges[c,f]	70	60	–	–
Sports locker rooms[b]	–	–	–	0.5
University/college laboratories[g]	25	10	0.18	–
Wood/metal shops[b]	20	10	0.18	0.5
Food and Beverage Service				
Bars, cocktail lounges	100	7.5	0.18	–
Cafeteria, fast food	100	7.5	0.18	–
Dining rooms	70	7.5	0.18	–
Kitchens (cooking)[f]	20[g]	7.5[g]	0.12[g]	0.7[c]
Hospitals, Nursing and Convalescent Homes				
Autopsy rooms[f]	–	–	–	0.7
Medical procedure rooms	20	15	–	–
Operating rooms	20	30	–	–
Patient rooms	10	25	–	–
Physical therapy	20	15	–	–
Recovery and ICU	10	15	–	–
Hotel, Motels, Resorts, and Dormitories				
Barracks sleeping areas[g]	20	5	0.06	–
Bathrooms/toilet—private[b]	–	–	–	25/50[h]
Bedroom/living room	10[g]	5	0.06	–
Conference/meeting[c]	–	5	0.06	–
Dormitory sleeping areas[c]	–	5	0.06	–
Gambling casinos	–	7.5	0.18	–
Laundry rooms, central[g]	10	5	0.12	–
Laundry rooms, within dwelling units[g]	10	5	0.12	–
Lobbies/prefunction	30[g]	7.5	0.06	–
Multipurpose assembly	120[g]	5	0.06	–

(Continued)

MINIMUM VENTILATION RATES (*Continued*)

Occupancy Classification	Occupant Density People/1,000 SF[a]	CFM per Person	CFM per SF[a]	Exhaust Airflow Rate CFM/SF[a]
Miscellaneous spaces[g]				
Banks or bank lobbies[g]	15	7.5	0.06	–
Freezer and refrigerated spaces (<50°F)[g]	–	10	–	–
General manufacturing (excludes heavy industrial and processes using chemicals)[g]	7	10	0.18	–
Janitor closets, trash rooms, recycling[g]	–	–	–	1.00
Kitchenettes[g]	–	–	–	0.30
Shipping/receiving[g]	2	10	0.12	–
Sorting, packing, light assembly[g]	7	7.5	0.12	–
Transportation waiting[g]	100	7.5	0.06	–
Offices				
Breakrooms[g]	50	5	0.12	–
Conference rooms	50	5	0.06	–
Main entry lobbies	10	5	0.06	–
Occupiable storage rooms for dry materials[g]	2	5	0.06	–
Office spaces	5	5	0.06	–
Reception areas	30	5	0.06	–
Telephone/data entry	60	5	0.06	–
Private Dwellings, Single and Multiple				
Common corridors[g]	–	–	0.06	–
Garages, common for multiple units[c,f]	–	–	–	0.75
Garages, separate for each dwelling[f]	–	–	–	100 cfm per car
Kitchens[f]	–	–	–	25/100[h]
Living areas[c,i]	Based upon number of bedrooms. First bedroom, 2; each additional bedroom, 1	0.35 ACH but not less than 15 cfm/person[d], 5[e]	–[d], 0.06[e]	–
Toilet rooms and bathrooms[b,c]	–	–	–	20/50[h]
Public Spaces				
Breakrooms[g]	25	5	0.06	–
Corridors	–	–	0.06	–
Courtrooms	70	5	0.06	–
Elevator car[c]	–	–	–	1.0
Legislative chambers	50	5	0.06	–
Libraries	10	5	0.12	–
Lobbies[g]	150	5	0.06	–
Museums (children's)	40	7.5	0.12	–
Museums/galleries	40	7.5	0.06	–
Occupiable storage rooms for liquids or gels[g]	2	5	0.12	–
Places of religious worship	120	5	0.06	–
Shower room (per shower head)[b,c]	–	–	–	50/20[h]
Smoking lounges[c,f]	70	60	–	–
Toilet rooms—public[b]	–	–	–	50/70[j]
Retail Stores, Sales Floors and Showroom Floors				
Dressing rooms	–	–	–	0.25
Mall common areas	40	7.5	0.06	–
Sales (except as below)	15	7.5	0.12	–
Shipping and receiving[c]	–	–	0.12	–
Smoking lounges[c,k]	70	60	–	–
Storage rooms[c]	–	–	0.12	–
Warehouses (see storage)	–	–	–	–
Specialty Shops				
Automotive motor-fuel dispensing stations[f]	–	–	–	1.5
Auto repair rooms[g]	–	–	–	1.50
Barber	25	7.5	0.06	0.5
Beauty salons[f]	25	20	0.12	0.6
Embalming rooms[c,f]	–	–	–	2.0

(*Continued*)

MINIMUM VENTILATION RATES (*Continued*)

Occupancy Classification	Occupant Density People/1,000 SF[a]	CFM per Person	CFM per SF[a]	Exhaust Airflow Rate CFM/SF[a]
Nail salons[f,k]	25	20	0.12	0.6
Pet shops (animal areas)	10	7.5	0.18	0.9
Supermarkets	8	7.5	0.06	–
Sports and Amusement				
Bowling alleys (seating area)	40	10	0.12	–
Disco/dance floors	100	20	0.06	–
Game arcades	20	7.5	0.18	–
Gym, stadium (play area)	–[d], 7[e]	–[d], 20[e]	0.30[d], 0.18[e]	–
Health club/aerobics room	40	20	0.06	–
Health club/weight room	10	20	0.06	–
Ice arenas without combustion engines	–	–	0.30[c]	0.5
Spectator areas	150	7.5	0.06	–
Swimming pools (pool and deck area)	–	–	0.48	–
Storage				
Repair garages, enclosed parking garages[f,l]	–	–	–	0.75
Soiled laundry storage rooms[g]	–	–	–	1.00
Storage rooms, chemical[g]	–	–	–	1.50
Warehouses	–	10[g]	0.06	–
Theaters				
Auditoriums (see education)	–	–	–	–
Lobbies[c]	150	5	0.06	–
Stages, studios	70	10	0.06	–
Ticket booths[c]	60	5	0.06	–
Transportation				
Platforms[c]	100	7.5	0.06	–
Transportation waiting[c]	100	7.5	0.06	–
Workrooms				
Bank vaults/safe deposit	5	5	0.06	–
Computer (without printing)	4	5	0.06	–
Copy, printing rooms	4[c]	5[c]	0.06[c]	0.5
Darkrooms	–	–	–	1.0
Meat processing[c,i]	10	15	–	–
Pharmacy (prep. area)	10	5	0.18	–
Photo studios	10	5	0.12	–

Notes:

a. Based on *net occupiable floor area.*

b. Mechanical exhaust is required and recirculation from such spaces is prohibited except that recirculation shall be permitted where the resulting supply airstream consists of not more than 10 percent air recirculated from these spaces. Recirculation of air that is contained completely within such spaces shall not be prohibited (see 2015 IMC Section 403.2.1, Items 2 and 4).

c. 2015 IMC only.

d. 2015 IMC.

e. ASHRAE Standard 62.1-2013.

f. Mechanical exhaust required and the recirculation of air from such spaces is prohibited. Recirculation of air that is contained completely within such spaces shall not be prohibited (see 2015 IMC Section 403.2.1, Item 3).

g. ASHRAE Standard 62.1-2013 only.

h. Rates are per room unless otherwise indicated. The higher rate shall be provided where the exhaust system is designed to operate intermittently. The lower rate shall be permitted only where the exhaust system is designed to operate continuously while occupied.

i. Spaces unheated or maintained below 50°F are not covered by these requirements unless the occupancy is continuous.

j. Rates are per water closet or urinal. The higher rate shall be provided where the exhaust system is designed to operate intermittently. The lower rate shall be permitted only where the exhaust system is designed to operate continuously while occupied.

k. For nail salons, each manicure and pedicure station shall be provided with a *source capture system* capable of exhausting not less than 50 cfm per station. Exhaust inlets shall be located in accordance with Section 502.20. Where one or more required source capture systems operate continuously during occupancy, the exhaust rate from such systems shall be permitted to be applied to the exhaust flow rate required by Table 403.3.1.1 for the nail salon.

l. Ventilation systems in enclosed parking garages shall comply with 2015 IMC Section 404.

A. Breathing zone outdoor airflow volumes must be corrected as follows:

$V_{BZ} = R_P P_Z + R_A A_Z$ Breathing zone outdoor airflow for each zone.

where:

A_Z = area of the zone.
P_Z = people per zone.
R_P = outdoor airflow rate for people.
R_A = outdoor airflow rate per area.

B. Single Zone Systems:

$V_{OT} = V_{OZ} = V_{BZ}/E_Z$ Outdoor air intake flow rate for single zone systems.

where:

V_{OT} = system outdoor air intake flow rate
V_{OZ} = zone outdoor airflow rate
E_Z = zone air distribution effectiveness factor from the table below.

C. 100-Percent Outdoor Air Systems:

$V_{OT} = V_{OZ1} + V_{OZ2} + \ldots$ Outdoor air intake flow rate for 100-percent outdoor air systems.

D. Multiple Zone Recirculating Systems:

$Z_p = V_{OZ}/V_{PZ}$ Primary outdoor air fraction for each zone—OA corrected for zone air distribution effectiveness divided by the primary airflow rate supplied to the zone (zone with highest primary outdoor air fraction shall be used in selection of E_V).

where:

V_{PZ} = primary airflow rate supplied to the zone. For variable volume supply, V_{PZ} shall be the lowest expected primary airflow rate to the zone when it is fully occupied.
E_V = system ventilation efficiency from table below.

E. Uncorrected Outdoor Air Intake for Multiple Zone Recirculating Systems:

$V_{OU} = D \, \Sigma_{\text{all zones}} R_P P_Z + \Sigma_{\text{all zones}} R_A A_Z$

where:

D = Occupant diversity: the ratio of the system population to the sum of the zone populations, determined in accordance with the following equation:

$D = P_S/\Sigma_{\text{all zones}} P_Z$

where:

P_S = System population: the total number of occupants in the area served by the system. For design purposes, P_S shall be the maximum number of occupants expected to be concurrently in all zones served by the system.

F. Corrected Outdoor Air Intake for Multiple Zone Recirculating Systems:

$V_{OT} = V_{OU}/E_V$ Outdoor air intake flow rate for multiple zone systems corrected for ventilation effectiveness.

Zone Air Distribution Effectiveness	
Air Distribution Configuration	E_Z
Ceiling supply of cool air.	1.0
Ceiling supply of warm air and floor return.	1.0
Ceiling supply of warm air at least 15°F above space temperature and ceiling return.	0.8
Ceiling supply of warm air less than 15°F above space temperature and ceiling return provided that the 150 fpm supply air jet reaches to within 4.5 feet of the floor level.	1.0
Ceiling supply of warm air less than 15°F above space temperature and ceiling return provided that the supply air jet is less than 150 fpm.	0.8

(Continued)

Zone Air Distribution Effectiveness	
Air Distribution Configuration	E_Z
Floor supply of cool air and ceiling return provided that the 150 fpm supply jet reaches at least 4.5 feet above the floor. *Note*: Most underfloor air distribution systems comply with this provision.	1.0
Floor supply of cool air and ceiling return, provided low velocity displacement ventilation achieves unidirectional flow and thermal stratification.	1.2
Floor supply of warm air and floor return.	1.0
Floor supply of warm air and ceiling return.	0.7
Makeup supply drawn in on the opposite side of the room from the exhaust and/or return.	0.8
Makeup supply drawn in near to the exhaust and/or return location.	0.5

System Ventilation Efficiency Table	
Max Z_p Zone with Max % OA	E_V
≤0.15	1
≤0.25	0.9
≤0.35	0.8
≤0.45	0.7
≤0.55	0.6
≤0.65	0.5
≤0.75	0.4
>0.75	0.3

8.02 *ASHRAE Standard 62.1-2013*: Return Air, Transfer Air, or Exhaust Air Classifications

A. **Class 1: Air with low contaminant concentration, low sensory-irritation intensity, and inoffensive odor. Class 1 air may be recirculated or transferred to any space. This includes:**

1. Offices.
2. Reception/waiting areas.
3. Telephone/data entry.
4. Lobbies.
5. Conference/meeting rooms.
6. Corridors.
7. Storage rooms.
8. Break rooms.
9. Coffee stations.
10. Equipment rooms.
11. Mechanical rooms.
12. Electrical/telephone closets.
13. Elevator machine rooms.
14. Laundry rooms within dwelling units.
15. Sports arena.
16. Correctional facility day room and guard station.
17. Educational facilities: classrooms, lecture classrooms, lecture halls, computer lab, media center, music/theater/dance studios, multiuse assembly.
18. Hotels, motels, resorts, dormitories: bedrooms, living rooms, barracks, sleeping quarters, lobbies, prefunction spaces, multipurpose assembly.
19. Computer rooms.

20. Photo studios.
21. Shipping/receiving rooms.
22. Transportation waiting rooms.
23. Public assembly spaces: auditorium seating area, places of religious worship, courtrooms, legislative chambers, libraries, lobbies, museums/galleries (all types).
24. Mall common areas.
25. Supermarkets.
26. Sports and entertainment: sports arena (play area), spectator areas, disco/dance floors, bowling alleys, gambling casinos, game arcades, stages, studios.

B. **Class 2: Air with moderate contaminant concentration, mild sensory-irritation intensity, or mildly offensive odors. Air that is not harmful or objectionable but is inappropriate for transfer or recirculation to spaces used for different purposes. Class 2 air may be recirculated within the space of origin but may not be recirculated or transferred to Class 1 spaces. Class 2 air may be recirculated or transferred to other Class 2 or Class 3 spaces with the same occupancy and use, or where contaminants are from similar sources and will not react to form more hazardous contaminants. Class 2 air may be recirculated or transferred to Class 4 spaces. This includes:**

1. Kitchens (commercial) and kitchenettes.
2. Toilet/bath rooms (public and private).
3. Locker rooms.
4. Locker/dressing rooms.
5. Central laundry rooms.
6. Science laboratories.
7. University and college laboratories.
8. Art classrooms.
9. Retail sales areas.
10. Barber shops.
11. Beauty and nail salons.
12. Prison cells with toilets.
13. Darkrooms.
14. Pet shops (animal areas).
15. Copy printing rooms.
16. Wood/metal shop classrooms.
17. Correctional facility booking/waiting areas.
18. Food and beverage services: restaurant dining rooms, cafeterias, fast food establishments, bars, cocktail lounges.
19. Bank vaults/safe deposit vaults.
20. Pharmacy preparation areas.
21. Warehouses.
22. Coin-operated laundries.
23. Gym/stadium (play areas).
24. Swimming pools and decks.
25. Health club/aerobics rooms.
26. Health club/weight rooms.
27. Hydraulic elevator machine rooms.

C. **Class 3: Air with significant contaminant concentration, significant sensory-irritation intensity, or offensive odor. Class 3 air may be recirculated within the space of origin only and cannot be recirculated to any other space. This includes:**

1. Commercial kitchen hoods other than grease hoods.
2. Residential kitchen vented hoods.
3. Refrigeration machinery rooms.

4. Boiler rooms.
5. Soiled laundry storage areas.
6. Janitor closets.
7. Trash/recycle rooms.
8. General chemical/biological laboratories.
9. Daycare sick rooms.

D. Class 4: Air with highly objectionable fumes or gases or with potentially danger-ous particles, bio-aerosols, or gases, at such high concentrations as to pose a health hazard. Class 4 air shall not be recirculated or transferred to any space or recirculated within the space of origin. This includes:

1. Commercial kitchen grease hoods.
2. Laboratory hoods.
3. Paint spray booths.
4. Diazo printing equipment discharges.
5. Chemical storage rooms.
6. Auto repair rooms.
7. Parking garages.

8.03 ASHRAE Standard 62.2-2013

A. Outdoor air must be provided to each dwelling unit in accordance with the following table:

Floor Area Square Feet	Number of Bedrooms				
	1	2	3	4	5
<500	30	38	45	53	60
501–1,000	45	53	60	68	75
1,001–1,500	60	68	75	83	90
1,501–2,000	75	83	90	98	105
2,001–2,500	90	98	105	113	120
2,501–3,000	105	113	120	128	135
3,001–3,500	120	128	135	143	150
3,501–4,000	135	143	150	158	165
4,001–4,500	150	158	165	173	180
4,501–5,000	165	173	180	188	195

Notes:

1 In lieu of the preceding table, the following equation may be used to determine the minimum outdoor air quantity.
$Q_{OA} = 0.03 \times A_{FLOOR} + 7.5 \times (N_{BR} + 1)$.
Q_{OA} = Quantity of Outdoor Air—CFM.
A_{FLOOR} = Floor Area of Residence—Square Feet.
N_{BR} = Number of Bedrooms—Minimum of 1.

2 Exhaust requirements:
  a. Intermittent:
    1. Kitchen: 100 CFM.
    2. Bathroom: 50 CFM.
  b. Continuous:
    1. Kitchen: 5.0 AC/h.
    2. Bathroom: 20 CFM.

8.04 ASHRAE Standard 170-2013 *Ventilation of Health Care Facilities* (incorporated as Part 4 of the 2014 Facility Guidelines Institute *Guidelines for Design and Construction of Hospitals and Outpatient Facilities*)

Area Designation	Pressure Relationship	Minimum OA AC/h	Minimum Total AC/h	All Air Exhaust to Outdoors
Surgery and Critical Care				
Operating room (Classes B and C)	Pos	4	20	NR
Operating/surgical cystoscopic rooms	Pos	4	20	NR
Delivery room (Caesarean)	Pos	4	20	NR
Substerile service area	NR	2	6	NR
Recovery room	NR	2	6	NR
Critical and intensive care	NR	2	6	NR
Intermediate care	NR	2	6	NR
Wound intensive care (burn unit)	NR	2	6	NR
Newborn intensive care	Pos	2	6	NR
Treatment room	NR	2	6	NR
Trauma room (crisis or shock)	Pos	3	15	NR
Medical/anesthesia gas storage	Neg	NR	8	Yes
Laser eye room	Pos	3	15	NR
ER waiting rooms	Neg	2	12	Yes
Triage	Neg	2	12	Yes
ER decontamination	Neg	2	12	Yes
Radiology waiting rooms	Neg	2	12	Yes
Procedure room (Class A surgery)	Pos	3	15	NR
Emergency department exam/treat-ment room	NR	2	6	NR
Inpatient Nursing				
Patient room	NR	2	4	NR
Nourishment area or room	NR	NR	2	NR
Toilet room	Neg	NR	10	Yes
Newborn nursery suite	NR	2	6	NR
Protective environment room	Pos	2	12	NR
Airborne Infectious Isolation (AII) room	Neg	2	12	Yes
Combination AII/Protective Environ-ment (PE) room	Pos	2	12	Yes
AII anteroom	Neg	NR	10	Yes
PE anteroom	Neg	NR	10	NR
Combination AII/PE anteroom	Neg	NR	10	Yes
Labor/delivery/recovery/postpartum (LDRP)	NR	2	6	NR
Labor/delivery/recovery (LDR)	NR	2	6	NR
Patient Corridor	NR	NR	2	NR
Nursing Facility				
Resident room	NR	2	2	NR
Resident gathering/activity/dining	NR	4	4	NR
Resident unit corridor	NR	NR	4	NR

(Continued)

Area Designation	Pressure Relationship	Minimum OA AC/h	Minimum Total AC/h	All Air Exhaust to Outdoors
Physical therapy	Neg	2	6	NR
Occupational therapy	NR	2	6	NR
Bathing room	Neg	NR	10	Yes
Radiology				
X-ray (diagnostic and treatment)	NR	2	6	NR
X-ray (surgery/critical care and catheterization)	Pos	3	15	NR
Darkroom	Neg	2	10	Yes
Diagnostic and Treatment				
Bronchoscopy, sputum collection, and pentamidine administration	Neg	2	12	Yes
Laboratory, general	Neg	2	6	NR
Laboratory, bacteriology	Neg	2	6	Yes
Laboratory, biochemistry	Neg	2	6	Yes
Laboratory, cytology	Neg	2	6	Yes
Laboratory, glass washing	Neg	2	10	Yes
Laboratory, histology	Neg	2	6	Yes
Laboratory, microbiology	Neg	2	6	Yes
Laboratory, nuclear medicine	Neg	2	6	Yes
Laboratory, pathology	Neg	2	6	Yes
Laboratory, serology	Neg	2	6	Yes
Laboratory, sterilizing	Neg	2	10	Yes
Laboratory, media transfer	Pos	2	4	NR
Nonrefrigerated body-holding room	Neg	NR	10	Yes
Autopsy room	Neg	2	12	Yes
Pharmacy	Pos	2	4	NR
Examination room	NR	2	6	NR
Medication room	NR	2	4	NR
Gastrointestinal endoscopy procedure room	NR	2	6	NR
Endoscope cleaning	Neg	2	10	Yes
Treatment room	NR	2	6	NR
Hydrotherapy	Neg	2	6	NR
Physical therapy	Neg	2	6	NR
Dialysis treatment area	NR	2	6	NR
Dialyzer reprocessing room	Neg	NR	10	Yes
Nuclear medicine hot lab	Neg	NR	6	Yes
Nuclear medicine treatment room	Neg	2	6	Yes
Sterilizing				
Sterilizer equipment room	Neg	NR	10	Yes
Central Medical and Surgical Supply				
Soiled or decontamination room	Neg	2	6	Yes
Clean workroom	Pos	2	4	NR
Sterile storage	Pos	2	4	NR

(Continued)

Area Designation	Pressure Relationship	Minimum OA AC/h	Minimum Total AC/h	All Air Exhaust to Outdoors
Service				
Food preparation center	NR	2	10	NR
Warewashing	Neg	NR	10	Yes
Dietary storage	NR	NR	2	NR
Laundry, general	Neg	2	10	Yes
Soiled linen sorting and storage	Neg	NR	10	Yes
Clean linen storage	Pos	NR	2	NR
Linen and trash chute room	Neg	NR	10	Yes
Bedpan room	Neg	NR	10	Yes
Bathroom	Neg	NR	10	Yes
Janitor's closet	Neg	NR	10	Yes
Support Space				
Soiled workroom or soiled holding	Neg	2	10	Yes
Clean workroom or clean holding	Pos	2	4	NR
Hazardous material storage	Neg	2	10	Yes

Notes:
Pos = Positive Pressure Relationship
Neg = Negative Pressure Relationship
NR = No Requirement

8.05 Enclosed Parking Garages

A. 2015 IMC

1. Ventilation rates:
    a. Minimum: 0.05 CFM/SF.
    b. Design: 0.75 CFM/SF.
    c. Mechanical ventilation systems may reduce the 0.75 CFM/SF ventilation requirement when the system operates automatically by means of carbon monoxide detectors applied in conjunction with nitrogen dioxide detectors. Such detectors shall be installed in accordance with their manufacturers' recommendations.

B. Enclosed Parking Garage Design Recommendations

1. Exhaust 0.75 CFM/SF at one end of the garage on each floor using a masonry plenum or ductwork (a floor-to-floor exhaust plenum is normally easier because floor-to-floor heights are generally limited in a garage and ductwork does not fit). Exhaust 1/2 of the air high and 1/2 of the air low. This will remove contaminants that are heavier than air (flammable vapors) and contaminants that are lighter than air (carbon monoxide).
2. Supply approximately 0.75 CFM/SF at the other end of the garage on each floor using a masonry plenum or ductwork (a floor-to-floor supply plenum is normally easier because floor-to-floor heights are generally limited in a garage and ductwork does not fit). Supply 1/2 of the air high and 1/2 of the air low. This exhaust and supply design will provide a sweeping air motion through the garage. Depending on the location of the entrances and exits to the garage, the supply quantity may be reduced to allow air to enter through the entrances and exits provided that short circuiting of the supply air is prevented.
3. Utilize VFDs to control the speed and the airflow of the fan based on vehicle operation and the presence of occupants, or carbon monoxide detectors applied in conjunction with nitrogen dioxide detectors. Note that the minimum garage ventilation rate is only 8 percent of the design airflow (0.05 CFM/SF divided by 0.75 CFM/SF). A single fan operated by a VFD will only turn down to about 25 percent. Use at least two fans with

VFDs; this will permit a turndown of 12.5 percent and will allow for partial capacity in the event of fan failure.

4. Garages should not be heated. The volume of air, even under code minimum airflow requirements, has a substantial impact and is a waste of energy.

8.06 Outside Air Intake and Exhaust Locations

A. 2015 IMC

1. Intakes or exhausts—10 feet from lot lines, buildings on same lot or center line of street or public way.
2. Intakes—10 feet horizontally from any hazardous or noxious contaminant (plumbing vents, chimneys, vents, stacks, alleys, streets, parking lots, loading docks). When within 10 feet horizontally, intake must be a minimum of 3 feet below or 25 feet above any source of contaminant.
3. Exhausts—shall not create a public nuisance or be directed onto walkways. For environmental air exhaust systems, outlets shall be 3 feet from property lines, 3 feet from operable openings into buildings for all occupancies other than Group U, and 10 feet from mechanical air intakes.
4. Opening protection:
   a. Protect intake and exhaust openings with corrosion resistant screens, louvers, or grilles.
   b. Exhaust openings: between 1/4″ and 1/2″ opening screens.
   c. Intake openings—residential: between 1/4″ and 1/2″ opening screens.
   d. All other intake openings: between 1/4″ and 1″ opening screens.

B. NFPA 90A-2015

1. Outside air intakes shall be located to avoid drawing in combustible materials and toxic or hazardous vapors.
2. Outside air intakes shall be protected with corrosion resistant screens not larger than 1/2″ mesh.
3. Outside air intakes shall be located to minimize the hazard from fires in other structures. Intakes shall be equipped with a fire damper when protection from fire hazards is required.
4. Outside air intake shall be located so as to minimize the introduction of smoke into the building. Intakes shall be equipped with a smoke damper when protection from smoke hazards is required.

C. ASHRAE Standard 62.1-2013—Air Intake Minimum Separation Distances

1. Significantly contaminated exhaust (high contaminant concentration, significant sensory-irritation intensity, offensive odor): 15 feet.
2. Noxious or dangerous exhaust air with highly objectionable fumes or gases and or exhaust air with potentially dangerous contaminants (laboratory exhaust, fumes, gases, potentially dangerous particles, bio-aerosols, gases at high concentrations to be harmful): 30 feet.
3. Plumbing vents terminating less than 3 feet above the level of the outdoor air intake: 10 feet.
4. Plumbing vents terminating at least 3 feet above the level of the outdoor air intake: 3 feet.
5. Vents, chimneys, flues, and other combustion appliance discharge: 15 feet.
6. Garage entry, automobile loading area, drive-in queue: 15 feet.
7. Truck loading area or dock, bus parking/idling area: 25 feet.
8. Driveway, street, or parking area: 5 feet.
9. Street or thoroughfare with high traffic volume: 25 feet.

10. Roof, landscaped grade or other surface directly below intake: 1 foot (or expected average snow depth, whichever is greater).
11. Garbage storage/pickup area, dumpsters: 15 feet.
12. Cooling tower intake or basin: 15 feet.
13. Cooling tower exhaust: 25 feet.
14. Class 1 air: 10 feet (the author's interpretation of Class 1 air).
15. Class 2 air: 15 feet (the author's interpretation of Class 2 air).
16. Class 3 air: 15 feet (see item number 1 preceding the definition of Class 3 air).
17. Class 4 air: 30 feet (see item number 2 preceding the definition of Class 4 air).

D. ASHRAE Standard 170-2013 *Ventilation of Health Care Facilities* **(incorporated as Part 4 of the 2014 Facility Guidelines Institute** *Guidelines for Design and Construction of Hospitals and Outpatient Facilities*

1. Outdoor air intakes shall be located at least 25 feet from cooling towers and all exhaust outlets of ventilating systems, combustion equipment stacks, medical-surgical vacuum systems, plumbing vents, or areas that may collect vehicular exhaust or other noxious fumes. Prevailing winds and/or proximity to other structures may require greater clearances.
2. The bottom of outdoor air intakes serving central systems shall be as high as practical, but at least 6 feet above ground level, or if installed above the roof, 3 feet above roof level.
3. Relief air is exempt from the 25 foot separation requirement. Relief air is defined as air that otherwise could be returned to an air handling unit from the occupied space but is being discharged to the outdoors to maintain building pressure, such as during outside air economizer operation.
4. Exhaust outlets from areas that may be contaminated shall discharge in a vertical direction at least 10 feet above roof level and shall be located not less than 10 feet horizontally from air intakes, openable windows/doors, or areas that are normally accessible to the public or maintenance personnel and that are higher in elevation than the exhaust discharge.
5. Exhaust outlets from areas that may be contaminated shall be arranged to minimize recirculation of exhaust air into the building.

8.07 Indoor Air Quality (IAQ)

A. Causes of Poor IAQ

1. Inadequate ventilation—50 percent of all IAQ problems are due to lack of ventilation.
2. Poor intake/exhaust locations.
3. Inadequate filtration or dirty filters.
4. Intermittent airflow.
5. Poor air distribution.
6. Inadequate operation.
7. Inadequate maintenance.

B. IAQ Control Methods

1. Control temperature and humidity.
2. Ventilation—dilution.
3. Remove pollution source.
4. Filtration.

C. IAQ Factors

1. Thermal environment.
2. Smoke.
3. Odors.
4. Irritants—dust.
5. Stress problems (perceptible, nonperceptible).

6. Toxic gases—carbon monoxide, carbon dioxide.
7. Allergens—pollen.
8. Biological contaminants—bacteria, mold, pathogens, legionella, micro-organisms, fungi.

D. CO_2 Levels and IAQ

1. Outdoor background level: 500–700 PPM CO_2 avg.
2. *ASHRAE Standard 62.1* recommends: 1000–1200 PPM CO_2 max.
3. OSHA and U.S. Air Force standard: 650 PPM CO_2 max.
4. Human discomfort begins: 800–1000 PPM CO_2.
5. Long-term health effects: >12,000 PPM CO_2.

8.08 Effects of Carbon Monoxide

A. Effects of Various Concentrations of Carbon Monoxide with Respect to Time are shown in the following table.

Hours of Exposure	Concentration of Carbon Monoxide in PPM ±		
	Barely Perceptible	Sickness	Deadly
0.5	600	1000	2000
1.0	200	600	1600
2	100	300	1000
3	75	200	700
4	50	150	400
5	35	125	300
6	25	120	200
7	25	100	200
8	25	100	150

B. Carbon Monoxide Concentration versus Time versus Symptoms are shown in the following table.

Concentration of CO in the Air	Inhalation Time	Toxic Symptoms Developed
9 PPM	Short-term exposure	ASHRAE recommended maximum allowable concentration for short term exposure in living area.
35 PPM	8 hours	The maximum allowable concentration for a continuous exposure, in any 8-hour period, according to federal law.
200 PPM	2–3 hours	Slight headache, tiredness, dizziness, nausea; maximum CO concentration exposure at any time as prescribed by OSHA
400 PPM	1–2 hours	Frontal headaches
	after 3 hours	Life threatening
	–	Maximum PPM in flue gas (on a free air basis) according to EPA and AGA
800 PPM	45 minutes	Dizziness, nausea, and convulsions
	2 hours	Unconscious
	2–3 hours	Death
1,600 PPM	20 minutes	Headache, dizziness, nausea
	1 hour	Death
3,200 PPM	5–10 minutes	Headache, dizziness, nausea
	30 minutes	Death

Concentration of CO in the Air	Inhalation Time	Toxic Symptoms Developed
6,400 PPM	1–2 minutes	Headache, dizziness, nausea
	10–15 minutes	Death
12,800 PPM	1–3 minutes	Death

C. Carbon monoxide is lighter than air (specific gravity is 0.968).

8.09 Toilet Rooms

A. *ASHRAE Standard 62.1-2013*

1. Private: 50 CFM/room for intermittent operation, 25 CFM/room for continuous operation. For toilet rooms intended to be occupied by one person at a time.
2. Public: 70 CFM/water closet and urinal where periods of heavy use are expected to occur, 50 CFM/water closet and urinal otherwise.

B. *2015 IMC*

1. Private: 50 CFM/room for intermittent operation, 25 CFM/room for continuous operation. For toilet rooms intended to be occupied by one person at a time.
2. Public: 70 CFM/water closet and urinal for intermittent operation, 50 CFM/water closet and urinal for continuous operation.

C. Recommended Design Requirements

1. 2.0 CFM/sq.ft.
2. 10 AC/h
3. 100 CFM/water closet and urinal.
4. Toilet room ventilation:
   a. For toilet rooms with high fixture densities (stadiums, auditoriums), the 50 CFM/water closet and urinal dictates.
   b. For toilet rooms with ceiling heights over 12 feet, the 10 AC/h dictates.
   c. For toilet rooms with ceiling heights 12 feet and under, the 2.0 CFM/sq.ft. dictates.
   d. If toilet rooms are designed for a 100 CFM/water closet or urinal, you will always meet the 2.0 CFM/sq.ft. and the 10 AC/h recommended airflow requirements.

8.10 Electrical Rooms

A. Recommended Minimum Ventilation Rate

1. 2.0 CFM/sq.ft.
2. 10.0 AC/h
3. 5 CFM/KVA of transformer.

B. Electrical Room Design Guidelines

1. Determine heat gain from transformers, panelboards, and other electrical equipment contained in the electrical room. Then, determine required airflow for ventilation or tempering of space.
2. Generally, electrical equipment rooms only require ventilation to keep equipment from overheating. Most electrical rooms are designed for 95°F to 104°F; however, consult the electrical engineer for equipment temperature tolerances. If space temperatures 90°F and below are required by equipment, air conditioning (tempering) of the space will be required.
3. If outside air is used to ventilate the electrical room, the electrical room design temperature will be 10°F to 15°F above outside summer design temperatures.
4. If conditioned air from an adjacent space is used to ventilate the electrical room, the electrical room temperature can be 10°F to 20°F above the adjacent spaces.

8.11 Mechanical Rooms

A. Recommended Minimum Ventilation Rate

1. 2.0 CFM/sq.ft.
2. 10.0 AC/h

B. Mechanical Equipment Room Design Guidelines

1. Determine heat gain from motors, pumps, fans, transformers, panelboards, and other mechanical and electrical equipment contained in the mechanical room. Then, determine the required airflow for the ventilation or tempering of space.
2. Generally, mechanical equipment rooms only require ventilation. Most mechanical rooms are designed for 95°F to 104°F; however, verify mechanical equipment temperature tolerances. If space temperatures below 90°F are required by mechanical equipment, air conditioning (tempering) of the space will be required.
3. A number of products (DDC control panels, variable frequency drives, other electronic components) will perform better if the mechanical room is tempered in lieu of just ventilating the room.
4. If outside air is used to ventilate the mechanical room, the mechanical room design temperature will be 10°F to 15°F above outside summer design temperatures.
5. If conditioned air from an adjacent space is used to ventilate the mechanical room, the mechanical room temperature can be 10°F to 20°F above the adjacent spaces.

C. Boiler Rooms—Cleaver Brooks 10 CFM/BHP

1. 8 CFM/BHP combustion air.
2. 2 CFM/BHP ventilation.
3. 1 BHP = 33,500 Btuh.

D. Chiller Rooms—*ASHRAE Standard* 15-2013 and *ASHRAE Standard* 34-2013

1. See *ASHRAE Standard 15-2013* and *ASHRAE Standard 34-2013* for complete refrigeration system requirements.
2. Scope:
   a. To establish safeguards for life, limb, health, and property.
   b. To define practices that are consistent with safety.
   c. To prescribe safety standards.
3. Application: The standard applies to all mechanical and absorption refrigerating systems and heat pumps used in institutional, public assembly, residential, commercial, large mercantile, industrial, and mixed-use occupancies; to parts and components added after adoption of this code; and to substitutions of refrigerant having a different designation.
4. Refrigerant classification is shown in the following table:

	Safety Group	
Higher Flammability	A3	B3
Lower Flammability	A2	B2 Ammonia
No Flame Propagation	A1 R-11, R-12, R-22, R-134a, R-410a	B1 R-123
	Lower toxicity	Higher toxicity

5. Requirements for refrigerant use:
   a. Requirements for refrigerant use are based on the probability that the refrigerant will enter an occupied space and on the type of occupancy (institutional, public assembly, residential, commercial, large mercantile, industrial, and mixed-use).

   b. The total amount of refrigerant permitted to be installed in a system is determined by the type of occupancy, the refrigerant group, and the probability that refrigerant will enter the occupied space.

   c. Refrigerant piping shall not be installed in an enclosed stairways, stair landings, or means of egress.

   d. Refrigeration system components shall not interfere with free passage through public hallways, and limitations regarding size are based on refrigerant type.

6. Service provisions:

   a. All serviceable components of refrigerating systems shall be safely accessible.

   b. Properly located stop valves, liquid and vapor transfer valves, refrigerant storage tanks, and adequate venting are required when needed for safe servicing of equipment.

   c. Refrigerant systems with more than 6.6 lbs. of refrigerant require stop valves at:

     1) The suction inlet of each compressor, compressor unit, or condensing unit.

     2) The discharge outlet of each compressor, compressor unit, or condensing unit.

     3) The outlet of each liquid receiver.

   d. Refrigerant systems with more than 110 lbs. of refrigerant require stop valves at:

     1) The suction inlet of each compressor, compressor unit, or condensing unit.

     2) The discharge outlet of each compressor, compressor unit, or condensing unit.

     3) The inlet of each liquid receiver, except for self-contained systems or where the receiver is an integral part of the condenser or condensing unit.

     4) The outlet of each liquid receiver.

     5) The inlet and outlet of condensers when more than one condenser is used in parallel.

   e. Stop valves shall be suitably labeled.

7. Installation requirements:

   a. Air ducts passing through machinery rooms shall be of tight construction and shall have no openings in such rooms. Access doors and panels in ductwork and air handling units shall be gasketed and tight fitting.

   b. Refrigerant piping crossing an open space that affords passageway in any building shall not be less than 7'-3" above the floor.

   c. Passages shall not be obstructed by refrigerant piping.

   d. Refrigerant piping shall not be placed in, or pass through, any elevator, dumb-waiter, or other shaft containing moving objects or in any shaft that has openings to living quarters or main exits.

   e. Refrigerant piping shall not be installed vertically through floors from one story to another except as follows:

     1) Basement to first floor, top floor to mechanical equipment penthouse or roof.

     2) Adjacent floors served by the refrigerating system.

     3) Where the refrigerant concentration does not exceed that listed in Table 4-1 or Table 4-2 of ASHRAE Standard 34 for the smallest occupied space through which the refrigerant piping passes.

     4) For the purpose of interconnecting separate pieces of equipment in other than industrial occupancies and where the refrigerant concentration exceeds that listed in Table 4-1 or Table 4-2 of ASHRAE Standard 34 for the smallest occupied space. The piping may be carried in an approved, rigid and tight, continuous fire-resistive pipe, duct, or shaft having no openings into floors not served by the refrigerating system or carried exposed on the outer wall of the building. Or the piping may be located on the exterior wall of a building when vented to the outdoors or to the space served by the system and not used as an air shaft, closed court, or similar space.

8. Refrigeration equipment room requirements:

   a. Provide proper space for service, maintenance, and operation.

   b. Minimum clear headroom shall be 7'-3".

   c. Doors shall be outward opening, self-closing, fire-rated, tight fitting, and adequate in number to ensure freedom for persons to escape in an emergency. No other openings shall be permitted in equipment rooms that will permit passage of refrigerant to other parts of the building.

d. Refrigeration equipment rooms require a refrigerant detector located in the equipment room in an area where refrigerant from a leak will concentrate, set to alarm and start the ventilation system when the level reaches the refrigerant's toxicity level. The alarm shall annunciate visual and audible alarms inside the refrigerating machinery room and outside each entrance to the refrigerating machinery room. The alarm shall be of the manual reset type with the reset located inside the refrigeration equipment room.

e. Periodic test of alarm and sensors are required.

f. Mechanical rooms shall be vented to the outdoors.

g. Mechanical ventilation shall be capable of exhausting the air quantity determined by the formula in Section 8.11.5. The exhaust quantity depends on the amount of refrigerant contained in the system. To obtain a reduced airflow for normal ventilation, multiple fans, multispeed fans, or fans with variable frequency drives may be used. Provision shall be made for inlet air to replace that being exhausted. Openings for inlet air shall be positioned to avoid recirculation.

h. Minimum ventilation rate shall be 0.5 CFM per square foot of machine room area or 20 CFM per person.

i. No open flames that use combustion air from the machinery room shall be installed where any refrigerant other than carbon dioxide (R-744), water (R-718), or ammonia (R-717) is used. A sealed air duct may be used to supply combustion air to fuel-burning appliances in the machinery room, or a refrigerant detector may be used to automatically shut down the combustion process in the event of refrigerant leakage.

j. There shall be no flame-producing device or continuously operating hot surface over 800°F permanently installed in the room.

k. Walls, floors, and ceilings shall be tight and of non-combustible construction with a minimum 1-hour fire resistance rating.

l. The machinery room shall have a door that opens directly to the outside or through a vestibule equipped with self-closing, tight-fitting doors.

m. All machinery room wall, floor, and ceiling penetrations shall be sealed.

n. Where Groups A2, A3, B2, and B3 refrigerants are used, the machinery room shall conform to Class I, Division 2 of the National Electric Code. Groups A1 and B1 are exempt from this requirement.

o. Emergency shutdown of the refrigeration equipment shall be provided immediately outside the machinery room door.

p. Ventilation fans shall be on a separate electrical circuit and shall have a control switch located immediately outside the machinery room door so they can be activated in an emergency.

q. Refrigeration compressors, piping, equipment, valves, switches, ventilation equipment, and associated appurtenances shall be labeled in accordance with *ANSI/ASME A13.1*.

8.12 Combustion Air

A. 2015 IMC

1. Oil-fired appliances shall be provided with combustion air in accordance with NFPA 31.
2. The requirements for combustion and dilution air for gas-fired appliances shall be in accordance with the *International Fuel Gas Code*.

B. NFPA 54—2015 National Fuel Gas Code

1. Inside air:
   a. Minimum required space volume: 50 ft.^{3} per 1,000 Btu/h of the combined fuel-burning appliance input capacity.
   b. Number of openings: Two openings are required—one within 1 foot of the ceiling of the room, and one within 1 foot of the floor.

    c. Opening size on the same story: The net free area of each opening shall be equal to 1.0 square inch for each 1,000 Btu/h of the combined fuel-burning appliance input rating (the sum of all appliances within the room), 100 square inches minimum. The minimum dimension of air openings shall not be less than 3 inches.

    d. Opening size on the different stories: The net free area of each opening shall be equal to 2.0 square inches for each 1,000 Btu/h of the combined fuel-burning appliance input rating (the sum of all appliances within the room).

2. Outdoor air:

    a. Two permanent opening methods:

      1) Number of openings: Two openings are required—one within 1 foot of the ceiling of the room and one within 1 foot of the floor.

      2) Direct opening size: The net free area of each opening shall be equal to 1.0 square inch for each 4,000 Btu/h of the combined fuel-burning appliance input rating (the sum of all appliances within the room).

      3) Horizontal duct opening size: The net free area of each opening shall be equal to 1.0 square inch for each 2,000 Btu/h of the combined fuel-burning appliance input rating (the sum of all appliances within the room).

      4) Vertical opening size: The net free area of each opening shall be equal to 1.0 square inch for each 4,000 Btu/h of the combined fuel-burning appliance input rating (the sum of all appliances within the room).

    b. One permanent opening method:

      1) Number of openings: One opening is required—one within 1 foot of the ceiling.

      2) The appliance will have at least 1 inch clearance on the sides and back of the appliance, and 6 inches in front of the appliance.

      3) The opening shall directly communicate with the outdoors or shall communicate through vertical or horizontal ducts to the outdoors.

      4) Opening size: The net free area of each opening shall be equal to 1.0 square inch for each 3,000 Btu/h of the combined fuel-burning appliance input rating (the sum of all appliances within the room). Not less than the sum of the areas of all vent connectors in the space.

3. Combination indoor and outdoor combustion air:

    a. Indoor openings shall comply with the indoor air requirements listed above.

    b. Outdoor openings shall comply with the outdoor air requirements listed above.

    c. The outdoor opening shall be sized to compensate for the deficiency in available volume of all communicating interior spaces. The minimum size of outdoor opening(s) shall be the full size of the outdoor opening(s) calculated in accordance with the requirements listed above, multiplied by 1 minus the ratio of available interior volume divided by the required interior volume.

4. Forced combustion air supply:

    a. Where combustion air is provided by mechanical means, the system shall deliver a minimum of 0.35 CFM per 1,000 Btu/h of the combined fuel-burning appliance input rating (the sum of all appliances within the room).

    b. Appliances shall be interlocked with a makeup air unit to prevent operation if the makeup air unit is not operating.

5. Direct vent appliances:

    a. Appliances must be listed and labeled for a direct combustion air connection.

    b. Appliances must be installed in accordance with the manufacturers' installation instructions.

6. Combustion air ducts:

    a. Galvanized steel construction.

    b. Unobstructed termination.

    c. Same cross-sectional area as the free area of the openings.

    d. Serves a single appliance enclosure.

    e. Separate ducts must be provided for the upper and lower combustion air openings. The separation between these ducts shall be maintained from source to discharge.

    f. Ducts that serve the upper combustion air opening cannot slope downward toward the source of the combustion air.

    g. The bottom of the combustion air opening shall be a minimum of 12 inches above grade.

    h. Ducts shall not be screened where terminating in an attic space.

    i. The remaining space within a chimney surrounding a chimney liner, gas vent, special gas vent, or plastic piping shall not be used to supply combustion air.

7. Opening protection:

    a. Metal louver: Maximum 75 percent free area.

    b. Wood louvers: Maximum 25 percent free area.

    c. Dampers (fire, smoke, control): Dampers shall be interlocked to operate with the appliance. Manually operated dampers are not permitted.

8.13 Hazardous Locations

A. Hazardous location requirements for electrical and electronic equipment are defined in the *2014 National Electrical Code (NEC NFPA 70)*, Articles 500 through 510.

B. Hazardous Classifications

1. Class I: Class I locations are those spaces where flammable gases or vapors are, or where they may be present in the air in quantities sufficient to produce explosive or ignitable mixtures.

    a. Class I locations are subdivided into four groups based on the type of flammable gases or vapors:

      1) Group A: Acetylene.

      2) Group B: Flammable gas (hydrogen, ethylene oxide, propylene oxide); flammable liquid-produced vapor, or combustible liquid-produced vapor mixed with air that may burn or explode, having either a maximum experimental safe gap (MESG) value less than or equal to 0.45 mm or a minimum igniting current ratio (MIC ratio) less than or equal to 0.40.

      3) Group C: Flammable gas (Ethyl Ether, Ethylene); flammable liquid-produced vapor, or combustible liquid-produced vapor mixed with air that may burn or explode, having either a maximum experimental safe gap (MESG) value greater than 0.45 mm and less than or equal to 0.75 mm, or a minimum igniting current ratio (MIC ratio) greater than 0.40 and less than or equal to 0.80.

      4) Group D: Flammable gas (Acetone, Ammonia, Butane, Gasoline, Propane); flammable liquid-produced vapor, or combustible liquid-produced vapor mixed with air that may burn or explode, having either a maximum experimental safe gap (MESG) value greater than 0.75 mm or a minimum igniting current ratio (MIC ratio) greater than 0.80.

    b. Class I locations are also subdivided into two divisions:

      1) Class I, Division 1:

        a) Locations where ignitable concentrations of flammable gases or vapors can exist under normal operating conditions; or

        b) Locations where ignitable concentrations of flammable gases or vapors may exist frequently because of repair or maintenance operations or because of leakage; or

        c) Locations where breakdown or faulty operation of equipment or processes might release ignitable concentrations of flammable gases or vapors, and might cause the simultaneous failure of electric equipment.

      2) Class I, Division 2:

        a) Locations where volatile flammable liquids or flammable gases are handled, processed, or used, but in which the liquids, vapors, or gases will normally be confined within closed containers or closed systems where they can escape only in case of an accidental rupture or breakdown of such containers or systems, or in the case of abnormal operation or equipment; or

      b) Locations where ignitable concentrations of gases or vapors are normally prevented by positive mechanical ventilation, and have the potential to become hazardous through failure or abnormal operation of the ventilating equipment; or

      c) Locations that are adjacent to Class I, Division 1 locations, and to which ignitable concentrations of gases or vapors might occasionally be communicated unless such communication is prevented by adequate positive pressure ventilation from a source of clean air, and effective safeguards against ventilation failure are provided.

2. Class II: Class II locations are spaces or areas that contain combustible dusts.

  a. Class II locations are subdivided into three groups based on the type of combustible dusts:

    1) Group E: Atmospheres containing combustible metal dusts, including aluminum, magnesium, and their commercial alloys, or other combustible dusts whose particle size, abrasiveness, and conductivity present similar hazards in the use of electrical equipment.

    2) Group F: Atmospheres containing combustible carbonaceous dusts that have more than 8 percent total entrapped volatiles or have been sensitized by other materials so that they present an explosion hazard (coal, carbon black, charcoal, and coke dust).

    3) Group G: Atmospheres containing combustible dusts not included in Group E or F, such as flour, grain, wood, plastic, and chemicals.

  b. Class II, Division 1:

    1) Locations in which combustible dust is in the air under normal operating conditions in quantities sufficient to produce explosive or ignitable mixtures; or

    2) Locations where mechanical failure or abnormal operation of machinery or equipment might cause such explosive or ignitable mixtures to be produced, and might also provide a source of ignition through the simultaneous failure of electrical equipment, through the operation of protection devices, or from other causes; or

    3) Locations in which Group E combustible dusts may be present in quantities sufficient to be hazardous.

  c. Class II, Division 2:

    1) Locations in which combustible dust due to abnormal operations may be present in the air in quantities sufficient to produce explosive or ignitable mixtures; or

    2) Locations where combustible dust accumulations are present but are normally insufficient to interfere with the normal operation of electrical equipment or other apparatus, but could as a result of infrequent malfunctioning of handling or processing equipment become suspended in the air; or

    3) Locations in which combustible dust accumulations on, in, or in the vicinity of the electrical equipment could be sufficient to interfere with the safe dissipation of heat from electrical equipment, or could be ignitable by abnormal operation or the failure of electrical equipment.

3. Class III: Class II locations are spaces or areas that contain easily ignitable fibers or flyings, but where such fibers or flyings are not likely to be in suspension in the air in quantities sufficient to produce ignitable mixtures.

  a. Class III, Division 1: Locations in which easily ignitable fibers or materials producing combustible flyings are handled, manufactured, or used.

  b. Class III, Division 2: Locations in which easily ignitable fibers are stored or handled other than in the process of manufacturing.

C. Hazardous Location Protection Techniques

1. Purged and pressurized systems: Spaces and equipment are pressurized at pressures above the external atmosphere with noncontaminated air or other nonflammable gas to prevent explosive gases or vapors from entering the enclosure.

2. Intrinsically safe systems: Electrical circuits are designed so that they do not release sufficient energy to ignite an explosive atmosphere.

3. Explosion-proof equipment: Explosion-proof equipment is designed and built to withstand an internal explosion without igniting the surrounding atmosphere.
4. Nonincendive circuits, components, and equipment: Circuits designed to prevent any arc or thermal effect produced, under intended operating conditions of the equipment or produced by opening, shorting, or grounding of the field wiring, is not capable, under specified test conditions, of igniting the flammable gas, vapor, or dust-air mixtures.
5. Oil immersed equipment: The arcing portions of the equipment are immersed in an oil at a depth that the arc will not set off any hazardous gases or vapors above the surface of the oil.
6. Hermetically sealed equipment: The equipment is sealed against the external atmosphere to prevent the entry of hazardous gases or vapors.
7. Dust-ignition-proof equipment: Dust-ignition-proof equipment is designed and built to exclude dusts and, where installed and protected, will not permit arcs, sparks, or heat generated or liberated inside the enclosure to cause ignition of the exterior accumulations or atmospheric suspensions of a specified dust on or in the enclosure.
8. Dust-tight equipment: Dust-tight equipment is designed to prevent the entrance of dust into equipment.
9. Combustible gas detection system: Gas detection equipment shall be listed for detection of the specific gas or vapor to be encountered.
10. Classification versus Protection Techniques is shown in the following table:

Protection Techniques	Class I		Class II		Class III	
	Div 1	Div 2	Div 1	Div 2	Div 1	Div 2
Purged and Pressurized	X	X	X	X	X	X
Intrinsically Safe Systems	X	X	X	X	X	X
Explosion-Proof Equipment	X	X	N/A	N/A	N/A	N/A
Nonincendive Circuits, Components, and Equipment	N/A	X	N/A	X	X	X
Hermetically Sealed Equipment	N/A	X	N/A	X	X	X
Oil Immersed Equipment	N/A	X	N/A	N/A	N/A	N/A
Dust-Ignitionproof Equipment	N/A	N/A	X	X	N/A	N/A
Dusttight Equipment	N/A	N/A	N/A	X	X	X
Combustible Gas Detection Systems	X	X	N/A	N/A	N/A	N/A

Notes:
X = Appropriate to the classification.
N/A = Not acceptable to the classification.

D. Ventilation Requirements

1. Ventilation, natural or mechanical, must be sufficient to limit the concentrations of flammable gases or vapors to a maximum level of 25 percent of their Lower Flammable Limit/Lower Explosive Limit (LFL/LEL).
2. Minimum ventilation required: 1.0 CFM/sq.ft. of floor area or 6.0 air changes per hour, whichever is greater. If a reduction in the classification is desired, the airflow must be four times the airflow just specified.
3. Recommendation: Ventilate all hazardous locations with 2.0 CFM/sq.ft. of floor area or 12 air changes per hour minimum with half the airflow supplied and exhausted high (within 6 inches of the ceiling or structure) and half the airflow supplied and exhausted low (within 6 inches of the floor).
4. A ventilation rate that is a minimum of four times the ventilation rate required to prevent the space from exceeding the maximum level of 25 percent LFL/LEL using fugitive emissions calculations.
5. Ventilate the space so accumulation pockets for lighter-than-air or heavier-than-air gases or vapors are eliminated.

6. Monitoring of the space is recommended to ensure that the 25 percent LFL/LEL is not exceeded.

E. Hazardous Location Definitions

1. *Boiling Point.* The temperature at which the vapor pressure of a liquid equals the atmospheric pressure of 14.7 pounds per square inch absolute.
2. *Combustible Liquids.* Liquids having flash points at or above 100°F. Combustible liquids shall be subdivided as Class II or Class III liquids as follows:
   a. Class II. Liquids having flash points at or above 100°F and below 140°F.
   b. Class IIIA. Liquids having flash points at or above 140°F and below 200°F.
   c. Class IIIB. Liquids having flash points at or above 200°F.
3. *Explosion.* An effect produced by the sudden violent expansion of gases, which can be accompanied by a shockwave or disruption, or both, of enclosing materials or structures. An explosion might result from chemical changes such as rapid oxidation, deflagration, or detonation; decomposition of molecules, and runaway polymerization; or physical changes such as pressure tank ruptures.
4. *Explosive.* Any chemical compound, mixture, or device, the primary or common purpose of which is to function by explosion.
5. *Flammable.* Any material capable of being ignited from common sources of heat or at a temperature of 600°F or less.
6. *Flammable Compressed Gas.* An air/gas mixture that is flammable when the gas is 13 percent or less by volume or when the flammable range of the gas is wider than 12 percent regardless of the lower limitation determined at atmospheric temperature and pressures.
7. *Flammable Liquids.* Liquids having flash points below 100°F and having vapor pressures not exceeding 40 pounds per square inch absolute at 100°F. Flammable liquids shall be subdivided as Classes IA, IB, and IC as follows:
   a. Class IA. Liquids having flash points below 73°F and having boiling points below 100°F.
   b. Class IB. Liquids having flash points below 73°F and having boiling points above 100°F.
   c. Class IC. Liquids having flash points at or above 73°F and below 100°F.
8. *Flammable Solids.* A solid, other than a blasting agent or explosive, that is capable of causing a fire through friction, absorption of moisture, spontaneous chemical change, or retaining heat from manufacturing or processing, or which has an ignition temperature below 212°F, or which burns so vigorously and persistently when ignited as to create a serious hazard.
9. *Flash Point.* The minimum temperature in °F at which a flammable liquid will give off sufficient vapors to form an ignitable mixture with air near the surface or in the container, but will not sustain combustion.
10. *Noncombustible.* A material that, in the form in which it is used and under the conditions anticipated, will not ignite, burn, support combustion, or release flammable vapors when subject to fire or heat.
11. *Pyrophoric.* A material that will spontaneously ignite in air at or below 130°F.

PART 9

Humidification
Rules of Thumb

9.01 Window Types and Space Humidity Values

A. **Single Pane Windows**   **±10 percent RH Maximum**

B. **Double Pane Windows**   **±30 percent RH Maximum**

C. **Triple Pane Windows**   **±40 percent RH Maximum**

D. **The preceding numbers are based on the following:**

1. 0°F, outside design temperature.
2. 72°F, inside design temperature.
3. $R_{\text{INSIDE AIR FILM}} = 0.680$ $U_{\text{INSIDE AIR FILM}} = 1.471$
4. $R_{\text{SINGLE GLASS}} = 0.909$ $U_{\text{SINGLE GLASS}} = 1.100$
5. $R_{\text{DOUBLE GLASS}} = 1.667$ $U_{\text{DOUBLE GLASS}} = 0.600$
6. $R_{\text{TRIPLE GLASS}} = 2.000$ $U_{\text{TRIPLE GLASS}} = 0.500$
7. Standard air at sea level.
8. The relative humidity numbers presented earlier in this list are rounded for ease of remembrance.
9. The glass R-values and U-values are for average glass construction. Modern glass construction can achieve higher R-values/lower U-values.
10. For additional information on moisture condensation on glass, see the tables at the end of this chapter.

9.02 Proper Vapor Barriers

A. **Proper vapor barriers and moisture control must be provided to prevent moisture condensation in walls and to prevent mold, fungi, bacteria, and other plant and micro-organism growth.**

9.03 Human Comfort

A. **30–60 percent RH**

9.04 Electrical Equipment, Computers

A. **35–55 percent RH**

9.05 Winter Design Relative Humidities

A. **Outdoor Air Below 32°F**

1. 70–80 percent RH
2. Design Wet Bulb Temperatures 2 to 4°F below Design Dry Bulb Temperatures

B. **Outdoor Air 32–60°F: 50 percent RH**

9.06 Optimum Relative Humidity Ranges for Health

Health Aspect	Optimum Relative Humidity Range for Controlling Health Aspect
Bacteria	20–70%
Viruses	40–78%
Fungi	0–70%
Mites	0–60%
Respiratory Infections (1)	40–50%
Allergic Rhinitis and Asthma	40–60%
Chemical Interactions	0–40%
Ozone Production	75–100%
Combined Health Aspects	40–60%

Note:
(1) **Insufficient data above 50 percent RH.**

9.07 Moisture Condensation on Glass

A. The subsequent moisture condensation tables are based on the following:

1. $R_{\text{INSIDE AIR FILM}} = 0.680$ $\quad$ $U_{\text{INSIDE AIR FILM}} = 1.471$
2. $R_{\text{SINGLE GLASS}} = 0.909$ $\quad$ $U_{\text{SINGLE GLASS}} = 1.100$
3. $R_{\text{DOUBLE GLASS}} = 1.818$ $\quad$ $U_{\text{DOUBLE GLASS}} = 0.550$
4. $R_{\text{TRIPLE GLASS}} = 2.500$ $\quad$ $U_{\text{TRIPLE GLASS}} = 0.400$
5. Standard air at sea level.

B. The glass surface temperatures, which are also the space dewpoint temperatures, listed in the moisture condensation tables that follow, were developed using the equations in Part 3.

Temp. Room °F	Temp. Outside °F	Single Pane Glass		Double Pane Glass		Triple Pane Glass	
		T_{GLASS} / T_{DEWPOINT}	% R.H.	T_{GLASS} / T_{DEWPOINT}	% R.H.	T_{GLASS} / T_{DEWPOINT}	% R.H.
	−30	−6.1	4.5	29.5	25.9	39.2	38.5
	−25	−2.3	5.6	31.3	27.9	40.5	40.5
	−20	1.4	6.9	33.2	30.2	41.9	42.8
	−15	5.2	8.4	35.1	32.6	43.2	45.0
	−10	8.9	10.1	36.9	35.1	44.6	47.5
	−5	12.6	12.1	38.8	37.9	46.0	50.1
	0	16.4	14.5	40.7	40.8	47.3	52.7
65	5	20.1	17.2	42.6	44.0	48.7	55.5
	10	23.9	20.3	44.4	47.1	50.0	58.3
	15	27.6	23.9	46.3	50.7	51.4	61.4
	20	31.3	27.9	48.2	54.5	52.8	64.7
	25	35.1	32.6	50.0	58.3	54.1	67.9
	30	38.8	37.9	51.9	62.6	55.5	71.4
	35	42.6	44.0	53.8	67.1	56.8	74.9
	40	46.3	50.7	55.6	71.7	58.2	78.3

(Continued)

Temp. Room °F	Temp. Outside °F	Single Pane Glass		Double Pane Glass		Triple Pane Glass	
		T_{GLASS} / $T_{DEWPOINT}$	% R.H.	T_{GLASS} / $T_{DEWPOINT}$	% R.H.	T_{GLASS} / $T_{DEWPOINT}$	% R.H.
66	−30	−5.8	4.4	30.1	25.6	39.9	38.2
	−25	−2.1	5.5	32.0	27.7	41.2	40.2
	−20	1.7	6.7	33.8	29.9	42.6	42.5
	−15	5.4	8.2	35.7	32.3	44.0	44.8
	−10	9.1	9.9	37.6	34.9	45.3	47.1
	−5	12.9	11.8	39.4	37.4	46.7	49.7
	0	16.6	14.1	41.3	40.4	48.0	52.2
	5	20.4	16.8	43.2	43.5	49.4	55.1
	10	24.1	19.8	45.1	46.8	50.8	58.0
	15	27.8	23.3	46.9	50.1	52.1	60.9
	20	31.6	27.3	48.8	53.8	53.5	64.1
	25	35.3	31.8	50.7	57.8	54.8	67.2
	30	39.1	37.0	52.5	61.8	56.2	70.8
	35	42.8	42.8	54.4	66.3	57.6	74.4
	40	46.6	49.5	56.3	71.0	58.9	78.0
67	−30	−5.6	4.3	30.7	25.4	40.6	37.9
	−25	−1.8	5.4	32.6	27.5	42.0	40.1
	−20	1.9	6.6	34.5	29.7	43.3	42.2
	−15	5.7	8.0	36.3	32.0	44.7	44.5
	−10	9.4	9.7	38.2	34.5	46.1	46.9
	−5	13.1	11.6	40.1	37.2	47.4	49.3
	0	16.9	13.8	41.9	39.9	48.8	52.0
	5	20.6	16.4	43.8	43.0	50.1	54.6
	10	24.4	19.4	45.7	46.2	51.5	57.5
	15	28.1	22.7	47.6	49.7	52.9	60.6
	20	31.8	26.6	49.4	53.2	54.2	63.5
	25	35.6	31.1	51.3	57.1	55.6	66.9
	30	39.3	36.0	53.2	61.3	56.9	70.1
	35	43.1	41.8	55.0	65.4	58.3	73.7
	40	46.8	48.2	56.9	70.1	59.7	77.5
68	−30	−5.3	4.3	31.3	25.1	41.3	37.7
	−25	−1.6	5.3	33.2	27.2	42.7	39.8
	−20	2.2	6.5	35.1	29.4	44.1	42.0
	−15	5.9	7.8	37.0	31.8	45.4	44.2
	−10	9.7	9.5	38.8	34.1	46.8	46.6
	−5	13.4	11.3	40.7	36.8	48.1	48.9
	0	17.1	13.5	42.6	39.6	49.5	51.6
	5	20.9	16.0	44.4	42.5	50.9	54.4
	10	24.6	18.9	46.3	45.7	52.2	57.0
	15	28.4	22.2	48.2	49.1	53.6	60.1
	20	32.1	26.0	50.0	52.6	54.9	63.0
	25	35.8	30.3	51.9	56.4	56.3	66.3
	30	39.6	35.2	53.8	60.5	57.7	69.7
	35	43.3	40.7	55.7	64.8	59.0	73.0
	40	47.1	47.1	57.5	69.2	60.4	76.7
69	−30	−5.1	4.2	32.0	25.0	42.1	37.6
	−25	−1.3	5.2	33.8	26.9	43.4	39.5
	−20	2.4	6.3	35.7	29.1	44.8	41.7
	−15	6.2	7.7	37.6	31.4	46.2	44.0
	−10	9.9	9.2	39.5	33.9	47.5	46.2
	−5	13.6	11.1	41.3	36.4	48.9	48.7
	0	17.4	13.2	43.2	39.2	50.2	51.2
	5	21.1	15.6	45.1	42.2	51.6	53.9
	10	24.9	18.5	46.9	45.2	53.0	56.8
	15	28.6	21.7	48.8	48.6	54.3	59.5
	20	32.3	25.3	50.7	52.1	55.7	62.7
	25	36.1	29.6	52.5	55.7	57.0	65.7
	30	39.8	34.3	54.4	59.8	58.4	69.1
	35	43.6	39.8	56.3	64.0	59.8	72.6
	40	47.3	45.9	58.2	68.6	61.1	76.0

(Continued)

Temp. Room °F	Temp. Outside °F	Single Pane Glass		Double Pane Glass		Triple Pane Glass	
		T_{GLASS} / $T_{DEWPOINT}$	% R.H.	T_{GLASS} / $T_{DEWPOINT}$	% R.H.	T_{GLASS} / $T_{DEWPOINT}$	% R.H.
70	−30	−4.8	4.1	32.6	24.8	42.8	37.3
	−25	−1.1	5.0	34.5	26.8	44.2	39.4
	−20	2.7	6.2	36.3	28.8	45.5	41.4
	−15	6.4	7.5	38.2	31.1	46.9	43.7
	−10	10.2	9.1	40.1	33.6	48.2	45.9
	−5	13.9	10.8	41.9	36.0	49.6	48.3
	0	17.6	12.9	43.8	38.8	51.0	51.0
	5	21.4	15.3	45.7	41.7	52.3	53.5
	10	25.1	18.0	47.6	44.8	53.7	56.3
	15	28.9	21.2	49.4	48.0	55.0	59.0
	20	32.6	24.8	51.3	51.5	56.4	62.1
	25	36.3	28.8	53.2	55.3	57.8	65.3
	30	40.1	33.6	55.0	59.0	59.1	68.4
	35	43.8	38.8	56.9	63.2	60.5	71.9
	40	47.6	44.8	58.8	67.7	61.8	75.3
71	−30	−4.6	4.0	33.2	23.6	43.5	37.0
	−25	−0.8	5.0	35.1	26.5	44.9	39.1
	−20	2.9	6.0	37.0	28.7	46.2	41.1
	−15	6.7	7.4	38.8	30.8	47.6	43.3
	−10	10.4	8.8	40.7	33.2	49.0	45.7
	−5	14.1	10.6	42.6	35.8	50.3	48.0
	0	17.9	12.6	44.4	38.4	51.7	50.5
	5	21.6	14.9	46.3	41.3	53.0	53.0
	10	25.4	17.6	48.2	44.3	54.4	55.8
	15	29.1	20.7	50.1	47.6	55.8	58.7
	20	32.8	24.1	51.9	50.9	57.1	61.6
	25	36.6	28.2	53.8	54.6	58.5	64.7
	30	40.3	32.7	55.7	58.5	59.8	67.8
	35	44.1	37.9	57.5	62.5	61.2	71.3
	40	47.8	43.7	59.4	66.9	62.6	74.9
72	−30	−4.3	4.0	33.8	24.3	44.3	36.9
	−25	−0.6	4.8	35.7	26.3	45.6	38.8
	−20	3.2	5.9	37.6	28.4	47.0	41.0
	−15	6.9	7.2	39.5	30.6	48.3	43.0
	−10	10.7	8.7	41.3	32.9	49.7	45.3
	−5	14.4	10.4	43.2	35.4	51.1	47.8
	0	18.1	12.3	45.1	38.1	52.4	50.1
	5	21.9	14.6	46.9	40.8	53.8	52.8
	10	25.6	17.2	48.8	43.8	55.1	55.3
	15	29.4	20.5	50.7	47.1	56.5	58.2
	20	33.1	23.6	52.6	50.5	57.9	61.2
	25	36.8	27.5	54.4	54.0	59.2	64.2
	30	40.6	32.0	56.3	57.8	60.6	67.4
	35	44.3	36.9	58.2	61.9	61.9	70.6
	40	48.1	42.7	60.0	66.0	63.3	74.2
73	−30	−4.1	3.8	34.5	24.2	45.0	36.7
	−25	−0.3	4.8	36.3	26.0	46.3	38.6
	−20	3.4	5.8	38.2	28.1	47.7	40.7
	−15	7.2	7.1	40.1	30.3	49.1	42.9
	−10	10.9	8.5	42.0	32.7	50.4	45.0
	−5	14.7	10.2	43.8	35.0	51.8	47.4
	0	18.4	12.1	45.7	37.7	53.1	49.7
	5	22.1	14.3	47.6	40.5	54.5	52.4
	10	25.9	16.9	49.4	43.3	55.9	55.1
	15	29.6	19.7	51.3	46.5	57.2	57.4
	20	33.4	23.1	53.2	49.9	58.6	60.7
	25	37.1	26.9	55.0	53.3	59.9	63.6
	30	40.8	31.2	56.9	57.1	61.3	66.8
	35	44.6	36.1	58.8	61.2	62.7	70.2
	40	48.3	41.6	60.7	65.4	64.0	73.5

(Continued)

Temp. Room °F	Temp. Outside °F	Single Pane Glass		Double Pane Glass		Triple Pane Glass	
		T_{GLASS} / $T_{DEWPOINT}$	% R.H.	T_{GLASS} / $T_{DEWPOINT}$	% R.H.	T_{GLASS} / $T_{DEWPOINT}$	% R.H.
74	−30	−3.8	3.8	35.1	24.0	45.7	36.4
	−25	−0.1	4.7	37.0	25.9	47.1	38.4
	−20	3.7	5.7	38.8	27.8	48.4	40.4
	−15	7.4	6.9	40.7	30.0	49.8	42.5
	−10	11.2	8.3	42.6	32.3	51.2	44.8
	−5	14.9	9.9	44.5	34.8	52.5	47.0
	0	18.6	11.8	46.3	37.3	53.9	49.5
	5	22.4	14.0	48.2	40.1	55.2	51.9
	10	26.1	16.4	50.1	43.0	56.6	54.6
	15	29.9	19.3	51.9	46.0	58.0	57.5
	20	33.6	22.6	53.8	49.3	59.3	60.2
	25	37.3	26.2	55.7	52.9	60.7	63.3
	30	41.1	30.5	57.5	56.4	62.0	66.2
	35	44.8	35.2	59.4	60.4	63.4	69.6
	40	48.6	40.7	61.3	64.6	64.8	73.1
75	−30	−3.5	3.7	35.7	23.8	46.4	36.2
	−25	0.2	4.6	37.6	25.6	47.8	38.2
	−20	3.9	5.6	39.5	27.7	49.2	40.3
	−15	7.7	6.8	41.3	29.7	50.5	42.2
	−10	11.4	8.1	43.2	32.0	51.9	44.5
	−5	15.2	9.7	45.1	34.4	53.2	46.7
	0	18.9	11.6	46.9	36.9	54.6	49.1
	5	22.6	13.6	48.8	39.6	56.0	51.7
	10	26.4	16.1	50.7	42.6	57.3	54.2
	15	30.1	18.9	52.6	45.7	58.7	57.0
	20	33.9	22.1	54.4	48.8	60.0	59.7
	25	37.6	25.7	56.3	52.3	61.4	62.7
	30	41.3	29.7	58.2	56.0	62.8	65.9
	35	45.1	34.4	60.0	59.7	64.1	69.0
	40	48.8	39.6	61.9	63.8	65.5	72.4
76	−30	−3.3	3.6	36.4	23.6	47.2	36.1
	−25	0.4	4.5	38.2	25.4	48.5	37.9
	−20	4.2	5.5	40.1	27.4	49.9	39.9
	−15	7.9	6.6	42.0	29.5	51.2	41.9
	−10	11.7	8.0	43.8	31.7	52.6	44.2
	−5	15.4	9.5	45.7	34.1	54.0	46.5
	0	19.1	11.3	47.6	36.6	55.3	48.8
	5	22.9	13.4	49.4	39.2	56.7	51.3
	10	26.6	15.7	51.3	42.1	58.0	53.8
	15	30.4	18.5	53.2	45.1	59.4	56.5
	20	34.1	21.5	55.1	48.4	60.8	59.4
	25	37.8	25.0	56.9	51.7	62.1	62.2
	30	41.6	29.1	58.8	55.3	63.5	65.3
	35	45.3	33.6	60.7	59.2	64.8	68.3
	40	49.1	38.8	62.5	63.1	66.2	71.7
77	−30	−3.0	3.6	37.0	24.4	47.9	35.8
	−25	0.7	4.4	38.8	25.2	49.3	37.8
	−20	4.4	5.3	40.7	27.2	50.6	39.7
	−15	8.2	6.5	42.6	29.3	52.0	41.8
	−10	11.9	7.8	44.5	31.5	53.3	43.8
	−5	15.7	9.3	46.3	33.7	54.7	46.1
	0	19.4	11.1	48.2	36.3	56.1	48.6
	5	23.1	13.0	50.1	38.9	57.4	50.9
	10	26.9	15.4	51.9	41.6	58.8	53.5
	15	30.6	18.0	53.8	44.6	60.1	56.0
	20	34.4	21.1	55.7	47.9	61.5	58.9
	25	38.1	24.5	57.6	51.3	62.9	61.9
	30	41.8	28.4	59.4	54.7	64.2	64.7
	35	45.6	32.8	61.3	58.5	65.6	68.0
	40	49.3	37.8	63.2	62.5	66.9	71.1

(Continued)

Temp. Room °F	Temp. Outside °F	Single Pane Glass		Double Pane Glass		Triple Pane Glass	
		T_{GLASS} / $T_{DEWPOINT}$	% R.H.	T_{GLASS} / $T_{DEWPOINT}$	% R.H.	T_{GLASS} / $T_{DEWPOINT}$	% R.H.
78	−30	−2.8	3.5	37.6	23.2	48.6	35.6
	−25	0.9	4.3	39.5	25.1	50.0	37.5
	−20	4.7	5.3	41.3	26.9	51.3	39.4
	−15	8.4	6.3	43.2	29.0	52.7	41.5
	−10	12.2	7.6	45.1	31.2	54.1	43.7
	−5	15.9	9.1	47.0	33.5	55.4	45.8
	0	19.7	10.8	48.8	35.9	56.8	48.2
	5	23.4	12.8	50.7	38.5	58.1	50.5
	10	27.1	15.0	52.6	41.3	59.5	53.1
	15	30.9	17.7	54.4	44.2	60.9	55.8
	20	34.6	20.6	56.3	47.3	62.2	58.4
	25	38.4	24.0	58.2	50.7	63.6	61.3
	30	42.1	27.8	60.0	54.0	64.9	64.2
	35	45.8	32.0	61.9	57.8	66.3	67.4
	40	49.6	37.0	63.8	61.8	67.7	70.7
79	−30	−2.5	3.5	38.2	23.0	49.4	35.5
	−25	1.2	4.2	40.1	24.8	50.7	37.3
	−20	4.9	5.1	42.0	26.8	52.1	39.3
	−15	8.7	6.2	43.8	28.7	53.4	41.2
	−10	12.4	7.5	45.7	30.9	54.8	43.4
	−5	16.2	8.9	47.6	33.2	56.2	45.6
	0	19.9	10.6	49.5	35.6	57.5	47.8
	5	23.6	12.5	51.3	38.1	58.9	50.3
	10	27.4	14.7	53.2	40.9	60.2	52.7
	15	31.1	17.3	55.1	43.8	61.6	55.3
	20	34.9	20.2	56.9	46.8	63.0	58.1
	25	38.6	23.4	58.8	50.1	64.3	60.8
	30	42.3	27.1	60.7	53.6	65.7	63.9
	35	46.1	31.3	62.5	57.1	67.0	66.8
	40	49.8	36.0	64.4	61.0	68.4	70.1
80	−30	−2.3	3.4	38.9	22.9	50.1	35.3
	−25	1.5	4.2	40.7	24.6	51.4	37.0
	−20	5.2	5.0	42.6	26.5	52.8	39.0
	−15	8.9	6.1	44.5	28.5	54.2	41.0
	−10	12.7	7.3	46.3	30.6	55.5	43.0
	−5	16.4	8.7	48.2	32.8	56.9	45.3
	0	20.2	10.4	50.1	35.3	58.2	47.4
	5	23.9	12.2	51.9	37.7	59.6	49.9
	10	27.6	14.4	53.8	40.5	61.0	52.4
	15	31.4	16.9	55.7	43.4	62.3	54.9
	20	35.1	19.7	57.6	46.4	63.7	57.6
	25	38.9	22.9	59.4	49.5	65.0	60.3
	30	42.6	26.5	61.3	53.0	66.4	63.3
	35	46.3	30.6	63.2	56.6	67.8	66.4
	40	50.1	35.3	65.0	60.3	69.1	69.5

People/Occupancy Rules of Thumb

10.1 2015 IMC and ASHRAE Standard 62.1-2013

OCCUPANCY SCHEDULE

Occupancy Classification	Max. Occupant Load	
	People per 1,000 SF	SF/Person
Correctional Facilities		
Booking/waiting	50	20
Cell—with plumbing fixtures	25	40
Cell—without plumbing fixtures	25	40
Day room	30	33
Dining halls (see food and beverage service)		
Guard stations	15	67
Dry Cleaners, Laundries		
Coin-operated dry cleaner	20	50
Coin-operated laundries	20	50
Commercial dry cleaner	30	33
Commercial laundry	10	100
Storage, pick-up	30	33
Education		
Art classroom	20	50
Auditoriums	150	6
Classrooms (ages 5 to 8)	25	40
Classrooms (ages 9 plus)	35	28
Computer lab	25	40
Corridors (see public spaces)	–	–
Daycare (through age 4)	25	40
Daycare sickroom	25	40
Lecture classroom	65	15
Lecture hall (fixed seats)	150	6
Locker/dressing rooms	–	–
Media center	25	40
Multiuse assembly	100	10
Music/theater/dance	35	28
Science laboratories	25	40
Smoking lounges	70	14
Sports locker rooms	–	–
University/college laboratories	25	40
Wood/metal shops	20	50
Food and Beverage Service		
Bars, cocktail lounges	100	10
Cafeteria, fast food	100	10
Dining rooms	70	14
Kitchens (cooking)	20	–
Hospitals, Nursing and Convalescent Homes		
Autopsy rooms	–	–
Medical procedure rooms	20	50
Operating rooms	20	50
Patient rooms	10	100
Physical therapy	20	50
Recovery and ICU	10	100
Hotels, Motels, Resorts, and Dormitories		
Barracks sleeping areas	20	50
Bathrooms/toilet—private	–	–
Bedroom/living room	10	100
Laundry rooms, central	10	100
Laundry rooms, within dwelling units	10	100

(Continued)

OCCUPANCY SCHEDULE (*Continued*)

Occupancy Classification	Max. Occupant Load	
	People per 1,000 SF	SF/Person
Conference/meeting	–	–
Dormitory sleeping areas	–	–
Gambling casinos	–	–
Lobbies/prefunction	30	33
Multipurpose assembly	120	8
Miscellaneous Spaces		
Banks or bank lobbies	15	67
Freezer and refrigerated spaces (<50°F)	–	–
General manufacturing (excludes heavy industrial and processes using chemicals)	7	140
Janitor closets, trash rooms, recycling	–	–
Kitchenettes	–	–
Shipping/receiving	2	500
Sorting, packing, light assembly	7	140
Transportation waiting	100	10
Offices		
Conference rooms	50	20
Main entry lobbies	10	100
Occupiable storage rooms for dry materials	2	500
Office spaces	5	200
Reception areas	30	33
Telephone/data entry	60	16
Private Dwellings, Single and Multiple		
Common corridors	–	–
Garages, common for multiple units	–	–
Garages, separate for each dwelling	–	–
Kitchens	–	–
Living areas	Based upon number of bedrooms. First bedroom, 2; each additional bedroom, 1	–
Toilet rooms and bathrooms	–	–
Public spaces		
Breakrooms	25	40
Corridors	–	–
Courtrooms	70	14
Elevator car	–	–
Legislative chambers	50	20
Libraries	10	100
Lobbies	150	6
Museums (children's)	40	25
Museums/galleries	40	25
Occupiable storage rooms for liquids or gels	2	500
Places of religious worship	120	8
Shower room (per shower head)	–	–
Smoking lounges	70	14
Toilet rooms—public	–	–
Retail Stores, Sales Floors, and Showroom Floors		
Dressing rooms	–	–
Mall common areas	40	25
Sales (except as below)	15	66
Shipping and receiving	–	–
Smoking lounges	70	14
Storage rooms	–	–
Warehouses (see storage)	–	–

(*Continued*)

OCCUPANCY SCHEDULE (*Continued*)

Occupancy Classification	Max. Occupant Load	
	People per 1,000 SF	SF/Person
Specialty Shops		
Automotive motor-fuel dispensing stations	–	–
Barber	25	40
Beauty salons	25	40
Nail salons	25	40
Embalming rooms	–	–
Pet shops (animal areas)	10	100
Supermarkets	8	125
Sports and Amusement		
Bowling alleys (seating area)	40	25
Disco/dance floors	100	10
Game arcades	20	50
Gym, stadium (play area)	7	–
Health club/aerobics room	40	25
Health club/weight room	10	100
Ice arenas without combustion engines	–	–
Spectator areas	150	6
Swimming pools (pool and deck area)	–	–
Storage		
Repair garages, enclosed parking garages	–	–
Soiled laundry storage rooms	–	–
Storage rooms, chemical	–	–
Warehouses	–	–
Theaters		
Auditoriums (see education)	–	–
Lobbies	150	6
Stages, studios	70	14
Ticket booths	60	16
Transportation		
Platforms	100	10
Transportation waiting	100	10
Workrooms		
Bank vaults/safe deposit	5	200
Computer (without printing)	4	250
Copy, printing rooms	4	250
Darkrooms	–	–
Meat processing	10	100
Pharmacy (prep. area)	10	100
Photo studios	10	100

Lighting Rules of Thumb

11.01 Code Lighting Power Level Requirements—Building Area Method

Building Area	Watts/Square Foot
	2015 IECC and ASHRAE Std 90.1-2013
Auditorium	–
Automotive Facility	0.80
Bank/Financial Institution	–
Classroom/Lecture Hall	–
Convention Center	1.01
Corridor, Restroom, Support Area	–
Courthouse	1.01
Dining-Bar, Lounge Leisure	1.01
Dining-Cafeteria, Fast Food	0.90
Dining-Family	0.95
Dormitory	0.57
Exercise Center	0.84
Exhibition Hall	–
Grocery Store	–
Fire Station	0.67
Gymnasium	0.94
Healthcare Clinic	0.90
Hospital	1.05
Hotel/Motel	0.87
Industrial Work, <20' Ceiling Height	–
Industrial Work, ≥20' Ceiling Height	–
Kitchen	–
Library	1.19
Lobby–Hotel	–
Lobby–Other	–
Mall, Arcade, or Atrium	–
Multifamily	0.51
Manufacturing Facility	1.17
Museum	1.02
Office	0.82
Parking Garage	0.21
Penitentiary	0.81
Police Station	0.87
Post Office	0.87
Religious Building	1.00
Restaurant	–
Retail	1.26
School/University	0.87
Sports Arena	0.91
Storage, Industrial and Commercial	–
Theater-Motion Picture	0.76
Theater-Performance	1.39
Town Hall	0.89
Transportation	0.70
Warehouse	0.66
Workshop	1.19
Other	–

11.02 Code Lighting Power Level Requirements—Space-by-Space Method

	Watts/Square Foot
Common Space Type	2015 IECC and ASHRAE Std 90.1-2013
Atrium	
That is 40 feet or less in height	0.03/ft. total height
That is greater than 40 feet in height	0.04 + 0.02/ft. total height
Audience Seating Area	
In a Gymnasium	0.65
In an Auditorium	0.63
In a Convention Center	0.82
In a Penitentiary	0.28
In Religious Buildings	1.53
In a Sports Arena	0.43
In a Performing Arts Theater	2.43
In a Motion Picture Theater	1.14
All Other Audience Seating Areas	0.43
Automotive-Service Repair	0.67
Banking Activity Area	1.01
Classroom/Lecture/Training	
In a Penitentiary	1.34
All Other Classroom/Lecture/Training	1.24
Computer Rooms	1.71
Conference/Meeting/Multipurpose	1.23
Confinement Cells	-[a], 0.81[b]
Convention Center-Exhibit Space	1.45
Copy/Print Rooms	0.72
Corridor	
In a Facility for the Visually Impaired	0.92
In a Hospital	0.79[a], 0.99[b]
In a Manufacturing Facility	0.41
All Other Corridors	0.66
Courtroom	1.72
Dining Areas	
In a Penitentiary	0.96
In a Facility for the Visually Impaired	1.90[a], 2.65[b]
In a Bar/Lounge/Leisure Dining	1.07
In a Cafeteria or Fast Food Dining	0.65
In Family Dining	0.89
All Other Dining Areas	0.65
Dormitory-Living Quarters	0.38
Dressing Rooms for Performing Arts Theater	0.61
Electrical/Mechanical Rooms	0.95[a], 0.42[b]
Emergency Vehicle Garage	0.56
Facility for the Visually Impaired	
In a Chapel	2.21
In a Recreation Room/Common Living Room	2.41
Fire Station-Sleeping Quarters	0.22
Food Preparation	1.21
Guest Rooms	0.47[a], 0.91[b]
Gymnasium/Fitness Centers	
In an Exercise Area	0.72
In a Playing Area	1.20

(Continued)

Common Space Type	Watts/Square Foot
	2015 IECC and ASHRAE Std 90.1-2013
Healthcare Facility	
In an Exam/Treatment Room	1.66
In an Imaging Room	1.51
In a Medical Supply Room	0.74
In a Nursery	0.88
In a Nurse's Station	0.71
In an Operating Room	2.48
In a Patient Room	0.62
In a Physical Therapy Room	0.91
In a Recovery Room	1.15
Laboratory	
In or as a classroom	1.43
All Other Laboratories	1.81
Laundry/Washing Area	0.60
Library	
In Reading Area	1.06
In the Stacks	1.71
Loading Dock, Interior	0.47
Lobby	
In a Facility for the Visually Impaired	1.80
For an Elevator	0.64
In a Hotel	1.06
In a Motion Picture Theater	0.59
In a Performing Arts Theater	2.00
All Other Lobbies	0.90
Locker Room	0.75
Lounge/Breakroom	
In a Healthcare Facility	0.92
All Other Lounges/Breakrooms	0.73
Manufacturing Facility	
In a Detailed Manufacturing Area	1.29
In an Equipment Room	0.74
In an Extra High Bay Area (>50 feet Floor-to-Ceiling Height)	1.05
In a High Bay Area (25–50 feet Floor-to-Ceiling Height)	1.23
In a Low Bay Area (<25 feet Floor-to-Ceiling Height)	1.19
Museum	
In a General Exhibition Area	1.05
In a Restoration Area	1.02
Office	
Enclosed	1.11
Open Plan	0.98
Parking Area, Interior	0.19
Pharmacy Area	1.68
Post Office-Sorting Area	0.94
Religious Buildings	
In a Fellowship Hall	0.64
In a Worship/Pulpit/Choir Area	1.53
Fellowship Hall	0.9
Restrooms	
In a Facility for the Visually Impaired	1.21
All Other Restrooms	0.98

(Continued)

Common Space Type	Watts/Square Foot
	2015 IECC and ASHRAE Std 90.1-2013
Retail Facilities	
In a Sales Area	1.59[a], 1.44[b]
In a Dressing/Fitting Room	0.71
In a Mall Concourse	1.10
Seating Area, General	0.54
Sports Arena	
For a Class I Facility	3.68
For a Class II Facility	2.40
For a Class III Facility	1.80
For a Class IV Facility	1.20
Stairwell	0.69
Storage Room	0.63[a]
$< 50\ ft^2$	1.24[b]
All Other Storage Rooms	0.63[b]
Transportation Facility	
In a Baggage/Carousel Area	0.53
In an Airport Concourse	0.36
At a Terminal Ticket Counter	0.80
Warehouse	
For Medium to Bulky, Palletized Items	0.58
For Smaller, Hand-Carried Items	0.95
Workshop	1.59

Notes:
 a. 2015 IECC.
 b. ASHRAE Standard 90.1-2013.

Appliance/Equipment Rules of Thumb

12.01 Offices and Commercial Spaces

A. Total Appliance/Equipment Heat Gain   0.5–8.0 watt/sq.ft.

B. Computer equipment loads for office spaces range between 0.5 watt/sq.ft. and
    3.5 watts/sq.ft. (2.0 watts/sq.ft. is recommended). If actual computer equipment
    loads are available, they should be used in lieu of the values listed here.

C. Depending on the facility, the appliance/equipment diversity factors can range
    from 25 to 75 percent (recommend diversities of 50 percent are recommended).

12.02 Computer Rooms, Data Centers, and Internet Host Sites

A. 2.0–500.0 watts/sq.ft. (Recommend 300 watts/sq.ft. Minimum)

12.03 Telecommunication Rooms

A. 50.0–120.0 watts/sq.ft.

12.04 Electrical Equipment Heat Gain

A. **Transformers**

1. 150 KVA and smaller   50 watts/KVA
2. 151–500 KVA   30 watts/KVA
3. 501–1000 KVA   25 watts/KVA
4. 1001–2500 KVA   20 watts/KVA
5. Larger than 2500 KVA   15 watts/KVA

B. **Switchgear**

1. Low voltage breaker 0–40 amps   10 watts
2. Low voltage breaker 50–100 amps   20 watts
3. Low voltage breaker 225 amps   60 watts
4. Low voltage breaker 400 amps   100 watts
5. Low voltage breaker 600 amps   130 watts
6. Low voltage breaker 800 amps   170 watts
7. Low voltage breaker 1,600 amps   460 watts
8. Low voltage breaker 2,000 amps   600 watts
9. Low voltage breaker 3,000 amps   1,100 watts
10. Low voltage breaker 4,000 amps   1,500 watts
11. Medium voltage breaker/switch 600 amps   1,000 watts
12. Medium voltage breaker/switch 1,200 amps 1,500 watts
13. Medium voltage breaker/switch 2,000 amps 2,000 watts
14. Medium voltage breaker/switch 3,000 amps 2,500 watts

C. **Panelboards**

1. 2 watts per circuit.

D. Motor Control Centers

1. 500 watts per section—each section is approximately 20" wide × 20" deep × 84" high

E. Starters

1.	Low voltage starters size 00	50 watts
2.	Low voltage starters size 0	50 watts
3.	Low voltage starters size 1	50 watts
4.	Low voltage starters size 2	100 watts
5.	Low voltage starters size 3	130 watts
6.	Low voltage starters size 4	200 watts
7.	Low voltage starters size 5	300 watts
8.	Low voltage starters size 6	650 watts
9.	Medium voltage starters size 200 amps	400 watts
10.	Medium voltage starters size 400 amps	1,300 watts
11.	Medium voltage starters size 700 amps	1,700 watts

F. Variable Frequency Drives

1. 2–6% of the KVA rating: 3% is most common

G. Miscellaneous Equipment

1.	Bus duct	0.015 watts/ft./amp
2.	Capacitors	2 watts/KVAR

Notes:

1 Actual electrical equipment heat gain values will vary from one manufacturer to another—use actual values when available.

2 In the past, electrical equipment rooms only required ventilation to keep equipment from overheating. Most electrical rooms were designed for 95°F to 104°F, although electrical equipment used today may require a maximum design temperature of 90°F because of electronic components and controls. Consult your electrical engineer for equipment temperature tolerances. If space temperatures below 90°F are required by the equipment, the air conditioning (tempering) of space will be required.

3 If outside air is used to ventilate the electrical room, the electrical room design temperature will be 10°F to 15°F above outside summer design temperatures.

4 If conditioned air from an adjacent space is used to ventilate the electrical room, the electrical room temperature can be 10°F to 20°F above the adjacent spaces.

5 Elevator machine rooms require 90°F space temperature (maximum) due to electronic components of elevator equipment. Therefore, elevator machine rooms must be air conditioned (tempered).

12.05 Motor Heat Gain

A. Motors Only

1.	Motors 0–2 hp	190 watts/hp
2.	Motors 3–20 hp	110 watts/hp
3.	Motors 25–200 hp	75 watts/hp
4.	Motors 250 hp and Larger	60 watts/hp

B. **Motors and driven equipment are shown in the following table:**

| Motor Horsepower | The Location of Motor and Driven Equipment with Respect to a Conditioned Space or Airstream | | |
	Motor In, Driven Equipment In Btu/h	Motor Out, Driven Equipment In Btu/h	Motor In, Driven Equipment Out Btu/h
1/20	360	130	240
1/12	580	200	380
1/8	900	320	590
1/6	1,160	400	760
1/4	1,180	640	540
1/3	1,500	840	660
1/2	2,120	1,270	850
3/4	2,650	1,900	740
1	3,390	2,550	850
1-½	4,960	3,820	1,140
2	6,440	5,090	1,350
3	9,430	7,640	1,790
5	15,500	12,700	2,790
7-½	22,700	19,100	3,640
10	29,900	24,500	4,490
15	44,400	38,200	6,210
20	58,500	50,900	7,610
25	72,300	63,600	8,680
30	85,700	76,300	9,440
40	114,000	102,000	12,600
50	143,000	127,000	15,700
60	172,000	153,000	18,900
75	212,000	191,000	21,200
100	283,000	255,000	28,300
125	353,000	318,000	35,300
150	420,000	382,000	37,800
200	569,000	509,000	50,300
250	699,000	636,000	62,900

12.06 Miscellaneous Guidelines

A. **Actual equipment layouts and information should be used for calculating equipment loads.**

B. **Movie projectors, slide projectors, overhead projectors, and similar types of equipment can generally be ignored because lights are off when being used and the lighting load will normally be larger than this equipment heat gain.**

C. **Items such as coffee pots, microwave ovens, refrigerators, food warmers, etc., should be considered when calculating equipment loads.**

D. **Kitchen, laboratory, hospital, computer room, and process equipment should be obtained from the owner, architect, engineer, or consultant due to the extreme variability of equipment loads.**

PART 13

Cooling Load Factors

13.01 Diversity Factors

A. *Diversity factors* are an engineer's judgment applied to various people, lighting, equipment, and total loads to consider actual usage. Actual diversities may vary depending on building type and occupancy. Diversities listed here are for office buildings and similar facilities.

B. **Room/Space Peak Loads**

1. People $1.0 \times$ Calculated Load.
2. Lights $1.0 \times$ Calculated Load.
3. Equipment $1.0 \times$ Calculated Load.*

C. **Floor/Zone Block Loads**

1. People $0.90 \times$ Sum of Peak Room/Space People Loads.
2. Lights $0.95 \times$ Sum of Peak Room/Space Lighting Loads.
3. Equipment $0.90 \times$ Sum of Peak Room/Space Equipment Loads.
4. Floor/Zone Total Loads $0.90 \times$ Sum of Peak Room/Space Total Loads.

D. **Building Block Loads**

1. People $0.75 \times$ Sum of Peak Room/Space People Loads.
2. Lighting $0.95 \times$ Sum of Peak Room/Space Lighting Loads.
3. Equipment $0.75 \times$ Sum of Peak Room/Space Equipment Loads.
4. Building Total Load $0.85 \times$ Sum of Peak Room/Space Total Loads.

13.02 Safety Factors

A. **Room/Space Peak Loads** $1.1 \times$ **Calc. Load.**

B. **Floor/Zone Loads (Sum of Peak)** $1.0 \times$ **Calc. Load.**

C. **Floor/Zone Loads (Block)** $1.1 \times$ **Calc. Load.**

D. **Building Loads (Sum of Peak)** $1.0 \times$ **Calc. Load.**

E. **Building Loads (Block)** $1.1 \times$ **Calc. Load.**

13.03 Cooling Load Factors

A. **Lighting Load Factors**

1. Existing lighting fixtures
   a. Fluorescent lights $1.25 \times$ Bulb Watts.
   b. Incandescent lights $1.00 \times$ Bulb Watts.
   c. HID lighting $1.25 \times$ Bulb Watts.
2. New lighting fixtures
   a. Fluorescent lights $0.85–1.15 \times$ Bulb Watts.
   b. Incandescent lights $1.00 \times$ Bulb Watts.
   c. HID lighting $0.85–1.15 \times$ Bulb Watts.
   d. Electronic ballasts have provided better energy performance. Electronic ballast factors are very dependent on the manufacturer, type of lighting fixture, and type of lamp. Consult electrical engineer for exact lighting watts required for fixtures.

---

*Calculated Load may have a diversity factor that has been calculated using individual pieces of equipment, equipment as a group, or incorporating no equipment at all.

B. Return Air Plenum (RAP) Factors

1. Heat of lights to space with RAP $0.76 \times$ Lighting Load.
2. Heat of lights to RAP $0.24 \times$ Lighting Load.
3. Heat of roof to space with RAP $0.30 \times$ Roof Load.
4. Heat of roof to RAP $0.70 \times$ Roof Load.

C. Ducted Exhaust or Return Air (DERA) Factors

1. Heat of lights to space with DERA $1.00 \times$ Lighting Load.
2. Heat of roof to space with DERA $1.00 \times$ Roof Load.

D. Other Cooling Load Factors (CLFs) are in accordance with ASHRAE recommendations.

1. CLF $\times$ Other Loads.

Heating Load Factors

14.01 Safety Factors

A.	Room/Space Peak Loads	1.1 × Calc. Load
B.	Floor/Zone Loads (Sum of Peak)	1.0 × Calc. Load
C.	Floor/Zone Loads (Block)	1.1 × Calc. Load
D.	Building Loads (Sum of Peak)	1.0 × Calc. Load
E.	Building Loads (Block)	1.1 × Calc. Load
F.	Generally: Sum of Peak Loads	1.1 × Block Loads

14.02 Heating Load Credits

A. **Solar.** Credit for solar gains should not be taken unless the building is specifically designed for solar heating. Solar gain is not a factor at night when design temperatures generally reach their lowest point.

B. **People.** Credit for people should not be taken. People gain is not a factor at night when design temperatures generally reach their lowest point because buildings are generally unoccupied at night.

C. **Lighting.** Credit for lighting should not be taken. Lighting is an inefficient means to heat a building and lights are generally off at night when design temperatures generally reach their lowest point.

D. **Equipment.** Credit for equipment should not be taken unless a reliable source of heat is generated 24 hours a day (e.g., computer facility, industrial process). Only a portion of this load should be considered (50 percent) and the building heating system should be able to keep the building from freezing if these equipment loads are shut down for extended periods of time. Consider what would happen if the system or process shut down for extended periods of time.

14.03 Heating System Selection Guidelines

A. If heat loss exceeds 450 Btu/h per lineal feet of wall, heat should be provided from under the window or from the base of the wall to prevent downdrafts.

B. If heat loss is between 250 and 450 Btu/h per lineal feet of wall, heat should be provided from under the window or from the base of the wall, or it may be provided from overhead diffusers, located adjacent to the perimeter wall, discharging air directly downward, and blanketing the exposed wall and window areas.

C. If heat loss is less than 250 Btu/h per lineal feet of wall, heat should be provided from under the window or from the base of the wall, or it may be provided from overhead diffusers, located adjacent to or slightly away from the perimeter wall, discharging air directed at, or both directed at and directed away from, the exposed wall and window areas.

Design Conditions and Energy Conservation

15.01 Design Conditions

A. Outside Design Conditions

1. Outdoor design conditions should be taken from either the *ASHRAE Handbook of Fundamentals*, local weather data, or some other recognized source.
2. ASHRAE summer design conditions are based on the following:
   a. Total yearly hours: 8,760 hours.
   b. Annual extreme values represent maximum summer design conditions.
   c. 0.4 percent design values represent summer design conditions that, on average, are exceeded fewer than 35 hours annually above stated conditions.
   d. 1.0 percent design values represent summer design conditions that, on average, are exceeded fewer than 88 hours annually above stated conditions.
   e. 2.0 percent design values represent summer design conditions that, on average, are exceeded fewer than 175 hours annually above stated conditions.
   f. 5.0 percent design values represent summer design conditions that, on average, are exceeded fewer than 438 hours annually above stated conditions.
3. ASHRAE winter design conditions are based on the following:
   a. Total yearly hours: 8,760 hours.
   b. Annual extreme values represent maximum winter design conditions.
   c. 99.6 percent design values represent winter design conditions that, on average, are exceeded fewer than 35 hours annually below stated conditions.
   d. 99.0 percent design values represent winter design conditions that, on average, are exceeded fewer than 88 hours annually below stated conditions.
4. Outside design condition example: Ambient weather conditions are based on Pittsburgh (Allegheny County Airport), Pennsylvania weather data from the American Society of Heating, Refrigerating, and Air Conditioning Engineers, Inc. (ASHRAE) *Handbook of Fundamentals, 2005 Edition* (see Fig. 15.1).
   a. Abbreviations:
      1) db = dry bulb temperature.
      2) wb = wet bulb temperature.
      3) dp = dewpoint temperature.
   b. Pittsburgh, PA summer design conditions:
      1) Average annual extreme value: 92.6°F db/81.3°F wb.
      2) Annual extreme value (20 years): 99.1°F db/81.3°F wb.
      3) 0.4% values: 89.1°F db/74.9°F wb.
      4) 1.0% values: 86.2°F db/73.3°F wb.
      5) 2.0% values: 83.8°F db/71.9°F wb.
   c. Pittsburgh, PA winter design conditions:
      1) Average annual extreme value: −4.6°F db.
      2) Annual extreme value (20 years): −19.3°F db.
      3) 99.6% values: 1.8°F db/−7.1°F dp.
      4) 99.0% values: 7.5°F db/−2.3°F dp.
5. Recommended outside design conditions values:
   a. General facilities and spaces—office buildings, schools, commercial and industrial facilities, other noncritical temperature and humidity control facilities:
      1) Summer: 1.0% values.
      2) Winter: 99.0% values.
   b. Semicritical facilities and spaces—hospitals, medical facilities, laboratories, other semicritical temperature and humidity control facilities:
      1) Summer: 0.4% values.
      2) Winter: 99.6% values.
   c. Critical facilities and spaces—laboratories, computer facilities, high-tech industrial, other critical temperature and humidity control facilities:
      1) Summer: 0.4% values.
      2) Winter: 99.6% values.
      3) Annual extreme values may even be appropriate here.

2005 ASHRAE Handbook - Fundamentals (IP)

Design conditions for PITTSBURGH, PA, USA

Station Information

Station name		WMO#	Lat	Long	Elev	StdP	Hours +/- UTC	Time zone code	Period
1a		1b	1c	1d	1e	1f	1g	1h	1i
PITTSBURGH		725200	40.50N	80.22W	1224	14.058	-5.00	NAE	7201

Annual Heating and Humidification Design Conditions

Coldest month	Heating DB		Humidification DP/MCDB and HR							Coldest month WS/MCDB				MCWS/PCWD to 99.6% DB	
			99.6%			99%				0.4%		1%			
	99.6%	99%	DP	HR	MCDB	DP	HR	MCDB	WS	MCDB	WS	MCDB	MCWS	PCWD
2	3a	3b	4a	4b	4c	4d	4e	4f	5a	5b	5c	5d	6a	6b
1	1.8	7.5	-7.1	3.9	4.3	-2.3	5.1	9.7	28.5	25.6	25.9	24.6	9.6	260

Annual Cooling, Dehumidification, and Enthalpy Design Conditions

Hottest month	Hottest month DB range	Cooling DB/MCWB						Evaporation WB/MCDB						MCWS/PCWD to 0.4% DB	
		0.4%		1%		2%		0.4%		1%		2%			
		DB	MCWB	DB	MCWB	DB	MCWB	WB	MCDB	WB	MCDB	WB	MCDB	MCWS	PCWD
7	8	9a	9b	9c	9d	9e	9f	10a	10b	10c	10d	10e	10f	11a	11b
7	18.9	89.1	72.5	86.2	70.9	83.8	69.3	74.9	85.0	73.3	82.5	71.9	80.3	10.3	240

Dehumidification DP/MCDB and HR									Enthalpy/MCDB					
0.4%			1%			2%			0.4%		1%		2%	
DP	HR	MCDB	DP	HR	MCDB	DP	HR	MCDB	Enth	MCDB	Enth	MCDB	Enth	MCDB
12a	12b	12c	12d	12e	12f	12g	12h	12i	13a	13b	13c	13d	13e	13f
71.8	123.0	80.1	70.3	116.9	78.2	69.0	111.6	76.8	31.5	85.1	30.0	82.7	28.7	80.5

Extreme Annual Design Conditions

Extreme Annual WS			Extreme Max WB	Extreme Annual DB				n-Year Return Period Values of Extreme DB							
				Mean		Standard deviation		n=5 years		n=10 years		n=20 years		n=50 years	
1%	2.5%	5%		Max	Min	Max	Min	Max	Min	Max	Min	Max	Min	Max	Min
14a	14b	14c	15	16a	16b	16c	16d	17a	17b	17c	17d	17e	17f	17g	17h
23.5	19.8	17.9	81.3	92.6	-4.6	3.5	7.9	95.1	-10.3	97.2	-14.9	99.1	-19.3	101.7	-25.1

Monthly Design Dry Bulb and Mean Coincident Wet Bulb Temperatures

%	Jan		Feb		Mar		Apr		May		Jun	
	DB	MCWB	DB	MCWB	DB	MCWB	DB	MCWB	DB	MCWB	DB	MCWB
	18a	18b	18c	18d	18e	18f	18g	18h	18i	18j	18k	18l
0.4%	61.9	55.8	64.5	53.3	76.4	58.8	82.1	61.8	85.9	68.3	90.5	72.2
1%	58.0	52.8	61.6	52.4	73.0	56.7	79.9	61.4	84.2	66.8	88.2	71.1
2%	55.1	50.4	58.1	49.1	70.0	55.6	77.1	60.3	82.4	65.6	86.4	70.4

%	Jul		Aug		Sep		Oct		Nov		Dec	
	DB	MCWB	DB	MCWB	DB	MCWB	DB	MCWB	DB	MCWB	DB	MCWB
	18m	18n	18o	18p	18q	18r	18s	18t	18u	18v	18w	18x
0.4%	93.4	74.1	92.4	73.6	87.2	70.4	77.9	64.0	71.5	59.5	63.9	56.4
1%	91.4	73.5	90.2	73.4	85.1	70.1	76.1	62.8	69.1	58.2	61.3	54.5
2%	89.5	73.1	88.0	72.3	83.0	68.8	74.1	61.4	66.6	56.5	58.7	53.2

Monthly Design Wet Bulb and Mean Coincident Dry Bulb Temperatures

%	Jan		Feb		Mar		Apr		May		Jun	
	WB	MCDB	WB	MCDB	WB	MCDB	WB	MCDB	WB	MCDB	WB	MCDB
	19a	19b	19c	19d	19e	19f	19g	19h	19i	19j	19k	19l
0.4%	56.4	60.8	56.0	62.5	61.2	72.0	65.2	76.3	71.9	81.5	75.3	86.1
1%	54.0	57.4	53.4	58.8	59.4	68.2	63.7	75.1	70.6	80.0	74.2	84.6
2%	50.9	55.1	51.1	55.9	57.3	66.3	62.2	73.5	68.9	78.4	73.2	83.1

%	Jul		Aug		Sep		Oct		Nov		Dec	
	WB	MCDB	WB	MCDB	WB	MCDB	WB	MCDB	WB	MCDB	WB	MCDB
	19m	19n	19o	19p	19q	19r	19s	19t	19u	19v	19w	19x
0.4%	77.3	88.5	77.0	87.8	74.0	82.9	66.9	73.6	61.8	67.7	58.4	62.1
1%	76.2	86.9	75.6	85.9	72.9	81.0	65.4	72.2	60.3	66.1	56.7	60.3
2%	75.2	85.4	74.5	83.9	71.5	78.9	64.0	70.6	59.0	64.7	54.2	57.6

Monthly Mean Daily Temperature Range

Jan	Feb	Mar	Apr	May	Jun	Jul	Aug	Sep	Oct	Nov	Dec
20a	20b	20c	20d	20e	20f	20g	20h	20i	20j	20k	20l
14.1	15.6	18.2	20.3	20.3	19.9	18.9	18.7	19.0	18.8	15.4	13.5

WMO#	World Meteorological Organization number	Lat	Latitude, °
Elev	Elevation, ft	StdP	Standard pressure at station elevation, psi
DB	Dry bulb temperature, °F	DP	Dew point temperature, °F
WS	Wind speed, mph	Enth	Enthalpy, Btu/lb
MCDB	Mean coincident dry bulb temperature, °F	MCDP	Mean coincident dew point temperature, °F
MCWS	Mean coincident wind speed, mph	PCWD	Prevailing coincident wind direction, °, 0 = North, 90 = East

Long — Longitude, °
WB — Wet bulb temperature, °F
HR — Humidity ratio, grains of moisture per lb of dry air
MCWB — Mean coincident wet bulb temperature, °F

FIGURE 15.1

d. Mechanical equipment rooms, electrical equipment rooms, and similar spaces:
   1) Summer: 2.0% values.
   2) Winter: 99.0% values.

B. **Indoor Design Conditions**

INDOOR DESIGN CONDITIONS

Facility Type/Spaces (1)	Cooling		Heating	
	Temperature °F Dry Bulb	Humidity %RH	Temperature °F Dry Bulb	Humidity %RH
Office Buildings, Commercial Facilities				
Offices	75	50	70	–
Conference Rooms, Meeting Rooms, Classrooms, Training Rooms	75	50	70	–
Reception Areas, Lobbies, Corridors	75	50	70	–
Network Computer Rooms	72 ± 2	–	72 ± 2	–
Telecommunication Areas/Data Entry	75	50	70	–
General Spaces				
Mechanical/Electrical/Equip. Rooms	85–90	–	60	–
Elevator Machine Rooms	85–90	–	65	–
Telecommunication Rooms	72 ± 2	–	72 ± 2	–
Toilet Rooms, Locker Rooms, Shower Rooms	78	50	70	–
Janitor Closets, Housekeeping	85	–	60	–
Lobbies, Corridors, Elevator Lobbies, Atriums	75	50	70	–
Vestibules, Entrances, Stairs	–	–	60	–
Stairs on Perimeter of Building or w/Glass	80	–	60	–
Storage	78	50	70	–
Garages—Open	–	–	–	–
Garages—Enclosed	–	–	50 (9)	–
Educational Facilities				
Auditoriums, Gymnasiums, Multipurpose Rooms	75	50	70	–
Classrooms, Lecture Halls	75	50	70	–
Laboratories—High School (4)	75	50	70	–
Laboratories—College or University (4)	75 ± 2	45 ± 5	72 ± 2	45 ± 5
Libraries	75	50	70	–
Music Rooms, Art Rooms	75	50	70	–
Training Shops	75	50	70	–
Food and Beverage Service				
Restaurants, Dining Rooms, Bars, Lounges, Cocktail Lounges	75	50	70	–
Cafeteria, Fast Food	75	50	70	–
Kitchens, Dishwashing	80	–	68	–
Hotels, Motels, Resorts, and Dormitories				
Conference Rooms, Meeting Rooms, Ballrooms	75	50	70	–
Bedrooms, Bathrooms	75	50	70	–
Dormitory Sleeping Areas	75	50	70	–
Living Rooms, Dining Rooms	75	50	70	–
Gambling Casinos, Gaming Rooms	75	50	70	–
Correctional Facilities				
Prison Cells	75	50	70	–
Dining Halls, Day Rooms	75	50	70	–
Guard Stations	75	50	70	–
Retail Stores and Specialty Shops				
Malls and Arcades	75	50	70	–
Department Stores, Supermarkets, Showroom Floors, Clothiers, Furniture	75	50	70	–
Dressing Rooms	75	50	70	–
Shipping and Receiving	–	–	60	–
Storage Rooms	78	50	70	–
Warehouses (7)	78	50	70	–

(Continued)

INDOOR DESIGN CONDITIONS (*Continued*)

Facility Type/Spaces (1)	Cooling		Heating	
	Temperature °F Dry Bulb	Humidity %RH	Temperature °F Dry Bulb	Humidity %RH
Barber Shop, Beauty Shop, Nail Salons	75	50	70	–
Hardware, Drug Stores, Fabric Stores, Specialty Stores	75	50	70	–
Pet Stores	75	50	70	–
Automobile Showrooms	75	50	70	–
Theaters				
Auditoriums, Concert Halls, Performing Arts Centers—Seating Area (5)	75	50	70	–
Stages, Performing Arts Studios—Performance Areas (5)	72 ± 2	40 ± 5	72 ± 2	40 ± 5
3D Theaters (6)	72	50	70	–
Lobbies, Ticket Booths, On-Call Windows	75	50	70	–
Sports and Entertainment				
Ballrooms, Disco, Dance Establishments	75	50	70	–
Bowling Alleys, Game Rooms, Arcades	75	50	70	–
Firing Ranges	75	50	70	–
Gymnasiums, Playing Floors	75	50	65	–
Swimming Pools, Natatoriums	75	50	70	–
Spectator Areas	75	50	70	–
Transportation				
Bus Stations, Airports	75	50	70	–
Waiting Areas	75	50	70	–
Storage Facilities				
Repair Garages (8)	–	–	65	–
Warehouses (7)	78	50	70	–
Workrooms				
Bank Vaults	75	50	70	–
Darkrooms	75	50	70	–
Duplicating, Printing	75	50	70	–
Pharmacy	75	50	70	–
Photo Studios	75	50	70	–
Private Dwelling—Single and Multiple				
Living Rooms, Dining Rooms, Bedrooms, Kitchens	75	50	70	–
Toilet Rooms, Bathrooms	78	50	70	–
Garages	–	–	50 (9)	–
Laboratories, Computer Facilities, High-Tech Industrial, Special Facilities (4)				
Laboratories—Research	72 ± 2	45 ± 5	72 ± 2	45 ± 5
Computer Rooms, Data Centers, Internet Host Sites, Server Rooms, Demarc Rooms	70 ± 2	45 ± 5	70 ± 2	45 ± 5
High Tech	68 ± 2	45 ± 5	68 ± 2	45 ± 5
High Tech (low humidity)	68 ± 2	35 ± 5	68 ± 2	35 ± 5
Animal Research Rooms	68–84 ± 2	40–70 ± 5	68–84 ± 2	40–70 ± 5
Museums, Galleries, Rare Document Libraries, Archives	72 ± 2	40 ± 5	72 ± 2	40 ± 5
Surgical and Critical Care (2)				
Operating room (Class B and C)	75	max 60	68	min 20
Operating/surgical cystoscopic rooms	75	max 60	68	min 20
Delivery room (Caesarean)	75	max 60	68	min 20
Substerile service area	–	–	–	–
Recovery room	75	max 60	70	min 20
Critical and intensive care	75	max 60	70	min 30
Intermediate care	75	max 60	70	–
Wound intensive care (burn unit)	75	max 60	70	min 40
Newborn intensive care	78	max 60	72	min 30
Treatment room	75	max 60	70	min 20
Trauma room (crisis or shock)	75	max 60	70	min 20
Medical/anesthesia gas storage	–	–	–	–
Laser eye room	75	max 60	70	min 20

(*Continued*)

INDOOR DESIGN CONDITIONS (*Continued*)

Facility Type/Spaces (1)	Cooling		Heating	
	Temperature °F Dry Bulb	Humidity %RH	Temperature °F Dry Bulb	Humidity %RH
ER waiting rooms	75	max 65	70	–
Triage	75	max 60	70	–
ER decontamination	75	–	–	–
Radiology waiting rooms	75	max 60	70	–
Procedure room (Class A surgery)	75	max 60	70	min 20
Emergency department exam/treatment room	75	max 60	70	–
Inpatient Nursing				
Patient room	75	max 60	70	–
Nourishment area or room	–	–	–	–
Toilet room	–	–	–	–
Newborn nursery suite	78	max 60	72	min 30
Protective environment room	75	max 60	70	–
Airborne Infectious Isolation (AII) room	75	max 60	70	–
Combination AII/Protective Environment (PE) room	75	max 60	70	–
AII anteroom	–	–	–	–
PE anteroom	–	–	–	–
Combination AII/PE anteroom	–	–	–	–
Labor/delivery/recovery/postpartum (LDRP)	75	max 60	70	–
Labor/delivery/recovery (LDR)	75	max 60	70	–
Patient Corridor	–	–	–	–
Nursing Facility				
Resident room	75	–	70	–
Resident gathering/activity/dining	75	–	70	–
Resident unit corridor	–	–	–	–
Physical therapy	75	–	70	–
Occupational therapy	75	–	70	–
Bathing room	75	–	70	–
Radiology				
X-ray (diagnostic and treatment)	78	max 60	72	–
X-ray (surgery/critical care and catheterization)	75	max 60	70	–
Darkroom	–	–	–	–
Diagnostic and Treatment				
Bronchoscopy, sputum collection, and pentamidine administration	73	–	68	–
Laboratory, general	75	–	70	–
Laboratory, bacteriology	75	–	70	–
Laboratory, biochemistry	75	–	70	–
Laboratory, cytology	75	–	70	–
Laboratory, glass washing	–	–	–	–
Laboratory, histology	75	–	70	–
Laboratory, microbiology	75	–	70	–
Laboratory, nuclear medicine	75	–	70	–
Laboratory, pathology	75	–	70	–
Laboratory, serology	75	–	70	–
Laboratory, sterilizing	75	–	70	–
Laboratory, media transfer	75	–	70	–
Nonrefrigerated body-holding room	75	–	70	–
Autopsy room	75	–	68	–
Pharmacy	–	–	–	–
Examination room	75	max 60	70	–
Medication room	75	max 60	70	–
Gastrointestinal endoscopy procedure room	73	max 60	68	min 20
Endoscope cleaning	–	–	–	–
Treatment room	75	max 60	70	–
Hydrotherapy	80	–	72	–
Physical therapy	80	max 65	72	–
Dialysis treatment area	78	–	72	–
Dialyzer reprocessing room	–	–	–	–

(Continued)

INDOOR DESIGN CONDITIONS (*Continued*)

Facility Type/Spaces (1)	Cooling		Heating	
	Temperature °F Dry Bulb	Humidity %RH	Temperature °F Dry Bulb	Humidity %RH
Nuclear medicine hot lab	75	–	70	–
Nuclear medicine treatment room	75	–	70	–
Sterilizing				
Sterilizer equipment room	–	–	–	–
Central Medical and Surgical Supply				
Soiled or decontamination room	78	–	72	–
Clean workroom	78	max 60	72	–
Sterile storage	78	max 60	72	–
Service				
Food preparation center	78	–	72	–
Warewashing	–	–	–	–
Dietary storage	78	–	72	–
Laundry, general	–	–	–	–
Soiled linen sorting and storage	–	–	–	–
Clean linen storage	78	–	72	–
Linen and trash chute room	–	–	–	–
Bedpan room	–	–	–	–
Bathroom	78	–	72	–
Janitor's closet	–	–	–	–
Support Space				
Soiled workroom or soiled holding	–	–	–	–
Clean workroom or clean holding	–	–	–	–
Hazardous material storage	–	–	–	–

Notes:

1 Indoor design conditions are recommendations that may be used when owners of the facility used have no specific criteria. When codes, processes, or other criteria require different design conditions than those listed here, they should be used for design purposes.

2 ASHRAE Standard 170-2013 *Ventilation of Health Care Facilities* (incorporated as Part 4 of the 2014 Facility Guidelines Institute *Guidelines for Design and Construction of Hospitals and Outpatient Facilities*) provides ranges for the spaces. The systems shall be capable of maintaining the rooms within the temperature and relative humidity ranges listed in the table during normal operation. Some specialty procedures may require special temperatures and relative humidity levels. The hospital staff should be consulted to determine any special requirements.

3 Surgeons or procedures may require room temperatures and relative humidities that exceed the minimum indicated ranges.

4 With laboratories, computer facilities, high-tech industrial, special facilities, museums, and other spaces requiring critical temperature and relative humidity control, the owners and their staff should be consulted to verify the specific requirements. Requirements provided here are for general guidance only.

5 Some performances will require strict temperature and relative humidity control during the performance. Careful consideration must be given to the types of performances and their requirements when designing these facilities.

6 3D type theaters (IMAX and OMNIMAX) should be maintained at temperatures at or below 72°F because the incidents of motion sickness increase considerably above these temperatures.

7 Warehouse temperature and relative humidity requirements can vary considerably depending on the materials being stored.

8 Repair garages are generally not air conditioned. If air conditioning is desired, the cooling design temperature recommendation would be 80 to 85°F.

9 Heating parking garages is not recommended because minimum-code-required ventilation rates require heating of the makeup air for the heating to become effective.

15.02 General Energy Conservation Requirements

A. **The major energy codes referenced by the 2015 International Building Code (2015 IBC) and the 2015 International Mechanical Code (2015 IMC) are:**

1. 2015 International Energy Conservation Code (2015 IECC is based on and references ASHRAE Standard 90.1-2013).
2. ASHRAE Standard 90.1-2013.

B. ASHRAE Standard 90.1-2013 requires glazing systems to be certified in accordance with National Fenestration Rating Council (NFRC) Standards 100, 200, 300, and 400 and tested by a laboratory accredited by a nationally recognized accreditation organization for the following:

1. Air leakage.
2. U-values.
3. Emissivity coatings.
4. Solar Heat Gain Coefficients (SHGCs).
5. Shading Coefficients (SCs).
6. Visible Light Transmittance (VT).
7. Condensation Resistance Ratings (CRs).

C. National Fenestration Rating Council Product Certification Program (NFRC 700-2015)

1. Paragraph 4 reads " . . . products may be authorized for certification only if they have been rated in accordance with NFRC-approved procedures, computer programs, and test methods."
2. Paragraph 4.3.2 reads "Product Evaluation. A Licensee shall obtain from an NFRC-accredited laboratory NFRC required ratings for each product to be authorized for certification."
3. Paragraph 4.3.2.1 reads "Valid Computational Procedure."
   a. Paragraph 4.3.2.1.A reads "This procedure is used for obtaining U-factor ratings. The licensee shall obtain a simulation report from an NFRC-accredited simulation laboratory for each product line to be authorized for certification."
   b. Paragraph 4.3.2.1.B reads "The licensee shall then obtain a physical test report from an NFRC-accredited test laboratory. The test report shall contain the test results of the baseline product (the representative product of the product line) chosen by the licensee in order to validate the simulations conducted for the product line."
4. Paragraph 4.3.2.2 reads "Computational Procedure. This procedure is used for obtaining SHGC, VT, and Condensation Resistance ratings. Under this procedure, the licensee shall obtain a simulation report from an NFRC-accredited simulation laboratory for each product line to be authorized for certification. The Testing Alternative procedure for these ratings is to be used only if the product cannot be simulated."

D. As can be seen in paragraphs 15.02.B and C, shown earlier, glazing systems must be certified and the certification must involve a physical product test. I include this information to emphasize this building envelope requirement, because many design professionals are unaware of the magnitude of these requirements and the potential construction cost impacts. Recommended specification text is indicated in the following.

1. A representative sample of each glazing product to be installed on the project shall be tested in a certified, independent, testing laboratory in accordance with NFRC testing procedures.
2. Glazing products shall include, but are not limited to, curtain wall assemblies, field-assembled units, factory-assembled units, spandrel glazing units, operable units, fixed units, glazing with and without frit, clear units, tinted/colored units, low e-units, low iron units, metal-/aluminum-framed units, wood-framed units, sealed units, and all other glazing units.
3. U-values and solar heat gain coefficients (SHGCs) shall be determined by an NFRC test method at a certified, independent testing laboratory and shall be expressed as Btu/(h ft.^{2} °F) for each glazing product. Test glazing products with a 15-mph wind (6.7 m/s), 0°F (−18°C) cold side temperature, and 70°F (21°C) warm side temperature. The NFRC testing procedure shall include both the thermal computer model and the physical product test.
4. Maximum U-values must be specified and include the effects of the framing system.
   a. For example: Maximum U-values including the frame shall be 0.46 Btu/(h ft.^{2} °F).
   b. Include the maximum U-values obtained from ASHRAE Std 90.1 or the IECC, or values used in the COMcheck program or the Energy Model that shows compliance.

5. The maximum SHGC must be specified.
   a. For example: the maximum SHGC shall be 0.39.
   b. The maximum SHGC values obtained from ASHRAE Std 90.1 or the IECC, or values used in the COMcheck program or the Energy Model that shows compliance.

E. Prescriptive Code Approach

1. The prescriptive approach is an explicitly defined design approach based on design, construction, and maintenance requirements for all building systems.
2. The prescriptive approach is less flexible.
3. The prescriptive approach requires less design effort.
4. The following is an illustration of the ASHRAE Standard 90.1 prescriptive approach.

THE ASHRAE STANDARD 90.1 PRESCRIPTIVE APPROACH

Architectural Building Envelope		Mechanical HVAC		Plumbing Water Heating	Electrical Lighting		Electrical Power/ Other
Mandatory Requirements		Mandatory Requirements		Mandatory Requirements	Mandatory Requirements		Mandatory Requirements
Prescriptive Method	Building Envelope Trade-off Method	Simple Method	Prescriptive Method	Prescriptive Method	Space-by-Space Method	Overall Building Method	Prescriptive Method

F. Performance Code Approach

1. The performance approach is a design approach based on performance goals, objectives, and criteria.
2. The performance approach is more flexible.
3. The performance approach requires a much greater design effort.
4. The following is an illustration of ASHRAE Standard 90.1 performance approach.

ASHRAE STANDARD 90.1 PERFORMANCE APPROACH

Architectural Building Envelope	Mechanical HVAC	Plumbing Water Heating	Electrical Lighting	Electrical Power/ Other
Mandatory Requirements	Mandatory Requirements	Mandatory Requirements	Mandatory Requirements	Mandatory Requirements
Energy Cost Budget Method "The Energy Model"				

G. As can be seen in the previous illustrations, all disciplines (architectural, mechanical, plumbing, and electrical) must be actively involved in the energy conservation aspects of building design and construction. Even the contractors must be cognizant of how their construction activities and policies affect the end product with respect to energy conservation. This book will concentrate on the HVAC aspects of energy conservation. However, some of the lighting energy conservation requirements are contained in Part 11.

H. It's always surprising when a design colleague or someone else involved in the design process raises the question(s): "Who is going to enforce these energy conservation requirements (implying who understands these requirements well enough to enforce them)? or Who are the "Energy Police" who will enforce these requirements? The answer should roughly be: "It is the design professional's responsibility to enforce these requirements, especially if they have

RA (registered architect) or PE (professional engineer) after their names." Your professional license as an architect or engineer requires you to uphold the laws and codes of the jurisdictions in which you are licensed, in addition to protecting the public health and safety. Therefore, it is ultimately the design professionals' responsibility.

I. Energy Performance Computer Programs are quite useful in determining compliance. The Department of Energy (DOE) has a free residential compliance program (REScheck) and a free commercial compliance program (COMcheck) that can be downloaded from their web site. It is easy to use, and many municipalities accept the output from these programs as certification of compliance. These programs also provide checklists that assist in the design of the building or structure. These programs utilize the prescriptive requirements to determine compliance and generally cannot be used to determine the compliance for the performance approach.

J. Summaries of the energy codes outlining/highlighting the more frequently uses requirements are indicated in the following. For detailed design requirements, consult the official code text.

15.03 2015 IECC

A. Scope and Intent

1. Minimum prescriptive and performance requirements for building energy conservation.
2. Applicable buildings: residential and commercial.
3. Systems covered—provide effective use of energy in buildings:
   a. Architectural—Building envelope.
   b. Mechanical—HVAC and plumbing.
   c. Electrical—Power and lighting.
4. The code is not intended to limit flexibility or negate the safety, health, or environmental requirements of building design and construction.

B. Design Conditions

1. The United States is categorized into eight climate zones with Climate Zone 1 at the tip of Florida and Climate Zone 8 in Alaska.
2. Outdoor design conditions—2013 ASHRAE Handbook of Fundamentals.
3. Indoor design conditions.
   a. Heating: 72°F maximum.
   b. Cooling: 75°F minimum.

C. Residential Energy Efficiency—Mandatory Requirements

1. Duct insulation
   a. Supply and return ducts in attics 3 inches in diameter and greater: Insulation with R-value = 8.0.
   b. Supply and return ducts in attics less than 3 inches in diameter: Insulation with R-value = 6.0.
   c. Supply and return ducts in other portions of the building 3 inches in diameter and greater: Insulation with R-value = 6.0.
   d. Supply and return ducts in other portions of the building less than 3 inches in diameter: Insulation with R-value = 4.2.
   e. Exception: ducts completely inside the building envelope do not require insulation.
2. Pipe insulation
   a. Pipe fluid above 105°F: Insulation with R-value = 3.0.
   b. Pipe fluid below 55°F: Insulation with R-value = 3.0.
   c. Domestic hot water: Insulation with R-value = 3.0.

D. Commercial Energy Efficiency Requirements

1. The 2015 IECC references ASHRAE Standard 90.1—Energy Standard for Buildings Except Low Rise Residential Buildings, but also provides its own prescriptive and performance requirements.
2. Mandatory requirements:
   a. Heating and cooling load calculations shall be performed in accordance with *ASHRAE Standard 183* and equipment size based on these calculations.
   b. Heating and cooling loads shall be adjusted to account for energy recovery systems.
   c. Minimum HVAC equipment performance requirements are defined in the code. These requirements are not included here. Chiller performance requirements can be found in Part 28.
   d. Thermostatic controls shall be provided for all HVAC systems. Humidity controls shall be provided for HVAC systems where humidification or dehumidification, or both are provided.
   e. A 5°F deadband is required between heating and cooling setpoints.
   f. Off-hour setback controls are required.
   g. Damper controls: Outdoor air and exhaust air systems (gravity vents and louvers) shall be provided with motorized shutoff dampers when the system is not in use.
      1) Exception: Gravity dampers are permitted in buildings that are two stories and less in height.
      2) Exception: Gravity dampers are permitted in buildings of any height located in Climate Zones 1, 2, or 3.
      3) Exception: Gravity dampers are permitted in systems with airflows of 300 CFM and less.
   h. Energy recovery systems shall be provided when the following conditions are met:
      1) HVAC systems where the supply airflow rate of a fan system exceeds the values listed in 2015 IECC, Tables C403.2.7(1) and C403.2.7(2) based on climate zone and percentage of outdoor airflow rate at design conditions.
      2) The energy recovery system shall recover at least 50 percent of the energy difference between the enthalpy difference of the outside air and the room air (50-percent minimum efficiency).
         a) Exceptions:
            1. Where energy recovery systems are prohibited by the *International Mechanical Code*.
            2. Laboratory fume hood systems that include at least one of the following:
               a. VAV exhaust hoods with supply air and exhaust airflow reduction to 50 percent or less of the system design airflow.
               b. Direct makeup air delivered to the laboratory fumes hoods equal to at least 75 percent of the exhaust flow rate, heated no warmer than 2°F below the room design temperature, cooled to no cooler than 3°F above the room design temperature, no humidification added, and no simultaneous heating and cooling used for dehumidification control.
            3. Heating only systems with design space temperatures less than 60°F.
            4. Where 60 percent of the outdoor heating energy is provided by site recovered energy or site solar energy.
            5. Heating energy recovery in Climate Zones 1 and 2.
            6. Cooling energy recovery in Climate Zones 3C, 4C, 5B, 5C, 6B, 7, and 8.
            7. Systems requiring dehumidification that employ energy recovery in series with the cooling coil.
            8. Where the largest exhaust source is less than 75 percent of the design outdoor airflow.
            9. Systems expected to operate less than 20 hours per week at the outdoor air percentage covered by Table C403.2.7(1).
            10. Hazardous exhaust systems: See Part 17.
            11. Commercial kitchen hoods.

3) Energy recovery control or bypass must be provided to permit the operation of airside economizers.

   i. Ductwork:
      1) Construction and materials: See Part 17.
      2) Insulation: See Part 35.
   j. Piping:
      1) Construction and materials: See Part 18 through Part 22.
      2) Insulation: See Part 35.
   k. HVAC system balancing is required for both the air systems and the hydronic systems.
   l. Operation and maintenance manuals are required to be turned over to the owner.

3. Economizers—prescriptive approach:
   a. See 2015 IECC, Sec. C403.3 for detailed requirements.
   b. Economizer systems shall be integrated with the mechanical cooling system and be capable of providing partial cooling even where additional mechanical cooling is required to provide the remainder of the cooling load.
   c. HVAC system design and economizer controls shall be such that economizer operation does not increase building heating energy during normal operation.
   d. Air economizers:
      1) Air economizer systems shall be capable of modulating outdoor air and return air dampers to provide up to 100 percent of the design supply air quantity as outdoor air for cooling.
      2) Economizer dampers shall be capable of being sequenced with the mechanical cooling equipment and shall not be controlled by only mixed-air temperature.
      3) Air economizers shall be capable of automatically reducing outdoor air intake to the design minimum outdoor air quantity when outdoor air intake will no longer reduce cooling energy usage.
      4) Systems shall be capable of relieving excess outdoor air during air economizer operation to prevent overpressurizing the building. The relief air outlet shall be located to avoid recirculation into the building.
   e. Water-side economizers:
      1) Water-side economizers shall be capable of providing 100 percent of the cooling load at outside air temperatures of 50°F db/45°F wb and below.

4. Hydronic and multiple-zone HVAC systems controls and equipment—prescriptive approach:
   a. See 2015 IECC, Sec. C403.4 for detailed requirements.
   b. Fan control:
      1) Each cooling system shall be designed to vary the indoor fan airflow as a function of load.
      2) Static pressure sensors used to control VAV fans shall be located such that the controller setpoint is not greater than 1.2 inches w.c.
      3) For systems with direct digital control of individual zones reporting to the central control panel, the static pressure setpoint shall be reset based on the zone requiring the most pressure. In such case, the setpoint is reset lower until one zone damper is nearly wide open.
   c. Hydronic systems controls:
      1) Three-pipe hydronic systems are prohibited.
      2) Two-pipe changeover hydronic systems shall be provided with a deadband of at least 15°F, operation of one mode for at least four hours before changing to the other mode, and automatic controls that allow heating and cooling supply temperatures at the changeover point to be no more than 30°F apart.
      3) Hydronic heat pump systems shall be designed with a 20°F deadband between the removal of heat and the addition of heat to the loop. A bypass around the closed circuit evaporative cooler or the cooling tower is required for climate zones 3 and 4 to prevent heat loss, except for minimum flow to prevent freezing. For climate zones 3 and 4 where a separate heat exchanger is used to isolate the cooling tower from the heat pump loop, heat loss shall be controlled by shutting down

the circulation pump on the cooling tower loop. For climate zones 5 through 8, where a closed circuit evaporative cooler or cooling tower is used, a separate heat exchanger shall be used to isolate the cooling tower from the heat pump loop, and heat loss shall be controlled by shutting down the circulation pump on the cooling tower loop. All water-source heat pumps shall be provided with two-position control valves on all heat pump systems with total pump energy exceeding 10 horsepower to stop flow when the heat pump has cycled off (variable flow water distribution system).

4) Hydronic heating and cooling systems 500,000 Btu/h capacity or greater shall have automatic supply-water temperature reset controls using coil valve position, zone-return water temperature, building-return water temperature, or outside air temperature as an indicator of building heating or cooling demand. The temperature shall be capable of being reset by not less than 25 percent of the design supply-to-return water temperature difference.

5) Hydronic heating and cooling systems 500,000 Btu/h capacity or greater with a combined motor capacity of 10 horsepower or greater with three or more control valves shall reduce system pump flow by at least 50 percent of the design flow rate utilizing adjustable speed drives on pumps or multiple-staged pumps where at least 1/2 of the total pump horsepower can be turned off or where control valves are designed to modulate closed as a function of load. Pump flow shall be controlled to maintain one control valve nearly wide open or to satisfy the minimum differential pressure.

6) Boiler systems with a design input greater than 1,000,000 Btu/h and less than or equal to 5,000,000 Btu/h shall have a minimum turndown ratio of 3 to 1; greater than 5,000,000 Btu/h and less than or equal to 10,000,000 Btu/h shall have a minimum turndown ratio of 4 to 1; and greater than 10,000,000 Btu/h shall have a minimum turndown ratio of 5 to 1.

7) Chilled water and boiler plants having more than one chiller and/or boiler shall have the capability to reduce flow automatically through the chiller/boiler plant when a chiller/boiler is shut down.

d. Heat rejection equipment fans 7.5 horsepower and greater shall have automatic controls in order to reduce fan speed to 2/3 of full speed or less.

e. Complex mechanical systems serving multiple zones:

1) Supply air systems serving multiple zones shall be variable air volume systems that, during periods of occupancy, are designed and capable of being controlled to reduce primary air supply to each zone to 30 percent of maximum supply air, 300 CFM or less, minimum ventilation requirements, higher rate if approved by the code official, or the airflow rate required to maintain pressure relationships or minimum air change rates before reheating, recooling, or mixing takes place.

2) Fractional horsepower fan motors that are greater than or equal to 1/12 horsepower and less than 1 horsepower shall be electronically commutated motors or shall have a minimum motor efficiency of 70 percent.

3) Multiple-zone HVAC systems shall include controls that automatically reset the supply air temperature not less than 25 percent of the difference between the design supply air temperature and the design room temperature in response to representative building loads or to outdoor air temperature.

4) Multiple-zone HVAC systems with direct digital control of individual zone boxes reporting to a central control panel shall have automatic controls configured to reduce outdoor air intake flow below design rates in response to changes in system ventilation efficiency as defined by the *International Mechanical Code*.

f. Condenser water heat recovery is required for the heating or reheating of service hot water, provided the facility operates 24 hours per day and the total water-cooled system exceeds 6,000,000 Btu/h capacity of heat rejection and the design service water heating load exceeds 1,000,000 Btu/h.

g. Cooling systems shall not use hot gas bypass or other evaporator pressure control systems unless the system is designed with multiple steps of unloading or continuous capacity modulation. The capacity of the hot gas bypass shall be limited to 50 percent of total capacity for systems with a rated capacity greater than or equal to 240,000 Btu/h, and 25 percent of total capacity for systems with a rated capacity less than 240,000 Btu/h.

E. Performance Approach

1. When using a performance-based energy compliance approach for either a residential or commercial facility, the energy model will have to be performed at least twice: once for the standard reference design using the standard energy performance values defined in the code and once using the proposed design performance values as defined on the construction documents.
2. Energy analysis shall be performed over a full calendar year (8,760 hours) using climatic data, energy rates, building envelope data, occupancy schedules, and simulated loads as applicable to the location.
3. The heating and cooling system zoning, orientation, and other building features for the standard building shall be the same as the proposed building except:
    a. The window area of the standard design shall be the same as the proposed design, or 35 percent of the above-grade wall area, whichever is less.
    b. The skylight area of the standard design shall be the same as the proposed design, or 3 percent of the gross roof area, whichever is less.

15.04 ASHRAE Standard 90.1-2013

A. Purpose: To provide minimum requirements for the energy-efficient design of buildings except low-rise residential buildings.

B. Scope

1. Application:
    a. New buildings and their systems.
    b. New portions of buildings and their systems.
    c. New systems and equipment in existing buildings.
    d. New equipment or building systems specifically identified as part of industrial or manufacturing processes.
2. Building elements and systems:
    a. Building envelope.
    b. HVAC systems.
    c. Service water heating systems.
    d. Electrical power distribution and metering systems.
    e. Electric motors, belts, and drives.
    f. Lighting.
3. The code is not intended to limit flexibility or to negate safety, health, or environmental requirements of building design and construction.

C. Mandatory Provisions

1. Mechanical equipment shall meet minimum equipment efficiencies. These requirements were not included here. Chiller performance requirements can be found in Part 28.
2. Heating and cooling load calculations shall be performed in accordance with ASHRAE Standard 183-2007 (RA2014) and equipment size based on these calculations.
3. The supply of heating and cooling energy shall be individually controlled by a thermostat responding to temperatures within the zone.
4. A 5°F deadband is required between heating and cooling setpoints. Where heating and cooling systems serving a zone are controlled by separate thermostats, provisions in the control system shall prevent the simultaneous heating and cooling of the zone.

5. HVAC systems shall be provided with off-hour controls:
   a. Exceptions:
      1) HVAC systems intended to operate continuously.
      2) HVAC systems with design heating and cooling capacities less than 15,000 Btu/h and with accessible on/off controls.
   b. Automatic shutdown shall be equipped with at least one of the following:
      1) Time schedules for seven different day types:
         a) Capable of retaining control programs for a minimum of 10 hours during a power outage.
         b) Accessible manual override for system operation up to two hours.
      2) Occupant sensor.
      3) Manually operated timer with 2-hour operation limitation.
      4) Interlock with security system—security system activation shuts down the HVAC system.
   c. Setback controls:
      1) Heating systems: Provide controls with a setback temperature of at least 10°F lower than the occupied heating setpoint (or 4°F lower than the occupied heating setpoint for radiant heating systems).
      2) Cooling systems: Provide controls with a setup temperature of at least 5°F higher than the occupied cooling setpoint.
   d. Optimum start controls (system with setback controls and DDC).
   e. Zone isolation controls for HVAC systems serving zones that are intended to operate or be occupied nonsimultaneously.
6. Ventilation control:
   a. Stair and elevator shaft vents shall be equipped with motorized dampers that are closed during normal building operation and are opened as required by fire and smoke detection systems.
   b. Damper controls: Outdoor air and exhaust air systems (gravity vents and louvers) shall be provided with motorized shutoff dampers when the system is not in use or during preoccupancy building warm-up, cool-down, and setback.
      1) Exception: Gravity dampers are permitted for exhaust and relief in buildings three stories and less in height and in buildings of any height in Climate Zones 1, 2, and 3.
      2) Exception: Gravity dampers are permitted in systems with an airflow of 300 CFM or less.
      3) Exception: Ventilation systems serving unconditioned spaces.
   c. Recommendation: Provide all dampers with a maximum damper leakage rate of 4.0 CFM per square foot of damper area at 1.0 inches water column differential.
   d. Ventilation fan controls: Fans greater than 3/4 horsepower shall have automatic shutdown controls that shut off fans when not required.
   e. Demand control ventilation is required for HVAC systems serving spaces larger than 500 square feet with an occupant density exceeding 25 people per 1,000 square feet (some exceptions apply; see ASHRAE Standard 90.1-2013, Sec. 6.4.3.8). The HVAC systems must also have one or more of the following:
      1) Air-side economizer.
      2) Automatic modulating control of the outdoor air damper.
      3) Design outdoor airflow greater than 3,000 CFM.
   f. Ductwork:
      1) Construction and materials: See Part 17.
      2) Insulation: See Part 35.
   g. Piping:
      1) Construction and materials: See Part 18 through Part 22.
      2) Insulation: See Part 35.
   h. Construction requirements:
      1) Record drawings shall be required.
      2) Operation and maintenance manuals shall be required.

3) Air distribution and hydronic systems shall be balanced.
4) HVAC control systems shall be commissioned for projects larger than 50,000 square feet.

D. Simplified Prescriptive Approach

1. Building meeting the following:
   a. Two stories or less.
   b. 25,000 square feet or less.
2. HVAC system serves a single zone.
3. HVAC system varies indoor fan airflow.
4. HVAC provided by packaged or split system equipment.
5. Airside economizer shall be provided.
6. System changeover shall be by a manual changeover or by a dual setpoint thermostat.
7. The system does not permit simultaneous heating and cooling.

E. Prescriptive Approach

1. Economizers—either an airside or a waterside economizer is required as indicated in the following tables.

Climate Zones	Comfort Cooling Capacity Requiring Economizer
1A, 1B	No economizer required
2A, 2B, 3A, 4A, 5A, 6A, 3B, 3C, 4B, 4C, 5B, 5C, 6B, 7, 8	≥ 54,000 Btu/h

Climate Zones	Computer Room Cooling Capacity Requiring Economizer
1A, 1B, 2A, 3A, 4A	No economizer required
2A, 5A, 6A, 7, 8	≥ 135,000 Btu/h
3B, 3C, 4B, 4C, 5B, 5C, 6B	≥ 65,000 Btu/h

a. Exceptions:
   1) Systems smaller than those indicated in the tables under the design conditions listed.
   2) In hospitals and ambulatory surgery centers where 75 percent or more of the HVAC system serve spaces that require humidification levels above a 35°F dewpoint temperature.
   3) Where 25 percent or more of the HVAC system serves spaces that require humidification levels above a 35°F dewpoint temperature.
   4) Systems that utilize condenser heat recovery.
   5) Residential HVAC systems where the capacity is less than five times the requirements listed in the preceding table.
   6) Systems that serve spaces whose sensible cooling load at design conditions, excluding transmission and infiltration loads, is less than or equal to transmission and infiltration losses at an outdoor temperature of 60°F.
   7) Systems expected to operate less than 20 hours per week.
   8) Where the use of outdoor air for cooling will affect supermarket open refrigerated casework systems.
   9) Other exceptions apply; see ASHRAE Standard 90.1-2013, Sec. 6.5.1.
2. Airside economizers:
   a. Design capacity: 100 percent of the supply air quantity.
   b. Controls: Must be sequenced with mechanical cooling systems and shall not be controlled by only mixed air temperature.
   c. Minimum position: System shall reduce outside airflow to the minimum position when the outside air will no longer reduce cooling energy usage.

    d. Relief: HVAC systems must provide a means to relieve excess outside air.

    e. Sensors must meet certain accuracies (see ASHRAE Standard 90.1-2013, Sec. 6.5.1.1.6).

3. Waterside economizers:

    a. Design capacity: 100 percent of the expected system cooling load at outside air temperatures of 50°F db/45°F wb and below.

    b. Maximum water pressure drop: 15 feet of water.

4. Economizer control:

    a. Economizers shall provide partial cooling even when additional mechanical cooling is required to meet the load.

    b. Economizers shall not increase the building heating energy use during normal operation.

    c. Economizer control methods.

Climate Zones	Allowed Control Types	Prohibited Control Types
1B, 2B, 3B, 3C, 4B, 4C, 5A, 5B, 5C, 6A, 6B, 7, 8	Fixed dry bulb Differential dry bulb Fixed enthalpy Differential enthalpy	
1A, 2A, 3A, 4A	Fixed dry bulb Fixed enthalpy Differential enthalpy	Differential dry bulb

5. Thermostatic zone controls shall reduce the following:

    a. Reheating.

    b. Recooling.

    c. Mixing.

    d. Simultaneous heating and cooling.

    e. Exceptions: When the quantity of air to be reheated, recooled, or mixed is no greater than the following:

      1) The prescribed code ventilation requirements.

      2) Zones where special pressure relationships are required to prevent cross-contamination.

      3) Code required minimum air change rates—hospitals are an example.

      4) Where 75 percent or more of the energy required for reheating is provided by an energy recovery system.

      5) Laboratory exhaust systems.

      6) Other exceptions apply; see ASHRAE Standard 90.1-2013, Sec. 6.5.2.1.

6. Humidistatic zone controls shall reduce the following:

    a. Reheating.

    b. Recooling.

    c. Mixing.

    d. Simultaneous heating and cooling.

    e. Exceptions:

      1) Systems that reduce supply air quantities to 50 percent or lower.

      2) Individual cooling systems with capacity less than or equal to 80,000 Btu/h and reduces cooling capacity to 50 percent before reheating.

      3) Individual cooling systems with capacity of 40,000 Btu/h or less.

      4) HVAC systems serving process needs and requirements.

      5) Where 75 percent or more of the energy required for reheating is provided by an energy recovery system.

7. Hydronic systems controls

    a. Three-pipe hydronic systems are prohibited.

    b. Two-pipe changeover hydronic systems shall be provided with a deadband of at least 15°F, operation of one mode for at least four hours before changing to the other mode, and automatic control that allows heating and cooling supply temperatures at the changeover point to be no more than 30°F apart.

    c. Hydronic heat pump systems shall be design with a 20°F deadband between the removal of heat and the addition of heat to the loop. Bypass around the closed circuit evaporative cooler or the cooling tower is required for Climate Zones 3 through 8 with heating degree days in excess of 1800 to prevent heat loss, except for minimum flow to prevent freezing.

    d. Hydronic systems having 10 horsepower or more of total pump system power:

        1) Provide control valves to modulate or step closed as a function of load to reduce water flow to 50 percent or less of the design flow rate.

        2) Chilled water pumps serving variable-flow systems having motors exceeding 5 horsepower shall have controls and/or devices (such as variable-speed control) that will result in pump motor demand of not more than 30 percent of their design wattage at 50 percent of the design water flow.

           a) Exception: Where minimum flow required is less than the minimum flow required by the equipment manufacturer for proper equipment operation and where the total pump system power is 75 horsepower or less.

           b) Exception: Systems with no more than three control valves.

    e. Hydronic heating and cooling systems 300,000 Btu/h capacity or greater shall have automatic supply-water temperature reset controls using zone-return water temperature, building-return water temperature, or outside air temperature as an indicator of building heating or cooling demand.

    f. Heat rejection equipment: Heat rejection equipment fans 7.5 horsepower and greater shall have automatic controls to be able to reduce fan speed to 2/3 of full speed or less and as a function of leaving water temperature or condensing temperature/pressure of the heat rejection device.

    g. Condenser water heat recovery is required for heating or reheating of service hot water provided the facility operates 24 hours per day and the total water cooled system exceeds the 6,000,000 Btu/h capacity of heat rejection and the design service water heating load exceeds 1,000,000 Btu/h.

  8. Fan system power and efficiency.

    a. See ASHRAE Standard 90.1-2013, Sec. 6.5.3.1.

  9. Fan control.

    a. See ASHRAE Standard 90.1-2013, Sec. 6.5.3.2.

  10. Multiple-zone VAV system ventilation optimization control.

    a. See ASHRAE Standard 90.1-2013, Sec. 6.5.3.3.

  11. Supply air temperature reset controls.

    a. See ASHRAE Standard 90.1-2013, Sec. 6.5.3.4.

  12. Fractional horsepower fan motors.

    a. See ASHRAE Standard 90.1-2013, Sec. 6.5.3.5.

  13. Energy recovery systems are required for individual fan systems with a design supply fan capacity that exceeds the values listed in ASHRAE Standard 90.1-2013, Tables 6.5.6.1-1 and 6.5.6.1-2 based on climate zone and percentage of outdoor airflow rate at design conditions. The energy recovery system will have a minimum energy recovery effectiveness of 50 percent (50 percent of the difference between the outside air enthalpy and the return air enthalpy at design conditions).

    a. Exceptions:

        1) Laboratory exhaust systems as defined in the following.

        2) Commercial kitchen hoods.

        3) Hazardous exhaust systems: See Part 17.

        4) Heating only systems with design space temperatures less than 60°F.

        5) Where 60 percent of the outdoor heating energy is provided by site recovered energy or site solar energy.

        6) Heating energy recovery in Climate Zones 1 and 2.

        7) Cooling energy recovery in Climate Zones 3C, 4C, 5B, 5C, 6B, 7, and 8.

        8) Where the largest exhaust source is less than 75 percent of the design outdoor airflow.

9) Systems requiring dehumidification that employ energy recovery in series with the cooling coil.

10) Systems expected to operate less than 20 hours per week at the outdoor air percentage covered by Table 6.5.6.1-1.

14. Commercial kitchen hood exhaust systems.
   a. See ASHRAE Standard 90.1-2013, Sec. 6.5.7.1.

15. Buildings with laboratory exhaust systems having a total exhaust airflow rate greater than 5,000 CFM shall have at least one of the following features:
   a. VAV laboratory exhaust and room supply system capable of reducing exhaust and makeup airflow rates and/or incorporate a heat recovery system to precondition makeup air from laboratory exhaust that shall meet the following:

   $$A + B \times (E/M) \geq 50\%$$

   where
   A = percentage that the exhaust and makeup airflow rates can be reduced from design conditions
   B = percentage sensible recovery effectiveness
   E = exhaust airflow rate through the heat recovery device at design conditions
   M = makeup airflow rate of the system at design conditions

   b. VAV exhaust hoods with supply air and exhaust airflow reduction to 50 percent or less of the system design airflow.
   c. Direct makeup air delivered to the laboratory fumes hoods equal to at least 75 percent of the exhaust flow rate, heated no warmer than 2°F below the room design temperature, cooled to no cooler than 3°F above the room design temperature, no humidification added, and no simultaneous heating and cooling used for dehumidification control.
   d. Heat recovery systems to precondition makeup air without using any exceptions.

16. Hot gas bypass:
   a. Hot gas bypass shall not be used unless the system has multiple steps of capacity control or continuous modulation capacity control.
   b. Hot gas bypass shall be limited as follows:
      1) System capacity ≤ 240,000 Btu/h: 15% of total system capacity.
      2) System capacity > 240,000 Btu/h: 10% of total system capacity.

F. Performance Approach—Energy Cost Budget Method (Compliance Only)

1. Mandatory energy conservation requirements must still be met using the performance approach.
2. The energy cost budget for the proposed building must be less than or equal to the energy cost budget for the budget building design for compliance.
3. When using a performance-based energy compliance approach for a commercial facility, the energy model will have to be performed at least twice: once for the budget building design using the budget energy performance values defined in the code, and once using the proposed building design values as defined on the construction documents.
4. See ASHRAE Standard 90.1-2013, Sec. 11 for complete requirements of the energy cost budget method.

G. Performance Rating Method—Normative Appendix G

1. This performance rating method is intended for use in rating the energy efficiency of building designs that exceed the minimum requirements of this code.
2. Mandatory energy conservation requirements must still be met using the performance rating method.
3. The improved performance of the proposed building design is calculated in accordance with provisions of this appendix using the following formula:

Percentage improvement = 100 × (Baseline building performance − Proposed building performance)/Baseline building performance

4. See ASHRAE Standard 90.1-2013, Normative Appendix G for complete requirements of the performance rating method.

15.05 ASHRAE Standard 90.2-2007

A. Purpose: To provide minimum requirements for the energy-efficient design of residential buildings.

B. Duct Insulation

1. All portions of the air distribution system used for heating or cooling shall be insulated with R-8 insulation.
2. Ducts are not required to be insulated:
   a. When supply and return ductwork are within the conditioned space.
   b. If it is exhaust ductwork.

C. Pipe Insulation

1. Piping shall be insulated as follows:

Fluid Design Operating Temperature	Conductivity Btu in./h ft.2 °F	Nominal Pipe or Tube Diameter				
		<1"	1" to 1-1/4"	1-1/2" to 3-1/2"	4" to 6"	≥8"
Heating Systems—Hot Water and Steam Condensate						
201–250°F	0.27–0.30	1.5	1.5	2.0	2.0	2.0
141–200°F	0.25–0.29	1.0	1.0	1.0	1.5	1.5
105–140°F	0.22–0.28	0.5	0.5	1.0	1.0	1.0
Heating Systems—Steam						
212–250°F 0–15 Psig	0.27–0.30	1.5	1.5	2.0	2.0	2.0
Cooling Systems—Chilled Water, Glycol, Brine, and Refrigerant						
40–55°F	0.22–0.28	0.5	0.5	1.0	1.0	1.0
< 40°F	0.22–0.28	0.5	1.0	1.0	1.0	1.5

D. Ventilation: See Part 8 or ASHRAE Standard 62.2.

PART 16

HVAC System Selection Criteria

16.01 HVAC System Selection Criteria

A. Building Type

1. Institutional: hospital, prisons, nursing homes, education.
2. Commercial: offices, stores.
3. Residential: hotel, motel, apartments.
4. Industrial: manufacturing.
5. Research and development: laboratories.

B. Owner Type

1. Government.
2. Developer.
3. Business.
4. Private.

C. Performance Requirements

1. Supporting a process: computer facility, telephone facility.
2. Promoting a germ-free environment.
3. Increasing sales and rental income.
4. System efficiency.
5. Increasing property salability.
6. Standby and reserve capacity.
7. Reliability, life expectancy: frequency of maintenance and failure.
8. How will equipment failures affect the building? Owner operations?

D. Capacity Requirements

1. Cooling loads: magnitude and characteristics.
2. Heating loads: magnitude and characteristics.
3. Ventilation.
4. Zoning requirements:
   a. Occupancy.
   b. Solar exposure.
   c. Special requirements.
   d. Space temperature and humidity tolerances.

E. Spatial Requirements

1. Architectural constraints:
   a. Aesthetics.
   b. Structural support.
   c. Architectural style and function.
2. Space available to house equipment and location.
3. Space available for distribution of ducts and pipes.
4. Acceptability of components obtruding into occupied space, physically and visually.
5. Furniture placement.
6. Flexibility.
7. Maintenance accessibility.
8. Roof.
9. Available space constraints.
10. Are mechanical rooms/shafts required?

F. Comfort Considerations

1. Control options.
2. Noise and vibration control.
3. Heating, ventilating, and air conditioning.
4. Filtration.
5. Air quality control.

G. First Cost

1. System cost. Return on investment.
2. Cost to add zones.
3. Ability to increase capacity.
4. Contribution to life safety needs.
5. Air quality control.
6. Future cost to replace and/or repair.

H. Operating Costs

1. Energy costs.
2. Energy type:
   a. Electricity. Voltage available, rate schedule.
   b. Gas.
   c. Oil.
   d. District steam.
   e. District chilled water.
   f. Other sources.
3. Energy types available at project site.
4. Equipment selection.

I. Maintenance Cost

1. Cost to repair.
2. Capabilities of owner's maintenance personnel.
3. Cost of system failure on productivity.
4. Economizer cycle:
   a. Airside economizer.
   b. Waterside economizer.
5. Heat recovery.
6. Future cost to replace.
7. Ease and quickness of servicing.
8. Ease and quickness of adding zones.
9. Extent and frequency of maintenance.

J. Codes

1. Codes govern HVAC and other building systems design.
2. Most building codes are adopted and enforced at the local level.
3. Most of the states have adopted the International Series of Codes.
4. Codes are not enforceable unless adopted by municipality, borough, county, state, etc.
5. Codes regulate:
   a. Design and construction.
   b. Allowable construction types.
   c. Building height.
   d. Egress requirements.
   e. Structural components.
   f. Light and ventilation requirements.
   g. Material specifications.
6. Code approaches:
   a. Prescriptive. Dictate specific materials and methods (ASTM A53, Steel Pipe, Welded).
   b. Performance. Dictate desired results (HVAC system to provide and maintain a design temperature of 70°F winter and 75°F/50 percent RH summer).
7. Codes developed because of:
   a. Loss of life.
   b. Loss of property.
   c. Pioneered by insurance industry.

8. 2015 International Code Council Series of Codes (ICC):
   a. 2015 International Building Code.
   b. 2015 International Mechanical Code.
   c. 2015 International Energy Conservation Code.
   d. 2015 International Plumbing Code.
   e. 2015 International Fire Code.
   f. 2015 International Fuel Gas Code.
   g. 2015 International Residential Code.
   h. 2015 International Existing Building Code.
   i. 2015 International Performance Code for Buildings and Facilities.
   j. 2015 International Private Sewage Disposal Code.
   k. 2015 International Property Maintenance Code.
   l. 2015 International Zoning Code.
   m. 2015 International Wildland-Urban Interface Code.

16.02 Heating System Selection Guidelines

A. If heat loss exceeds 450 Btu/h per lineal feet of wall, heat should be provided from under the window or from the base of the wall to prevent downdrafts.

B. If heat loss is between 250 and 450 Btu/h per lineal feet of wall, heat should be provided from under the window or from the base of the wall, or it may be provided from overhead diffusers, located adjacent to the perimeter wall, discharging air directly downward, and blanketing the exposed wall and window areas.

C. If heat loss is less than 250 Btu/h per lineal feet of wall, heat should be provided from under the window or from the base of the wall, or it may be provided from overhead diffusers, located adjacent to or slightly away from the perimeter wall, discharging air directed at, or both directed at and directed away from, the exposed wall and window areas.

Air Distribution Systems

17.01 Ductwork Systems

A. Ductwork Sizing Criteria Table

DUCTWORK SIZING CRITERIA

System Type	Maximum Friction Rate in. W.G./100 ft.	Minimum Velocity ft./min.	Maximum Velocity ft./min.	Comments/Reasons
General Air Handling Systems				
Low Pressure Ducts	0.10 (0.15)	–	1,500–1,800	When CFM > 6,000 velocity governs; when CFM < 6,000 friction rate governs; applicable for supply, return, exhaust, and outside air systems
Medium Pressure Ducts	0.20 (0.25)	–	2,000–2,500	When CFM > 6,000 velocity governs; when CFM < 6,000 friction rate governs; applicable for supply systems only
High Pressure Ducts	0.40 (0.45)	–	2,500–3,500	When CFM > 5,000 velocity governs; when CFM < 5,000 friction rate governs; applicable for supply systems only
Transfer Air Ducts	0.03–0.05	–	1,000	When CFM > 3,200 velocity governs; when CFM < 3,200 friction rate governs
Outside Air Shafts	0.05–0.10	–	1,000	When CFM > 1,200 velocity governs; when CFM < 1,200 friction rate governs
Gravity Relief Air Shafts	0.03–0.05	–	1,000	When CFM > 3,200 velocity governs; when CFM < 3,200 friction rate governs
General Exhaust and Special Exhaust Systems				
General Exhaust Ducts	0.10 (0.15)	–	1,500–1,800	When CFM > 6,000 velocity governs; when CFM < 6,000 friction rate governs
Toilet Exhaust Ducts	0.10 (0.15)	–	1,500–1,800	When CFM > 6,000 velocity governs; when CFM < 6,000 friction rate governs
Kitchen Hood Exhaust Ducts	–	1,500	2,200	2015 IMC: 500 FPM min.; NFPA 96-2014: 500 FPM min.
Dishwasher Exhaust Ducts	0.10 (0.15)	1,500	2,200	
Acid, Ammonia, and Solvent Mains	0.50 (0.60)	1,000	3,000	Mains and risers 1,500–3,000 FPM; Branches and lateral 1,000–2,000 FPM
Acid, Ammonia, and Solvent Stacks	–	3,000	4,000	
Silane Ducts	–	250	–	Velocity across the neck of the cylinder or cabinet window or access port
Louvers				
Intake	–	–	500	Maximum velocity through free area; assuming 50% free area—max. velocity 250 FPM through gross louver area
Exhaust or Relief	–	–	700	Maximum velocity through free area; assuming 50% free area—max. velocity 350 FPM through gross louver area

Notes:
1. Friction rates in parenthesis should only be used when space constraints dictate.
2. Maximum aspect ratio 4:1; unless space constraints dictate greater aspect ratios.
3. When diffusers, registers, and grilles are mounted to supply, return, and exhaust ducts, duct velocities should not exceed 1,500 FPM or noise will result.

B. Ductwork System Sizing

1. Low pressure: 0.10 (0.15) in. W.G./100 ft.;   1,500–1,800 FPM maximum.
2. Medium pressure: 0.20 (0.25) in. W.G./100 ft.;   2,000–2,500 FPM maximum.
3. High pressure: 0.40 (0.45) in. W.G./100 ft.;   2,500–3,500 FPM maximum.
4. Transfer ducts: 0.03–0.05 in. W.G./100 ft.;   1,000 FPM maximum.

5. Transfer grilles: 0.03–0.05 in. W.G. pressure drop.
6. Outside air shafts: 0.05–0.10 in. W.G./100 ft.;   1,000 FPM maximum.
7. Gravity relief air shafts: 0.03–0.05 in. W.G./100 ft.;   1,000 FPM Maximum.
8. Decrease or increase duct size whenever the duct changes by 4" or more in one or two dimensions. Do *NOT* use fractions of an inch for duct sizes.
9. Try to change only one duct dimension at a time because it is easier to fabricate fittings and therefore generally less expensive—that is, 36×12 to 30×12 in lieu of 36×12 to 32×10.
10. Duct taps should be 2" smaller than the main duct to properly construct and seal the duct. The duct size should be 2" wider than diffusers, registers, and grilles.
11. All 90-degree square elbows should be provided with double radius turning vanes. Elbows in dishwasher, kitchen, and laundry exhausts should be of unvaned smooth radius construction with a radius equal to 1-½ times the width of the duct.
12. Provide flexible connections at the point of connection to equipment in all ductwork systems (supply, return, and exhaust) connected to air handling units, fans, and other equipment.
13. Provide access doors to access all fire dampers, smoke dampers, smoke detectors, volume dampers, motor-operated dampers, humidifiers, coils (steam, hot water, chilled water, electric), and other items located in ductwork that requires service and/ or inspection.
14. All rectangular duct taps should be made with shoe (45 degree) fittings. Do *NOT* use splitter dampers or extractors.
15. NFPA 90A-2015:
    a. Service openings shall be located at approximately 20-foot intervals in horizontal ducts and at the base of each vertical riser to facilitate cleaning unless the ductwork can be accessed through removable diffusers, registers, and grilles.
       1) Exception: Service openings are not required where all of the following can be met:
          a) The occupancy has no process that produces combustible material such as dust, lint, or greasy vapors (banks, offices, churches, hotels, and health care facilities, except kitchens, laundries, and manufacturing portions of such facilities).
          b) The air inlets are at least 7 feet above the floor and are protected by metal screens (registers, grilles) that prevent paper, refuse, or other combustible solids from entering the system.
          c) The minimum return duct design velocity is 1,000 FPM.
    b. Air outlets and inlets shall be located at least 3" above the floor unless provisions have been made to prevent dirt and dust from entering the system. Where outlets are located less than 7 feet above the floor, outlet openings shall be protected by a grille or screen with a maximum ½" opening size (register or grille).
16. Maximum ductwork hanger spacing:
    a. SMACNA minimum requirements:
       1) Horizontal: 8 to 10 feet maximum.
       2) Vertical: One- or two-story intervals—12 to 24 feet.
    b. Recommended:
       1) Horizontal ducts less than 4 square feet: 8 feet maximum.
       2) Horizontal ducts 4 to 10 square feet: 6 feet maximum.
       3) Horizontal ducts greater than 10 square feet: 4 feet maximum.
       4) Vertical round ducts: 12 feet maximum.
       5) Vertical rectangular ducts: 10 feet maximum.

DUCTWORK SUPPORT

Ductwork Type	Maximum Hanger Spacing Feet
Horizontal Ducts Less than 4 Square Feet	8
Horizontal Ducts 4 to 10 Square Feet	6
Horizontal Ducts Greater than 10 Square Feet	4
Vertical Round Ducts	12
Vertical Rectangular Ducts	10

C. **Friction Loss Estimate**

1. 1.5 × System Length (ft./100) × Friction Rate (in. W.G./100 ft.).

D. **Ductwork Sizes**

1. 4" × 4" smallest rectangular size.
2. 8" × 4" smallest recommended size.
3. Rectangular ducts: Use even duct sizes—that is, 24 × 12, 10 × 6, 72 × 36, 48 × 12.
4. 4:1 maximum recommended aspect ratio.
5. 3" smallest round size, odd and even sizes available.
6. Round ducts available in 0.5-inch increments for duct sizes through 5.5-inch diameter, 1-inch increments for duct sizes 6 inches through 20 inches, and 2-inch increments for duct sizes 22 inches and greater.

17.02 Duct Construction

A. **Sheet Metal and Air Conditioning Contractors' National Association (SMACNA) Duct Construction Manuals:**

1. *SMACNA—HVAC Duct Construction Standards—Metal and Flexible*, Third Edition, referred to herein as SMACNA-HVAC.
2. *SMACNA—Fibrous Glass Duct Construction Standards*, Seventh Edition, referred to herein as SMACNA-FG.
3. *SMACNA—Rectangular Industrial Duct Construction Standard*, Second Edition, referred to herein as SMACNA-IDC.
4. *SMACNA—Round Industrial Duct Construction Standard*, Second Edition, referred to herein as SMACNA-RIDC.
5. *SMACNA—Thermoplastic Duct (PVC) Construction Manual*, Second Edition, referred to herein as SMACNA-PVC.

B. *SMACNA-HVAC* **Pressure Ratings**

1. ±½; ±1; ±2; ±3; ±4; ±6; ±10.

C. *SMACNA-IDC* **and** *SMACNA-RIDC* **Pressure Ratings**

1. +12" to + 100" by multiples of 2".
2. −4" to −100" by multiples of −2".

D. **SMACNA Ductwork Testing**

1. −4" W.G. and lower: 1.5 × Pressure Rating.
2. −3" to +3" W.G.:   Generally not tested.
3. +4" W.G. and higher: 1.5 × Pressure Rating.

E. **ASHRAE Standard 90.1-2013: A minimum of 25 percent of duct systems designed to operate at static pressures exceeding 3" WC and all ductwork located outdoors shall be leak tested according to industry-accepted procedures.**

F. **Recommended Testing**

1. All supply duct systems operating at static pressures 3" and higher must be leak tested from air the handling unit to the vertical riser and from the vertical riser to 5 feet beyond shaft penetration on each and every floor (ductwork hidden in shaft construction).
2. All return duct systems operating at static pressures 3" and higher must be leak tested from the air handling unit to the vertical riser and from the vertical riser to 5 feet beyond shaft penetration on each and every floor (the ductwork is hidden in the shaft construction).
3. Leak test a representative sample of duct systems designed to operate at static pressures exceeding 3" WC on each floor to complete the minimum 25-percent leak testing required by ASHRAE Standard 90.1 and other energy conservation codes.

G. *SMACNA-HVAC* Ductwork Seal Classes

1. Seal Class A: 2–5 percent total system leakage (seal all transverse joints, longitudinal seams, and duct penetrations).
2. Seal Class B: 3–10 percent total system leakage (seal all transverse joints and longitudinal seams).
3. Seal Class C: 5–20 percent total system leakage (seal all transverse joints).
4. Unsealed: 10–40 percent total system leakage.
5. SMACNA recommended seal classes.

SMACNA DUCTWORK SEAL CLASSES

Seal Class	Applicable Static Pressure Construction Class
A	4" WC and higher
B	3" WC
C	2" WC
C	½" WC and higher for all ductwork upstream of VAV terminal units

H. ASHRAE Standard 90.1-2013 Ductwork Seal Class

1. ASHRAE Standard 90.1-2013 requires ductwork and all plenums with a pressure rating to be constructed to Seal Class A (seal all transverse joints, longitudinal seams, and duct penetrations).

I. Recommended Ductwork Seal Classes

1. Seal Class A: Seal all transverse joints, longitudinal seams, and duct penetrations.
2. Seal Class B: Seal all transverse joints and longitudinal seams.
3. Seal Class C: Seal all transverse joints.

RECOMMENDED DUCTWORK SEAL CLASSES

	SMACNA Pressure Class (in. WC)						
	±½	±1	±2	±3	±4	±6	±10
Supply Ductwork							
Outdoors	A	A	A	A	A	A	A
Unconditioned Space	B	B	B	A	A	A	A
Conditioned Space	B	B	B	A	A	A	A
Return Ductwork							
Outdoors	A	A	A	A	A	A	A
Unconditioned Space	B	B	B	B	A	A	A
Conditioned Space	B	B	B	B	A	A	A
	SMACNA Pressure Class (in. WC)						
	±½	±1	±2	±3	±4	±6	±10
Exhaust Ductwork							
Outdoors	B	B	B	B	A	A	A
Unconditioned Space	B	B	B	B	A	A	A
Conditioned Space	B	B	B	B	A	A	A

J. Ductwork Materials

1. Galvanized Steel: HVAC Applications; Most Common; Galvanized steel sheets meeting *ASTM A90, A525*, and *A527, Lock Forming Quality*.

ASTM Galvanized Coating Designations	Minimum Coating Weight oz./sq.ft.		
	Triple Spot Test Average Total Both Sides	Single Spot Test	
		One Side	Total Both Sides
G210	2.10	0.72	1.80
G185	1.85	0.64	1.60
G165	1.65	0.56	1.40
G140	1.40	0.48	1.20
G115	1.15	0.40	1.00
G90	0.90	0.32	0.80
G60	0.60	0.20	0.50
G40	0.40	0.12	0.30
G30	0.30	0.10	0.25

2. Carbon steel: Breechings, flues, and stacks; carbon steel meeting *ASTM A569* for stacks and breechings 24" and larger; galvanized sheet steel meeting *ASTM A527* with *ANSI/ASTM A525 G90* zinc coating for stacks and breechings less than 24".
3. Aluminum: Moisture laden air streams; aluminum base alloy sheets meeting *ASTM B209*, Lock Forming Quality.
4. Stainless steel: Kitchen hood and fume hood exhaust; stacks and breechings (prefabricated); Type 304, 304L, 316, or 316L stainless steel sheets meeting *ASTM A167*:
   a. 304 and 316: Non-welded applications.
   b. 304L and 316L: Welded applications.
   c. Kitchen exhaust finish:
      1) Concealed: None.
      2) Exposed: No. 2B, No. 4, or match equipment (No. 4 preferred).
   d. Lab fume exhaust finish:
      1) Concealed: No. 2B.
      2) Exposed: No. 2B.
5. Fiberglass: HVAC applications; 1"-thick glass duct board meeting U.L. 181.
6. Fiberglass reinforced: Chemical exhaust; plastic (FRP).
7. Polyvinyl chloride (PVC): Chemical exhaust, underground ducts; PVC conforming to *NFPA 91, ASTM D1784, D1785, D1927*, and *D2241*.
8. Concrete: Underground ducts, air shafts; reinforced concrete pipe meeting *ASTM C76, Class IV*.
9. Gypsum: Air shafts (generally provided by architects).
   a. 2015 IMC:
      1) Temperature shall not exceed 125°F.
      2) Gypsum board surface temperature must be maintained above the dewpoint.
      3) Gypsum board ducts shall not be used for supply air.
   b. NFPA 90A-2015: Gypsum board ducts shall be permitted to be used for negative pressure exhaust and return ducts where the temperature of the conveyed air does not exceed 125°F.
10. Copper: ornamental.
11. Polyvinyl Steel and Stainless Steel (PVS and PVSS): Chemical exhaust; common type: Halar-coated stainless steel, Teflon-coated stainless steel.
12. Sheet metal gauges (applies to preceding item numbers 1, 3, 4, and 10):
    a. 16, 18, 20, 22, 24, 26 SMACNA or welded construction.
    b. 10, 11, 12, 13, 14 welded construction only.

K. Flexible Duct

1. 5–8 ft. maximum recommended length.
2. Insulated, uninsulated.
3. NFPA 90A-2015: 14 feet maximum.

L. Ductwork Sheet Metal Gauges and Weights

SMACNA HVAC DUCTWORK SHEET METAL GAUGES

Maximum Duct Dimension	SMACNA Pressure Class													
	±½		±1		±2		±3		±4		±6		±10	
	A	B	A	B	A	B	A	B	A	B	A	B	A	B
4"–8"	26	–	26	–	26	–	24	26	24	26	24	26	22	24
9"–10"	26	–	26	–	26	–	24	26	22	26	24	24	20	22
11"–12"	26	–	26	–	26	–	24	26	22	26	20	24	18	22
13"–14"	26	–	26	–	24	26	22	24	20	24	20	22	18	20
15"–16"	26	–	26	–	24	26	22	24	20	24	18	22	16	20
17"–18"	26	–	24	26	22	26	20	24	18	24	18	22	16	20
19"–20"	24	26	24	26	20	26	18	24	18	24	16	22	–	18
21"–22"	22	26	22	26	18	26	18	24	18	24	16	22	–	18
23"–24"	22	26	22	26	18	26	18	24	18	22	16	22	–	18
25"–26"	20	26	20	26	18	26	18	24	16	22	–	20	–	18
27"–28"	18	26	18	26	18	24	18	22	16	22	–	20	–	18
29"–30"	18	26	18	26	18	24	18	22	16	22	–	18	–	18
31"–36"	181	26	18	24	16	24	16	20	–	20	–	18	–	16
37"–42"	6	26	16	24	–	22	–	20	–	18	–	16	–	16
43"–48"	16	26	16	22	–	22	–	18	–	18	–	16	–	16
49"–54"	–	26	–	22	–	20	–	18	–	18	–	16	–	16
55"–60"	–	24	–	22	–	20	–	18	–	16	–	16	–	16
61"–72"	–	22	–	18	–	18	–	16	–	16	–	16	–	16
73"–84"	–	22	–	18	–	16	–	16	–	16	–	16	–	16
85"–96"	–	20	–	18	–	16	–	16	–	16	–	16	–	16
97"–108"	–	18	–	16	–	16	–	16	–	16	–	16	–	16
109"–120"	–	16	–	16	–	16	–	16	–	16	–	16	–	16

Notes:
1 The table is based on the following:
   a. Column A: Duct gauge requirement with no reinforcement.
   b. Column B: Duct gauge with reinforcement as indicated below.
   c. ±½" Pressure Class: 5 feet reinforcing spacing for 19"–120".
   d. ±1" Pressure Class: 5 feet reinforcing spacing for 17"–108" and 4 feet spacing for 109"–120".
   e. ±2" Pressure Class: 5 feet reinforcing spacing for 13"–84", 4 feet spacing for 85"–108", and 3 feet spacing for 109"–120".
   f. ±3" Pressure Class: 5 feet reinforcing spacing for 4"–84", 4 feet spacing for 85"–96", and 3 feet spacing for 97"–120".
   g. ±4" Pressure Class: 5 feet reinforcing spacing for 4"–60", 4 feet spacing for 61"–72", and 3 feet spacing for 73"–120".
   h. ±6" Pressure Class: 5 feet reinforcing spacing for 4"–48", 4 feet spacing for 49"–60", and 3 feet spacing for 61"–120".
   i. ±10" Pressure Class: 5 feet reinforcing spacing for 4"–42", 4 feet spacing for 43"–54", 3 feet spacing for 55"–72", and 2 feet spacing for 73"–120".
2 Lighter sheet metal gauges may be used with additional reinforcing, and heavier gauges may be used with less reinforcing (see the SMACNA manuals).
3 Commercial installations recommend a 24-gauge minimum.

SHEET METAL GAUGES AND WEIGHTS

Sheet Metal Gauge	Material Weight lbs./sq.ft.		
	Galvanized Steel	300 Series Stainless Steel	Aluminum
26	0.906	0.748	0.224
24	1.156	0.987	0.282
22	1.406	1.231	0.352
20	1.656	1.491	0.451
18	2.156	2.016	0.563
16	2.656	2.499	0.718
14	3.281	3.154	0.901
12	4.531	4.427	1.141
10	5.781	5.670	1.436

SHEET METAL GAUGES

Sheet Metal Gauge	Thickness Inches	Remarks	Sheet Metal Gauge	Thickness Inches	Remarks
0	0.3125	Welded Ductwork Only	19	0.0437	SMACNA Ductwork Construction
1	0.2810		20	0.0375	
2	0.2650		21	0.0343	
3	0.2500		22	0.0312	
4	0.2340		23	0.0280	
5	0.2187		24	0.0250	
6	0.2030		25	0.0218	
7	0.1875		26	0.0187	
8	0.1720		27	0.0170	Gauges Not Permitted for Ductwork Construction
9	0.1560		28	0.0156	
10	0.1400		29	0.0140	
11	0.1250		30	0.0125	
12	0.1090		31	0.0109	
13	0.0937		32	0.0100	
14	0.0780		33	0.0093	
15	0.0700		34	0.0085	
16	0.0625	SMACNA Ductwork Construction	35	0.0078	
17	0.0560		36	0.0070	
18	0.0500				

17.03 Kitchen Exhaust Ducts and Hoods

A. For examples of kitchen hood exhaust systems, see Figs. 17.1 through 17.3.

B. *2015 IMC*

1. Exhaust/makeup air:
   a. Exhaust systems: 500 ft./min. minimum duct velocity.
   b. Type I hood exhaust systems shall be independent of all other exhaust systems. Combining Type I systems permitted if all of the following are met:
      1) Hoods are located on the same floor.
      2) Hoods located in the same room or adjoining rooms.
      3) Interconnecting ducts do not penetrate fire rated assemblies.
      4) Solid fuel appliances must have separate exhaust system.
   c. Type II hood exhaust systems shall be independent of all other exhaust systems. Combining Type II hoods is permitted following the same rules as listed for Type I hoods.
   d. Hoods serving solid fuel cooking appliances must have separate exhaust systems from all the other hoods.
   e. Makeup air systems: ΔT shall not be greater than 10°F, unless it is part of the AC system or will not cause a decrease in comfort conditions.
   f. Supply air shall be approximately equal to the exhaust air.
   g. The exhaust shall terminate a minimum of 40" above the roof.
2. Duct sheet metal construction:
   a. 16 ga. steel.
   b. 18 ga. 304 stainless steel.
   c. Type I hood exhaust ducts shall be all welded or brazed construction.
   d. Type I hood horizontal duct slope:
      1) Horizontal ducts 75 feet or less in length: 1/4" per foot.
      2) Horizontal ducts greater than 75 feet in length: 1" per foot.

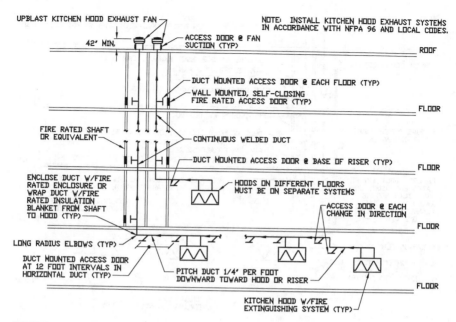

FIGURE 17.1 KITCHEN HOOD EXHAUST SYSTEM—UPBLAST FAN.

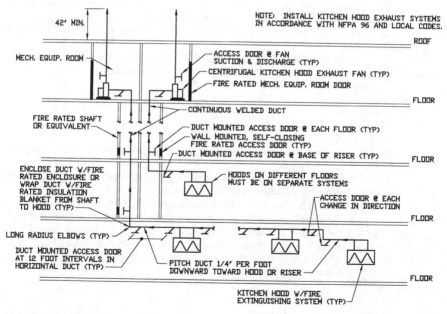

FIGURE 17.2 KITCHEN HOOD EXHAUST SYSTEM—UTILITY SET FAN.

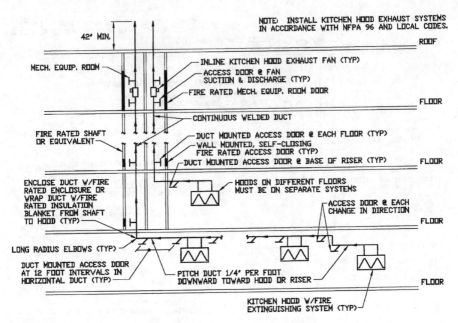

FIGURE 17.3   KITCHEN HOOD EXHAUST SYSTEM—INLINE FAN.

e. Type I hood exhaust ducts shall be enclosed in a fire rated enclosure from the penetration of the ceiling, wall, or floor to the point of the outlet terminal. The rating of the enclosure shall not be less than that of the assembly penetrated and not less than 1 hour.
1) Horizontal (in kitchen): Fire rated duct wrap recommended.
2) Horizontal (shaft offsets): Shaft enclosure recommended.
3) Vertical: Shaft enclosure recommended.
3. Cleanouts:
a. Base of riser.
b. Horizontal:
1) Every 20 feet.
2) Not more than 10 feet from changes in direction that are greater than 45 degrees.
4. Hoods:
a. Type I hoods: Serve appliances that produce grease or smoke—such as griddles, fryers, broilers, ovens, ranges, and wok ranges.
1) Type I hood exhaust system shall operate automatically through an interlock with the cooking appliances, by means of heat sensors, or by other approved methods.
b. Type II hoods: Serve appliances that produce heat or steam but do not produce grease or smoke—such as steamers, kettles, pasta cookers, and dishwashers.
c. Domestic appliances used for commercial purposes shall be provided with Type I or Type II hoods as applicable.
d. Hood construction:
1) Type I hoods:
    Steel: 18 gauge
    Stainless steel: 20 gauge
2) Type II hoods:
    Steel   22 gauge
    Stainless steel: 24 gauge
e. Hood exhaust:

Type of Hood	Minimum CFM per Lineal Foot of Hood			
	Type of Cooking Appliances			
	Extra-Heavy Duty	Heavy Duty	Medium Duty	Light Duty
Wall-Mounted Canopy	550	400	300	200
Single Island Canopy	700	600	500	400
Double Island Canopy (per side)	550	400	300	250
Backshelf/Pass-Over	Not permitted	400	300	250
Eyebrow	Not permitted	Not permitted	250	250

Notes:

1 Airflows indicated in the table are net quantity of exhaust air and shall be calculated by subtracting any airflow supplied directly to a hood cavity from the total exhaust flow rate of the hood.

2 Where more than one type of appliance is located under a single hood, the highest exhaust rate shall be used.

3 Extra-heavy duty cooking appliances: Cooking appliances using solid fuel as the primary source of heat for cooking, such as wood, charcoal, briquettes, and mesquite. Type I hoods serving barbeque pits, barbeque cooking appliances, solid fuel burning stoves and ovens, hickory grilles, charbroilers, and charcoal grilles. Hoods serving these systems must have separate exhaust systems from all the other hoods.

4 Heavy duty cooking appliances: Type I hoods serving electric under-fired broilers, electric chain (conveyor) broilers, gas open-burner ranges (with or without oven), electric and gas wok ranges, and electric and gas over-fired (upright) broilers and salamanders.

5 Medium duty cooking appliances: Type I hoods serving electric discrete element ranges (with or without oven), electric and gas hot-top ranges, electric and gas griddles, electric and gas double-sided griddles, electric and gas fryers (open deep fat fryers, donut fryers, kettle fryers, and pressure fryers), electric and gas pasta cookers, electric and gas conveyor pizza ovens, electric and gas tilting skillets (braising pans), and electric and gas rotisseries.

6 Light duty cooking appliances: Type I hoods serving electric and gas ovens (standard, bake, roasting, revolving, retherm, convection, combination convection/steamer, conveyor, deck or deck style pizza, and pastry), electric and gas steam-jacketed kettles, electric and gas compartment steamers (both pressure and atmospheric), and electric and gas cheesemelters.

C. NFPA 96-2014

1. Exhaust/makeup air:
   a. Exhaust systems: 500 ft./min. minimum duct velocity.
   b. Supply air shall be adequate to prevent negative pressures from exceeding 0.02" WC.
   c. Exhaust shall terminate a minimum of 40" above the roof.
   d. Exhaust ducts shall not pass through fire walls.
   e. All ducts shall lead directly to the exterior of the building to reduce the risk of fire hazard.
   f. Exhaust ducts shall be independent of all other exhaust systems.
   g. Hoods serving solid fuel cooking appliances must have separate exhaust systems from all the other hoods.
2. Duct sheet metal construction:
   a. Carbon steel: 16 gauge
   b. Stainless steel: 18 gauge
   c. Exhaust ducts shall be all welded construction.
   d. Horizontal duct slope:
      1) All ducts shall be installed without forming drips or traps that might collect residues.
      2) All duct runs up to 75 feet in length shall be installed with a minimum of 2 percent slope. Duct runs greater than 75 feet in length shall be installed with a minimum of 8 percent slope.
   e. Exhaust ducts shall be enclosed in a fire rated enclosure from the penetration of the ceiling, wall, or floor to the point of the outlet terminal.
      1) Horizontal (in kitchen): Fire rated duct wrap recommended.
      2) Horizontal (shaft offsets): Shaft enclosure recommended.
      3) Vertical: Shaft enclosure recommended.
      4) 1 hour rating minimum for buildings less than four stories.
      5) 2 hour rating minimum for buildings four stories or more.

   f. Exhaust duct enclosures shall be vented to the exterior of the building through weather-protected openings.
   g. Each exhaust duct system shall constitute an individual system serving only exhaust hoods in one fire zone on one floor.
   h. Common duct (manifold) systems: Master kitchen exhaust ducts that serve multiple tenants shall include provisions to bleed air from outdoors or from adjacent spaces into the master exhaust duct to maintain the necessary minimum air velocity in the master exhaust duct.
      1) The bleed air duct shall have a fire damper at least 12" from the master exhaust duct connection.
      2) The bleed air duct shall have a volume balancing damper upstream of the fire damper.
      3) The bleed air duct cannot be used for exhaust of grease-laden vapors and shall be labeled as such.
      4) The bleed air duct shall have the same construction requirements as the exhaust duct.
   i. Dampers shall not be installed in exhaust ducts or exhaust duct systems.
3. Cleanouts:
   a. Horizontal: Every 12 feet.
   b. Vertical: Every floor.
4. Hoods:
   a. Steel: 18 gauge.
   b. Stainless steel: 20 gauge.
5. Hood exhaust: Exhaust air volumes for hoods shall be of sufficient level to provide for capture and removal of grease-laden cooking vapors.
6. Fire damper: A fire damper with a 286°F fusible link is required at each supply air connection to the hood.
   a. Exception: If the supply air connection discharges air out the face of the hood rather than the bottom or into the hood and is isolated from the exhaust hood by continuously welded construction, it does not require a fire damper.

D. ASHRAE Standard 154-2011

1. Exhaust/makeup air:
   a. Exhaust systems: 500 ft./min. minimum duct velocity.
   b. The commercial kitchen ventilation system shall provide pressure differentials to control odor migration and to control dust, dirt, and insects.
      1) Kitchen—negative (maximum 0.02 in. w.c.) with respect to dining and other adjacent areas.
      2) Negative with respect to outdoors when the food-service facility shares a wall with an adjacent non-food-service facility, such as a retail center.
   c. Exhaust discharge shall be designed to prevent re-entrainment into air intakes.
   d. The minimum horizontal distance between intakes and discharge shall be 10 feet.
2. Hoods:
   a. Type I hoods: A hood designed to capture heat, smoke and/or grease-laden vapor produced by a cooking process, incorporating listed grease-removal devices and fire suppression equipment. Equipment requiring Type I hoods—ranges, fryers, griddles, broilers, and ovens that produce smoke or grease-laden vapors.
   b. Type II hoods: A hood designed to capture heat, odors, products of combustion, and/or moisture where smoke or grease laden vapor is not present. A Type II hood may or may not have filters or baffles and does not have a fire-suppression system. Equipment requiring Type II hoods—dishwashers, microwave ovens, toasters, steam tables, popcorn poppers, hot dog cookers, coffee makers, rice cookers, egg cookers, and holding/warming ovens.
   c. Mounting heights and overhang requirements:

Type of Hood	Mounting Height	End Overhang	Front Overhang	Rear Overhang
Wall-Mounted Canopy	78"	6"	12"	N/A
Single Island Canopy	78"	12"	12"	12"
Double Island Canopy	78"	12"	12"	N/A
Eyebrow	78"	N/A	12"	N/A
Backshelf/Pass-over	24"	6"	10"	N/A

Note:

1 Mounting heights are minimum dimensions and are listed with respect to the finished floor except the backshelf/pass-over hoods, which are the maximum dimensions above the cooking surface.

d. Hood exhaust:

Type of Hood	Type II Minimum Net Exhaust Flow Rate CFM/Lineal Foot of Hood Length	
	Light Duty	Medium Duty
Wall-Mounted Canopy	200	300
Single Island Canopy	400	500
Double Island Canopy	250	250
Eyebrow	250	250
Backshelf/Pass-over	200	300

e. Appliance duty level:

1) Light duty: A cooking process requiring an exhaust airflow rate of less than 200 CMF/ft. for capture, containment, and removal of the cooking effluent and products of combustion. Gas and electric ovens (standard, bake, roasting, revolving, rethermalizer, convection, combination convection/steamer, conveyor, deck or deck style pizza and pastry ovens, electric and gas steam–jacketed kettles, electric and gas compartment steamers, electric and gas cheesemelters, and electric and gas rethermalizers).

2) Medium duty: A cooking process requiring an exhaust airflow rate of 200 to 300 CMF/ft. for capture, containment, and removal of the cooking effluent and products of combustion. Electric discrete element ranges, electric and gas hot-top ranges, electric and gas griddles, electric and gas double-sided griddles, electric and gas fryers (open deep fat fryers, donut fryers, kettle fryers, and pressure fryers), electric and gas pasta cookers, electric and gas conveyor (pizza) ovens, electric and gas tilting skillets/braising pans, and electric and gas rotisseries.

3) Heavy duty: A cooking process requiring an exhaust airflow rate of 300 to 400 CMF/ft. for capture, containment, and removal of the cooking effluent and products of combustion. Electric and gas underfired broilers, electric and gas chain (conveyor) broilers, gas open-burner ranges (with or without oven), electric and gas wok ranges, electric and gas overfired (upright) broilers, and salamanders.

4) Extra-heavy duty: A cooking process requiring an exhaust airflow rate of greater than 400 CMF/ft. for capture, containment, and removal of the cooking effluent and products of combustion. Appliances using solid fuel such as wood, charcoal, briquettes, and mesquite.

E. **Figures 17.4 and 17.5 are photographs of an upblast kitchen hood exhaust fan and makeup air unit in their installed conditions.**

FIGURE 17.4   PHOTOGRAPH OF AN UPBLAST KITCHEN HOOD EXHAUST FAN.

FIGURE 17.5   PHOTOGRAPH OF A KITCHEN HOOD MAKEUP AIR UNIT.

17.04 Louvers

A. **Louvers: Use stationary louvers only. Do not use operable louvers because they become rusty or become covered with snow and ice and may not operate:**
1. Intake (outdoor air): 500 ft./min. maximum velocity through free area.
2. Exhaust or relief: 700 ft./min. maximum velocity through free area.
3. Free area range:
   a. Metal: 40–70 percent of gross area. Recommend using 50 percent free area.
   b. Wood: 20–25 percent of gross area.
4. Pressure loss: 0.01–0.10" W.G.

17.05 Volume Dampers (Manual or Balancing Dampers)/ Motor Operated Dampers (Control Dampers)

A. **Damper Characteristics**
1. Opposed blade: Balancing, mixing, modulating, and 2-position control applications.
2. Parallel blade: Two-position applications (open/closed).
3. Pressure Loss: 0.15" W.G. @ 2000 FPM (full open)
4. Size dampers at a flow rate of approximately 1,200–1,500 CFM/sq.ft. (1,200–1,500 FPM) rather than on duct size.
5. Linkage type:
   a. Concealed—inside duct. When specifying concealed linkage, be careful of duct air temperatures and actuator ratings (e.g., generator radiator exhaust can reach temperatures in excess of some actuator ratings).
   b. Exposed—outside duct.
6. Dampers may be specified with integral insulation.

B. **Damper Leakage Classes (AMCA Certified)**
1. Class I dampers:
   4.0 CFM/sq.ft. @ 1" W.G. differential.
   8.0 CFM/sq.ft. @ 4" W.G. differential.
   11.0 CFM/sq.ft. @ 8" W.G. differential.
   14.0 CFM/sq.ft. @12" W.G. differential.
2. Class II dampers:
   10.0 CFM/sq.ft. @ 1" W.G. differential.
   20.0 CFM/sq.ft. @ 4" W.G. differential.
   28.0 CFM/sq.ft. @ 8" W.G. differential.
   35.0 CFM/sq.ft. @ 12" W.G. differential.
3. Class III dampers:
   40.0 CFM/sq.ft. @ 1" W.G. differential.
   80.0 CFM/sq.ft. @ 4" W.G. differential.
   112.0 CFM/sq.ft. @ 8" W.G. differential.
   140.0 CFM/sq.ft. @ 12" W.G. differential.

C. **Damper Types**
1. Standard V-groove blade—approximately 2,000 FPM maximum velocity.
2. Airfoil blade—approximately 4,000 FPM maximum velocity.

D. **Recommended**
1. Two-position ducted applications: AMCA certified Ultra-low Leakage Class with a maximum 8.0 CFM/sq.ft. leakage rate at a 4" WC pressure differential, airfoil-parallel blade, motor-operated damper.

2. All other ducted applications: AMCA certified Ultra-low Leakage Class with a maximum 8.0 CFM/sq.ft. leakage rate at 4" WC pressure differential, airfoil-opposed blade, motor operated damper.
3. Non-ducted applications: AMCA certified Ultra-low Leakage Class with a maximum 8.0 CFM/sq.ft. leakage rate at 4" WC pressure differential, insulated-airfoil-opposed blade, motor-operated damper.

17.06 Fire Dampers, Smoke Dampers, and Combination Fire/Smoke Dampers

A. Fire, Smoke, and Combination Damper Classifications

1. Damper type:
   a. Expanding curtain type (fire damper only):
      1) Type A: Frame and damper storage are located in the airstream.
      2) Type B: Damper storage is totally recessed out of the airstream.
      3) Type C: Frame and damper storage are totally recessed out of the airstream.
      4) Recommend using Type C in ducted and ducted transfer applications and Type A in transfer grille applications (to fit within the grille dimension, must oversize the grille to account for the frame and blades).
   b. Opposed blade type:
      1) V-groove blades: Maximum velocity of 2,000 FPM.
      2) Airfoil blades: Maximum velocity of 4,000 FPM.
      3) Blades and frame are located in the airstream. Must account for the pressure drop of the damper and frame in static pressure calculations.
      4) Leakage class:
         a) Leakage Class I:
            4.0 CFM/sq.ft. @ 1" WC pressure differential.
            8.0 CFM/sq.ft. @ 4" WC pressure differential.
            11.0 CFM/sq.ft. @ 8" WC pressure differential.
            14.0 CFM/sq.ft. @ 12" WC pressure differential.
         b) Leakage Class II:
            10.0 CFM/sq.ft. @ 1" WC pressure differential.
            20.0 CFM/sq.ft. @ 4" WC pressure differential.
            28.0 CFM/sq.ft. @ 8" WC pressure differential.
            35.0 CFM/sq.ft. @ 12" WC pressure differential.
         c) Leakage Class III: (Not Permitted by IMC Code)
            40.0 CFM/sq.ft. @ 1" WC pressure differential.
            80.0 CFM/sq.ft. @ 4" WC pressure differential.
            112.0 CFM/sq.ft. @ 8" WC pressure differential.
            140.0 CFM/sq.ft. @ 12" WC pressure differential.
         d) Leakage Class IV: (Not Permitted by IMC Code)
            60.0 CFM/sq.ft. @ 1" WC pressure differential.
            120.0 CFM/sq.ft. @ 4" WC pressure differential.
            168.0 CFM/sq.ft. @ 8" WC pressure differential.
            210.0 CFM/sq.ft. @ 12" WC pressure differential.
2. Fire rating:
   a. 1-½ hour.
   b. 3 hour.
3. Closure rating:
   a. *U.L. 555* and UL 555S require fire, smoke, and fire/smoke dampers to bear an affixed label stating whether the damper is static or dynamic rated.
   b. Dynamic Rating: Dynamic rated dampers must be U.L. tested and show airflow and maximum static pressure against which the damper will operate (fully close).

Dampers are tested to 4" static pressure for "no duct" applications and 8" static pressure for "in duct" applications.

c. Static Rating: Static rated dampers have not been U.L. tested against airflow and may not close under medium-to-high airflow conditions that may be encountered in HVAC systems that do not shut down in the event of fire.

d. Recommend using dynamically rated fire/smoke dampers in all applications.

4. Temperature rating of fusible links:
   a. Standard: 165°F.
   b. Optional expanding curtain type (see code requirements): 212°F, 285°F.
   c. Optional blade type (see code requirements): 212°F, 250°F, 285°F, 350°F, 450°F.
   d. Smoke control requirements:
      1) Primary: 285°F (can be overridden by the fire department).
      2) Secondary: 350°F (cannot be overridden by fire department).

B. Fire/Smoke Damper Recommendations

1. Fire dampers (HVAC applications):
   a. Curtain type: Type C, 1-½ or 3 hours to match wall construction, Expanding Curtain Type Fire Damper with 165°F fusible link for all applications (including transfer duct applications) except transfer grille applications shall be Type A.
   b. Blade type: 3,000 FPM minimum velocity, Airfoil Blade, Leakage Class I at 4" WC pressure differential, 1-½ or 3 hours to match wall construction, Dynamic Fire Damper at 8" WC closure rating with 165°F fusible link.
2. Smoke Dampers and Combination Fire/Smoke Dampers (HVAC Applications):
   a. Blade type: 3,000 FPM minimum velocity, Airfoil Blade, Leakage Class I at 4" WC pressure differential, 1-½ or 3 hours to match wall construction, Dynamic Fire Damper at 8" WC closure rating with 250°F primary fusible link and 350°F secondary fusible link.
3. Fire dampers, smoke dampers, and combination fire/smoke dampers (smoke control applications):
   a. Blade type: 3,000 FPM minimum velocity, Airfoil Blade, Leakage Class I at 4" WC pressure differential, 1-½ or 3 hours to match wall construction, Dynamic Fire Damper at 8" WC closure rating with 285°F primary fusible link and 350°F secondary fusible link.
4. Fire dampers, smoke dampers, and fire/smoke dampers: Blowout panels should be considered for ductwork systems under the following circumstances:
   a. Whenever, the potential exists for fire, smoke, and/or fire/smoke dampers to close suddenly and cause system pressures to exceed construction pressures of the ductwork especially in systems utilizing dynamic rated dampers.
   b. Whenever human operation of fire, smoke, and/or fire/smoke dampers is required by code, by local authorities, or for smoke evacuation systems, in the event that the fire department personnel or owner's operating personnel inadvertently close all the dampers, and system pressures exceed construction pressures of the ductwork.

C. 2015 IMC

1. Installation shall comply with the IMC and manufacturer's installation instructions and listing.
2. Testing procedures:
   a. Fire dampers: UL 555.
   b. Smoke dampers: UL 555S.
   c. Combination fire/smoke dampers: UL 555 and UL 555S.
   d. Ceiling dampers: UL 555C.
   e. Actuators: UL 555 and UL 555S.
3. Fire protection rating:
   a. Less than 3-hour rated assemblies: 1-½ hours
   b. Three hours and above rated assemblies: 3 hours

4. Fire damper actuating devices:
    a. HVAC systems: 50°F above the normal operating temperature within the duct system, but not less than 160°F.
    b. Smoke control systems: 350°F maximum.
5. Smoke damper actuating devices:
    a. Elevated temperature rating: 250°F minimum, 350°F maximum.
    b. Duct mounted smoke damper: Provide duct mounted smoke detector located within 5 feet with no inlet/outlets between damper and detector.
    c. Unducted smoke damper: Provide space-mounted smoke detector located within 5 feet horizontally of wall opening with damper.
    d. Smoke dampers may be controlled by the smoke detection system where a smoke detection system is installed in all areas served by the duct in which the damper will be located.
    e. Smoke damper leakage rating shall be Class I or II.
6. Combination fire/smoke damper actuating devices:
    a. Smoke control system: 50°F above smoke control design temperature, but not less than 160°F or more than 350°F.
    b. Smoke detectors as indicated under smoke damper actuating devices.
7. Access: Fire, smoke, and fire/smoke dampers shall be provided with an approved means of access. Access doors shall be labeled with 0.5"-high letters minimum reading: "FIRE DAMPER," "SMOKE DAMPER," or "FIRE/SMOKE DAMPER," respectively.
8. Fire dampers are required at duct and transfer openings at the following locations:
    a. Fire walls.
    b. Fire barriers:
        1) Exception: Dampers are not required in penetrations of walls with a required 1-hour fire-resistance rating or less by a ducted HVAC system that is of sheet steel not less than 26 gauge and is continuous from the air-handling equipment to the air outlet or inlet terminals in areas of other than Use Group H where the building is equipped throughout with an automatic sprinkler system.
        2) Exception: Dampers are not required in ducts used as an approved smoke control system *where the damper would interfere with the operation of the smoke control system.*
    a. Fire partitions:
        1) Exception: Dampers are not required in penetrations of tenant separation and corridor walls in buildings of other than Use Group H where the building is equipped throughout with an automatic sprinkler system.
        2) Exception: Dampers are not required in duct systems constructed of code-approved materials that meet all of the following:
            a) Duct size 100 sq. in. or less.
            b) Duct constructed of a minimum of 24 gauge steel.
            c) Duct cannot have openings that communicate the corridor with adjoining rooms or spaces.
            d) Duct is installed above a ceiling.
            e) Duct shall not terminate at a fire rated wall with a register.
            f) A minimum 12" long × 16 gauge sleeve shall be centered at each duct opening.
        3) Exception: Dampers are not required in penetrations of walls with a required 1-hour fire-resistance rating or less by a ducted HVAC system that is of sheet steel not less than 26 gauge and is continuous from the air-handling equipment to the air outlet or inlet terminals in areas of other than Use Group H where the building is equipped throughout with an automatic sprinkler system.
9. Smoke dampers are required at duct and transfer openings at the following locations:
    a. Smoke barriers and corridors with smoke and draft controls.
        1) Exception: Dampers are not required at corridor penetrations where the building is equipped throughout with an approved smoke control system.
        2) Exception: Ducts penetrating smoke barriers where the duct serves a single smoke compartment and are constructed of steel.
        3) Exception: Dampers are not required in ducts that do not serve the corridor and are constructed of minimum 26 gauge steel.

10. Fire and smoke dampers or combination fire/smoke dampers are required at duct and transfer openings at the following locations:
   a. Shaft enclosures:
      1) Exception: Fire dampers are not required in exhaust systems equipped with steel exhaust air subducts extending at least 22" vertically in an exhaust shaft and where there is continuous airflow upward to the outside.
      2) Exception: Fire dampers are not required in penetrations tested in accordance with ASTM E 119 or UL 263 as part of the fire-resistance-rated assembly.
      3) Exception: Smoke dampers are not required in bathroom, toilet, kitchen, and clothes dryer exhaust openings equipped with 26 gauge minimum steel exhaust air subducts extending at least 22" vertically in an exhaust shaft and where there is continuous airflow upward to the outside in Groups B and R occupancies equipped throughout with an automatic sprinkler system.
      4) Exception: Fire dampers and smoke dampers are not required in ducts used as an approved smoke control system where the damper would interfere with the operation of the smoke control system.
      5) Exception: Fire dampers and smoke dampers are not required in parking garage exhaust ducts that are separated from other building shafts by not less than 2-hour fire-resistance-rated assemblies.
      6) Exception: Fire dampers and smoke dampers are not required in kitchen and clothes dryer exhaust systems.
   b. Horizontal Assemblies (floor, floor/ceiling, roof ceiling): Horizontal assemblies shall be protected by shaft enclosures.
      1) Exception: Fire dampers may be permitted to be installed at each floor provided the duct does not connect more than two floors in occupancies other than I-2 (Hospital) Occupancies and I-3 (Prison) Occupancies.
   c. Fire/smoke dampers may be an individual fire damper and smoke damper in series or a combination fire/smoke damper.

D. NFPA 90A-2015

1. Installation shall comply with the manufacturer's installation instructions and UL listing.
2. Testing procedures:
   a. Fire dampers: UL 555.
   b. Smoke dampers: UL 555S.
   c. Combination fire/smoke dampers: UL 555 and UL 555S.
   d. Ceiling dampers: UL 555C.
   e. Actuators: UL 555 and UL 555S.
3. Fire protection rating:
   a. Less than 3-hour-rated assemblies: 1-½ hours
   b. Three-hour and above rated assemblies: 3 hours
4. Fire damper actuating devices:
   a. HVAC systems: 50°F above ambient temperature, but not less than 160°F.
   b. Smoke control systems: 50°F above smoke control design temperature, but not more than 350°F.
5. Smoke damper actuating devices:
   a. Duct Mounted Smoke Damper: Provide duct mounted smoke detector located within 5 feet with no inlet/outlets between damper and detector.
   b. Unducted Smoke Damper: Provide space mounted smoke detector located within 5 feet of wall opening with damper.
   c. Smoke dampers may be controlled by area smoke detectors at smoke doors, corridors, or where total coverage smoke detection system is employed.
6. Combination fire/smoke damper actuating devices:
   a. Smoke Control System: 50°F above smoke control design temperature, but not more than 350°F.
   b. Smoke detectors as indicated under smoke damper actuating devices.

7. Access: A service opening shall be provided adjacent to each fire damper, smoke damper, fire/smoke damper, and smoke detector. Service openings shall be identified with letters 0.5" high minimum to indicate the type and location of the fire protection device.
8. Fire dampers shall be installed at the following penetration locations:
   a. Fire-rated walls and partitions with a 2-hour rating or more.
   b. Air transfer openings in partitions that are required to have a fire resistance rating and in which other openings are required to be protected.
   c. Fire-rated floors: Where air ducts extend through only one floor and serve only two adjacent floors, the ducts may be enclosed or provided with a fire damper at each floor penetration.
   d. Shafts:
      1) Less than four stories: One-hour rating.
      2) Four stories or more: Two-hour rating.
      3) Shafts that constitute air ducts or that enclose air ducts used for movement of environmental air shall not enclose the following:
         a) Kitchen hood exhaust ducts.
         b) Ducts used to remove flammable vapors.
         c) Ducts used for moving, conveying, or transporting stock, vapor, or dust.
         d) Ducts used for the removal of nonflammable corrosive fumes and vapors.
         e) Refuse or linen chutes.
         f) Piping containing hazardous materials or combustible piping.
         g) Combustible storage.
      4) Exception: A fire damper is not required where the following occur:
         a) Branch ducts connected to enclosed exhaust risers enclosed in shafts.
         b) The airflow moves upward.
         c) Steel subducts at least 22" in length are carried up inside the riser from each inlet.
         d) The riser is appropriately sized to accommodate the flow restriction created by the subduct.
9. Smoke dampers shall be installed at the following penetration locations:
   a. Smoke Barriers: Damper shall be installed within 2 feet of the smoke barrier and prior to any air inlet or outlet.
      1) Exception: Smoke dampers shall not be required on air systems other than where necessary for the proper function of that system where the system is designed specifically to accomplish the following:
         a) Function as an engineered smoke control system.
         b) Provide air to other areas of the building during a fire emergency.
         c) Provide pressure differentials during a fire emergency.
      2) Exception: Smoke dampers shall not be required where ducts serve a single smoke compartment and no other smoke compartment.
   b. Smoke dampers shall be installed in air handling systems with a capacity greater than 15,000 CFM to isolate air handling equipment (supply and return).
      1) Exception: Air handling units located on the floor they serve and serving only that floor do not require smoke dampers.
      2) Exception: Air handling units located on the roof and serving only the floor immediately below the roof do not require smoke dampers.
10. Fire/smoke dampers shall be installed at the following penetration locations:
    a. Fire-rate and smoke-rated walls and partitions.
11. Maintenance: At least every 4 years the following shall be performed:
    a. Fusible links shall be removed.
    b. All dampers shall be operated to verify that they close fully.
    c. The latch, if provided, shall be checked.
    d. Moving parts shall be lubricated as necessary.

17.07 HVAC Smoke Detection Systems Control

A. 2015 IMC

1. HVAC systems shall be equipped with smoke detectors listed and labeled for installation in air distribution systems.
2. Smoke detectors shall be installed in accordance with NFPA 72 and manufacturer's installation instructions.
3. Smoke detectors are required at the following locations:
   a. Return Air Systems: Smoke detectors are required in return air systems with design air capacity greater than 2,000 CFM (upstream of filters, exhaust connections, outdoor air connections, etc.).
   b. Common Supply and Return Systems: Smoke detectors are required in the return air system where multiple air handling systems share common supply or return air ducts or plenums with a combined capacity greater than 2,000 CFM.
   c. Return Air Risers: Smoke detectors are required in systems where the return air riser serves two or more floors and serves any portion of a return air system having a design capacity greater than 15,000 CFM. Smoke detectors shall be installed at each floor where the return air duct connects to the riser.
   d. Fan Powered Boxes: Smoke detectors are required for fan-powered boxes with a capacity greater than 2,000 CFM.
   e. Exception: Smoke detectors are not required where air distribution systems are incapable of spreading smoke beyond the enclosing walls, floors, and ceilings of the room or space in which smoke is generated.
   f. Exception: Smoke detectors are not required where the building is equipped throughout with area smoke detectors connected to a fire alarm system.
4. Control/supervision:
   a. Upon detection of smoke, the air distribution system shall be shutdown. Air distribution systems that are part of a smoke control system shall switch to smoke control operation.
   b. All smoke detectors shall be connected to the fire alarm system where a fire alarm system is required.

B. NFPA 90A-2015

1. Smoke detectors shall be installed in accordance with NFPA 72 and the manufacturer's installation instructions.
2. Smoke dampers installed to isolate the air handling system shall be arranged to close automatically when the system is not in operation.
3. Supply Air System: Smoke detectors are required in supply air systems with design air capacity greater than 2,000 CFM (downstream of filters, upstream of supply connections).
4. Return Air Risers: Smoke detectors are required in systems where the return air riser serves two or more floors and serves any portion of a return air system having a design capacity greater than 15,000 CFM. Smoke detectors shall be installed at each floor where the return air duct connects to the riser.
   a. Exception: Return air smoke detectors are not required where the entire space served by the air distribution system is protected by an area smoke detection system.
5. Exception: Smoke detectors are not required for fan units whose sole function is to remove air from the inside of the building to the outside of the building.
6. Smoke detectors shall automatically stop their respective fans.
7. Where the system is functioning as an engineered smoke control system, the smoke detectors are not required to stop the air handling system.

C. Because the IMC and NFPA requirements are different, I recommend meeting both codes by providing smoke detectors in both the supply and return systems with a capacity greater than 2,000 CFM.

17.08 Sound Attenuators

A. Types

1. Rectangular: 3-, 5-, 7-, and 10-foot lengths.
2. Round: Two or three times the diameter.

B. Locating

1. Centrifugal and axial fans:
   a. Discharge: 1 duct diameter from discharge for every 1,000 FPM.
   b. Intake: 0.75 duct diameters from intake for every 1,000 FPM.
2. Elbows: 3 duct diameters up and down stream.
3. Terminal Boxes: 1 duct diameter down stream.
4. Mechanical Equipment Rooms: Install in or close to mechanical equipment room wall opening.

17.09 Terminal Units

A. For diagrammatic examples of air terminal units, see Fig. 17.6.

B. Variable Air Volume (VAV) Terminal Units

1. VAV w/o reheat:
   a. Controls space temperature by varying the quantity of supply air.
   b. Supply temperature is constant.
   c. The energy savings is due to reduced supply air quantities and therefore reduced horsepower.
2. VAV w/reheat:
   a. Integrates heating at the VAV terminal unit to offset heating load, limit maximum humidity, provide reasonable air movement, and provide ventilation air.
3. Minimum CFM for VAV boxes:
   a. Dictated by ASHRAE Standard 62.1.
   b. Typical office building range: 30 percent to 50 percent of design flow.
   c. When interior spaces are occupied or lights are on, the VAV terminal unit will maintain a minimum flow to offset the heat gain. Therefore, the only time a VAV terminal unit serving an interior space will be closed is when the space is unoccupied and the lights are off.

C. Fan-Powered Terminal Units

1. Parallel fan-powered terminal units:
   a. Primary air is modulated in response to cooling demands and the fan is energized at a predetermined reduced primary airflow.
   b. The fan is the first stage of heating by utilizing plenum air for return. The second stage of heating is the reheat coil.
   c. Fan is located outside the primary airstream to allow intermittent fan operation.
2. Series fan-powered terminal units:
   a. A constant volume fan mixes primary air with a varying amount of air from the ceiling plenum.
   b. The fan is located within the primary airstream and runs continuously.
   c. Series fan-powered boxes are generally used with low temperature supply air from the air handling unit.

D. Induction Terminal Units

1. Reduces cooling capacity by reducing primary air and inducing room or ceiling plenum air.
2. Incorporates reduced supply air quantity energy savings of the VAV system and air volume to space is constant to reduce the effect of stagnant air.

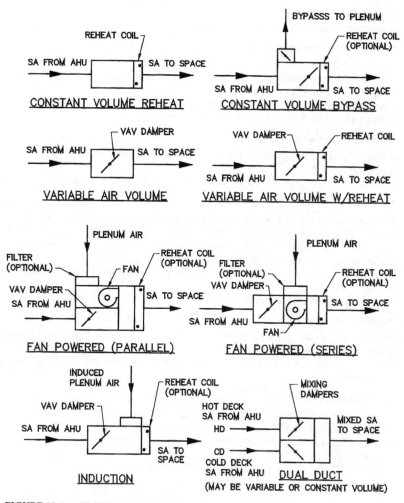

FIGURE 17.6 AIR TERMINAL UNITS.

E. Constant Volume Reheat (CVR) Terminal Units

1. CVR terminal units provide zone/space control for areas of unequal loading, simultaneous cooling/heating, and close tolerance of temperature control.
2. Conditioned air is delivered to each terminal unit at a fixed temperature, and is then reheated to control space temperature.
3. Energy inefficient system.
4. Energy codes restrict the use of these systems.

F. Constant Volume Bypass Terminal Units

1. Variation of CVR system. Constant volume primary air system with VAV secondary system.
2. Supply air to space varied by dumping air to return air plenum.
3. Energy codes restrict the use of these systems.

G. Dual Duct Terminal Units

1. A constant volume of supply air is delivered to the space.
2. Space temperature is maintained by mixing varying amounts of hot and cold air.

3. Energy inefficient system.
4. Energy codes restrict the use of these systems.

H. VAV Dual Duct Terminal Units

1. A variable volume of supply air is delivered to space.
2. Space temperature is maintained by supplying either hot or cold air in varying amounts and limiting the amount of hot and cold air mixing.
3. More energy efficient than standard dual duct systems.
4. Energy codes restrict the use of these systems.

I. Single Zone Systems

1. Supply unit serves single temperature zone and varies supply air temperature to control space temperature.
2. Single zone systems are generally small capacity systems or serve large open areas.

J. Multizone Systems

1. Supply unit serves two or more temperature zones and varies supply air temperature to each zone by mixing hot and cold air with zone dampers at the unit to control space temperature.
2. Each zone is served by a separate ductwork system.
3. Similar to dual duct systems, but where mixing occurs at the unit.
4. Limited number of zones, inflexible system, energy inefficient, and not a recommended system.
5. Multizone systems are essentially obsolete.

K. Terminal Unit Types

1. Pressure-independent terminal units: Terminal unit airflow is independent of pressure upstream of the box. Recommend using pressure-independent terminal units.
2. Pressure-dependent terminal units: Terminal unit airflow is dependent on pressure upstream of box.

L. Terminal Unit Installation

1. Locate all terminal units for unobstructed access to unit access panels, controls, and valving.
2. Minimum straight duct length upstream of terminal units:
   a. Manufacturers generally recommend 1.5 duct diameters based on terminal unit inlet size.
   b. 2.0 duct diameters are the recommended minimum.
   c. 3.0–5.0 duct diameters are preferred.
   d. Best to use 3 feet of straight duct upstream of terminal units because you do not have to concern yourself with box size when producing ductwork layout (the maximum terminal unit inlet size is 16 inches with 2 duct diameters, which results in 32 inches, and most of the time you are not using 16-inch terminal units).
3. Duct runout to the terminal unit should never be smaller than the terminal unit's inlet size; it may be larger than the inlet size, though. Terminal unit inlet and discharge ductwork should be sized based on ductwork sizing criteria and not the terminal unit inlet and discharge connection sizes. The transition from the inlet and discharge connection sizes to the air terminal unit should be made at the terminal unit. A minimum of 3 feet of straight duct should be provided upstream of all terminal units.

M. Zoning

1. Partitioned offices:
   a. One, two, three, or four offices/terminal unit.
   b. Two or three offices/terminal unit most common.
   c. One office/terminal unit; most desirable, also most expensive.
2. Open offices:
   a. 400–1,200 sq.ft./terminal unit.
3. Perimeter and interior spaces should be zoned separately.
4. Group spaces/zones/rooms/areas of similar thermal occupancy:

a. For example, group offices with offices.
b. Don't put offices with conference rooms or other dissimilar rooms.
c. Don't put east offices with south offices, etc.
d. Corner offices or spaces should be treated separately.

17.10 Process Exhaust Systems

A. Ductwork material must be selected to suit the material or chemical being exhausted—carbon steel, 304 or 316 stainless steel, Teflon- or Halar-coated stainless steel, fiberglass reinforced plastic (FRP), and polyvinyl chloride (PVC) are some examples. Sprinklers are generally required in FRP and PVC ductwork systems in all sizes larger than 8 inches in diameter.

B. Process exhaust ductwork cannot penetrate fire walls, fire separation assemblies, or smoke walls.

C. Process exhaust systems should be provided with a blast gate or butterfly damper at each tap for a hood or equipment, at each lateral, and at each submain. At all fans, large laterals, and submains, a tight shutoff—style butterfly damper should be provided for balancing and positive shutoff in addition to the blast gate. Blast gates should be specified with a wiper gasket, of EPDM or other suitable material, to provide as tight a seal as possible for blast gates; otherwise, blast gates tend to experience high leakage rates. Wind loading on blast gates installed on the roof or outside the building must be considered, especially in large blast gates. Blast gate blades will act as a sail in the wind and cause considerable stress on the ductwork system.

D. Process exhaust ductwork should be sloped a minimum of 1/8 inch per foot with a drain provided at the low point. The drain should be piped to the appropriate waste system.

E. Process exhaust systems are required, in most cases, to undergo a treatment process—scrubbing, abatement, burning, or filtering.

F. Duct sizing must be based on capture velocities and entrainment velocities of the material or chemical being exhausted. For most chemical or fume exhaust systems, the mains, risers, submains, and large laterals should be sized for 2,000 to 3,000 feet per minute, and small laterals and branches should be sized for 1,500 to 2,500 feet per minute. Discharge stacks should be sized for 3,000 to 4,000 feet per minute discharge velocity and should terminate a minimum of 8 feet above the roof and a minimum of 10 feet from any openings or intakes. Properly locate discharge stacks and coordinate discharge height to prevent contamination of outside air intakes, cooling tower intakes, and combustion air intakes. Clearly indicate termination heights.

G. The connection to a fume hood or other piece of equipment will generally require between 1.0 and 3.0 inches WC negative pressure.

H. Branches and laterals should be connected above duct centerline. If branches and laterals are connected below the duct centerline, drains will be required at the low point. Hoods, tools, and equipment must be protected from the possibility of drainage contaminating or entering equipment when taps are connected below the centerline.

I. Specify proper pressure class upstream and downstream of scrubbers and other abatement equipment.

J. When ductwork is installed outside or in unconditioned spaces, verify if condensation will occur on the outside or the inside of this duct. Insulate the duct and/or heat trace if required.

K. Process exhaust fans are required to be on emergency power by code.

L. Process exhaust ductwork cannot penetrate fire-rated construction. Fire dampers are generally not desirable. If penetrating fire-rated construction cannot be avoided, process exhaust ductwork must be enclosed in a fire-rated enclosure until it exits the building, or sprinkler protection located inside the duct may be used if approved by authority having jurisdiction.

M. Provide pressure ports at the end of all laterals, submains, and mains.

N. Generally, drains are required in fan scroll, scrubber, and other abatement equipment.

O. Provide flexible connections at fans and specify flexible connection material suitable for application.

P. If adjustable or variable frequency drives are required or used, locate and coordinate them with the electrical engineer. Use direct drive fans with adjustable or variable frequency drives.

17.11 Hazardous Exhaust Systems

A. Hazardous exhaust systems as defined in the 2015 IMC.

B. A hazardous exhaust system shall include exhaust systems containing:
1. Flammable vapors.
2. Gases.
3. Fumes.
4. Mists.
5. Dusts.
6. Paint residue.
7. Corrosive fumes.
8. Volatile or airborne materials posing a health hazard.

C. Hazardous Exhaust System Concerns:
1. Combustibility.
2. Flammability.
3. Toxicity.
4. Corrosiveness.
5. Explosiveness.
6. Microbial.
7. Pathogenic.

D. Hospital and research laboratory exhaust systems are designed to exhaust different substances. However, these substances may or may not be flammable, toxic, corrosive, or pathogenic. For the classification and identification of hazardous substances, see NFPA 704. NFPA 704 covers the concerns of combustibility, flammability, toxicity, corrosiveness, and explosiveness, but this standard does not address microbial, pathogenic, and other hospital or research exhaust

hazards. Laboratory exhaust systems involve the use of chemicals and other hazardous materials for:

1. Testing.
2. Analysis.
3. Teaching.
4. Research.
5. Development.
6. Nonproduction purposes.
7. 2015 IMC: Laboratory exhaust systems do not have to be independent of other exhaust systems provided that all of the following conditions are met.
   a. All hazardous exhaust ductwork and other laboratory exhaust ductwork within both the occupied space and the shaft is under negative pressure while in operation.
   b. All hazardous exhaust ductwork manifolded together within the occupied space must originate in the same fire area.
   c. Hazardous exhaust ductwork originating in different fire areas and manifolded together in a common shaft shall be equipped with steel exhaust air subducts extending at least 22" vertically in exhaust shafts where there is continuous airflow upward to the outside.
   d. Each control branch has a flow regulating device.
   e. Perchloric acid hoods must have a separate exhaust system and cannot be manifolded together.
   f. Radioisotope hoods are properly filtered.
   g. Biological safety cabinets are properly filtered.
   h. A provision is made for continuous operation of the negative static pressure in the ductwork with standby fans.

E. **Hazardous exhaust systems are required wherever hazardous materials are present to create any one of the following conditions. The criteria is based on the normal operating conditions and not the conditions that would exist in an accident or unusual condition.**

1. Materials are present in concentrations at room temperature that exceed 25 percent of the lower flammability limit of the substance.
2. Materials are present at any concentration with a health hazard of 4.
   a. Exception: Hazardous exhaust systems are required for laboratories where materials are present with a health hazard of 4 at concentrations exceeding 1 percent of the median lethal concentration for acute inhalation toxicity.
3. Materials are present with a health hazard of 1, 2, or 3 at concentrations exceeding 1 percent of the median lethal concentration for acute inhalation toxicity.

F. **Hazardous exhaust systems must be independent of all other exhaust systems.**

G. **Hazardous exhaust systems must be located in separate shafts from other HVAC duct systems and in separate shafts from other hazardous exhaust systems originating in different fire areas.**

H. **Hazardous exhaust systems must segregate compatible and incompatible material exhaust air streams.**

I. **Ductwork design methods:**

1. Vapors, gases, and smoke: Constant velocity or equal friction methods.
2. Dust, fibers, and particulate matter: Constant velocity method.

J. **Exhaust makeup air shall be delivered to the space with hazardous exhaust systems in quantities nearly equal to the exhaust air quantities. Normally, the makeup air is slightly less than the exhaust air quantity to help confine the contaminants.**

K. Hazardous exhaust systems that penetrate a fire-rated floor/ceiling assembly or fire-rated wall assembly must be enclosed in a fire-resistance-rated shaft enclosure, meeting the fire rating of the highest rated assembly penetrated, from where the exhaust system penetrates the rated enclosure until it terminates outdoors.

L. In lieu of enclosing the hazardous exhaust duct that penetrates a fire-rated wall assembly in a fire-rated enclosure, the interior of the duct may be equipped with an approved automatic fire suppression system suitable for the materials being exhausted. Hazardous exhaust systems that penetrate a fire-rated floor/ceiling assembly must be enclosed in a fire-rated shaft, regardless of whether the system is protected by a fire suppression system or not.

M. Ducts shall not penetrate a fire wall.

N. Fire dampers and smoke dampers are not permitted in hazardous exhaust systems.

O. Hazardous exhaust systems shall be protected by an approved automatic fire suppression system. The automatic fire suppression system must be compatible with the materials being exhausted (water, dry chemical, carbon dioxide).

1. Except hazardous exhaust systems conveying nonflammable and noncombustible materials at all concentrations.
2. Except in metallic and noncombustible, nonmetallic exhaust ducts in semiconductor fabrication facilities.
3. Except in ducts where the cross-sectional duct diameter is less than 10 inches.
4. Except in laboratory hoods or laboratory exhaust systems.

P. Ductwork materials for hazardous exhaust systems:

1. G90 galvanized steel.
2. 304 or 316 stainless steel.
3. Fiberglass reinforced: Chemical exhaust; plastic (FRP).
4. Polyvinyl chloride (PVC): Chemical exhaust, underground ducts; PVC conforming to *NFPA 91, ASTM D1784, D1785, D1927,* and *D2241.*
5. Polyvinyl steel and stainless steel (PVS and PVSS): Chemical exhaust; common type: Halar-coated stainless steel, Teflon-coated stainless steel.
6. Nonmetal ducts must meet the ASTM E 84 flame spread index of 25 or less and a smoke developed index of 50 or less.
7. Ducts shall be constructed of materials that are compatible with the exhaust.
8. Minimum hazardous exhaust duct thickness:

Diameter of Duct or Maximum Side Dimension	Minimum Nominal Thickness		
	Nonabrasive Materials (Gauge)	Nonabrasive/ Abrasive Materials (Gauge)	Abrasive Materials (Gauge)
0–8 inches	24	22	20
9–18 inches	22	20	18
19–30 inches	20	18	16
Over 30 inches	18	16	14

Q. Hazardous exhaust ducts shall be supported at intervals not exceeding 10 feet. Supports shall be constructed of noncombustible materials.

17.12 Galvanized Rectangular Ductwork Weights—Pound per Lineal Foot

GALVANIZED RECTANGULAR DUCT WEIGHT

Width + Depth Inches	26 (12")	24 (24")	22 (48")	20 (60")	18 (60+")	16	Surface Area sq.ft./ ln.ft.
38	–	9.15	11.13	13.11	17.07	21.03	6.34
39	–	9.39	11.42	13.46	17.52	21.58	6.50
40	–	9.63	11.72	13.80	17.97	22.13	6.67
41	–	9.87	12.01	14.15	18.42	22.69	6.83
42	–	10.12	12.30	14.49	18.87	23.24	7.00
43	–	10.36	12.60	14.84	19.31	23.79	7.17
44	–	10.60	12.89	15.18	19.76	24.35	7.34
45	–	10.84	13.18	15.53	20.21	24.90	7.50
46	–	11.08	13.47	15.87	20.66	25.45	7.67
47	–	11.32	13.77	16.22	21.11	26.00	7.83
48	–	11.56	14.06	16.56	21.56	26.56	8.00
49	–	11.80	14.35	16.91	22.01	27.11	8.17
50	–	12.04	14.65	17.25	22.46	27.67	8.34
51	–	12.28	14.94	17.60	22.91	28.22	8.50
52	–	12.52	15.23	17.94	23.36	28.77	8.67
53	–	12.76	15.52	18.29	23.81	29.32	8.83
54	–	13.01	15.82	18.63	24.26	29.88	9.00
55	–	13.25	16.11	18.98	24.70	30.43	9.17
56	–	13.49	16.40	19.32	25.15	30.99	9.34
57	–	13.73	16.70	19.67	25.60	31.54	9.50
58	–	13.97	16.99	20.01	26.05	32.09	9.67
59	–	14.21	17.28	20.36	26.50	32.65	9.83
60	–	14.45	17.58	20.70	26.95	33.20	10.00
61	–	–	17.87	21.05	27.40	33.75	10.17
62	–	–	18.16	21.39	27.85	34.31	10.34
63	–	–	18.45	21.74	28.30	34.86	10.50
64	–	–	18.75	22.08	28.75	35.41	10.67
65	–	–	19.04	22.43	29.20	35.97	10.83
66	–	–	19.33	22.77	29.65	36.52	11.00
67	–	–	19.63	23.12	30.09	37.07	11.17
68	–	–	19.92	23.46	30.54	37.63	11.34
69	–	–	20.21	23.81	30.99	38.18	11.50
70	–	–	20.50	24.15	31.44	38.73	11.67
71	–	–	20.80	24.50	31.89	39.29	11.83
72	–	–	21.09	24.84	32.34	39.84	12.00
73	–	–	21.38	25.19	32.79	40.39	12.17
74	–	–	21.68	25.53	33.24	40.95	12.34
75	–	–	21.97	25.88	33.69	41.50	12.50
76	–	–	22.26	26.22	34.14	42.05	12.67
77	–	–	22.55	26.57	34.59	42.61	12.83
78	–	–	22.85	26.91	35.04	43.16	13.00
79	–	–	23.14	27.26	35.48	43.71	13.17
80	–	–	23.43	27.60	35.93	44.27	13.34
81	–	–	23.73	27.95	36.38	44.82	13.50
82	–	–	24.02	28.29	36.83	45.37	13.67
83	–	–	24.31	28.64	37.28	45.93	13.83
84	–	–	24.61	28.98	37.73	46.48	14.00
85	–	–	24.90	29.33	39.18	47.03	14.17
86	–	–	25.19	29.67	38.63	48.59	14.34
87	–	–	25.48	30.02	39.08	48.14	14.50
88	–	–	25.78	30.36	39.53	48.69	14.67
89	–	–	26.07	30.71	39.98	49.25	14.83
90	–	–	26.36	31.05	40.43	49.80	15.00
91	–	–	26.66	31.40	40.87	50.35	15.17
92	–	–	26.95	31.74	41.32	50.91	15.34

(Continued)

GALVANIZED RECTANGULAR DUCT WEIGHT (Continued)

Width + Depth Inches	Sheet Metal Gauge						Surface Area sq.ft./ln.ft.
	26 (12")	24 (24")	22 (48")	20 (60")	18 (60+")	16	
93	–	–	27.24	32.09	41.77	51.46	15.50
94	–	–	27.53	32.43	42.22	52.01	15.67
95	–	–	27.83	32.78	42.67	52.57	15.83
96	–	–	28.12	33.12	43.12	53.12	16.00
97	–	–	28.41	33.47	43.57	53.67	16.17
98	–	–	28.71	33.81	44.02	54.23	16.34
99	–	–	29.00	34.16	44.47	54.78	16.50
100	–	–	29.29	34.50	44.92	55.33	16.67
101	–	–	29.58	34.85	45.37	55.89	16.83
102	–	–	29.88	35.19	45.82	56.44	17.00
103	–	–	30.17	35.54	46.26	56.99	17.17
104	–	–	30.46	35.88	46.71	57.55	17.34
105	–	–	30.76	36.23	47.16	58.10	17.50
106	–	–	31.05	36.57	47.61	58.65	17.67
107	–	–	31.34	36.92	48.06	59.21	17.83
108	–	–	31.64	37.26	48.51	59.76	18.00
109	–	–	31.93	37.61	48.96	60.31	18.17
110	–	–	32.22	37.95	49.41	60.87	18.34
111	–	–	32.51	38.30	49.86	61.42	18.50
112	–	–	32.81	38.64	50.31	61.97	18.67
113	–	–	33.10	38.99	50.76	62.53	18.83
114	–	–	33.39	39.33	51.21	63.08	19.00
115	–	–	33.69	39.68	51.65	63.63	19.17
116	–	–	33.98	40.02	52.10	64.19	19.34
117	–	–	34.27	40.37	52.55	64.74	19.50
118	–	–	34.56	40.71	53.00	65.29	19.67
119	–	–	34.86	41.06	53.45	65.85	19.83
120	–	–	35.15	41.40	53.90	66.40	20.00
121	–	–	35.44	41.75	54.35	66.95	20.17
122	–	–	35.74	42.09	54.80	67.51	20.34
123	–	–	36.03	42.44	55.25	68.06	20.50
124	–	–	36.32	42.78	55.70	68.61	20.67
125	–	–	36.61	43.13	56.15	69.17	20.83
126	–	–	36.91	43.47	56.60	69.72	21.00
127	–	–	37.20	43.82	57.04	70.27	21.17
128	–	–	37.49	44.16	57.49	70.83	21.34
129	–	–	37.79	44.51	57.94	71.38	21.50
130	–	–	38.08	44.85	58.39	71.93	21.67
131	–	–	38.37	45.20	58.84	72.49	21.83
132	–	–	38.67	45.54	59.29	73.04	22.00
133	–	–	38.96	45.89	59.74	73.59	22.17
134	–	–	39.25	46.23	60.19	74.15	22.34
135	–	–	39.54	46.58	60.64	74.70	22.50
136	–	–	39.84	46.92	61.09	75.25	22.67
137	–	–	40.13	47.27	61.54	75.81	22.83
138	–	–	40.42	47.61	61.99	76.36	23.00
139	–	–	40.72	47.96	62.43	76.91	23.17
140	–	–	41.01	48.30	62.88	77.46	23.34
141	–	–	41.30	48.65	63.33	78.02	23.50
142	–	–	41.59	48.99	63.78	78.57	23.67
143	–	–	41.88	49.34	64.23	79.13	23.83
144	–	–	42.18	49.68	64.68	79.68	24.00
145	–	–	42.47	50.03	65.13	80.23	24.17
146	–	–	42.77	50.37	65.58	80.79	24.34
147	–	–	43.06	50.72	66.03	81.34	24.50
148	–	–	43.35	51.06	66.48	81.89	24.67
149	–	–	43.64	51.41	66.93	82.45	24.83
150	–	–	43.94	51.75	67.38	83.00	25.00
151	–	–	44.23	52.10	67.82	83.55	25.17
152	–	–	44.52	52.44	68.27	84.11	25.34

(Continued)

GALVANIZED RECTANGULAR DUCT WEIGHT (*Continued*)

Width + Depth Inches	Sheet Metal Gauge						Surface Area sq.ft./ ln.ft.
	26 (12")	24 (24")	22 (48")	20 (60")	18 (60+")	16	
153	–	–	44.82	52.79	68.72	84.66	25.50
154	–	–	45.11	53.13	69.17	85.21	25.67
155	–	–	45.40	53.48	69.62	85.77	25.83
156	–	–	45.70	53.82	70.07	86.32	26.00
157	–	–	45.99	54.17	70.52	86.87	26.17
158	–	–	46.28	54.51	70.97	87.43	26.34
159	–	–	46.57	54.86	71.42	87.98	26.50
160	–	–	46.87	55.20	71.87	88.53	26.67
161	–	–	47.16	55.55	72.32	89.09	26.83
162	–	–	47.45	55.89	72.77	89.64	27.00
163	–	–	47.75	56.24	73.21	90.19	27.17
164	–	–	48.04	56.58	73.66	90.75	17.34
165	–	–	48.33	56.93	74.11	91.30	27.50
166	–	–	48.62	57.27	74.56	91.85	27.67
167	–	–	48.92	57.62	75.01	92.41	27.83
168	–	–	49.21	57.96	75.46	92.96	28.00
169	–	–	49.50	58.31	75.91	93.51	28.17
170	–	–	49.80	58.65	76.36	94.07	28.34
171	–	–	50.09	59.00	76.81	94.62	28.50
172	–	–	50.38	59.34	77.26	95.17	28.67
173	–	–	50.67	59.69	77.71	95.73	28.83
174	–	–	50.97	60.03	78.16	96.28	29.00
175	–	–	51.26	60.38	78.60	96.83	29.17
176	–	–	51.55	60.72	79.05	97.39	29.34
177	–	–	51.85	61.07	79.50	97.94	29.50
178	–	–	52.14	61.41	79.95	98.49	29.67
179	–	–	52.43	61.76	80.40	99.05	29.83
180	–	–	52.73	62.10	80.85	99.60	30.00
181	–	–	53.02	62.45	81.30	100.15	30.17
182	–	–	53.31	62.79	81.75	100.71	30.34
183	–	–	53.60	63.14	82.20	101.26	30.50
184	–	–	53.90	63.48	82.65	101.81	30.67
185	–	–	54.19	63.83	83.10	102.37	30.83
186	–	–	54.48	64.17	83.55	102.92	31.00
187	–	–	54.78	64.52	83.99	103.47	31.17
188	–	–	55.07	64.86	84.44	104.03	31.34
189	–	–	55.36	65.21	84.89	104.58	31.50
190	–	–	55.65	65.55	85.34	105.13	31.67
191	–	–	55.95	65.90	85.79	105.69	31.83
192	–	–	56.24	66.24	86.24	106.24	32.00
193	–	–	56.53	66.59	86.69	106.79	32.17
194	–	–	56.83	66.93	87.14	107.35	32.34
195	–	–	57.12	67.28	87.59	107.90	32.50
196	–	–	57.41	67.62	88.04	108.45	32.67
197	–	–	57.70	67.97	88.49	109.01	32.83
198	–	–	58.00	68.31	88.94	109.56	33.00
199	–	–	58.29	68.66	89.38	110.11	33.17
200	–	–	58.58	69.00	89.83	110.67	33.34
201	–	–	58.88	69.35	90.28	111.22	33.50
202	–	–	59.17	69.69	90.73	111.77	33.67
203	–	–	59.46	70.04	91.18	112.33	33.83
204	–	–	59.76	70.38	91.63	112.88	34.00
205	–	–	60.05	70.73	92.08	113.43	34.17
206	–	–	60.34	71.07	92.53	113.99	34.34
207	–	–	60.63	71.42	92.98	114.54	34.50
208	–	–	60.93	71.76	93.43	115.09	34.67
209	–	–	61.22	72.11	93.88	115.65	34.83
210	–	–	61.51	72.45	94.33	116.20	35.00
211	–	–	61.81	72.80	94.77	116.75	35.17
212	–	–	62.10	73.14	95.22	117.31	35.34

(Continued)

GALVANIZED RECTANGULAR DUCT WEIGHT (*Continued*)

Width + Depth Inches	Sheet Metal Gauge						Surface Area sq.ft./ ln.ft.
	26 (12")	24 (24")	22 (48")	20 (60")	18 (60+")	16	
213	–	–	62.39	73.49	95.67	117.86	35.50
214	–	–	62.68	73.83	96.12	118.41	35.67
215	–	–	62.98	74.18	96.57	118.97	35.83
216	–	–	63.27	74.52	97.02	119.52	36.00
217	–	–	63.56	74.87	97.47	120.07	36.17
218	–	–	63.86	75.21	97.92	120.63	36.34
219	–	–	64.15	75.56	98.37	121.18	36.50
220	–	–	64.44	75.90	98.82	121.73	36.67
221	–	–	64.73	76.25	99.27	122.29	36.83
222	–	–	65.03	76.59	99.72	122.84	37.00
223	–	–	65.32	76.94	100.16	123.39	37.17
224	–	–	65.61	77.28	100.61	123.95	37.34
225	–	–	65.91	77.63	101.06	124.50	37.50
226	–	–	66.20	77.97	101.51	125.05	37.67
227	–	–	66.49	78.32	101.96	125.61	37.83
228	–	–	66.79	78.66	102.41	126.16	38.00
229	–	–	67.08	79.01	102.86	126.71	38.17
230	–	–	67.37	79.35	103.31	127.27	38.34
231	–	–	67.66	79.70	103.76	127.82	38.50
232	–	–	67.96	80.04	104.21	128.37	38.67
233	–	–	68.25	80.39	104.66	128.93	38.83
234	–	–	68.54	80.73	105.11	129.48	39.00
235	–	–	68.84	81.08	105.55	130.03	39.17
236	–	–	69.13	81.42	106.00	130.59	39.34
237	–	–	69.42	81.77	106.45	131.14	39.50
238	–	–	69.71	82.11	106.90	131.69	39.67
239	–	–	70.01	82.46	107.35	132.25	39.83
240	–	–	70.30	82.80	107.80	132.80	40.00

Notes:
1 Table includes 25 percent allowance for bracing, hangers, reinforcing, joints, and seams. Add 10 percent for insulated ductwork systems.
2 The first column is the sum of the width and depth of the duct (i.e., a 20 × 10 duct equals 30 inches).
3 Columns 2 through 7 give the weight of galvanized steel ducts in pounds per lineal foot.
4 Column 8 gives the ductwork surface area used for estimating insulation.
5 Numbers in parentheses below the sheet metal gauges indicate the maximum duct dimension for the indicated gauge.

17.13 Galvanized Round Ductwork Weights—Pound per Lineal Foot

GALVANIZED ROUND DUCT WEIGHT

Diameter	Gauge						Surface Area sq.ft./Lin.ft.
	26	24	22	20	18	16	
3	0.89	1.13	1.38	1.63	2.12	2.61	0.79
4	1.19	1.51	1.84	2.17	2.82	3.48	1.05
5	1.48	1.89	2.30	2.71	3.53	4.35	1.31
6	1.78	2.27	2.76	3.25	4.23	5.22	1.57
7	2.08	2.65	3.22	3.79	4.94	6.08	1.83
8	2.37	3.03	3.68	4.34	5.64	6.95	2.09
9	2.67	3.40	4.14	4.88	6.35	7.82	2.36
10	2.96	3.78	4.60	5.42	7.06	8.69	2.62
11	3.26	4.16	5.06	5.96	7.76	9.56	2.88
12	3.56	4.54	5.52	6.50	8.47	10.43	3.14

(Continued)

GALVANIZED ROUND DUCT WEIGHT (*Continued*)

| Diameter | Gauge | | | | | | Surface Area sq.ft./Lin.ft. |
	26	24	22	20	18	16	
14	4.15	5.30	6.44	7.59	9.88	12.17	3.67
16	4.74	6.05	7.36	8.67	11.29	13.91	4.19
18	5.34	6.81	8.28	9.75	12.70	15.65	4.71
20	5.93	7.57	9.20	10.84	14.11	17.38	5.24
22	6.52	8.32	10.12	11.92	15.52	19.12	5.76
24	7.12	9.08	11.04	13.01	16.93	20.86	6.28
26	7.71	9.84	11.96	14.09	18.34	22.60	6.81
28	8.30	10.59	12.88	15.17	19.76	24.34	7.33
30	8.89	11.35	13.80	16.26	21.17	26.08	7.85
32	9.49	12.11	14.72	17.34	22.58	27.81	8.38
34	10.08	12.86	15.64	18.43	23.99	29.55	8.90
36	10.67	13.62	16.56	19.51	25.40	31.29	9.42
38	11.27	14.38	17.48	20.59	26.81	33.03	9.95
40	11.86	15.13	18.40	21.68	28.22	34.77	10.47
42	12.45	15.89	19.32	22.76	29.63	36.51	11.00
44	13.05	16.65	20.24	23.84	31.04	38.24	11.52
46	13.64	17.40	21.17	24.93	32.46	39.98	12.04
48	14.23	18.16	22.09	26.01	33.87	41.72	12.57
50	–	18.92	23.01	27.10	35.28	43.46	13.09
52	–	19.67	23.93	28.18	36.69	45.20	13.61
54	–	20.43	24.85	29.26	38.10	46.94	14.14
56	–	21.18	25.77	30.35	39.51	48.67	14.66
58	–	21.94	26.69	31.43	40.92	50.41	15.18
60	–	22.70	27.61	32.52	42.33	52.15	15.71
62	–	23.45	28.53	33.60	43.74	53.89	16.23
64	–	24.21	29.45	34.68	45.16	55.63	16.76
66	–	24.97	30.37	35.77	46.57	57.37	17.28
68	–	25.72	31.29	36.85	47.98	59.10	17.80
70	–	26.48	32.21	37.93	49.39	60.84	18.33
72	–	27.24	33.13	39.02	50.80	62.58	18.85
74	–	27.99	34.05	40.10	52.21	64.32	19.37
76	–	28.75	34.97	41.19	53.62	66.06	19.90
78	–	29.51	35.89	42.27	55.03	67.80	20.42
80	–	30.26	36.81	43.35	56.44	69.53	20.94
82	–	31.02	37.73	44.44	57.86	71.27	21.47
84	–	31.78	38.65	45.52	59.27	73.01	21.99
86	–	32.53	39.57	46.61	60.68	74.75	22.51
88	–	33.29	40.49	47.69	62.09	76.49	23.04
90	–	34.05	41.41	48.77	63.50	78.23	23.56
92	–	34.80	42.33	49.86	64.91	79.96	24.09
94	–	35.56	43.25	50.94	66.32	81.70	24.61
96	–	36.32	44.17	52.02	66.73	83.44	25.13
98	–	37.07	45.09	53.11	69.14	85.18	25.66
100	–	37.83	46.01	54.19	70.55	86.92	26.18
102	–	38.59	46.93	55.28	71.97	88.66	26.70
104	–	39.34	47.85	56.36	73.38	90.39	27.23
106	–	40.10	48.77	57.44	74.79	92.13	27.75
108	–	40.86	49.69	58.53	76.20	93.87	28.27
110	–	41.61	50.61	59.61	77.61	95.61	28.80
112	–	42.37	51.53	60.70	79.02	97.35	29.32
114	–	43.13	52.45	61.78	80.43	99.09	29.85
116	–	43.88	53.37	62.86	81.84	100.82	30.37
118	–	44.64	54.29	63.95	83.25	102.56	30.89
120	–	45.40	55.21	65.03	84.67	104.30	31.42
122	–	46.15	56.13	66.11	86.08	106.04	31.94

(*Continued*)

GALVANIZED ROUND DUCT WEIGHT (*Continued*)

Diameter	26	24	22	20	18	16	Surface Area sq.ft./Lin.ft.
				Gauge			
124	–	46.91	57.05	67.20	87.49	107.78	32.46
126	–	47.67	57.97	68.28	88.90	109.52	32.99
128	–	48.42	58.89	69.37	90.31	111.25	33.51
130	–	49.18	59.81	70.45	91.72	112.99	34.03
132	–	49.94	60.73	71.53	93.13	114.73	34.56
134	–	50.69	61.66	72.62	94.54	116.47	35.08
136	–	51.45	62.58	73.70	95.95	118.21	35.60
138	–	52.21	63.50	74.79	97.37	119.95	36.12
140	–	52.96	64.42	75.87	98.78	121.68	36.65
142	–	53.72	65.34	76.95	100.19	123.42	37.18
144	–	54.48	66.26	78.04	101.60	125.16	37.70

Notes:
1 Table includes 25 percent allowance for bracing, hangers, reinforcing, joints, and seams. Add 10 percent for insulated ductwork systems.
2 Table gives weight of galvanized steel ducts in pounds per lineal foot.

17.14 Galvanized Flat Oval Ductwork Weights—Pounds per Lineal Foot

GALVANIZED FLAT OVAL DUCTWORK WEIGHT

Nominal Flat Oval Size	Equiv. Round	Cross Sectional Area sq.ft.	Surface Area sq.ft./ln.ft.	Gauge	Weight lbs./ln.ft.
3 × 8	5.1	0.15	1.57	24	2.3
3 × 9	5.6	0.18	1.83	24	2.6
3 × 11	6.0	0.22	2.09	24	3.1
3 × 12	6.4	0.25	2.36	24	3.4
3 × 14	6.7	0.29	2.62	24	3.8
3 × 15	7.0	0.32	2.88	24	4.2
3 × 17	7.3	0.36	3.14	24	4.5
3 × 19	7.5	0.39	3.40	24	4.9
3 × 22	8.0	0.46	3.93	24	5.7
4 × 7	5.7	0.18	1.57	24	2.3
4 × 9	6.2	0.22	1.83	24	2.6
4 × 10	6.7	0.26	2.09	24	3.1
4 × 12	7.2	0.31	2.36	24	3.4
4 × 13	7.6	0.35	2.62	24	3.8
4 × 15	8.0	0.40	2.88	24	4.2
4 × 17	8.4	0.44	3.14	24	4.5
4 × 18	8.5	0.48	3.40	24	4.9
4 × 20	9.0	0.52	3.68	24	5.3
4 × 21	9.5	0.57	3.93	24	5.7
5 × 8	6.6	0.25	1.83	24	2.6
5 × 10	7.3	0.30	2.09	24	3.0
5 × 11	7.9	0.35	2.36	24	3.4
5 × 13	8.4	0.41	2.62	24	3.8
5 × 14	8.8	0.46	2.88	24	4.2
5 × 16	9.3	0.52	3.14	24	4.5
5 × 18	9.5	0.57	3.40	24	4.9
5 × 19	10.0	0.63	3.66	24	5.3
5 × 21	10.5	0.68	3.93	24	5.7
6 × 8	6.9	0.26	1.83	24	2.6
6 × 9	7.7	0.33	2.09	24	3.0
6 × 11	8.4	0.39	2.36	24	3.4
6 × 12	8.9	0.46	2.62	24	3.8
6 × 14	9.6	0.53	2.88	24	4.2
6 × 15	10.1	0.59	3.14	24	4.5

(*Continued*)

GALVANIZED FLAT OVAL DUCTWORK WEIGHT (*Continued*)

Nominal Flat Oval Size	Equiv. Round	Cross Sectional Area sq.ft.	Surface Area sq.ft./ln.ft.	Gauge	Weight lbs./ln.ft.
6 × 17	10.5	0.65	3.40	24	4.9
6 × 19	11.0	0.72	3.66	24	5.3
6 × 20	11.5	0.79	3.93	24	5.7
6 × 22	11.8	0.85	4.18	24	6.0
6 × 23	12.0	0.92	4.45	24	6.4
6 × 25	12.5	0.98	4.71	22	8.3
6 × 28	13.2	1.11	5.23	22	9.2
6 × 30	13.5	1.18	5.50	22	9.7
6 × 31	13.8	1.24	5.76	22	10.1
6 × 33	14.0	1.31	6.02	22	10.6
6 × 34	14.3	1.38	6.28	22	11.0
6 × 36	14.5	1.44	6.54	22	11.5
6 × 37	14.9	1.50	6.80	22	12.0
6 × 39	15.0	1.57	7.07	22	12.4
6 × 41	15.4	1.64	7.33	22	12.9
6 × 44	15.9	1.77	7.85	22	13.8
6 × 45	16.0	1.83	8.12	22	14.3
6 × 52	17.0	2.09	9.16	20	19.0
6 × 59	18.0	2.42	10.47	20	21.7
7 × 10	8.7	0.42	2.36	24	3.4
7 × 12	9.4	0.50	2.62	24	3.8
7 × 13	10.1	0.57	2.88	24	4.2
7 × 15	10.7	0.65	3.14	24	4.5
7 × 16	11.0	0.73	3.40	24	4.9
7 × 18	11.7	0.80	3.67	24	5.3
7 × 20	12.0	0.88	3.93	24	5.7
7 × 21	12.5	0.95	4.19	24	6.1
7 × 23	13.0	1.03	4.45	24	6.4
8 × 10	9.0	0.44	2.36	24	3.4
8 × 11	9.8	0.53	2.62	24	3.8
8 × 13	10.6	0.62	2.88	24	4.2
8 × 14	11.2	0.70	3.14	24	4.5
8 × 16	11.5	0.79	3.40	24	4.9
8 × 17	12.0	0.87	3.67	24	5.3
8 × 18	12.4	0.90	3.80	24	5.5
8 × 19	13.0	0.96	3.93	24	5.7
8 × 21	13.5	1.05	4.18	24	6.1
8 × 22	14.0	1.13	4.45	24	6.4
8 × 24	14.4	1.23	4.71	24	6.8
8 × 27	15.2	1.40	5.23	22	9.2
8 × 30	15.9	1.57	5.76	22	10.2
8 × 33	16.6	1.74	6.28	22	11.0
8 × 35	17.0	1.83	6.54	22	11.5
8 × 36	17.3	1.92	6.80	22	12.0
8 × 39	17.9	2.09	7.33	22	12.9
8 × 43	18.6	2.27	7.85	22	13.8
8 × 46	19.1	2.44	8.37	22	14.7
8 × 49	19.6	2.62	8.89	20	18.4
8 × 50	20.0	2.71	9.16	20	19.0
8 × 52	20.2	2.80	9.42	20	19.5
8 × 58	21.0	3.14	10.47	20	21.7
8 × 65	22.0	3.49	11.52	20	23.8
8 × 71	23.0	3.84	12.57	18	33.9
8 × 77	24.0	4.19	13.61	18	36.7
9 × 12	10.8	0.64	2.88	24	4.2
9 × 14	11.5	0.74	3.14	24	4.6
9 × 15	12.0	0.83	3.40	24	4.9

(Continued)

GALVANIZED FLAT OVAL DUCTWORK WEIGHT (*Continued*)

Nominal Flat Oval Size	Equiv. Round	Cross Sectional Area sq.ft.	Surface Area sq.ft./ln.ft.	Gauge	Weight lbs./ln.ft.
9 × 17	12.9	0.93	3.67	24	5.3
9 × 18	13.5	1.03	3.93	24	5.7
9 × 20	14.0	1.13	4.19	24	6.1
9 × 22	14.5	1.23	4.45	24	6.4
9 × 23	15.0	1.33	4.71	24	6.8
10 × 12	11.0	0.66	2.88	24	4.2
10 × 13	11.9	0.77	3.14	24	4.5
10 × 15	12.5	0.87	3.40	24	4.9
10 × 16	13.4	1.00	3.66	24	5.3
10 × 18	14.0	1.09	3.93	24	5.7
10 × 19	14.5	1.20	4.19	24	6.1
10 × 20	14.7	1.25	4.18	24	6.1
10 × 21	15.0	1.31	4.45	24	6.4
10 × 23	15.7	1.42	4.71	24	6.8
10 × 24	16.0	1.53	4.97	24	7.2
10 × 26	16.7	1.63	5.23	22	9.2
10 × 27	17.0	1.75	5.50	22	9.7
10 × 29	17.7	1.86	5.76	22	10.2
10 × 30	18.0	1.96	6.02	22	10.6
10 × 32	18.5	2.07	6.28	22	11.1
10 × 34	19.0	2.18	6.54	22	11.5
10 × 35	19.3	2.29	6.80	22	12.0
10 × 38	20.1	2.51	7.33	22	12.9
10 × 41	20.8	2.73	7.85	22	13.8
10 × 43	21.0	2.84	8.12	22	14.3
10 × 45	21.5	2.95	8.37	22	14.7
10 × 48	22.1	3.16	8.89	22	15.6
10 × 51	22.8	3.39	9.42	20	19.5
10 × 52	23.0	3.49	9.69	20	20.1
10 × 54	23.3	3.60	9.95	20	20.6
10 × 57	23.8	3.82	10.56	20	21.9
10 × 60	24.4	4.04	11.00	20	22.8
10 × 63	25.0	4.25	11.52	20	23.8
10 × 67	25.5	4.47	12.05	20	24.9
10 × 70	26.0	4.69	12.51	20	25.9
10 × 73	26.4	4.91	13.10	18	35.3
10 × 76	27.0	5.13	13.61	18	36.7
11 × 14	13.0	0.90	3.40	24	4.9
11 × 16	13.6	1.02	3.67	24	5.3
11 × 17	14.0	1.14	3.93	24	5.7
11 × 19	15.0	1.26	4.19	24	6.1
11 × 22	16.3	1.50	4.71	24	6.8
11 × 24	17.0	1.62	4.97	24	7.2
12 × 14	13.0	0.92	3.40	24	4.9
12 × 15	13.8	1.05	3.67	24	5.3
12 × 17	14.5	1.18	3.93	24	5.7
12 × 18	15.3	1.31	4.19	24	6.1
12 × 20	16.0	1.44	4.45	24	6.4
12 × 21	16.7	1.57	4.71	24	6.8
12 × 25	18.0	1.83	5.24	22	9.2
12 × 28	19.1	2.09	5.76	22	10.1
12 × 31	20.1	2.36	6.28	22	11.1
12 × 34	20.9	2.62	6.81	22	12.0
12 × 37	21.9	2.88	7.33	22	12.9
12 × 40	22.7	3.14	7.85	22	13.8
12 × 42	23.0	3.27	8.12	22	14.3
12 × 43	23.5	3.40	8.37	22	14.7
12 × 45	24.0	3.53	8.64	22	15.2

(Continued)

GALVANIZED FLAT OVAL DUCTWORK WEIGHT (*Continued*)

Nominal Flat Oval Size	Equiv. Round	Cross Sectional Area sq.ft.	Surface Area sq.ft./ln.ft.	Gauge	Weight lbs./ln.ft.
12 × 47	24.3	3.67	8.89	22	15.6
12 × 50	25.0	3.93	9.42	20	19.5
12 × 53	25.7	4.19	9.95	20	20.6
12 × 56	26.3	4.45	10.56	20	21.9
12 × 59	26.9	4.71	11.00	20	22.8
12 × 62	27.5	4.98	11.52	20	23.8
12 × 65	28.1	5.23	12.05	20	24.9
12 × 69	28.7	5.51	12.57	20	26.0
12 × 72	29.2	5.76	13.10	18	35.3
12 × 78	30.0	6.28	14.14	18	38.1
12 × 81	31.0	6.54	14.66	18	39.5
14 × 17	16.0	1.37	4.19	24	6.1
14 × 19	17.0	1.53	4.45	24	6.4
14 × 20	17.5	1.68	4.71	24	6.8
14 × 22	18.0	1.83	4.97	24	7.2
14 × 23	18.9	1.98	5.23	24	7.6
14 × 27	20.2	2.30	5.76	22	10.1
14 × 28	21.0	2.44	6.02	22	10.6
14 × 30	21.3	2.60	6.28	22	11.0
14 × 31	22.0	2.75	6.54	22	11.5
14 × 33	22.4	2.91	6.80	22	12.0
14 × 34	23.0	3.05	7.07	22	12.4
14 × 36	23.4	3.21	7.33	22	12.9
14 × 38	24.0	3.36	7.59	22	13.3
14 × 39	24.4	3.51	7.85	22	13.8
14 × 41	25.0	3.67	8.12	22	14.3
14 × 42	25.3	3.84	8.37	22	14.7
14 × 45	26.1	4.12	8.89	22	15.6
14 × 49	26.9	4.43	9.42	20	19.5
14 × 52	27.7	4.74	9.95	20	20.6
14 × 55	28.4	5.04	10.56	20	21.9
14 × 58	29.1	5.35	11.00	20	22.8
14 × 61	29.8	5.65	11.52	20	23.9
14 × 64	30.5	5.96	12.05	20	24.9
14 × 67	31.1	6.27	12.57	20	26.0
14 × 71	31.7	6.57	13.10	18	35.9
14 × 77	33.0	7.18	14.14	18	38.1
16 × 19	18.0	1.75	4.71	24	6.8
16 × 21	19.0	1.92	4.97	24	7.2
16 × 22	19.5	2.08	5.23	24	7.6
16 × 24	20.0	2.27	5.50	24	7.9
16 × 25	20.9	2.44	5.76	22	10.2
16 × 29	22.3	2.79	6.28	22	11.0
16 × 30	23.0	2.97	6.54	22	11.5
16 × 32	23.5	3.13	6.80	22	12.0
16 × 33	24.0	3.32	7.07	22	12.4
16 × 35	24.7	3.48	7.33	22	12.9
16 × 36	25.0	3.67	7.59	22	13.3
16 × 38	25.7	3.84	7.85	22	13.8
16 × 41	26.8	4.19	8.38	22	14.7
16 × 44	27.7	4.53	8.89	22	15.6
16 × 46	28.0	4.71	9.16	22	16.1
16 × 47	28.6	4.88	9.42	22	16.6
16 × 49	29.0	5.06	9.69	20	20.1
16 × 51	29.4	5.23	9.95	20	20.6
16 × 54	30.2	5.59	10.47	20	21.7
16 × 57	31.0	5.93	11.00	20	22.8
16 × 60	31.8	6.28	11.52	20	23.8

(Continued)

GALVANIZED FLAT OVAL DUCTWORK WEIGHT (*Continued*)

Nominal Flat Oval Size	Equiv. Round	Cross Sectional Area sq.ft.	Surface Area sq.ft./ln.ft.	Gauge	Weight lbs./ln.ft.
16 × 63	32.5	6.61	12.05	20	24.9
16 × 66	33.3	6.98	12.57	20	26.0
16 × 69	34.0	7.33	13.09	20	27.1
16 × 76	35.0	8.03	14.14	18	38.1
16 × 79	36.0	8.38	14.66	18	39.5
18 × 21	19.9	2.16	5.23	24	7.6
18 × 23	21.0	2.36	5.50	24	7.9
18 × 24	21.6	2.56	5.76	24	8.3
18 × 26	22.0	2.75	6.02	22	10.6
18 × 27	23.1	2.95	6.28	22	11.0
18 × 29	24.0	3.14	6.54	22	11.5
18 × 31	24.5	3.35	6.80	22	12.0
18 × 32	25.0	3.53	7.07	22	12.4
18 × 34	25.7	3.73	7.33	22	12.9
18 × 37	27.0	4.13	7.85	22	13.8
18 × 40	28.1	4.53	8.37	22	14.7
18 × 43	29.1	4.92	8.89	22	15.6
18 × 46	30.2	5.31	9.42	22	16.6
18 × 49	31.1	5.70	9.95	20	20.6
18 × 53	32.0	6.10	10.56	20	21.9
18 × 56	32.9	6.49	11.00	20	22.8
18 × 59	33.7	6.88	11.52	20	23.8
18 × 62	34.5	7.26	12.05	20	24.9
18 × 65	35.3	7.67	12.51	20	25.9
18 × 68	36.0	8.07	13.10	20	27.1
18 × 71	37.0	8.44	13.61	18	36.7
18 × 78	38.0	9.23	14.66	18	39.5
20 × 26	23.6	3.05	6.28	22	11.0
20 × 29	25.2	3.49	6.81	22	12.0
20 × 31	26.0	3.71	7.07	22	12.4
20 × 33	26.6	3.93	7.33	22	12.9
20 × 34	27.0	4.15	7.59	22	13.3
20 × 36	28.0	4.36	7.85	22	13.8
20 × 39	29.2	4.81	8.37	22	14.7
20 × 40	30.0	5.02	8.64	22	15.2
20 × 42	30.3	5.23	8.89	22	15.6
20 × 44	31.0	5.45	9.16	22	16.1
20 × 45	31.4	5.67	9.42	22	16.6
20 × 47	32.0	5.89	9.69	22	17.0
20 × 48	32.5	6.11	9.95	22	17.5
20 × 51	33.4	6.55	10.56	20	21.9
20 × 55	34.4	6.98	11.00	20	22.8
20 × 58	35.3	7.41	11.52	20	23.8
20 × 61	36.2	7.86	12.05	20	24.9
20 × 64	37.1	8.29	12.57	20	26.0
20 × 67	37.9	8.71	13.10	20	27.1
20 × 77	40.0	10.04	14.66	18	39.5
22 × 25	23.9	3.12	6.28	22	11.0
22 × 28	25.6	3.60	6.81	22	12.0
22 × 31	27.2	4.08	7.33	22	12.9
22 × 35	28.7	4.56	7.85	22	13.8
22 × 38	30.0	5.04	8.38	22	14.7
22 × 39	31.0	5.28	8.64	22	15.2
22 × 41	31.3	5.52	8.90	22	15.6
22 × 42	32.0	5.76	9.16	22	16.1
22 × 44	32.5	6.00	9.42	22	16.6

(Continued)

GALVANIZED FLAT OVAL DUCTWORK WEIGHT (*Continued*)

Nominal Flat Oval Size	Equiv. Round	Cross Sectional Area sq.ft.	Surface Area sq.ft./ln.ft.	Gauge	Weight lbs./ln.ft.
22 × 46	33.0	6.24	9.69	22	17.0
22 × 47	33.7	6.48	9.95	22	17.5
22 × 50	34.8	6.96	10.47	20	21.7
22 × 53	35.8	7.44	11.00	20	22.8
22 × 57	36.7	7.92	11.52	20	23.8
22 × 60	37.8	8.40	12.04	20	24.9
22 × 63	38.7	8.88	12.57	20	26.0
22 × 66	39.6	9.36	13.09	20	27.1
22 × 69	40.4	9.84	13.61	20	28.2
22 × 75	42.0	10.80	14.66	18	39.5
22 × 82	44.0	11.76	15.71	18	42.3
24 × 27	25.9	3.66	6.81	22	12.0
24 × 30	28.1	4.19	7.33	22	12.9
24 × 33	29.3	4.71	7.85	22	13.8
24 × 37	30.8	5.23	8.38	22	14.7
24 × 40	32.2	5.76	8.90	22	15.6
24 × 41	33.0	6.02	9.16	22	16.1
24 × 43	33.5	6.28	9.42	22	16.6
24 × 44	34.0	6.54	9.69	22	17.1
24 × 46	34.7	6.80	9.95	22	17.5
24 × 49	35.9	7.33	10.47	20	21.7
24 × 52	37.0	7.85	11.00	20	22.8
24 × 55	38.1	8.38	11.52	20	23.8
24 × 59	39.2	8.90	12.04	20	24.9
24 × 62	40.1	9.42	12.57	20	26.0
24 × 65	41.1	9.95	13.09	20	27.1
24 × 68	42.0	10.47	13.61	20	28.2
24 × 74	44.0	11.52	14.66	18	39.5
26 × 29	27.9	4.25	7.33	22	12.9
26 × 32	29.7	4.82	7.85	22	13.8
26 × 35	31.3	5.39	8.38	22	14.7
26 × 39	32.8	5.96	8.90	22	15.6
26 × 42	34.3	6.52	9.42	22	16.6
26 × 45	35.6	7.09	9.95	22	17.5
26 × 48	36.9	7.66	10.47	22	18.4
26 × 51	38.1	8.22	11.00	20	22.8
26 × 54	39.3	8.79	11.52	20	23.8
26 × 57	40.4	9.36	12.04	20	24.9
26 × 61	41.5	9.93	12.57	20	26.0
26 × 64	42.5	10.49	13.09	20	27.1
26 × 67	43.5	11.06	13.61	20	28.2
26 × 70	44.4	11.63	14.14	20	29.3
28 × 31	29.9	4.88	7.85	22	13.8
28 × 34	31.7	5.50	8.38	22	14.7
28 × 37	33.4	6.11	8.90	22	15.6
28 × 41	34.9	6.72	9.42	22	16.6
28 × 44	36.4	7.33	9.95	22	17.5
28 × 47	37.8	7.94	10.47	22	18.4
28 × 50	39.1	8.55	11.00	20	22.8
28 × 53	40.3	9.16	11.52	20	23.8
28 × 56	41.5	9.77	12.04	20	24.9
28 × 59	42.6	10.38	12.57	20	26.0
28 × 63	43.8	10.99	13.09	20	27.1
28 × 66	44.8	11.60	13.61	20	28.2
28 × 69	45.8	12.22	14.14	20	29.3
30 × 33	32.0	5.56	8.38	22	14.7
30 × 36	33.7	6.22	8.90	22	15.6
30 × 39	35.4	6.87	9.42	22	16.6

(*Continued*)

GALVANIZED FLAT OVAL DUCTWORK WEIGHT (*Continued*)

Nominal Flat Oval Size	Equiv. Round	Cross Sectional Area sq.ft.	Surface Area sq.ft./ln.ft.	Gauge	Weight lbs./ln.ft.
30 × 43	37.0	7.53	9.95	22	17.5
30 × 46	38.5	8.18	10.47	22	18.4
30 × 49	39.9	8.84	11.00	20	22.8
30 × 52	41.2	9.49	11.52	20	23.8
30 × 55	42.5	10.15	12.06	20	25.0
30 × 58	43.7	10.80	12.57	20	26.0
30 × 61	44.9	11.46	13.09	20	27.1
30 × 64	46.0	12.11	13.61	20	28.2
30 × 68	47.1	12.77	14.14	20	29.3
30 × 71	48.2	13.42	14.66	18	39.5
32 × 35	34.0	6.28	8.90	22	15.6
32 × 38	35.8	6.98	9.42	22	16.6
32 × 41	37.4	7.68	9.95	22	17.5
32 × 45	39.0	8.38	10.47	22	18.4
32 × 48	40.5	9.08	11.00	22	15.3
32 × 51	42.0	9.77	11.52	20	23.8
32 × 54	43.3	10.47	12.04	20	24.9
32 × 57	44.6	11.17	12.57	20	26.0
32 × 60	45.9	11.87	13.09	20	27.1
32 × 63	47.1	12.57	13.61	20	28.2
32 × 67	48.3	13.26	14.14	20	29.3
32 × 70	49.4	13.96	14.66	20	30.3
34 × 37	36.0	7.05	9.42	22	16.6
34 × 40	37.8	7.79	9.95	22	17.5
34 × 43	39.5	8.52	10.47	22	18.4
34 × 47	41.1	9.27	11.00	22	19.3
34 × 50	42.6	10.01	11.52	20	23.8
34 × 53	44.1	10.75	12.04	20	24.9
34 × 56	45.5	11.50	12.57	20	26.0
34 × 59	46.8	12.24	13.09	20	27.1
34 × 62	48.1	12.98	13.61	20	28.2
34 × 65	49.3	13.72	14.14	20	29.3
34 × 69	50.5	14.46	14.66	20	30.3
34 × 72	51.6	15.20	15.18	18	31.4
36 × 39	38.0	7.85	9.95	22	17.5
36 × 42	39.8	8.64	10.47	22	18.4
36 × 45	41.5	9.42	11.00	22	19.4
36 × 49	43.1	10.21	11.52	20	23.8
36 × 52	44.7	11.00	12.04	20	24.9
36 × 55	46.2	11.78	12.57	20	26.0
36 × 58	47.6	12.57	13.09	20	27.1
36 × 61	48.9	13.35	13.61	20	28.2
36 × 64	50.2	14.14	14.14	20	29.3
36 × 67	51.1	14.92	14.66	20	30.3
36 × 71	52.7	15.71	15.18	18	40.9
38 × 41	40.0	8.70	10.47	22	18.4
38 × 44	41.8	9.53	11.00	22	19.3
38 × 47	43.5	10.36	11.52	22	20.3
38 × 51	45.2	11.19	12.04	20	24.9
38 × 54	46.7	12.02	12.57	20	26.0
38 × 57	48.2	12.85	13.09	20	27.1
38 × 60	49.7	13.68	13.61	20	28.2
38 × 63	51.0	14.51	14.14	20	29.3
38 × 66	52.4	15.34	14.66	20	30.3
38 × 69	53.7	16.16	15.18	20	31.4
40 × 43	42.0	9.60	11.00	22	19.3
40 × 46	43.8	10.47	11.52	22	20.3
40 × 49	45.6	11.34	12.04	20	24.9

(Continued)

GALVANIZED FLAT OVAL DUCTWORK WEIGHT (*Continued*)

Nominal Flat Oval Size	Equiv. Round	Cross Sectional Area sq.ft.	Surface Area sq.ft./ln.ft.	Gauge	Weight lbs./ln.ft.
40 × 53	47.2	12.21	12.57	20	26.0
40 × 56	48.8	13.09	13.09	20	27.1
40 × 59	50.4	13.96	13.61	20	28.2
40 × 62	51.8	14.83	14.14	20	29.3
40 × 65	53.2	15.71	14.66	20	30.3
40 × 68	54.5	16.58	15.18	20	31.4
40 × 71	55.8	17.45	15.71	18	42.3

Notes:
1 Equivalent round is the diameter of the round duct which will have the capacity and friction equivalent to the flat oval duct size.
2 To obtain the rectangular duct size, use the Trane Ductulator and equivalent round duct size.
3 Table includes 25 percent allowance for bracing, hangers, reinforcing, joints, and seams. Add 10 percent for insulated ductwork systems.
4 Table lists standard sizes as manufactured by United Sheet Metal, a division of United McGill Corporation.

17.15 Ductwork Cost Ratios

DUCTWORK COST RATIOS

SMACNA Pressure Class	Installed Cost Ratio
± ½"	1.00
± 1"	1.05
± 2"	1.15
± 3"	1.40
± 4"	1.50
± 6"	1.60
± 10"	1.80

Aspect Ratios	Installed Cost Ratio	Operating Cost Ratio
1:1	1.00	1.000
2:1	1.13	1.001
3:1	1.28	1.005
4:1	1.45	1.010
5:1	1.65	1.012
6:1	1.85	1.020
7:1	2.08	1.030

17.16 Friction Loss Correction Factors for Ducts

FRICTION LOSS CORRECTION FACTORS FOR DUCTS

Velocity FPM	Material								
	Galv. Steel Stainless Steel	Duct Liner	Aluminum	Carbon Steel	Fiberous Glass (2)	PVC	Concrete or Conc. Block	Drywall
500	1.00	1.25	0.98	0.93	1.25	0.93	1.5–1.9	1.25
600	1.00	1.28	0.98	0.92	1.27	0.92	1.5–1.9	1.27
700	1.00	1.30	0.98	0.92	1.30	0.92	1.5–2.0	1.30
800	1.00	1.31	0.97	0.91	1.31	0.91	1.5–2.0	1.31
900	1.00	1.32	0.97	0.90	1.31	0.90	1.5–2.0	1.31

(*Continued*)

FRICTION LOSS CORRECTION FACTORS FOR DUCTS (*Continued*)

Velocity FPM	Material								
	Galv. Steel Stainless Steel	Duct Liner	Aluminum	Carbon Steel	Fiberous Glass (2)	PVC	Concrete or Conc. Block	Drywall
1,000	1.00	1.33	0.97	0.90	1.32	0.90	1.6–2.1	1.32
1,200	1.00	1.36	0.97	0.89	1.34	0.89	1.6–2.1	1.34
1,400	1.00	1.38	0.96	0.88	1.36	0.88	1.6–2.1	1.36
1,600	1.00	1.40	0.96	0.87	1.38	0.87	1.6–2.2	1.38
1,800	1.00	1.41	0.96	0.86	1.39	0.86	1.6–2.3	1.39
2,000	1.00	1.42	0.96	0.85	1.40	0.85	1.7–2.3	1.40
2,500	1.00	1.45	0.95	0.84	1.42	0.84	1.7–2.3	1.42
3,000	1.00	1.47	0.95	0.83	1.43	0.83	1.7–2.3	1.43
3,500	1.00	1.49	0.95	0.83	1.44	0.83	1.8–2.4	1.44
4,000	1.00	1.50	0.94	0.82	1.45	0.82	1.8–2.4	1.45
4,500	1.00	1.52	0.94	0.81	1.46	0.81	1.8–2.4	1.46
5,000	1.00	1.54	0.94	0.80	1.48	0.80	1.8–2.4	1.48
5,500	1.00	1.55	0.93	0.79	1.49	0.79	1.8–2.4	1.49
6,000	1.00	1.56	0.93	0.78	1.50	0.78	1.8–2.4	1.50

Notes:
  1 First number indicated is for smooth concrete; second number indicated is for rough concrete.
  2 Flexible ductwork has a friction loss correction factor of 1.5–2.0 times the value read from friction loss tables, ductulators, etc.

17.17 Velocity Pressures

VELOCITIES VS. VELOCITY PRESSURES

Velocity FPM	Velocity Pressure in. W.G.	Velocity FPM	Velocity Pressure in. W.G.	Velocity FPM	Velocity Pressure in. W.G.
50	0.0002	2,050	0.262	4,050	1.023
100	0.0006	2,100	0.275	4,100	1.048
150	0.001	2,150	0.288	4,150	1.074
200	0.002	2,200	0.302	4,200	1.100
250	0.004	2,250	0.316	4,250	1.126
300	0.006	2,300	0.330	4,300	1.153
350	0.008	2,350	0.344	4,350	1.180
400	0.010	2,400	0.359	4,400	1.207
450	0.013	2,450	0.374	4,450	1.235
500	0.016	2,500	0.390	4,500	1.262
550	0.019	2,550	0.405	4,550	1.291
600	0.022	2,600	0.421	4,600	1.319
650	0.026	2,650	0.438	4,650	1.348
700	0.031	2,700	0.454	4,700	1.377
750	0.035	2,750	0.471	4,750	1.407
800	0.040	2,800	0.489	4,800	1.436
850	0.045	2,850	0.506	4,850	1.466
900	0.050	2,900	0.524	4,900	1.497
950	0.056	2,950	0.543	4,950	1.528
1,000	0.062	3,000	0.561	5,000	1.559
1,050	0.069	3,050	0.580	5,050	1.590
1,100	0.075	3,100	0.599	5,100	1.622
1,150	0.082	3,150	0.619	5,150	1.654
1,200	0.090	3,200	0.638	5,200	1.686
1,250	0.097	3,250	0.659	5,250	1.718

(Continued)

VELOCITIES VS. VELOCITY PRESSURES (*Continued*)

Velocity FPM	Velocity Pressure in. W.G.	Velocity FPM	Velocity Pressure in. W.G.	Velocity FPM	Velocity Pressure in. W.G.
1,300	0.105	3,300	0.679	5,300	1.751
1,350	0.114	3,350	0.700	5,350	1.784
1,400	0.122	3,400	0.721	5,400	1.818
1,450	0.131	3,450	0.742	5,450	1.852
1,500	0.140	3,500	0.764	5,500	1.886
1,550	0.150	3,550	0.786	5,550	1.920
1,600	0.160	3,600	0.808	5,600	1.955
1,650	0.170	3,650	0.831	5,650	1.990
1,700	0.180	3,700	0.853	5,700	2.026
1,750	0.191	3,750	0.877	5,750	2.061
1,800	0.202	3,800	0.900	5,800	2.097
1,850	0.213	3,850	0.924	5,850	2.134
1,900	0.225	3,900	0.948	5,900	2.170
1,950	0.237	3,950	0.973	5,950	2.207
2,000	0.249	4,000	0.998	6,000	2.244
6,050	2.282	8,050	4.040	10,050	6.297
6,100	2.320	8,100	4.090	10,100	6.360
6,150	2.358	8,150	4.141	10,150	6.423
6,200	2.397	8,200	4.192	10,200	6.486
6,250	2.435	8,250	4.243	10,250	6.550
6,300	2.474	8,300	4.295	10,300	6.614
6,350	2.514	8,350	4.347	10,350	6.678
6,400	2.554	8,400	4.399	10,400	6.743
6,450	2.594	8,450	4.452	10,450	6.808
6,500	2.634	8,500	4.504	10,500	6.873
6,550	2.675	8,550	4.558	10,550	6.939
6,600	2.716	8,600	4.611	10,600	7.005
6,650	2.757	8,650	4.665	10,650	7.071
6,700	2.799	8,700	4.719	10,700	7.138
6,750	2.841	8,750	4.773	10,750	7.205
6,800	2.883	8,800	4.828	10,800	7.272
6,850	2.925	8,850	4.883	10,850	7.339
6,900	2.968	8,900	4.938	10,900	7.407
6,950	3.011	8,950	4.994	10,950	7.475
7,000	3.055	9,000	5.050	11,000	7.544
7,050	3.099	9,050	5.106	11,050	7.612
7,100	3.143	9,100	5.163	11,100	7.681
7,150	3.187	9,150	5.220	11,150	7.751
7,200	3.232	9,200	5.277	11,200	7.820
7,250	3.277	9,250	5.334	11,250	7.890
7,300	3.322	9,300	5.392	11,300	7.961
7,350	3.368	9,350	5.450	11,350	8.031
7,400	3.414	9,400	5.509	11,400	8.102
7,450	3.460	9,450	5.567	11,450	8.173
7,500	3.507	9,500	5.627	11,500	8.245
7,550	3.554	9,550	5.686	11,550	8.317
7,600	3.601	9,600	5.746	11,600	8.389
7,650	3.649	9,650	5.807	11,650	8.461
7,700	3.696	9,700	5.866	11,700	8.534
7,750	3.745	9,750	5.927	11,750	8.607
7,800	3.793	9,800	5.988	11,800	8.681
7,850	3.842	9,850	6.049	11,850	8.755
7,900	3.891	9,900	6.110	11,900	8.829
7,950	3.940	9,950	6.172	11,950	8.903
8,000	3.990	10,000	6.234	12,000	8.978

Note:

1 Velocity Pressure $= VP = \left[\dfrac{V}{4005} \right]^2 = \dfrac{(V)^2}{(4005)^2}$

Refer to the online resource for Section 17.18 Equivalent Round/Rectangular Ducts.
www.mheducation.com/HVACequations

Piping Systems, General

18.01 Piping Materials and Properties

A. Steel pipe and Type L copper pipe are the most common pipe materials used in HVAC applications.

B. Steel Pipe

1. Standard steel pipe sizes: 1/2", 3/4", 1", 1-1/4", 1-1/2", 2", 2-1/2", 3", 4", 6", 8", 10", 12", 14", 16", 18", 20", 24", 30", 36", 42", 48", 54", 60", 72", 84", and 96".
2. Nonstandard steel pipe sizes: 5", 22", 26", 28", 32", and 34" are not standard sizes and not readily available in all locations.
3. Standard steel pipe is the most common steel pipe used in HVAC applications.
4. Standard and XS steel pipe are available in sizes through 96 inch.
5. XXS steel pipe is available in sizes through 12 inch.
6. Schedule 40 steel pipe is available in sizes through 96 inch.
7. Schedule 80 and 160 steel pipe are available in sizes through 24 inch.
8. Standard and Schedule 40 steel pipe have the same dimensions and flow for 10 inch and smaller.
9. XS and Schedule 80 steel pipe have the same dimensions and flow for 8 inch and smaller.
10. XXS and Schedule 160 have no relationship for dimensions or flow.
11. Steel pipe is manufactured in accordance with ASTM Standards A53 and A106.
12. The ASTM standards refer to steel pipe grades A and B. Grade A steel pipe has a lower tensile strength and is not generally used for HVAC applications.
13. The ASTM standards refer to steel pipe Types E, S, and F.
    a. Type E (also referred to as ERW) steel pipe refers to electric resistance welded steel pipe.
    b. Type S steel pipe refers to seamless steel pipe.
    c. Type F steel pipe refers to furnace-butt welded steel pipe. This type is generally not used in HVAC applications and is only available in Grade A.

C. Copper Pipe

1. Standard copper pipe sizes: 1/2", 3/4", 1", 1-1/4", 1-1/2", 2", 2-1/2", 3", 4", 6", 8", 10", and 12".
2. Copper pipe is available in Types K, L, and M.
3. Types K, L, and M copper may be hard drawn or annealed (soft) temper.
4. Hard drawn copper pipe has higher allowable stress than annealed copper pipe.
5. Types K, L, and M designate decreasing wall thicknesses (Type K copper pipe has the thickest wall, while type M copper pipe has the thinnest wall).
6. Type K is generally used for higher pressure/temperature applications and for direct burial.
7. Type L copper pipe is the most common copper pipe used in HVAC applications.
8. Type M copper pipe should not be used where subject to external damage.
9. Copper pipe is manufactured in accordance with ASTM Standard B88.

D. Stainless Steel Pipe

1. Standard stainless steel pipe sizes: 1/2", 3/4", 1", 1-1/4", 1-1/2", 2", 2-1/2", 3", 4", 6", 8", 10", 12", 14", 16", 18", 20", and 24".
2. Schedule 5 and 10 stainless steel pipe are available in sizes through 24 inch.

E. In the following piping tables . . .

1. Uninsulated piping: Add 20 percent for hangers and supports.
2. Insulated piping: Add 25 percent for hangers, supports, and insulation.

F. Piping installations are generally governed by one of the following three codes . . .

1. ASME B31.1-1998: Power Piping:
   a. Applicable to electric generating stations, industrial and institutional plants, central and district heating/cooling plants, and geothermal heating.

2. ASME B31.3-1999: Process Piping:
   a. Applicable to petroleum refineries, chemical, pharmaceutical, textile, paper, semi-conductor, and cryogenic plants.
3. ASME B31.9-1996: Building Services Piping:
   a. Applicable to industrial, institutional, commercial, and public buildings and multiunit residences.
   b. Most HVAC applications fall under ASME B31.9 requirements.

PROPERTIES OF COPPER PIPE

Pipe Size in.	Type	Inside Dia. in.	Wall Thick in.	Outside Dia. in.	Area, sq.in.	Weight (1) Pipe lbs./ft.	Water lbs./ft.	Total lbs./ft.	Water Volume gal./ft.
1/2	K	0.527	0.049	0.625	0.218	0.301	0.095	0.396	0.011
	L	0.545	0.040	0.625	0.233	0.250	0.101	0.351	0.012
	M	0.569	0.028	0.625	0.254	0.179	0.110	0.289	0.013
3/4	K	0.745	0.065	0.875	0.436	0.562	0.189	0.751	0.023
	L	0.785	0.045	0.875	0.484	0.399	0.210	0.609	0.025
	M	0.811	0.032	0.875	0.517	0.288	0.224	0.512	0.027
1	K	0.995	0.065	1.125	0.778	0.736	0.337	1.073	0.040
	L	1.025	0.050	1.125	0.825	0.574	0.357	0.932	0.043
	M	1.055	0.035	1.125	0.874	0.407	0.379	0.786	0.045
1-1/4	K	1.245	0.065	1.375	1.217	0.909	0.527	1.437	0.063
	L	1.265	0.055	1.375	1.257	0.775	0.545	1.320	0.065
	M	1.291	0.042	1.375	1.309	0.598	0.567	1.165	0.068
1-1/2	K	1.481	0.072	1.625	1.723	1.194	0.746	1.941	0.089
	L	1.505	0.060	1.625	1.779	1.003	0.771	1.774	0.092
	M	1.527	0.049	1.625	1.831	0.825	0.793	1.618	0.095
2	K	1.959	0.083	2.125	3.014	1.810	1.306	3.116	0.157
	L	1.985	0.070	2.125	3.095	1.536	1.341	2.877	0.161
	M	2.009	0.058	2.125	3.170	1.280	1.373	2.654	0.165
2-1/2	K	2.435	0.095	2.625	4.657	2.567	2.018	4.585	0.242
	L	2.465	0.080	2.625	4.772	2.174	2.068	4.242	0.248
	M	2.495	0.065	2.625	4.889	1.777	2.118	3.895	0.254
3	K	2.907	0.109	3.125	6.637	3.511	2.876	6.387	0.345
	L	2.945	0.090	3.125	6.812	2.917	2.951	5.868	0.354
	M	2.981	0.072	3.125	6.979	2.348	3.024	5.371	0.363
4	K	3.857	0.134	4.125	11.684	5.712	5.062	10.774	0.607
	L	3.905	0.110	4.125	11.977	4.717	5.189	9.906	0.622
	M	3.935	0.095	4.125	12.161	4.089	5.269	9.358	0.632
5	K	4.805	0.160	5.125	18.133	8.484	7.856	16.341	0.942
	L	4.875	0.125	5.125	18.665	6.675	8.087	14.762	0.970
	M	4.907	0.109	5.125	18.911	5.839	8.193	14.033	0.982
6	K	5.741	0.192	6.125	25.886	12.166	11.215	23.381	1.345
	L	5.845	0.140	6.125	26.832	8.949	11.625	20.574	1.394
	M	5.881	0.122	6.125	27.164	7.822	11.769	19.590	1.411
8	K	7.583	0.271	8.125	45.162	22.732	19.566	42.298	2.346
	L	7.725	0.200	8.125	46.869	16.928	20.306	37.234	2.435
	M	7.785	0.170	8.125	47.600	14.443	20.623	35.066	2.473
10	K	9.449	0.338	10.125	70.123	35.330	30.381	65.711	3.643
	L	9.625	0.250	10.125	72.760	26.367	31.523	57.890	3.780
	M	9.701	0.212	10.125	73.913	22.445	32.023	54.468	3.840
12	K	11.315	0.405	12.125	100.554	50.695	43.565	94.259	5.224
	L	11.565	0.280	12.125	105.046	35.422	45.511	80.933	5.457
	M	11.617	0.254	12.125	105.993	32.203	45.921	78.124	5.506

PROPERTIES OF STEEL PIPE

Pipe Size in.	Schedule		Inside Dia. in.	Wall Thick in.	Outside Dia. in.	Area sq.in.	Weight (1) Pipe lbs./ft.	Water lbs./ft.	Total lbs./ft.	Water Volume gal./ft.
	10	–	0.674	0.083	0.840	0.357	0.671	0.155	0.826	0.019
	40	STD	0.622	0.109	0.840	0.304	0.851	0.132	0.983	0.016
1/2	80	XS	0.546	0.147	0.840	0.234	1.088	0.101	1.189	0.012
	160	–	0.466	0.187	0.840	0.171	1.304	0.074	1.378	0.009
	–	XXS	0.252	0.294	0.840	0.050	1.714	0.022	1.736	0.003
	10	–	0.884	0.083	1.050	0.614	0.857	0.266	1.123	0.032
	40	STD	0.824	0.113	1.050	0.533	1.131	0.231	1.362	0.028
3/4	80	XS	0.742	0.154	1.050	0.432	1.474	0.187	1.661	0.022
	160	–	0.614	0.218	1.050	0.296	1.937	0.128	2.065	0.015
	–	XXS	0.434	0.308	1.050	0.148	2.441	0.064	2.505	0.008
	10	–	1.097	0.109	1.315	0.945	1.404	0.409	1.813	0.049
	40	STD	1.049	0.133	1.315	0.864	1.679	0.374	2.053	0.045
1	80	XS	0.957	0.179	1.315	0.719	2.172	0.312	2.483	0.037
	160	–	0.815	0.250	1.315	0.522	2.844	0.226	3.070	0.027
	–	XXS	0.599	0.358	1.315	0.282	3.659	0.122	3.781	0.015
	10	–	1.442	0.109	1.660	1.633	1.806	0.708	2.513	0.085
	40	STD	1.380	0.140	1.660	1.496	2.273	0.648	2.921	0.078
1-1/4	80	XS	1.278	0.191	1.660	1.283	2.997	0.556	3.552	0.067
	160	–	1.160	0.250	1.660	1.057	3.765	0.458	4.223	0.055
	–	XXS	0.896	0.382	1.660	0.631	5.214	0.273	5.487	0.033
	10	–	1.682	0.109	1.900	2.222	2.085	0.963	3.048	0.115
	40	STD	1.610	0.145	1.900	2.036	2.718	0.882	3.600	0.106
1-1/2	80	XS	1.500	0.200	1.900	1.767	3.631	0.766	4.397	0.092
	160	–	1.338	0.281	1.900	1.406	4.859	0.609	5.468	0.073
	–	XXS	1.100	0.400	1.900	0.950	6.408	0.412	6.820	0.049
	10	–	2.157	0.109	2.375	3.654	2.638	1.583	4.221	0.190
	40	STD	2.067	0.154	2.375	3.356	3.653	1.454	5.107	0.174
2	80	XS	1.939	0.218	2.375	2.953	5.022	1.279	6.301	0.153
	160	–	1.689	0.343	2.375	2.241	7.444	0.971	8.415	0.116
	–	XXS	1.503	0.436	2.375	1.774	9.029	0.769	9.798	0.092
	10	–	2.635	0.120	2.875	5.453	3.531	2.363	5.893	0.283
	40	STD	2.469	0.203	2.875	4.788	5.793	2.074	7.867	0.249
2-1/2	80	XS	2.323	0.276	2.875	4.238	7.661	1.836	9.497	0.220
	160	–	2.125	0.375	2.875	3.547	10.013	1.537	11.549	0.184
	–	XXS	1.771	0.552	2.875	2.463	13.695	1.067	14.762	0.128
	10	–	3.260	0.120	3.500	8.347	4.332	3.616	7.948	0.434
	40	STD	3.068	0.216	3.500	7.393	7.576	3.203	10.779	0.384
3	80	XS	2.900	0.300	3.500	6.605	10.253	2.862	13.115	0.343
	160	–	2.626	0.437	3.500	5.416	14.296	2.346	16.642	0.281
	–	XXS	2.300	0.600	3.500	4.155	18.584	1.800	20.384	0.216
	10	–	4.260	0.120	4.500	14.253	5.614	6.175	11.789	0.740
	40	STD	4.026	0.237	4.500	12.730	10.791	5.515	16.306	0.661
4	80	XS	3.826	0.337	4.500	11.497	14.984	4.981	19.965	0.597
	160	–	3.438	0.531	4.500	9.283	22.509	4.022	26.531	0.482
	–	XXS	3.152	0.674	4.500	7.803	27.541	3.381	30.922	0.405
	10	–	5.295	0.134	5.563	22.020	7.770	9.540	17.310	1.144
	40	STD	5.047	0.258	5.563	20.006	14.618	8.667	23.285	1.039
5	80	XS	4.813	0.375	5.563	18.194	20.778	7.882	28.661	0.945
	160	–	4.313	0.625	5.563	14.610	32.962	6.330	39.291	0.759
	–	XXS	4.063	0.750	5.563	12.965	38.553	5.617	44.170	0.674
	10	–	6.357	0.134	6.625	31.739	9.290	13.751	23.040	1.649
	40	STD	6.065	0.280	6.625	28.890	18.974	12.517	31.491	1.501
6	80	XS	5.761	0.432	6.625	26.067	28.574	11.293	39.867	1.354
	160	–	5.189	0.718	6.625	21.147	45.297	9.162	54.459	1.099
	–	XXS	4.897	0.864	6.625	18.834	53.161	8.160	61.321	0.978
	10	–	8.329	0.148	8.625	54.485	13.399	23.605	37.005	2.830
	20	–	8.125	0.250	8.625	51.849	22.362	22.463	44.825	2.693
	30	–	8.071	0.277	8.625	51.162	24.697	22.166	46.862	2.658
8	40	STD	7.981	0.322	8.625	50.027	28.554	21.674	50.228	2.599
	80	XS	7.625	0.500	8.625	45.664	43.388	19.784	63.172	2.372
	–	XXS	6.875	0.875	8.625	37.122	72.425	16.083	88.508	1.928
	160	–	6.813	0.906	8.625	36.456	74.691	15.794	90.485	1.894

(Continued)

PROPERTIES OF STEEL PIPE (*Continued*)

Pipe Size in.	Schedule		Inside Dia. in.	Wall Thick in.	Outside Dia. in.	Area sq.in.	Weight (1) Pipe lbs./ft.	Weight (1) Water lbs./ft.	Weight (1) Total lbs./ft.	Water Volume gal./ft.
10	10	–	10.420	0.165	10.750	85.276	18.653	36.945	55.599	4.430
	20	–	10.250	0.250	10.750	82.516	28.036	35.750	63.785	4.287
	30	–	10.136	0.307	10.750	80.691	34.241	34.959	69.200	4.192
	40	STD	10.020	0.365	10.750	78.854	40.484	34.163	74.647	4.096
	60	XS	9.750	0.500	10.750	74.662	54.736	32.347	87.083	3.879
	80	–	9.564	0.593	10.750	71.840	64.328	31.125	95.453	3.732
	140	XXS	8.750	1.000	10.750	60.132	104.132	26.052	130.184	3.124
	160	–	8.500	1.125	10.750	56.745	115.647	24.585	140.231	2.948
12	10	–	12.390	0.180	12.750	120.568	24.165	52.236	76.401	6.263
	20	–	12.250	0.250	12.750	117.859	33.376	51.062	84.438	6.123
	30	–	12.090	0.330	12.750	114.800	43.774	49.737	93.511	5.964
	–	STD	12.000	0.375	12.750	113.097	49.563	48.999	98.562	5.875
	40	–	11.938	0.406	12.750	111.932	53.526	48.494	102.020	5.815
	–	XS	11.750	0.500	12.750	108.434	65.416	46.979	112.395	5.633
	80	–	11.376	0.687	12.750	101.641	88.510	44.036	132.545	5.280
	120	XXS	10.750	1.000	12.750	90.763	125.492	39.323	164.815	4.715
	160	–	10.126	1.312	12.750	80.531	160.274	34.890	195.164	4.183
14	10	–	13.500	0.250	14.000	143.139	36.713	62.014	98.728	7.436
	20	–	13.376	0.312	14.000	140.521	45.611	60.880	106.492	7.300
	30	STD	13.250	0.375	14.000	137.886	54.569	59.739	114.308	7.163
	40	–	13.126	0.437	14.000	135.318	63.302	58.626	121.928	7.029
	–	XS	13.000	0.500	14.000	132.732	72.091	57.506	129.597	6.895
	80	–	12.500	0.750	14.000	122.718	106.134	53.167	159.302	6.375
	160	–	11.188	1.406	14.000	98.309	189.116	42.592	231.708	5.107
16	10	–	15.500	0.250	16.000	188.692	42.053	81.750	123.803	9.802
	20	–	15.376	0.312	16.000	185.685	52.276	80.447	132.723	9.646
	30	STD	15.250	0.375	16.000	182.654	62.579	79.134	141.714	9.489
	40	XS	15.000	0.500	16.000	176.715	82.772	76.561	159.333	9.180
	80	–	14.314	0.843	16.000	160.921	136.465	69.718	206.183	8.360
	160	–	12.814	1.593	16.000	128.961	245.114	55.872	300.986	6.699
18	10	–	17.500	0.250	18.000	240.528	47.393	104.208	151.601	12.495
	20	–	17.376	0.312	18.000	237.132	58.940	102.737	161.677	12.319
	–	STD	17.250	0.375	18.000	233.705	70.589	101.252	171.841	12.141
	30	–	17.126	0.437	18.000	230.357	81.971	99.802	181.772	11.967
	–	XS	17.000	0.500	18.000	226.980	93.452	98.338	191.790	11.791
	40	–	16.876	0.562	18.000	223.681	104.668	96.909	201.577	11.620
	80	–	16.126	0.937	18.000	204.241	170.755	88.487	259.242	10.610
	160	–	14.438	1.781	18.000	163.721	308.509	70.932	379.440	8.505
20	10	–	19.500	0.250	20.000	298.648	52.733	129.388	182.122	15.514
	20	STD	19.250	0.375	20.000	291.039	78.600	126.092	204.691	15.119
	30	XS	19.000	0.500	20.000	283.529	104.132	122.838	226.970	14.729
	40	–	18.814	0.593	20.000	278.005	122.911	120.445	243.356	14.442
	80	–	17.938	1.031	20.000	252.719	208.873	109.490	318.363	13.128
	160	–	16.064	1.968	20.000	202.674	379.008	87.808	466.816	10.529
22	10	–	21.500	0.250	22.000	363.050	58.074	157.290	215.364	18.860
	20	STD	21.250	0.375	22.000	354.656	86.610	153.654	240.263	18.424
	30	XS	21.000	0.500	22.000	346.361	114.812	150.060	264.872	17.993
	80	–	19.750	1.125	22.000	306.354	250.818	132.727	383.545	15.915
	160	–	17.750	2.125	22.000	247.450	451.072	107.207	558.278	12.855
24	10	–	23.500	0.250	24.000	433.736	63.414	187.915	251.328	22.532
	20	STD	23.250	0.375	24.000	424.557	94.620	183.938	278.558	22.055
	–	XS	23.000	0.500	24.000	415.476	125.492	180.003	305.496	21.583
	30	–	22.876	0.562	24.000	411.008	140.681	178.068	318.749	21.351
	40	–	22.626	0.687	24.000	402.073	171.054	174.197	345.251	20.887
	80	–	21.564	1.218	24.000	365.215	296.359	158.228	454.587	18.972
	160	–	19.314	2.343	24.000	292.978	541.938	126.932	668.870	15.220
26	10	–	25.376	0.312	26.000	505.750	85.598	219.115	304.713	26.273
	–	STD	25.250	0.375	26.000	500.740	102.630	216.944	319.574	26.012
	20	XS	25.000	0.500	26.000	490.874	136.173	212.670	348.842	25.500
28	10	–	27.376	0.312	28.000	588.613	92.263	255.015	347.277	30.577
	–	STD	27.250	0.375	28.000	583.207	110.640	252.673	363.313	30.296
	20	XS	27.000	0.500	28.000	572.555	146.853	248.058	394.910	29.743
	30	–	26.750	0.625	28.000	562.001	182.732	243.485	426.217	29.195

(*Continued*)

PROPERTIES OF STEEL PIPE (*Continued*)

Pipe Size in.	Schedule		Inside Dia. in.	Wall Thick in.	Outside Dia. in.	Area sq.in.	Weight (1)			Water Volume gal./ft.
							Pipe lbs./ft.	Water lbs./ft.	Total lbs./ft.	
30	10	–	29.376	0.312	30.000	677.759	98.927	293.637	392.564	35.208
	–	STD	29.250	0.375	30.000	671.957	118.650	291.123	409.774	34.907
	20	XS	29.000	0.500	30.000	660.520	157.533	286.168	443.701	34.313
	30	–	28.750	0.625	30.000	649.181	196.082	281.255	477.337	33.724
	40	–	28.500	0.688	29.876	637.940	214.473	276.385	490.858	33.140
32	10	–	31.376	0.312	32.000	773.188	105.591	334.981	440.573	40.166
	–	STD	31.250	0.375	32.000	766.990	126.660	332.296	458.957	39.844
	20	XS	31.000	0.500	32.000	754.768	168.213	327.001	495.214	39.209
	30	–	30.750	0.625	32.000	742.643	209.432	321.748	531.180	38.579
	40	–	30.624	0.688	32.000	736.569	230.080	319.116	549.196	38.263
34	10	–	33.376	0.312	34.000	874.900	112.256	379.048	491.304	45.449
	–	STD	33.250	0.375	34.000	868.307	134.671	376.191	510.862	45.107
	20	XS	33.000	0.500	34.000	855.299	178.893	370.555	549.449	44.431
	30	–	32.750	0.625	34.000	842.389	222.782	364.962	587.744	43.760
	40	–	32.624	0.688	34.000	835.919	244.776	362.159	606.935	43.424
36	10	–	35.376	0.312	36.000	982.895	118.920	425.836	544.757	51.060
	–	STD	35.250	0.375	36.000	975.906	142.681	422.808	565.489	50.696
	20	XS	35.000	0.500	36.000	962.113	189.574	416.832	606.406	49.980
	30	–	34.750	0.625	36.000	948.417	236.133	410.899	647.031	49.268
	40	–	34.500	0.750	36.000	934.820	282.358	405.008	687.366	48.562
42	–	STD	41.250	0.375	42.000	1336.404	166.711	578.993	745.704	69.424
	20	XS	41.000	0.500	42.000	1320.254	221.614	571.996	793.610	68.585
	30	–	40.750	0.625	42.000	1304.203	276.183	565.042	841.225	67.751
	40	–	40.500	0.750	42.000	1288.249	330.419	558.130	888.549	66.922
48	–	STD	47.250	0.375	48.000	1753.450	190.742	759.677	950.418	91.088
	20	XS	47.000	0.500	48.000	1734.945	253.655	751.659	1005.314	90.127
	30	–	46.750	0.625	48.000	1716.537	316.234	743.684	1059.918	89.171
	40	–	46.500	0.750	48.000	1698.227	378.480	735.751	1114.231	88.220
54	–	STD	53.250	0.375	54.000	2227.046	214.772	964.860	1179.632	115.691
	20	XS	53.000	0.500	54.000	2206.183	285.695	955.822	1241.517	114.607
	30	–	52.750	0.625	54.000	2185.419	356.285	946.826	1303.111	113.528
	40	–	52.500	0.750	54.000	2164.754	426.540	937.873	1364.413	112.455
60	–	STD	59.250	0.375	60.000	2757.189	238.803	1194.543	1433.346	143.231
	20	XS	59.000	0.500	60.000	2733.971	317.736	1184.484	1502.220	142.024
	30	–	58.750	0.625	60.000	2710.851	396.336	1174.467	1570.803	140.823
	40	–	58.500	0.750	60.000	2687.829	474.601	1164.493	1639.095	139.627
72	–	STD	71.250	0.375	72.000	3987.123	286.863	1727.408	2014.272	207.123
	20	XS	71.000	0.500	72.000	3959.192	381.817	1715.307	2097.124	205.672
	30	–	70.750	0.625	72.000	3931.360	476.437	1703.249	2179.686	204.226
	40	–	70.500	0.750	72.000	3903.625	570.723	1691.233	2261.956	202.786
84	–	STD	83.250	0.375	84.000	5443.251	334.924	2358.271	2693.195	282.766
	20	XS	83.000	0.500	84.000	5410.608	445.898	2344.128	2790.027	281.071
	30	–	82.750	0.625	84.000	5378.063	556.539	2330.029	2886.567	279.380
	40	–	82.500	0.750	84.000	5345.616	666.845	2315.971	2982.816	277.694
96	–	STD	95.250	0.375	96.000	7125.574	382.985	3087.132	3470.117	370.160
	20	XS	95.000	0.500	96.000	7088.218	509.980	3070.948	3580.927	368.219
	30	–	94.750	0.625	96.000	7050.961	636.640	3054.806	3691.446	366.284
	40	–	94.500	0.750	96.000	7013.802	762.967	3038.707	3801.674	364.353

PROPERTIES OF STAINLESS STEEL PIPE

Pipe Size in.	Schedule		Inside Dia. in.	Wall Thick in.	Outside Dia. in.	Area sq.in.	Weight (1)			Water Volume gal./ft.
							Pipe lbs./ft.	Water lbs./ft.	Total lbs./ft.	
1/2	5	–	0.710	0.065	0.840	0.396	0.549	0.172	0.720	0.021
	10	–	0.674	0.083	0.840	0.357	0.684	0.155	0.839	0.019
3/4	5	–	0.920	0.065	1.050	0.665	0.697	0.288	0.985	0.035
	10	–	0.884	0.083	1.050	0.614	0.874	0.266	1.140	0.032
1	5	–	1.185	0.065	1.315	1.103	0.885	0.478	1.363	0.057
	10	–	1.097	0.109	1.315	0.945	1.432	0.409	1.842	0.049

(*Continued*)

PROPERTIES OF STAINLESS STEEL PIPE (*Continued*)

Pipe Size in.	Schedule		Inside Dia. in.	Wall Thick in.	Outside Dia. in.	Area sq.in.	Weight (1)			Water Volume gal./ft.
							Pipe lbs./ft.	Water lbs./ft.	Total lbs./ft.	
1-1/4	5	–	1.530	0.065	1.660	1.839	1.129	0.797	1.926	0.096
	10	–	1.442	0.109	1.660	1.633	1.842	0.708	2.549	0.085
1-1/2	5	–	1.770	0.065	1.900	2.461	1.299	1.066	2.365	0.128
	10	–	1.682	0.109	1.900	2.222	2.127	0.963	3.089	0.115
2	5	–	2.245	0.065	2.375	3.958	1.636	1.715	3.351	0.206
	10	–	2.157	0.109	2.375	3.654	2.691	1.583	4.274	0.190
2-1/2	5	–	2.709	0.083	2.875	5.764	2.524	2.497	5.022	0.299
	10	–	2.635	0.120	2.875	5.453	3.601	2.363	5.964	0.283
3	5	–	3.334	0.083	3.500	8.730	3.090	3.782	6.872	0.454
	10	–	3.260	0.120	3.500	8.347	4.419	3.616	8.035	0.434
4	5	–	4.334	0.083	4.500	14.753	3.994	6.392	10.385	0.766
	10	–	4.260	0.120	4.500	14.253	5.726	6.175	11.901	0.740
5	5	–	5.345	0.109	5.563	22.438	6.476	9.721	16.197	1.166
	10	–	5.295	0.134	5.563	22.020	7.925	9.540	17.465	1.144
6	5	–	6.407	0.109	6.625	32.240	7.737	13.968	21.705	1.675
	10	–	6.357	0.134	6.625	31.739	9.475	13.751	23.226	1.649
8	5	–	8.407	0.109	8.625	55.510	10.112	24.050	34.162	2.884
	10	–	8.329	0.148	8.625	54.485	13.667	23.605	37.273	2.830
10	5	–	10.482	0.134	10.750	86.294	15.497	37.386	52.883	4.483
	10	–	10.420	0.165	10.750	85.276	19.026	36.945	55.972	4.430
12	5	–	12.438	0.156	12.750	121.504	21.403	52.641	74.044	6.312
	10	–	12.390	0.180	12.750	120.568	24.648	52.236	76.884	6.263
14	5	–	13.688	0.156	14.000	147.153	23.527	63.754	87.281	7.644
	10	–	13.624	0.188	14.000	145.780	28.287	63.159	91.446	7.573
16	5	–	15.670	0.165	16.000	192.854	28.463	83.553	112.016	10.018
	10	–	15.624	0.188	16.000	191.723	32.384	83.063	115.447	9.960
18	5	–	17.670	0.165	18.000	245.224	32.058	106.243	138.301	12.739
	10	–	17.624	0.188	18.000	243.949	36.480	105.690	142.170	12.673
20	5	–	19.624	0.188	20.000	302.458	40.576	131.039	171.615	15.712
	10	–	19.564	0.218	20.000	300.611	46.979	130.239	177.218	15.616
22	5	–	21.624	0.188	22.000	367.250	44.672	159.110	203.782	19.078
	10	–	21.564	0.218	22.000	365.215	51.729	158.228	209.957	18.972
24	5	–	23.564	0.218	24.000	436.102	56.479	188.940	245.418	22.655
	10	–	23.500	0.250	24.000	433.736	64.682	187.915	252.597	22.532

18.02 Pipe Support and Pipe Spacing

HORIZONTAL PIPE SUPPORT SPACING

	Maximum Horizontal Hanger Spacing Feet							
	Steel				Copper			
Pipe Size	Recommend	Water Systems	Vapor Systems	Recommend	Water Systems	Vapor Systems	Minimum Rod Size in.
1/2	6	7	8	5	5	6	3/8
3/4	6	7	9	5	5	7	3/8
1	6	7	9	6	6	8	3/8
1-1/4	6	7	9	6	7	9	3/8
1-1/2	6	9	12	6	8	10	3/8
2	7	10	13	7	8	11	3/8
2-1/2	10	11	14	8	9	13	1/2
3	10	12	15	10	10	14	1/2
4	10	14	17	10	12	16	5/8
5	10	16	19	10	13	18	5/8
6	10	17	21	10	14	20	3/4
8	12	19	24	10	16	23	7/8
10	12	22	26	10	18	25	7/8
12	12	23	30	10	19	28	7/8
14	12	25	32	–	–	–	1

(Continued)

HORIZONTAL PIPE SUPPORT SPACING (*Continued*)

	Maximum Horizontal Hanger Spacing Feet						
		Steel			Copper		
Pipe Size	Recommend	Water Systems	Vapor Systems	Recommend	Water Systems	Vapor Systems	Minimum Rod Size in.
16	12	27	35	–	–	–	1
18	12	28	37	–	–	–	1-1/4
20	12	30	39	–	–	–	1-1/4
22	12	30	39	–	–	–	1-1/2
24	12	32	42	–	–	–	1-1/2
26	12	32	42	–	–	–	1-1/2
28	12	32	42	–	–	–	1-1/2
30	12	33	44	–	–	–	1-1/2
32	12	33	44	–	–	–	1-1/2
34	12	33	44	–	–	–	1-1/2
36	12	33	44	–	–	–	1-1/2
42	12	33	44	–	–	–	1-1/2
48	12	32	42	–	–	–	1-3/4
54	12	33	44	–	–	–	1-3/4
60	12	33	44	–	–	–	2
72	12	33	44	–	–	–	2
84	12	33	44	–	–	–	2-1/2
96	12	33	44	–	–	–	2-1/2

Note:
1 Recommended pipe support spacing is less than the maximum to more evenly distribute weight over a building's structural system. Consult the structural engineer for additional guidance on pipe support spacing, especially with steel bar joist construction.

VERTICAL PIPE SUPPORT SPACING

	Maximum Vertical Support Spacing Feet		
Pipe Size	Steel	Copper	Support
8" and Smaller	Every other floor and base of all pipe risers	Every floor and base of all pipe risers	Steel extension pipe clamps
10"–12"	Every other floor and base of all pipe risers	Every floor and base of all pipe risers	Steel extension pipe clamps
14"–24"	Every other floor and base of all pipe risers	Not applicable	Steel extension pipe clamps
26"–96"	Every floor and base of all pipe risers	Not applicable	Steel extension pipe clamps

PIPE SPACING ON RACKS

	Minimum Centerline-to-Centerline Dimensions, Inches										
Pipe Size	Pipe Size										
	1/2	3/4	1	1-1/4	1-1/2	2	2-1/2	3	4	5	6
1/2	7.5	–	–	–	–	–	–	–	–	–	–
3/4	8.0	8.0	–	–	–	–	–	–	–	–	–
1	8.0	8.5	8.5	–	–	–	–	–	–	–	–
1-1/4	8.5	8.5	8.5	9.0	–	–	–	–	–	–	–
1-1/2	8.5	8.5	9.0	9.0	9.0	–	–	–	–	–	–
2	9.0	9.0	9.5	9.5	9.5	10.0	–	–	–	–	–
2-1/2	10.0	10.0	10.5	10.5	10.5	11.0	12.0	–	–	–	–
3	10.0	10.5	10.5	11.0	11.0	11.5	12.5	12.5	–	–	–
4	11.5	11.5	12.0	12.0	12.0	12.5	13.5	14.0	15.0	–	–
5	12.0	12.0	12.5	12.5	12.5	13.0	14.0	14.5	15.5	16.0	–
6	12.5	12.5	13.0	13.0	13.0	13.5	14.5	14.5	16.0	16.5	17.0
8	13.5	14.0	14.0	14.5	14.5	15.0	16.0	16.0	17.5	18.0	18.5
10	15.0	15.0	15.5	15.5	15.5	16.0	17.0	17.5	18.5	19.0	19.5
12	16.5	16.5	17.0	17.5	17.0	17.5	18.5	19.0	20.0	20.5	21.0
14	17.5	17.5	18.0	18.0	18.0	18.5	19.5	20.0	21.0	21.5	22.0

(*Continued*)

PIPE SPACING ON RACKS (*Continued*)

Pipe Size	Minimum Centerline-to-Centerline Dimensions, Inches										
	Pipe Size										
	1/2	3/4	1	1-1/4	1-1/2	2	2-1/2	3	4	5	6
16	18.5	19.0	19.0	19.0	19.5	20.0	21.0	21.0	22.5	23.0	23.5
18	19.5	19.5	20.0	20.0	20.0	20.5	21.5	22.0	23.0	23.5	24.0
20	20.5	21.0	21.0	21.0	21.5	22.0	23.0	23.0	24.5	25.0	25.5
22	22.0	22.0	22.0	22.0	22.5	23.0	24.0	24.0	25.5	26.0	26.5
24	23.0	23.5	23.5	23.5	23.5	24.0	25.0	25.5	26.5	27.0	27.5
26	24.0	24.5	24.5	24.5	25.0	25.0	26.0	26.5	28.0	28.0	29.0
28	25.0	25.5	25.5	25.5	26.0	26.5	27.5	27.5	29.0	29.5	30.0
30	26.5	27.0	27.0	27.0	27.0	27.5	28.5	29.0	30.0	30.5	31.0
32	28.0	28.0	28.0	28.0	28.5	29.0	29.5	30.0	31.5	32.0	32.5
34	29.0	29.0	29.0	29.0	29.5	30.0	31.0	31.0	32.5	33.0	33.5
36	30.0	30.5	30.5	30.5	30.5	31.0	32.0	32.5	33.5	34.0	34.5
42	33.5	34.0	34.0	34.0	34.0	34.5	35.5	36.0	37.0	37.5	38.0
48	36.5	37.0	37.0	37.5	37.5	38.0	39.0	39.0	40.5	41.0	41.5
54	40.0	40.0	40.5	40.5	41.0	41.5	42.5	42.5	44.0	44.5	45.0
60	43.5	43.5	44.0	44.0	44.0	44.5	45.5	46.0	47.0	47.5	48.0
72	50.0	50.5	50.5	51.0	51.0	51.5	52.5	52.5	54.0	54.5	55.0
84	57.0	57.0	57.5	57.5	57.5	58.0	59.0	59.5	60.5	61.0	61.5
96	63.5	64.0	64.0	64.5	64.5	65.0	66.0	66.0	67.5	68.0	68.5

PIPE SPACING ON RACKS

Pipe Size	Minimum Centerline-to-Centerline Dimensions, Inches										
	Pipe Size										
	8	10	12	14	16	18	20	22	24	26	28
1/2	–	–	–	–	–	–	–	–	–	–	–
3/4	–	–	–	–	–	–	–	–	–	–	–
1	–	–	–	–	–	–	–	–	–	–	–
1-1/4	–	–	–	–	–	–	–	–	–	–	–
1-1/2	–	–	–	–	–	–	–	–	–	–	–
2	–	–	–	–	–	–	–	–	–	–	–
2-1/2	–	–	–	–	–	–	–	–	–	–	–
3	–	–	–	–	–	–	–	–	–	–	–
4	–	–	–	–	–	–	–	–	–	–	–
5	–	–	–	–	–	–	–	–	–	–	–
6	–	–	–	–	–	–	–	–	–	–	–
8	19.5	–	–	–	–	–	–	–	–	–	–
10	21.0	22.0	–	–	–	–	–	–	–	–	–
12	22.5	23.5	25.0	–	–	–	–	–	–	–	–
14	23.5	24.5	26.0	27.0	–	–	–	–	–	–	–
16	24.5	26.0	27.5	28.5	30.0	–	–	–	–	–	–
18	25.5	26.5	28.0	29.0	30.5	31.0	–	–	–	–	–
20	26.5	28.0	29.5	30.5	31.5	32.5	33.5	–	–	–	–
22	27.5	29.0	30.5	31.5	32.5	33.5	34.5	35.5	–	–	–
24	29.0	30.0	31.5	32.5	34.0	34.5	36.0	37.0	38.0	–	–
26	30.0	31.0	33.0	34.0	35.0	36.0	37.0	38.0	39.5	40.5	–
28	31.0	32.5	34.0	35.0	36.0	37.0	38.0	39.0	40.5	41.5	42.5
30	32.5	33.5	35.0	36.0	37.5	38.0	39.5	40.5	41.5	42.5	44.0
32	34.0	35.0	36.5	37.4	39.0	39.5	41.0	42.0	43.0	44.0	45.5
34	35.0	36.0	37.5	38.5	40.0	40.5	42.0	43.0	44.0	45.0	46.5
36	36.0	37.0	38.5	39.5	41.0	41.5	43.0	44.0	45.0	46.5	47.5
42	39.5	40.5	42.0	41.0	44.5	45.0	46.5	47.5	48.5	50.0	51.0
48	42.5	44.0	45.5	46.5	47.5	48.5	49.5	51.0	52.0	53.0	54.0
54	46.0	47.5	49.0	50.0	51.0	52.0	53.0	54.0	55.5	56.5	57.5
60	49.5	50.5	52.0	53.0	54.5	55.0	56.5	57.5	58.5	60.0	61.0
72	56.0	57.5	59.0	60.0	61.0	62.0	63.0	64.5	65.5	66.5	67.5
84	63.0	64.0	65.5	66.5	68.0	68.5	70.0	71.0	72.0	73.5	74.5
96	69.5	71.0	72.5	73.5	74.5	75.5	76.5	78.0	79.0	80.0	81.0

PIPE SPACING ON RACKS

Pipe Size	Minimum Centerline-to-Centerline Dimensions, Inches										
	Pipe Size										
	30	32	34	36	42	48	54	60	72	84	96
1/2	–	–	–	–	–	–	–	–	–	–	–
3/4	–	–	–	–	–	–	–	–	–	–	–
1	–	–	–	–	–	–	–	–	–	–	–
1-1/4	–	–	–	–	–	–	–	–	–	–	–
1-1/2	–	–	–	–	–	–	–	–	–	–	–
2	–	–	–	–	–	–	–	–	–	–	–
2-1/2	–	–	–	–	–	–	–	–	–	–	–
3	–	–	–	–	–	–	–	–	–	–	–
4	–	–	–	–	–	–	–	–	–	–	–
5	–	–	–	–	–	–	–	–	–	–	–
6	–	–	–	–	–	–	–	–	–	–	–
8	–	–	–	–	–	–	–	–	–	–	–
10	–	–	–	–	–	–	–	–	–	–	–
12	–	–	–	–	–	–	–	–	–	–	–
14	–	–	–	–	–	–	–	–	–	–	–
16	–	–	–	–	–	–	–	–	–	–	–
18	–	–	–	–	–	–	–	–	–	–	–
20	–	–	–	–	–	–	–	–	–	–	–
22	–	–	–	–	–	–	–	–	–	–	–
24	–	–	–	–	–	–	–	–	–	–	–
26	–	–	–	–	–	–	–	–	–	–	–
28	–	–	–	–	–	–	–	–	–	–	–
30	45.0	–	–	–	–	–	–	–	–	–	–
32	46.5	48.0	–	–	–	–	–	–	–	–	–
34	47.5	49.0	50.0	–	–	–	–	–	–	–	–
36	48.5	50.0	51.0	52.0	–	–	–	–	–	–	–
42	52.0	53.5	54.5	55.5	59.0	–	–	–	–	–	–
48	55.5	57.0	58.0	59.0	62.5	65.5	–	–	–	–	–
54	58.5	60.0	61.0	62.5	66.0	69.0	72.5	–	–	–	–
60	62.0	63.5	64.5	65.5	69.0	72.5	76.0	79.0	–	–	–
72	69.0	70.5	71.5	72.5	76.0	79.0	82.5	86.0	92.5	–	–
84	75.4	77.0	78.0	79.0	82.5	86.0	89.5	92.5	99.5	106.0	–
96	82.5	84.0	85.0	86.0	89.5	92.5	96.0	99.5	106.0	113.0	119.5

Notes:

1. Table based on schedule 40 pipe and includes the outside dimensions for flanges, fittings, etc.
2. Insulation over flanges, fittings, etc., is as follows:
   Pipe sizes 2" and smaller: 1-1/2" Insulation
   Pipe sizes 2-1/2" and larger: 2" Insulation
3. The spaces between fittings are as follows:
   Space between two pipes 3" and smaller: 1"
   Space between one pipe 3" and smaller, and one pipe 4" and larger: 1-1/2"
   Space between two pipes 4" and larger: 2"
4. For schedule 80 and 160 pipe and 300 lb. fittings, add the following:
   Pipe sizes 4" and smaller: 1"
   Pipe sizes 5"–12": 1-1/2"
   Pipe sizes 14" and larger: 2"
5. Tables do not include space for valve handles and stems, expansion joints, expansion loops, or pipe guides.

18.03 Pipe Expansion

A. Expansion Loops (See Part 3)

1. L-Bends. Anchor force = 500 lbs./dia. in.
2. Z-Bends. Anchor force = 200–500 lbs./dia. in.
3. U-Bends. Anchor force = 200 lbs./dia. in.
4. Locate anchors at beam locations, and avoid anchor locations at steel bar joists if at all possible.
5. The following expansion tables were created using the equations in Part 3.

THERMAL EXPANSION OF METAL PIPE

Saturated Steam Pressure Psig	Temperature °F	Linear Thermal Expansion Inches/100 Feet		
		Carbon Steel	Stainless Steel	Copper
–	−30	−0.19	−0.30	−0.32
–	−20	−0.12	−0.20	−0.21
–	−10	−0.06	−0.10	−0.11
–	0	0	0	0
–	10	0.08	0.11	0.12
–	20	0.15	0.22	0.24
−14.6	32	0.24	0.36	0.37
−14.6	40	0.30	0.45	0.45
−14.5	50	0.38	0.56	0.57
−14.4	60	0.46	0.67	0.68
−14.3	70	0.53	0.78	0.79
−14.2	80	0.61	0.90	0.90
−14.0	90	0.68	1.01	1.02
−13.7	100	0.76	1.12	1.13
−13.0	120	0.91	1.35	1.37
−11.8	140	1.06	1.57	1.59
−10.0	160	1.22	1.79	1.80
−7.2	180	1.37	2.02	2.05
−3.2	200	1.52	2.24	2.30
0	212	1.62	2.38	2.43
2.5	220	1.69	2.48	2.52
10.3	240	1.85	2.71	2.76
20.7	260	2.02	2.94	2.99
34.6	280	2.18	3.17	3.22
52.3	300	2.35	3.40	3.46
75.0	320	2.53	3.64	3.70
103.3	340	2.70	3.88	3.94
138.3	360	2.88	4.11	4.18
181.1	380	3.05	4.35	4.42
232.6	400	3.23	4.59	4.87
294.1	420	3.41	4.83	4.91
366.9	440	3.60	5.07	5.15
452.2	460	3.78	5.32	5.41
551.4	480	3.97	5.56	5.65
666.1	500	4.15	5.80	5.91
797.7	520	4.35	6.05	6.15
947.8	540	4.54	6.29	6.41
1118	560	4.74	6.54	6.64
1311	580	4.93	6.78	6.92
1528	600	5.13	7.03	7.18
1772	620	5.34	7.28	7.43
2045	640	5.54	7.53	7.69
2351	660	5.75	7.79	7.95
2693	680	5.95	8.04	8.20
3079	700	6.16	8.29	8.47
–	720	6.37	8.55	8.71
–	740	6.59	8.81	9.00
–	760	6.80	9.07	9.26
–	780	7.02	9.33	9.53
–	800	7.23	9.59	9.79
–	820	7.45	9.85	10.07
–	840	7.67	10.12	10.31
–	860	7.90	10.38	10.61
–	880	8.12	10.65	10.97
–	900	8.34	10.91	11.16
–	920	8.56	11.18	11.42
–	940	8.77	11.45	11.71
–	960	8.99	11.73	11.98
–	980	9.20	12.00	12.27
–	1000	9.42	12.27	12.54

Notes:
1 Table based on ASTM A53, Grade B, steel pipe.
2 Temperature range applicable through 400°F.
3 Table also applicable to copper tube.
4 For equations and diagrams relating to pipe expansion, see Part 3 Equations.
5 L-bend, Z-bend, and U-bend or loop dimensions are minimum dimensions; we recommend rounding up to nearest 1/2 foot (H = 2W).

PIPE EXPANSION L-BENDS

Pipe Size	Expansion of Longest Leg							
	1"	1-1/2"	2"	2-1/2"	3"	4"	5"	6"
1/2	5'9"	7'0"	8'2"	9'2"	10'0"	11'6"	12'9"	14'0"
3/4	6'6"	8'4"	9'3"	10'4"	11'3"	13'0"	14'8"	16'0"
1	7'2"	8'9"	10'2"	11'4"	12'6"	14'4"	16'0"	17'6"
1-1/4	8'0"	9'10"	11'4"	12'8"	14'0"	16'2"	18'0"	19'8"
1-1/2	8'8"	10'6"	12'2"	13'8"	15'0"	17'2"	19'3"	21'0"
2	9'8"	11'9"	13'8"	15'2"	16'8"	19'3"	21'6"	23'6"
2-1/2	10'8"	13'0"	15'0"	16'9"	18'4"	21'2"	23'8"	26'0"
3	11'8"	14'4"	16'6"	18'6"	20'2"	23'4"	26'2"	28'8"
4	13'3"	16'2"	18'8"	21'0"	23'0"	26'6"	29'8"	32'6"
5	14'8"	18'0"	20'9"	23'3"	25'6"	29'6"	32'10"	36'0"
6	16'2"	19'8"	22'8"	25'4"	27'9"	32'2"	35'10"	39'3"
8	18'4"	22'6"	26'0"	29'0"	31'8"	36'8"	41'0"	44'10"
10	20'6"	25'0"	29'9"	32'4"	35'6"	40'10"	45'8"	50'0"
12	22'3"	27'3"	31'6"	35'2"	38'6"	44'6"	49'9"	54'6"
14	23'4"	28'8"	33'0"	36'10"	40'4"	46'8"	52'2"	57'2"
16	25'0"	30'6"	35'3"	39'6"	43'2"	50'0"	55'8"	61'0"
18	26'6"	32'4"	37'6"	41'9"	45'9"	52'10"	59'2"	64'10"
20	27'10"	34'2"	39'6"	44'0"	48'3"	55'8"	62'3"	68'3"
22	29'3"	35'9"	41'4"	46'2"	50'8"	58'6"	65'4"	71'8"
24	30'6"	37'6"	43'2"	48'3"	52'10"	61'0"	68'3"	74'9"
26	31'9"	39'0"	45'0"	50'3"	55'0"	63'6"	71'0"	77'9"
28	33'0"	40'4"	46'8"	52'2"	57'2"	66'0"	73'8"	80'9"
30	34'2"	41'9"	48'3"	54'0"	59'2"	68'3"	76'3"	83'8"
32	35'3"	43'2"	50'0"	55'8"	61'0"	70'6"	78'9"	86'4"
34	36'4"	44'6"	51'4"	57'6"	63'0"	72'8"	81'2"	89'0"
36	37'6"	45'9"	52'10"	59'2"	64'9"	74'9"	83'8"	91'6"
42	40'6"	49'6"	57'2"	63'10"	70'0"	80'9"	90'3"	99'10"
48	43'2"	52'10"	61'0"	68'3"	74'9"	86'4"	96'5"	105'8"
54	45'9"	56'1"	64'9"	72'4"	79'3"	91'6"	102'4"	112'1"
60	48'3"	59'1"	68'3"	76'3"	83'7"	96'6"	107'10"	118'2"
72	52'10"	64'9"	74'9"	83'7"	91'6"	105'8"	118'2"	129'5"
84	57'1"	69'11"	80'9"	90'3"	98'10"	114'2"	127'7"	140'0"
96	61'0"	74'9"	86'4"	96'6"	105'8"	122'0"	136'5"	149'6"

PIPE EXPANSION Z-BENDS

Pipe Size	Anchor-to-Anchor Expansion							
	1"	1-1/2"	2"	2-1/2"	3"	4"	5"	6"
1/2	3'8"	4'6"	5'2"	5'10"	6'5"	7'4"	8'2"	9'0"
3/4	4'2"	5'2"	6'2"	6'8"	7'3"	8'6"	9'4"	10'3"
1	4'8"	5'8"	6'6"	7'4"	8'0"	9'2"	10'4"	11'3"
1-1/4	5'2"	6'4"	7'4"	8'2"	9'0"	10'4"	11'8"	12'8"
1-1/2	5'6"	6'10"	7'10"	8'9"	9'7"	11'0"	12'4"	13'6"
2	6'2"	7'8"	8'9"	9'9"	10'8"	12'4"	13'10"	15'2"
2-1/2	6'10"	8'4"	9'8"	10'9"	11'9"	13'8"	15'2"	16'8"
3	7'6"	9'2"	10'8"	12'0"	13'0"	15'0"	16'9"	18'4"
4	8'6"	10'6"	12'0"	13'6"	14'9"	17'0"	19'0"	20'10"
5	9'6"	11'8"	13'4"	15'0"	16'6"	19'0"	21'2"	23'2"
6	10'4"	12'8"	14'6"	16'4"	18'0"	20'8"	23'0"	25'3"
8	11'9"	14'6"	16'8"	18'8"	20'4"	23'6"	26'4"	28'10"
10	13'2"	16'2"	18'6"	20'9"	22'9"	26'3"	29'4"	32'2"
12	14'4"	17'6"	20'3"	22'8"	24'9"	28'8"	32'0"	35'0"
14	15'0"	18'4"	21'2"	23'8"	26'0"	30'0"	33'6"	36'8"
16	16'0"	19'8"	22'8"	25'4"	27'9"	32'0"	35'9"	39'3"
18	17'0"	20'10"	24'0"	26'10"	29'6"	34'0"	38'0"	41'8"
20	18'0"	22'0"	25'3"	28'4"	31'0"	35'9"	40'0"	43'10"
22	18'10"	23'0"	26'8"	29'8"	32'6"	37'8"	42'0"	46'0"
24	19'8"	24'0"	27'9"	31'0"	34'0"	39'2"	43'10"	48'0"
26	20'6"	25'0"	28'10"	32'4"	35'4"	40'10"	45'8"	50'0"

(Continued)

PIPE EXPANSION Z-BENDS (*Continued*)

Pipe Size	Anchor-to-Anchor Expansion							
	1"	1-1/2"	2"	2-1/2"	3"	4"	5"	6"
28	21'2"	26'0"	30'0"	33'6"	36'8"	42'4"	47'4"	52'0"
30	22'0"	26'10"	31'0"	34'8"	38'0"	43'10"	49'0"	53'8"
32	22'8"	27'9"	32'0"	35'10"	39'3"	45'4"	50'8"	55'6"
34	23'4"	28'8"	33'0"	37'0"	40'6"	46'8"	52'2"	57'2"
36	24'0"	29'6"	34'0"	38'0"	41'8"	48'0"	53'8"	58'10"
42	26'0"	31'9"	36'8"	41'0"	45'0"	52'0"	58'0"	63'6"
48	27'9"	34'0"	39'3"	43'10"	48'0"	55'6"	62'0"	67'11"
54	29'5"	36'0"	41'7"	46'6"	50'11"	58'10"	65'9"	72'0"
60	31'0"	38'0"	43'10"	49'0"	53'8"	62'0"	69'4"	75'11"
72	34'0"	41'7"	48'0"	53'8"	58'10"	67'11"	75'11"	83'2"
84	36'8"	44'11"	51'11"	58'0"	63'6"	73'4"	82'0"	89'10"
96	39'3"	48'0"	55'6"	62'0"	67'11"	78'5"	87'8"	96'0"

PIPE EXPANSION U-BENDS OR LOOPS

Pipe Size	Anchor-to-Anchor Expansion							
	1"		1-1/2"		2"		2-1/2"	
	W	H	W	H	W	H	W	H
1/2	1'2"	2'4"	1'6"	3'0"	1'8"	3'4"	1'10"	3'8"
3/4	1'4"	2'8"	1'8"	3'4"	1'10"	3'8"	2'2"	4'4"
1	1'6"	3'0"	1'9"	3'6"	2'0"	4'0"	2'4"	4'8"
1-1/4	1'8"	3'4"	2'0"	4'0"	2'4"	4'8"	2'8"	5'4"
1-1/2	1'9"	3'6"	2'2"	4'4"	2'6"	5'0"	2'9"	5'6"
2	1'11"	3'10"	2'4"	4'8"	2'9"	5'6"	3'2"	6'4"
2-1/2	2'2"	4'4"	2'8"	5'4"	3'0"	6'0"	3'3"	6'6"
3	2'4"	4'8"	3'0"	6'0"	3'4"	6'8"	3'9"	7'6"
4	2'8"	5'4"	3'3"	6'6"	3'9"	7'6"	4'2"	8'4"
5	3'0"	6'0"	3'8"	7'4"	4'2"	8'4"	4'8"	9'4"
6	3'3"	6'6"	4'0"	8'0"	4'7"	9'2"	5'2"	10'4"
8	3'8"	7'4"	4'6"	9'0"	5'2"	10'4"	5'10"	11'8"
10	4'2"	8'4"	5'0"	10'0"	5'10"	11'8"	6'6"	13'0"
12	4'6"	9'0"	5'6"	11'0"	6'4"	12'8"	7'2"	14'4"
14	4'8"	9'4"	5'9"	11'6"	6'8"	13'4"	7'6"	15'0"
16	5'0"	10'0"	6'2"	12'4"	7'1"	14'2"	8'0"	16'0"
18	5'4"	10'8"	6'6"	13'0"	7'6"	15'0"	8'6"	17'0"
20	5'8"	11'4"	7'0"	14'0"	7'11"	15'9"	8'10"	17'8"
22	5'10"	11'8"	7'3"	14'6"	8'3"	16'6"	9'3"	18'6"
24	6'1"	12'2"	7'6"	15'0"	8'8"	17'4"	9'8"	19'4"
26	6'5"	13'0"	7'10"	15'8"	9'0"	18'0"	10'2"	20'4"
28	6'8"	13'4"	8'2"	16'4"	9'4"	18'8"	10'6"	21'0"
30	6'10"	13'8"	8'6"	17'0"	9'8"	19'4"	11'0"	21'8"
32	7'1"	14'2"	8'8"	17'4"	10'0"	20'0"	11'2"	22'4"
34	7'4"	14'8"	9'0"	18'0"	10'4"	20'8"	11'6"	23'0"
36	7'6"	15'0"	9'2"	18'4"	10'8"	21'4"	12'0"	23'8"
42	8'1"	16'2"	10'0"	20'0"	11'6"	23'0"	12'9"	25'6"
48	8'8"	17'4"	10'7	21'2"	12'3"	24'6"	13'8"	27'4"
54	9'2"	18'4"	11'3"	22'6"	13'0"	26'0"	14'6"	29'0"
60	9'8"	19'4"	11'10"	23'8"	13'8"	27'4"	15'3"	30'6"
72	10'7"	21'2"	13'0"	26'0"	15'0"	30'0"	16'9"	33'6"
84	11'5"	22'10"	14'0"	28'0"	16'2"	32'4"	18'1"	36'2"
96	12'3"	24'6"	15'0"	30'0"	17'3"	34'6"	19'4"	38'8"

PIPE EXPANSION U-BENDS OR LOOPS

Pipe Size	Anchor-to-Anchor Expansion							
	3"		4"		5"		6"	
	W	H	W	H	W	H	W	H
1/2	2'0"	4'0"	2'4"	4'8"	2'8"	5'4"	2'10"	5'8"
3/4	2'4"	4'8"	2'8"	5'4"	3'0"	6'0"	3'3"	6'6"
1	2'6"	5'0"	3'0"	6'0"	3'4"	6'8"	3'6"	7'0"
1-1/4	2'10"	5'8"	3'3"	6'6"	3'8"	7'4"	4'0"	8'0"
1-1/2	3'0"	6'0"	3'6"	7'0"	3'10"	7'8"	4'3"	8'6"
2	3'4"	6'8"	4'0"	8'0"	4'4"	8'8"	4'9"	9'6"
2-1/2	3'8"	7'4"	4'3"	8'6"	4'10"	9'10"	5'2"	10'4"
3	4'1"	8'2"	4'8"	9'4"	5'4"	10'8"	5'9"	11'8"
4	4'7"	9'2"	5'4"	10'8"	5'10"	11'8"	6'6"	13'0"
5	5'2"	10'4"	6'0"	12'0"	6'8"	13'4"	7'3"	14'6"
6	5'7"	11'2"	6'6"	13'0"	7'2"	14'4"	8'0"	16'0"
8	6'4"	12'8"	7'4"	14'8"	8'4"	16'8"	9'0"	18'0"
10	7'1"	14'2"	8'2"	16'4"	9'2"	18'4"	10'0"	20'0"
12	7'9"	15'6"	9'0"	18'0"	10'0"	20'0"	11'0"	22'0"
14	8'1"	16'2"	9'4"	18'8"	10'6"	21'0"	11'6"	23'0"
16	8'8"	17'4"	10'0"	20'0"	11'2"	22'4"	12'3"	24'6"
18	9'2"	18'4"	10'8"	21'4"	11'10"	23'8"	13'0"	26'0"
20	9'8"	19'4"	11'2"	22'4"	12'6"	25'0"	13'8"	27'4"
22	10'2"	20'4"	11'8"	23'4"	13'2"	26'4"	14'4"	28'8"
24	10'8"	21'4"	12'3"	24'6"	13'8"	27'4"	15'0"	30'0"
26	11'0"	22'0"	12'9"	25'6"	14'4"	28'8"	15'7"	31'2"
28	11'6"	23'0"	13'2"	26'4"	14'10"	29'8"	16'2"	32'4"
30	12'0"	23'8"	13'8"	27'4"	15'4"	30'8"	16'9"	33'6"
32	12'3"	24'6"	14'2"	28'4"	15'10"	31'8"	17'3"	34'6"
34	12'8"	25'4"	14'6"	29'0"	16'4"	32'8"	18'0"	36'0"
36	13'0"	26'0"	15'0"	30'0"	16'10"	33'8"	18'4"	36'8"
42	14'0"	28'0"	16'2"	32'4"	18'2"	36'4"	20'0"	40'0"
48	15'0"	30'0"	17'4"	34'8"	19'4"	38'8"	21'2"	42'4"
54	15'11"	31'10"	18'4"	36'8"	20'6"	41'0"	22'5"	44'10"
60	16'9"	33'6"	19'4"	38'8"	21'7"	43'2"	23'8"	47'4"
72	18'4"	36'8"	21'2"	42'4"	23'8"	47'4"	25'11"	51'10"
84	19'10"	39'8"	22'10"	45'8"	25'7"	51'2"	28'0"	56'0"
96	21'2"	42'4"	24'5"	48'10"	27'4"	54'8"	29'11"	59'10"

PIPE EXPANSION U-BENDS OR LOOPS

Pipe Size	Anchor-to-Anchor Expansion							
	7"		8"		10"		12"	
	W	H	W	H	W	H	W	H
1/2	3'2"	6'4"	3'3"	6'6"	3'8"	7'4"	4'0"	8'0"
3/4	3'6"	7'0"	3'8"	7'4"	4'2"	8'4"	4'6"	9'0"
1	3'10"	7'8"	4'0"	8'0"	4'7"	9'2"	5'0"	10'0"
1-1/4	4'4"	8'8"	4'7"	9'2"	5'1"	10'2"	5'7"	11'2"
1-1/2	4'8"	9'4"	5'0"	10'0"	5'6"	11'0"	6'0"	12'0"
2	5'2"	10'4"	5'6"	11'0"	6'1"	12'2"	6'8"	13'4"
2-1/2	5'8"	11'4"	6'0"	12'0"	6'8"	13'4"	7'4"	14'8"
3	6'2"	12'4"	6'8"	13'4"	7'6"	15'0"	8'1"	16'2"
4	7'0"	14'0"	7'6"	15'0"	8'6"	17'0"	9'2"	18'4"
5	7'10"	15'8"	8'4"	16'8"	9'4"	18'8"	10'2"	20'4"
6	8'6"	17'0"	9'2"	18'4"	10'2"	20'4"	11'2"	22'4"
8	9'8"	19'4"	10'4"	20'8"	11'7"	23'2"	12'8"	25'4"
10	10'10"	21'8"	11'7"	23'2"	13'0"	26'0"	14'2"	28'4"
12	11'10"	23'8"	12'7"	25'2"	14'0"	28'0"	15'6"	31'0"
14	12'4"	24'8"	13'3"	26'6"	14'9"	29'6"	16'2"	32'4"
16	13'2"	26'4"	14'2"	28'4"	15'9"	31'6"	17'3"	34'6"
18	14'0"	28'0"	15'0"	30'0"	16'9"	33'6"	18'4"	36'8"
20	14'10"	29'8"	15'9"	31'6"	17'8"	35'4"	19'4"	38'8"
22	15'6"	31'0"	16'7"	33'2"	18'6"	37'0"	20'3"	40'6"
24	16'2"	32'4"	17'4"	34'8"	19'4"	38'8"	21'2"	42'4"
26	16'10"	33'8"	18'0"	36'0"	20'0"	40'0"	22'0"	44'0"

(Continued)

PIPE EXPANSION U-BENDS OR LOOPS (*Continued*)

Pipe Size	Anchor-to-Anchor Expansion							
	7"		8"		10"		12"	
	W	H	W	H	W	H	W	H
28	17'6"	35'0"	18'8"	37'4"	21'0"	42'0"	23'0"	46'0"
30	18'2"	36'4"	19'4"	38'8"	21'7"	43'2"	23'8"	47'4"
32	18'8"	37'4"	20'0"	40'0"	22'4"	44'8"	24'6"	49'0"
34	19'4"	38'8"	20'8"	41'4"	23'0"	46'0"	25'2"	50'4"
36	19'10"	39'8"	21'2"	42'4"	23'8"	47'4"	26'0"	52'0"
42	21'6"	43'0"	23'0"	46'0"	25'6"	51'0"	28'0"	56'0"
48	22'10"	45'8"	24'5"	48'10"	27'4"	54'8"	30'0"	60'0"
54	24'3"	48'6"	25'11"	51'10"	29'0"	58'0"	31'9"	63'6"
60	25'7"	51'2"	27'4"	54'8"	30'6"	61'0"	33'6"	67'0"
72	23'8"	47'4"	29'11"	59'10"	33'5"	66'10"	36'8"	73'4"
84	30'3"	60'6"	32'4"	64'8"	36'1"	72'2"	39'7"	69'2"
96	32'4"	64'8"	34'7"	69'2"	38'8"	77'4"	42'4"	84'8"

18.04 *ASME B31 Piping Code* Comparison

ASME B31 PIPING CODE COMPARISON

Item	Power Piping ASME B31.1-1998	Process Piping ASME B31.3-1996	Building Services Piping ASME B31.9-1996
Application	Power and auxiliary piping for electric generating stations, industrial and institutional plants, central and district heating/cooling plants, and geothermal heating systems.	Petroleum refineries, chemical, pharmaceutical, textile, paper, semiconductor, and cryogenic plants.	Industrial, institutional, commercial, and public buildings and multiunit residences.
Services	Systems include, but are not limited to, steam, water, oil, gas, and air.	Systems include, but are not limited to, raw, intermediate, and finished chemicals, petroleum products, gas, steam air water, fluidized solids, refrigerants, and cryogenic fluids.	Systems include, but are not limited to, water for heating and cooling, condensing water, steam or other condensate, other nontoxic liquids, steam, vacuum, other nontoxic, nonflammable gases, and combustible liquids including fuel oil.
General Limitations	This code does not apply to building services piping within the property limits or buildings of industrial and institutional facilities, which is in the scope of ASME B31.9 except that piping beyond the limitations of material, size, temperature, pressure, and service specified in ASME B31.9 shall conform to the requirements of ASME B31.1. This code excludes power boilers in accordance with the ASME Boiler and Pressure Vessel Code (BPVC) Section I.	This code excludes piping systems for internal gauge pressures above zero but less than 15 psig, provided the fluid is nonflammable, nontoxic, and not damaging to human tissue and its temperature is from −20°F through 366°F. This code excludes power boilers in accordance with the ASME Boiler and Pressure Vessel Code Section I and boiler external piping, which is required to conform to ASME B31.1.	This code prescribes requirements for the design, materials, fabrication, installation, inspection, examination, and testing of piping systems for building services. It includes piping systems in the building or within the property limits. This code excludes power boilers in accordance with the ASME Boiler and Pressure Vessel Code Section I and boiler external piping, which is required to conform to ASME B31.1.
Pipe Size Limitations	No limit	No limit	Carbon steel 30" nominal pipe size and 0.5" wall (30" xs steel pipe) Copper—12" nominal pipe size Stainless steel—12" of nominal pipe size and 0.5" wall
Pressure Limitations	No limit	No limit	Steam and condensate—150 psig Liquids—350 psig Vacuum—1 atmosphere external pressure Compressed air and gas—150 psig

(*Continued*)

ASME B31 PIPING CODE COMPARISON (*Continued*)

Item	Power Piping ASME B31.1-1998	Process Piping ASME B31.3-1996	Building Services Piping ASME B31.9-1996
Temperature Limitations	No limit	No limit	Steam and condensate—366°F Maximum (150 psig) Other gases and vapors—200°F maximum Nonflammable liquids—250°F maximum Minimum temperature all services—0°F
Bypass Requirements	All bypasses must be in accordance with MSS-SP-45. Pipe weight shall be minimum Schedule 80.	Bypasses not addressed—recommend the following B31.1	Bypasses not addressed—recommend the following B31.1
Class I Boiler Systems— ASME BPVC Section I	Boiler external piping is governed by ASME B31.1. All other piping may be governed by this code within the limitations of the code.	Boiler external piping is governed by ASME B31.1. All other piping may be governed by this code within the limitations of the code.	Boiler external piping is governed by ASME B31.1. All other piping may be governed by this code within the limitations of the code.
Class IV Boiler Systems— ASME BPVC Section IV	All piping, including boiler external piping, may be governed by this code within the limitations of the code.	All piping, including boiler external piping, may be governed by this code within the limitations of the code.	All piping, including boiler external piping, may be governed by this code within the limitations of the code.

Class I Boiler Systems
  1 Class I Steam Boiler Systems are constructed for Working Pressures above 15 psig.
  2 Class I Hot Water Boiler Systems are constructed for Working Pressures above 160 psig and/or Working Temperatures above 250°F.
Class IV Boiler Systems
  1 Class IV Steam Boiler Systems are constructed with a maximum Working Pressure of 15 psig.
  2 Class IV Hot Water Boiler Systems are constructed with a maximum Working Pressure of 160 psig and a maximum Working Temperature of 250°F.
Class I Boiler External Piping
  1 Steam Boiler Piping—ASME Code piping is required from the boiler through the 1st stop check valve to the 2nd stop valve.
  2 Steam Boiler Feedwater Piping—ASME Code piping is required from the boiler through the 1st stop valve to the check valve for single boiler feedwater installations and from the boiler through the 1st stop valve and through the check valve to the 2nd stop valve at the feedwater control valve for multiple boiler installations.
  3 Steam Boiler Bottom Blowdown Piping—ASME Code Piping is required from the boiler through the 1st stop valve to the 2nd stop valve.
  4 Steam Boiler Surface Blowdown Piping—ASME Code Piping is required from the boiler to the 1st stop valve.
  5 Steam and Hot Water Boiler Drain Piping—ASME Code Piping is required from the boiler through the 1st stop valve to the 2nd stop valve.
  6 Hot Water Boiler Supply and Return Piping—ASME Code piping is required from the boiler through the 1st stop check valve to the 2nd stop valve on both the supply and return piping.
Class IV Boiler External Piping
  1 All Class IV Boiler External Piping is governed by the respective piping system code.

ASME B31 PIPING CODE COMPARISON

Item	Power Piping ASME B31.1-1998	Process Piping ASME B31.3-1996	Building Services Piping ASME B31.9-1996
Piping Classifications	No classifications required by this code. The code deals with and governs all piping under its jurisdiction the same.		No classifications required by this code. The code deals with and governs all piping under its jurisdiction the same.
Low Temp Chilled Water (0–40°F)		D	
Chilled Water (40–60°F)		D	

(*Continued*)

ASME B31 PIPING CODE COMPARISON (*Continued*)

Item	Power Piping ASME B31.1-1998	Process Piping ASME B31.3-1996	Building Services Piping ASME B31.9-1996
Condenser Water (60–110°F)		D	
Low Temp Heating Water (110–250°F)		N	
High Temp Heating Water (250–450°F)		N—Except Boiler Ext. Piping B31.1 applicable	
Low Press. Steam (15 psig and Less)		N	
High Press. Steam (Above 15 psig)		N—Except Boiler Ext. Piping B31.1 applicable	
Hydrostatic Pressure Testing	Test Medium—Water, unless subject to freezing	Test Medium—Water, unless subject to freezing	Test Medium—Water, unless subject to freezing
	Boiler External Piping—ASME BPVC Section I	N/A	N/A
	Nonboiler External Piping—1.5 times the design pressure but not to exceed the max. allowable system pressure for a minimum of 10 minutes.	Category D or N Fluid Service—1.5 times the design pressure but not to exceed the max. allowable system pressure for a minimum of 10 minutes.	Nonboiler External Piping—1.5 times the design pressure but not to exceed the max. allowable system pressure for a minimum of 10 minutes.
	All Other Services—1.5 times the design pressure but not to exceed the max. allowable system pressure for a minimum of 10 minutes.		All Other Services—1.5 times the design pressure but not to exceed the max. allowable system pressure for a minimum of 10 minutes.
Examination, Inspection, and Testing Requirements	The degree of examination, inspection, and testing, and the acceptance standards must be mutually agreed upon by the manufacturer, fabricator, erector, or contractor and the owner.	The degree of examination, inspection, and testing, and the acceptance standards must be mutually agreed upon by the manufacturer, fabricator, erector, or contractor and the owner.	The degree of examination, inspection, and testing, and the acceptance standards must be mutually agreed upon by the manufacturer, fabricator, erector, or contractor and the owner.
	Class I Steam & Hot Water Systems—Nondestructive testing and visual examinations are required by this code. Percentage and types of tests performed must be agreed upon.	Category D Fluid Service—Visual Examination.	All Services—Visual Examinations.
	Class IV Steam & Hot Water Systems—Visual Examination only. All other services—Visual Examination only.	Category N Fluid Service—Visual Examination, 5% Random Examination of components, fabrication, welds, and installation. Random radiographic or ultrasonic testing of 5% of circumferential butt welds.	If more rigorous examination or testing is required, it must be mutually agreed upon.
	If more rigorous examination or testing is required, it must be mutually agreed upon.	If more rigorous examination or testing is required, it must be mutually agreed upon.	
Nondestructive Testing	Radiographic Ultrasonic Eddy current Liquid penetrant Magnetic particle Hardness tests	Radiographic Ultrasonic Eddy current Liquid penetrant Magnetic particle Hardness tests	Radiographic Ultrasonic Eddy current Liquid penetrant Magnetic particle Hardness tests

ASME B31 Chilled Water System Decision Diagram Chilled Water Systems (0–60°F)

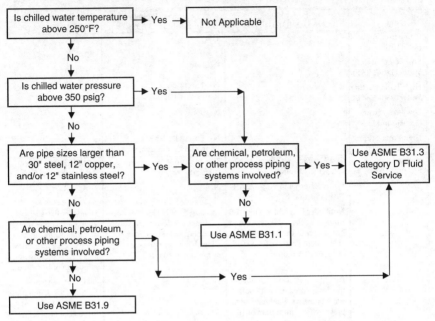

ASME B31 Condenser Water System Decision Diagram Condenser Water Systems (60–110°F)

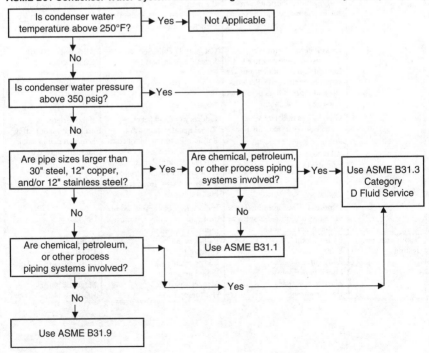

ASME B31 Heating Water System Decision Diagram Heating Water Systems (110–450°F)

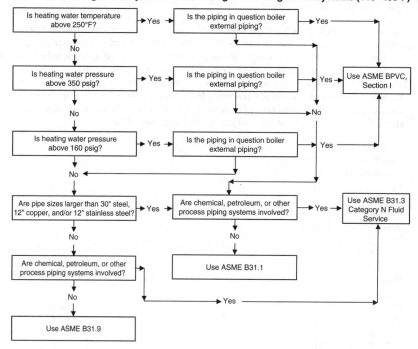

ASME B31 Steam and Steam Condensate System Decision Diagram
Steam and Steam Condensate Systems (0–300 psig)

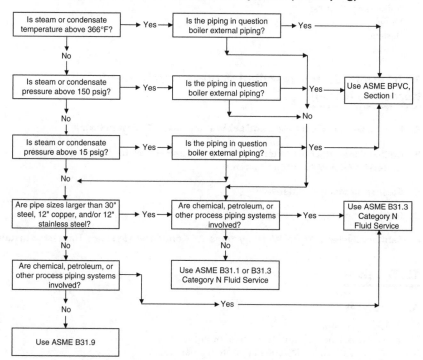

18.05 Galvanic Action

A. Galvanic action results from the electrochemical variation in the potential of metallic ions. If two metals of different potentials are placed in an electrolytic medium (i.e., water), the one with the higher potential will act as an anode and will corrode. The metal with the lower potential, being the cathode, will be unchanged. The greater the separation of the two metals on the following chart, the greater the speed and severity of the corrosion. The following list is in order of their anodic-cathodic characteristics (i.e., metals listed in the following will corrode those listed previously—for example, copper will corrode steel).

Magnesium Alloys
Alclad 3S
Aluminum Alloys
Low-Carbon Steel
Cast Iron
Stainless Steel, Type 410
Stainless Steel, Type 430
Stainless Steel, Type 404
Stainless Steel, Type 304
Stainless Steel, Type 316
Hastelloy A
Lead-Tin Alloys
Brass
Copper
Bronze
90/10 Copper-Nickel
70/30 Copper-Nickel
Inconel
Silver
Stainless Steel (passive)
Monel
Hastelloy C
Titanium

18.06 Piping System Installation Hierarchy (Easiest to Hardest to Install)

A. Natural Gas, Medical Gases, and Laboratory Gases, Easiest to Install

B. Chilled Water, Heating Water, Domestic Cold and Hot Water Systems, and Other Closed HVAC and Plumbing Systems

C. Steam and Steam Condensate

D. Refrigeration Piping Systems

E. Sanitary Systems, Storm Water Systems, AC Condensate Systems, Hardest to Install

18.07 Valves

A. Valve Types
1. Balancing valves:
   a. Duty: Balancing, shutoff (manual or automatic).
   b. A valve specially designed for system balancing.

2. Ball valves full port:
   a. Duty: Shutoff.
   b. A valve with a sphere-shaped internal flow device that rotates open and closed to permit flow, or obstruct flow, through the valve. The valve goes from full open to full close in a quarter turn. The opening in the spherical flow device is the same size or close to the same size as the pipe.
3. Ball valves, reduced port:
   a. Duty: Balancing, shutoff.
   b. A valve with a sphere-shaped internal flow device that rotates open and closed to permit flow, or obstruct flow, through the valve. The valve goes from full open to full close in a quarter turn. The opening in the spherical flow device is smaller than the pipe size.
4. Butterfly valves:
   a. Duty: Shutoff, balancing.
   b. A valve with a disc-shaped internal flow device that rotates open and closed to permit flow, or obstruct flow, through the valve. The valve goes from full open to full close in a quarter turn.
5. Check valves:
   a. Duty: Control flow direction.
   b. A valve opened by the flow of fluid in one direction that closes automatically to prevent flow in the reverse direction. (Types: Ball, Disc, Globe, Piston, Stop, Swing.)
6. Gate valves:
   a. Duty: Shutoff.
   b. A valve with a wedge- or gate-shaped internal flow device that moves on an axis perpendicular to the direction of flow.
   c. This valve is obsolete in today's hydronic piping systems (it is replaced by ball and butterfly valves).
7. Globe valves:
   a. Duty: Throttling.
   b. A valve with a disc or plug that moves on an axis perpendicular to the valve seat.
8. Plug valves:
   a. Duty: Shutoff, balancing.
   b. A valve with a cylindrical or cone-shaped internal flow device that rotates open and closed to permit flow, or obstruct flow, through the valve. The valve goes from full open to full close in a quarter turn.
9. Control Valves. Control valves are mechanical devices used to control the flow of steam, water, gas, and other fluids.
   a. *2-Way*. Temperature control, modulate flow to controlled device, variable flow system.
   b. *3-Way Mixing*. Temperature control, modulate flow to controlled device, constant flow system; two inlets and one outlet.
   c. *3-Way Diverting*. Used to divert flow; generally cannot modulate flow—Two positions; one inlet and two outlets.
   d. *Quick Opening Control Valves*. Quick opening control valves produce a wide free port area with a relatively small percentage of total valve stem stroke. The maximum flow is approached as the valve begins to open.
   e. *Linear Control Valves*. Linear control valves produce free port areas directly related to valve stem stroke. The opening and flow are related in direct proportion.
   f. *Equal Percentage Control Valves*. Equal percentage control valves produce an equal percentage increase in the free port area with each equal increment of valve stem stroke. Each equal increment of opening increases flow by an equal percentage over the previous value (most common HVAC control valve).
   g. Control valves are normally smaller than line size unless used in two-position applications (open/closed).

h. Control valves should normally be sized to provide 20–60 percent of the total system pressure drop.

   1) Water system control valves should be selected with a pressure drop equal to 2–3 times the pressure drop of the controlled device.
   OR
   Water system control valves should be selected with a pressure drop equal to 10 ft or the pressure drop of the controlled device, whichever is greater.
   OR
   Water system control valves for constant flow systems should be sized to provide 25 percent of the total system pressure drop.
   OR
   Water system control valves for variable flow systems should be sized to provide 10 percent of the total system pressure drop or 50 percent of the total available system pressure.

   2) Steam control valves should be selected with a pressure drop equal to 75 percent of the inlet steam pressure.

10. Specialty valves:
   a. Triple-duty valves: Combination check, balancing, and shutoff.
   b. Backflow preventer: prevent contamination of domestic water system. For HVAC applications, use reduced pressure zone backflow preventers.

11. Valves used for balancing need not be line size. Balancing valves should be selected for the midrange of its adjustment.

B. Valve Terms

1. *Actuator.* A mechanical, hydraulic, electric, or pneumatic device or mechanism used to operate a valve.
2. *Adjustable Travel Stop.* A mechanism used to limit the internal flow device travel.
3. *Back Face.* The side of the flange opposite the gasket.
4. *Blind Flange.* A flange with a sealed end to provide a pressure tight closure of a flanged opening.
5. *Body.* The pressure containing shell of a valve or fitting with ends for connection to the piping system.
6. *Bonnet.* A valve body component that contains an opening for the stem. The bonnet may be bolted (Bolted Bonnet), threaded (Threaded Bonnet), or a union (Union Bonnet).
7. *Bronze Mounted.* The seating surfaces of the valve are made of brass or bronze.
8. *Butt Welding Joints.* A joint made to pipes, valves, and fittings with ends adapted for welding by abutting the ends and welding them together.
9. *Chainwheel.* A manual actuator that uses a chain-driven wheel to turn the valve flow device by turning the stem, handwheel, or gearing.
10. *Cock.* A form of a plug valve.
11. *Cold Working Pressure.* Maximum pressure at which a valve or fitting is allowed to operate at ambient temperature.
12. *Concentric Reducer.* A reducer in which both openings are on the same centerline.
13. *Eccentric Reducer.* A reducer with the small end off-center.
14. *Elbow, Long Radius.* An elbow with a centerline turning radius of 1-1/2 times the nominal pipe size of the elbow.
15. *Elbow, Short Radius.* An elbow with a centerline turning radius of one times the nominal pipe size of the elbow.
16. *Face-to-Face Dimension.* The dimension from the face of the inlet to the face of the outlet of the valve or fitting.
17. *Female End.* Internally threaded portion of a pipe, valve, or fitting.
18. *Flanged Joint.* A joint made with an annular collar designed to permit a bolted connection.
19. *Grooved Joint* or *Mechanical Joint.* A joint made with a special mechanical device using a circumferential groove cut into or pressed into the pipes, valves, and fittings to retain a coupling member.

20. *Handwheel.* The valve handle shaped in the form of a wheel.
21. *Inside Screw.* The screw mechanism that moves the internal flow device located within the valve body.
22. *Insulating Unions (Dielectric Unions).* Used in piping systems to prevent dissimilar metals from coming into direct contact with each other. (See Galvanic Action Paragraph.)
23. *Male End.* Externally threaded portion of pipes, valves, or fittings.
24. *Memory Stop.* A device that allows for the repeatable operation of a valve at a position other than full open or full closed, often used to set or mark a balance position.
25. *Nipple.* A short piece of pipe with both ends externally threaded.
26. *Nominal Pipe Size (NPS).* The standard pipe size, but not necessarily the actual dimension.
27. *Nonrising Stem.* When the valve is operated, the stem does not rise through the bonnet; the internal flow device rises on the stem.
28. *Outside Screw and Yoke (OS&Y).* The valve packing is located between the stem threads and the valve body. The valve has a threaded stem that is visible.
29. *Packing.* A material that seals around the movable penetration of the valve stem.
30. *Rising Stem.* When the valve is operated, the stem rises through the bonnet and the internal flow device is moved up or down by the moving stem.
31. *Safety-Relief Valves.* A valve that automatically relieves the system pressure when the internal pressure exceeds a set value. Safety-relief valves may operate on pressure only or on a combination of pressure and temperature.
    a. *Safety Valve.* An automatic pressure relieving device actuated by the static pressure upstream of the valve and characterized by full opening pop action. A safety valve is used primarily for gas or vapor service.
    b. *Relief Valve.* An automatic pressure relieving device actuated by the static pressure upstream of the valve that opens further with the increase in pressure over the opening pressure. A relief valve is used primarily for liquid service.
    c. *Safety Relief Valve.* An automatic pressure actuated relieving device suitable for use either as a safety valve or relief valve, depending on application.
    d. Safety, Relief, and Safety Relief Valve testing is dictated by the Insurance Underwriter.
32. *Seat.* The portion of the valve that the internal flow device presses against to form a tight seal for shutoff.
33. *Slow Opening Valve.* A valve that requires at least five 360-degree turns of the operating mechanism to change from fully closed to fully open.
34. *Socket Welding Joint.* A joint made with a socket configuration to fit the ends of the pipes, valves, or fittings, which is then fillet welded in place.
35. *Soldered Joint.* A joint made with pipes, valves, or fittings in which the joining is accomplished by soldering or brazing.
36. *Stem.* A device that operates the internal flow control device.
37. *Threaded Joint.* A joint made with pipes, valves, or fittings in which the joining is accomplished by threading the components.
38. *Union.* A fitting that allows the assembly or disassembly of the piping system without rotating the piping.

C. Valve Abbreviations

TE	Threaded End
FE	Flanged End
SE	Solder End
BWE	Butt Weld End
SWE	Socket Weld End
TB	Threaded Bonnet
BB	Bolted Bonnet
UB	Union Bonnet
TC	Threaded Cap
BC	Bolted Cap

UC	Union Cap
IBBM	Iron Body, Bronze Mounted
DI	Ductile Iron
SB	Silver Brazed
DD	Double Disc
SW	Solid Wedge Disc
RWD	Resilient Wedge Disc
FW	Flexible Wedge
HW	Handwheel
NRS	Non-Rising Stem
RS	Rising Stem
OS&Y	Outside Screw & Yoke
ISNRS	Inside Screw NRS
ISRS	Inside Screw RS
FF	Flat Face
RF	Raised Face
HF	Hard Faced
MJ	Mechanical Joint
RJ	Ring Type Joint
F&D	Face and Drilled Flange
CWP	Cold Working Pressure
OWG	Oil, Water, Gas, Pressure
SWP	Steam Working Pressure
WOG	Water, Oil, Gas, Pressure
WWP	Water Working Pressure
FTTG	Fitting
FLG	Flange
DWV	Drainage-Waste-Vent Fitting
NPS	Nominal Pipe Size
IPS	Iron Pipe Size
NPT	National Standard Pipe Thread Taper

18.08 Strainers

A. Strainers shall be full line size.

B. Water Systems

1. Strainer type:
   a. 2" and smaller: "Y" Type.
   b. 2-1/2" to 16":   Basket type.
   c. 18" and larger: Multiple basket type.
2. Strainer perforation size:
   a. 4" and smaller: 0.057" dia. perforations.
   b. 5" and larger: 0.125" dia. perforations.
   c. Double perforation diameter for condenser water systems.

C. Steam Systems

1. Strainer type: "Y" Type.
2. Strainer perforation size:
   a. 2" and smaller: 0.033" dia. perforations.
   b. 2-1/2" and larger: 3/64" dia. perforations.

D. Strainer Pressure Drops, Water Systems: Pressure drops listed in the following are based on the GPM and pipe sizing of 4.0 ft./100 ft. pressure drop or 10 ft./sec. velocity.

1. 1-1/2" and smaller (Y type and Basket type):
    a. Pressure drop < 1.0 PSI, 2.31 ft. H_2O.
2. 2"–4" (Y type and Basket type):
    a. Pressure drop: 1.0 PSI, 2.31 ft. H_2O.
3. 5" and larger:
    a. Y-type pressure drop   1.5 PSI, 3.46 ft H_2O
    b. Basket-type pressure drop   1.0 PSI, 2.31 ft. H_2O

Hydronic (Water) Piping Systems

19.01 Hydronic Pipe Sizing

A. 4.0 ft./100 ft. Maximum pressure drop

B. 8 FPS Maximum velocity occupied areas

C. 10 FPS Maximum velocity unoccupied areas

D. Minimum pipe velocity 1.5 FPS, even under low load/flow conditions.

E. Pipe sizing tables are applicable to closed and open hydronic piping systems.

F. See the following pipe sizing tables for copper, steel, and stainless steel.

19.02 Friction Loss Estimate

A. 1.5 × System Length (ft.) × Friction Rate (ft./100 ft.)

B. Pipe Friction Estimate: 3.0 to 3.5 ft./100 ft.

19.03 Pipe Testing

A. 1.5 × System Working Pressure

B. 100 psi Minimum

19.04 Hydronic System Pipe Sizing Tables

HYDRONIC PIPING SYSTEMS—TYPE K COPPER PIPE

Pipe Size	Water Flow—GPM							
	Friction Rate—ft./100 ft.			Velocity—ft./sec.				
	2.0	3.0	4.0	4.0	5.0	6.0	8.0	10.0
1/2	1.0	1.2	1.4		Pressure	drop	governs	
3/4	2.4	3.0	3.5		with these	pipe	sizes	
1	5.2	6.4	7.4					
1-1/4	9	12	13					
1-1/2	15	18	21	21				
2	31	38	44	38				
2-1/2	55	67	78	58	73			
3	87	107	123	83	103	124		
4	183	224	258	146	182	219		
5	324	397	458	226	283	339	452	
6	515	631	729	323	403	484	645	
8	1,064	1,304		563	704	845	1,126	1,408
10	1,887	Velocity	governs	874	1,093	1,311	1,749	2,186
12	3,015	with these	pipe sizes	1,254	1,567	1,880	2,507	3,134

HYDRONIC PIPING SYSTEMS—TYPE L COPPER PIPE

Pipe Size	Water Flow—GPM							
	Friction Rate—ft./100 ft.			Velocity—ft./sec.				
	2.0	3.0	4.0	4.0	5.0	6.0	8.0	10.0
1/2	1.1	1.3	1.5		Pressure	drop	governs	
3/4	2.8	3.4	4.0		with these	pipe	sizes	
1	5.7	6.9	8.0					
1-1/4	10	12	14					
1-1/2	16	19	22	22				
2	32	39	45	39				
2-1/2	57	69	80	59	74			
3	90	111	128	85	106	127		
4	189	231	267	149	187	224		
5	337	412	476	233	291	349	465	
6	540	662	764	335	418	502	669	
8	1,117	1,368		584	730	877	1,169	1,461
10	1,980	Velocity	governs	907	1,134	1,361	1,814	2,268
12	3,191	with these	pipe sizes	1,310	1,637	1,965	2,619	3,274

HYDRONIC PIPING SYSTEMS—TYPE M COPPER PIPE

Pipe Size	Water Flow—GPM							
	Friction Rate—ft./100 ft.			Velocity—ft./sec.				
	2.0	3.0	4.0	4.0	5.0	6.0	8.0	10.0
1/2	1.2	1.5	1.7		Pressure	drop	governs	
3/4	3.1	3.7	4.3		with these	pipe	sizes	
1	6.1	7.5	8.6					
1-1/4	10	13	15	16				
1-1/2	16	20	23	23				
2	33	41	47	40				
2-1/2	58	72	83	61	76			
3	93	114	132	87	109	131		
4	192	236	272	152	190	227		
5	342	419	484	236	295	354	472	
6	549	672	776	339	423	508	677	
8	1,140	1,396		593	742	890	1,187	1,484
10	2,020	Velocity	governs	922	1,152	1,382	1,843	2,304
12	3,228	with these	pipe sizes	1,321	1,652	1,982	2,643	3,304

HYDRONIC PIPING SYSTEMS—STANDARD STEEL PIPE

Pipe Size	Water Flow—GPM							
	Friction Rate—ft./100 ft.			Velocity—ft./sec.				
	2.0	3.0	4.0	4.0	5.0	6.0	8.0	10.0
1/2	1.5	1.9	2.1		Pressure	drop	governs	
3/4	3.2	3.9	4.5		with these	pipe	sizes	
1	6.0	7.4	8.5					
1-1/4	12	15	18	19				
1-1/2	19	23	26	25				
2	36	44	51	42	52			
2-1/2	57	70	80	60	75			
3	100	123	142	92	115	138		
4	204	250	289	159	198	238		
5	368	451	521	249	312	374	499	
6	595	729	841	360	450	540	720	
8	1,216	1,489		624	780	936	1,247	1,559

(Continued)

HYDRONIC PIPING SYSTEMS—STANDARD STEEL PIPE (*Continued*)

Pipe Size	Friction Rate—ft./100 ft.			Water Flow—GPM Velocity—ft./sec.				
	2.0	3.0	4.0	4.0	5.0	6.0	8.0	10.0
10	2,198	governs	with	983	1,229	1,475	1,966	2,458
12	3,512	pipe	sizes	1,410	1,763	2,115	2,820	3,525
14				1,719	2,149	2,579	3,438	4,298
16	Velocity			2,277	2,847	3,416	4,554	5,693
18	these			2,914	3,642	4,371	5,827	7,284
20				3,629	4,536	5,443	7,257	9,071
22				4,422	5,527	6,633	8,843	11,054
24				5,293	6,616	7,940	10,586	13,233
26				6,243	7,804	9,364	12,486	15,607
28				7,271	9,089	10,907	14,542	18,178
30				8,378	10,472	12,566	16,755	20,944
32				9,562	11,953	14,344	19,125	23,906
34				10,826	13,532	16,238	21,651	27,064
36				12,167	15,209	18,251	24,334	30,418
42				16,662	20,827	24,992	33,323	41,654
48				21,861	27,327	32,792	43,722	54,653
54				27,766	34,707	41,649	55,532	69,414
60				34,375	42,969	51,563	68,751	85,938
72				49,710	62,137	74,564	99,419	124,274
84				67,864	84,830	101,796	135,728	169,660
96				88,838	111,048	133,257	177,677	222,096

HYDRONIC PIPING SYSTEMS—XS STEEL PIPE

Pipe Size	Friction Rate—ft./100 ft.			Water Flow—GPM Velocity—ft./sec.				
	2.0	3.0	4.0	4.0	5.0	6.0	8.0	10.0
1/2	1.1	1.3	1.5		Pressure	drop	governs	
3/4	2.4	3.0	3.4		with these	pipe	sizes	
1	4.7	5.8	6.7					
1-1/4	10	12	14					
1-1/2	15	19	22	22				
2	30	37	43	37				
2-1/2	48	59	69	53	66			
3	87	106	123	82	103	124		
4	179	219	253	143	179	215		
5	325	399	460	227	284	340	454	
6	520	637	736	325	406	487	650	
8	1,080	1,322		569	712	854	1,139	1,423
10	2,047	governs	with	931	1,164	1,396	1,862	2,327
12	3,325	pipe	sizes	1,352	1,690	2,028	2,704	3,380
14				1,655	2,069	2,482	3,310	4,137
16	Velocity			2,203	2,754	3,305	4,406	5,508
18	these			2,830	3,537	4,245	5,660	7,075
20				3,535	4,419	5,302	7,070	8,837
22				4,318	5,398	6,477	8,637	10,796
24				5,180	6,475	7,770	10,360	12,950
26				6,120	7,650	9,180	12,240	15,300
28				7,138	8,923	10,708	14,277	17,846
30				8,235	10,294	12,353	16,470	20,588
32				9,410	11,763	14,115	18,820	23,525
34				10,663	13,329	15,995	21,327	26,659
36				11,995	14,994	17,993	23,990	29,988
42				16,460	20,575	24,690	32,921	41,151
48				21,630	27,038	32,446	43,261	54,076
54				27,506	34,382	41,258	55,011	68,764
60				34,086	42,607	51,129	68,172	85,215
72				49,361	61,702	74,042	98,723	123,403
84				67,457	84,321	101,185	134,914	168,642
96				88,373	110,466	132,559	176,745	220,931

HYDRONIC PIPING SYSTEMS—XXS STEEL PIPE

Pipe Size	Water Flow—GPM							
	Friction Rate—ft./100 ft.			Velocity—ft./sec.				
	2.0	3.0	4.0	4.0	5.0	6.0	8.0	10.0
1/2	0.1	0.2	0.2		Pressure	drop	governs	
3/4	0.6	0.7	0.8		with these	pipe	sizes	
1	1.4	1.7	1.9					
1-1/4	4	5	6					
1-1/2	7	8	10					
2	15	19	22	22				
2-1/2	24	29	34	31				
3	47	58	67	52	65			
4	108	132	152	97	122	146		
5	209	256	296	162	202	242		
6	341	417	482	235	294	352	470	
8	825	1,010		463	579	694	926	1,157
10	1,545	Velocity	governs	750	937	1,125	1,499	1,874
12	2,639	with these	pipe sizes	1,132	1,414	1,697	2,263	2,829

HYDRONIC PIPING SYSTEMS—SCHEDULE 40 STEEL PIPE

Pipe Size	Water Flow—GPM							
	Friction Rate—ft./100 ft.			Velocity—ft./sec.				
	2.0	3.0	4.0	4.0	5.0	6.0	8.0	10.0
1/2	1.5	1.9	2.1		Pressure	drop	governs	
3/4	3.2	3.9	4.5		with these	pipe	sizes	
1	6.0	7.4	8.5					
1-1/4	12	15	18	19				
1-1/2	19	23	26	25				
2	36	44	51	42	52			
2-1/2	57	70	80	60	75			
3	100	123	142	92	115	138		
4	204	250	289	159	198	238		
5	368	451	521	249	312	374	499	
6	595	729	841	360	450	540	720	
8	1,216	1,489		624	780	936	1,247	1,559
10	2,198	governs	with	983	1,229	1,475	1,966	2,458
12	3, 65	pipe	sizes	1,396	1,744	2,093	2,791	3,489
14				1,687	2,109	2,531	3,374	4,218
16	Velocity			2,203	2,754	3,305	4,406	5,508
18	these			2,789	3,486	4,183	5,577	6,972
20				3,466	4,333	5,199	6,932	8,665
22				–	–	–	–	–
24				5,013	6,266	7,519	10,026	12,532
26				–	–	–	–	–
28				–	–	–	–	–
30				7,954	9,942	11,930	15,907	19,884
32				9,183	11,479	13,775	18,366	22,958
34				10,422	13,027	15,633	20,844	26,055
36				11,655	14,569	17,482	23,310	29,137
42				16,061	20,077	24,092	32,123	40,153
48				21,173	26,466	31,759	42,345	52,932
54				26,989	33,736	40,484	53,978	67,473
60				33,511	41,888	50,266	67,021	83,776
72				48,669	60,836	73,003	97,337	121,671
84				66,647	83,308	99,970	133,293	166,617
96				87,445	109,306	131,167	174,890	218,612

HYDRONIC PIPING SYSTEMS—SCHEDULE 80 STEEL PIPE

| Pipe Size | Water Flow—GPM | | | | | | | |
| | Friction Rate—ft./100 ft. | | | Velocity—ft./sec. | | | | |
	2.0	3.0	4.0	4.0	5.0	6.0	8.0	10.0
1/2	1.1	1.3	1.5		Pressure	drop	governs	
3/4	2.4	3.0	3.4		with	pipe	sizes	
1	4.7	5.8	6.7		these			
1-1/4	10	12	14					
1-1/2	15	19	22	22				
2	30	37	43	37				
2-1/2	48	59	69	53	66			
3	87	106	123	82	103	124		
4	179	219	253	143	179	215		
5	325	399	460	227	284	340	454	
6	520	637	736	325	406	487	650	
8	1,080	1,322		569	712	854	1,139	1,423
10	1,947	governs	with	896	1,120	1,344	1,791	2,239
12	3,057	pipe	sizes	1,267	1,584	1,901	2,534	3,168
14				1,530	1,912	2,295	3,060	3,825
16	Velocity			2,006	2,508	3,009	4,013	5,016
18	these			2,546	3,183	3,820	5,093	6,366
20				3,151	3,938	4,726	6,302	7,877
22				3,819	4,774	5,729	7,639	9,549
24				4,553	5,692	6,830	9,107	11,383

HYDRONIC PIPING SYSTEMS—SCHEDULE 160 STEEL PIPE

| | Water Flow—GPM | | | | | | | |
| | Friction Rate—ft./100 ft. | | | Velocity—ft./sec. | | | | |
Pipe Size	2.0	3.0	4.0	4.0	5.0	6.0	8.0	10.0
1/2	0.7	0.9	1.0		Pressure	drop	governs	
3/4	1.5	1.8	2.1		with	pipe	sizes	
1	3.1	3.8	4.4		these			
1-1/4	8	10	11					
1-1/2	11	14	16					
2	21	26	30	28				
2-1/2	38	47	54	44	55			
3	67	82	95	68	84			
4	135	166	191	116	145	174		
5	244	299	346	182	228	273		
6	396	485	560	264	330	395	527	
8	805	986		455	568	682	909	1,136
10	1,433	1,755	with	707	884	1,061	1,415	1,769
12	2,259		sizes	1,004	1,255	1,506	2,008	2,510
14	2,928			1,226	1,532	1,839	2,451	3,064
16	Velocity	governs		1,608	2,010	2,412	3,216	4,020
18	these	pipe		2,041	2,551	3,062	4,082	5,103
20				2,527	3,159	3,790	5,054	6,317
22				3,085	3,856	4,628	6,170	7,713
24				3,653	4,566	5,479	7,305	9,132

HYDRONIC PIPING SYSTEMS—SCHEDULE 5 STAINLESS STEEL PIPE

| | Water Flow—GPM | | | | | | | |
| | Friction Rate—ft./100 ft. | | | Velocity—ft./sec. | | | | |
Pipe Size	2.0	3.0	4.0	4.0	5.0	6.0	8.0	10.0
1/2	2.2	2.6	3.0		Pressure	drop	governs	
3/4	4.3	5.2	6.0		with	pipe	sizes	
1	8.3	10.2	11.7		these			
1-1/4	16	20	23	23				
1-1/2	24	29	34	31				
2	44	54	63	49	62			
2-1/2	73	89	103	72	90			
3	125	153	176	109	136	163		
4	248	303	350	184	230	276		
5	428	524	605	280	350	420	559	
6	686	840	970	402	502	603	804	
8	1,392	1,705		692	865	1,038	1,384	1,730
10	2,471	governs	with	1,076	1,345	1,614	2,152	2,690
12		pipe	sizes	1,515	1,894	2,272	3,030	3,787
14				1,835	2,293	2,752	3,669	4,587
16	Velocity			2,404	3,006	3,607	4,809	6,011
18	these			3,057	3,822	4,586	6,115	7,643
20				3,771	4,714	5,656	7,542	9,427
22				4,579	5,723	6,868	9,157	11,447
24				5,437	6,796	8,156	10,874	13,593

HYDRONIC PIPING SYSTEMS—SCHEDULE 10 STAINLESS STEEL PIPE

| | Water Flow—GPM | | | | | | | |
| | Friction Rate—ft./100 ft. | | | Velocity—ft./sec. | | | | |
Pipe Size	2.0	3.0	4.0	4.0	5.0	6.0	8.0	10.0
1/2	1.9	2.3	2.7		Pressure	drop	governs	
3/4	3.8	4.7	5.4		with	pipe	sizes	
1	6.8	8.3	9.6		these			
1-1/4	14	17	20	20				
1-1/2	21	25	29	28				
2	40	49	56	46	57			
2-1/2	67	83	95	68	85			
3	118	144	166	104	130	156		
4	237	290	335	178	222	267		
5	417	511	590	275	343	412	549	
6	672	823	951	396	495	594	791	
8	1,359	1,664		679	849	1,019	1,359	1,698
10	2,433	governs	with	1,063	1,329	1,595	2,126	2,658
12		pipe	sizes	1,503	1,879	2,255	3,006	3,758
14				1,818	2,272	2,726	3,635	4,544
16	Velocity			2,390	2,988	3,585	4,781	5,976
18	these			3,041	3,802	4,562	6,083	7,604
20				3,748	4,685	5,622	7,496	9,370
22				4,553	5,692	6,830	9,107	11,383
24				5,408	6,760	8,111	10,815	13,519

Notes:
1 Maximum recommended pressure drop: 4 ft./100 ft.
2 Maximum recommended velocity (occupied areas): 8 FPS.
3 Maximum recommended velocity (unoccupied areas, shafts, tunnels, etc.): 10 FPS.
4 Standard steel pipe and Type L copper pipe are the most common pipe materials used in HVAC applications.
5 Tables are applicable to closed and open hydronic piping systems.
6 Pipe sizes 5", 22", 26", 28", 32", and 34" are not standard sizes and are not readily available in all locations.
7 Types K, L, and M copper pipe are available in sizes up through 12 inch.
8 Standard and XS steel pipe are available in sizes through 96 inch.

9   XXS steel pipe is available in sizes through 12 inch.
10 Schedule 40 steel pipe is available in sizes through 96 inch.
11 Schedule 80 and 160 steel pipe are available in sizes through 24 inch.
12 Schedule 5 and 10 stainless steel pipe are available in sizes through 24 inch.
13 Standard and Schedule 40 steel pipe have the same dimensions and flow for 10 inch and smaller.
14 XS and Schedule 80 steel pipe have the same dimensions and flow for 8 inch and smaller.
15 XXS and Schedule 160 have no relationship for dimensions or flow.

19.05 Hydronic System Designs and Terminology

A. *Closed Piping Systems.* Piping systems with no more than one point of interface with a compressible gas (generally air). Examples: Chilled Water and Heating Water Systems.

B. *Open Piping Systems.* Piping systems with more than one point of interface with a compressible gas (generally air). Example: Condenser Water Systems

C. *Reverse Return Systems.* Where the length of supply and return piping is nearly equal. Reverse return systems are nearly self-balancing (see Figs. 19.1 through 19.5).

D. *Direct Return Systems.* Where the length of supply and return piping is unequal. Direct return systems are more difficult to balance (see Figs. 19.1 through 19.5).

E. One-Pipe Systems

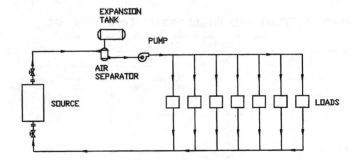

DIRECT RETURN HYDRONIC SYSTEM

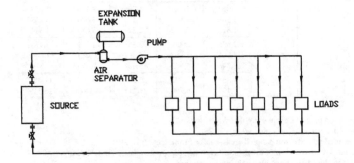

REVERSE RETURN HYDRONIC SYSTEM

FIGURE 19.1   HYDRONIC SYSTEM RETURN TYPES.

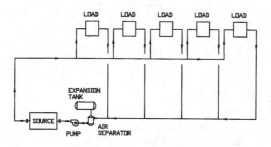

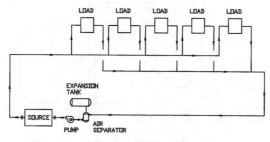

FIGURE 19.2   HYDRONIC SYSTEM RETURN TYPES.

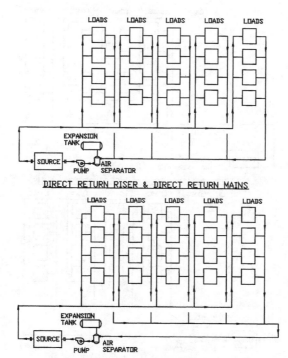

FIGURE 19.3   HYDRONIC SYSTEM RETURN TYPES.

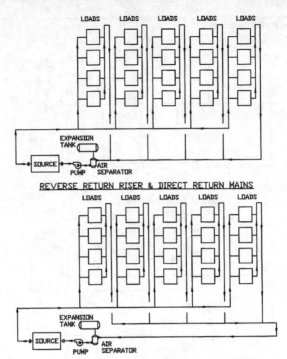

FIGURE 19.4 HYDRONIC SYSTEM RETURN TYPES.

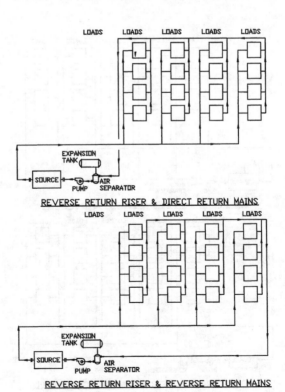

FIGURE 19.5 HYDRONIC SYSTEM RETURN TYPES.

1. One-pipe systems are constant volume flow systems.
2. *All Series Flow Arrangements.* Total circulation flows through every terminal user with lower inlet supply temperatures with each successive terminal device.
3. *Diverted Series Flow Arrangements.* Part of the flow goes through the terminal unit, while the remainder is diverted around the terminal unit using a resistance device (balancing valve, fixed orifice, diverting tees, or flow control devices).

F. Two-Pipe Systems (See Fig. 19.6)

1. The same piping is used to circulate chilled water and heating water.
2. Two-pipe systems are either constant volume flow or variable volume flow systems.
3. *Direct Return Systems.* In these systems, it is critical to provide proper balancing devices (balancing valves or flow control devices).
4. *Reverse Return Systems.* Generally limited to small systems, these systems will simplify balancing.

G. Three-Pipe Systems (Obsolete; See Fig. 19.7)

1. Separate chilled water and heating water supply piping; common return piping is used to circulate chilled water and heating water.

H. Four-Pipe Systems (See Figs. 19.8 and 19.9)

1. Separate supply and return piping (two separate systems) are used to circulate chilled water and heating water.
2. Four-pipe systems are either constant volume flow or variable volume flow systems.
3. *Direct Return Systems.* In these systems, it is critical to provide proper balancing devices (balancing valves or flow control devices).
4. *Reverse Return Systems.* Generally limited to small systems. These systems will simplify balancing.

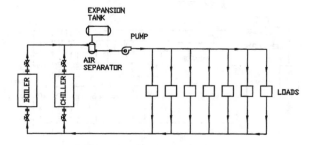

DIRECT RETURN HYDRONIC SYSTEM

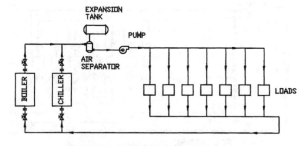

REVERSE RETURN HYDRONIC SYSTEM

FIGURE 19.6   TWO-PIPE HYDRONIC SYSTEMS.

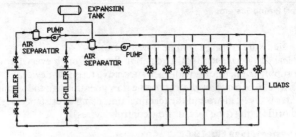

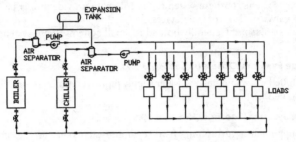

NOTE: 3-PIPE HYDRONIC SYSTEMS ARE OBSOLETE AND ARE NOT USED IN THE DESIGN OF HVAC SYSTEMS TODAY.

FIGURE 19.7 THREE-PIPE HYDRONIC SYSTEMS.

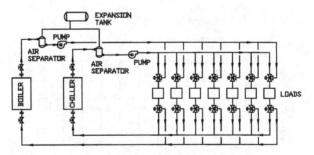

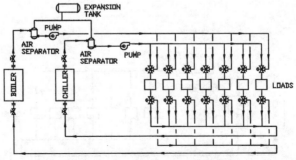

FIGURE 19.8 FOUR-PIPE HYDRONIC SYSTEMS COMMON LOAD SYSTEMS.

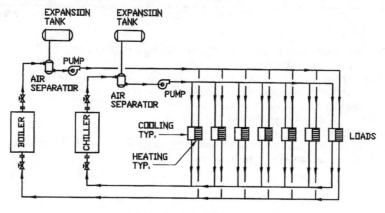

DIRECT RETURN HYDRONIC SYSTEM

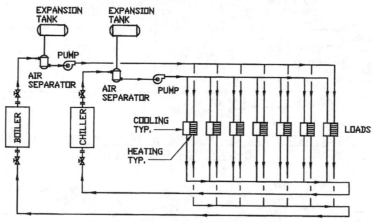

REVERSE RETURN HYDRONIC SYSTEM

FIGURE 19.9 FOUR-PIPE HYDRONIC SYSTEMS INDEPENDENT LOAD SYSTEMS.

I. Ring or Loop Type Systems

1. Piping systems that are laid out to form a loop with the supply and return mains parallel to each other.
2. Constant volume flow or variable volume flow systems.
3. They provide flexibility for future additions and provide service reliability.
4. These can be designed with better diversity factors.
5. During shutdown for emergency or scheduled repairs, maintenance, or modifications, loads, especially critical loads, can be fed from other direction or leg.
6. Isolation valves must be provided at critical junctions and between all major lateral connections so mains can be isolated and flow rerouted.
7. Flows and pressure distribution must be estimated by trial and error or by computer.

J. Constant Volume Flow Systems

1. *Direct Connected Terminals.* Flow created by a main pump through three-way valves.
2. *Indirect Connected Terminals.* Flow created by a separate pump with a bypass and without output controls.
   a. Permit variable volume flow systems.
   b. Subcircuits can be operated with high pump heads without penalizing the main pump.
   c. Require excess flow in the main circulating system.

221

3. Constant volume flow systems are limited to (see Figs. 19.10 and 19.11):
   a. Small systems with a single boiler or chiller.
   b. More than one boiler system if boilers are firetube or firebox boilers.
   c. Two chiller systems if chillers are connected in series.
   d. Small low-temperature heating water systems with 10–20°F delta T.
   e. Small chilled water systems with 7–10°F delta T.
   f. Condenser water systems.
   g. Large chilled water and heating water systems with primary/secondary pumping systems, constant flow primary circuits.

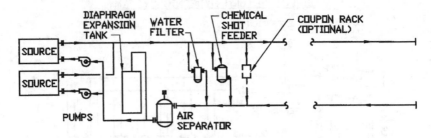

CONSTANT FLOW AT SOURCE & TERMINAL EQUIPMENT
(3-WAY CONTROL VALVES)

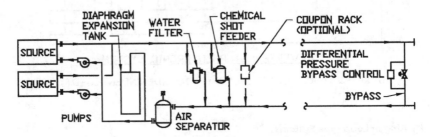

CONSTANT OR VARIABLE FLOW AT SOURCE
VARIABLE FLOW AT TERMINAL EQUIPMENT
(2-WAY CONTROL VALVES)

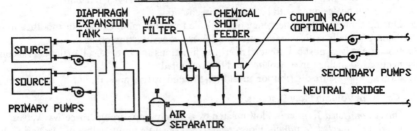

CONSTANT OR VARIABLE FLOW PRIMARY,
VARIABLE FLOW SECONDARY
(2-WAY CONTROL VALVES)

FIGURE 19.10   COUPLED CONSTANT vs. VARIABLE FLOW SYSTEMS.

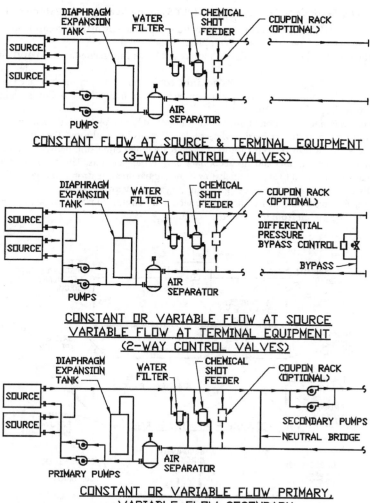

CONSTANT FLOW AT SOURCE & TERMINAL EQUIPMENT
(3-WAY CONTROL VALVES)

CONSTANT OR VARIABLE FLOW AT SOURCE
VARIABLE FLOW AT TERMINAL EQUIPMENT
(2-WAY CONTROL VALVES)

CONSTANT OR VARIABLE FLOW PRIMARY,
VARIABLE FLOW SECONDARY
(2-WAY CONTROL VALVES)

FIGURE 19.11 HEADERED CONSTANT vs. VARIABLE FLOW SYSTEMS.

4. Constant volume flow systems are not suited to (see Figs. 19.10 and 19.11):
   a. Multiple watertube boiler systems.
   b. Parallel chiller systems.
   c. Parallel boiler systems.
5. Constant volume flow systems are generally energy inefficient.

K. Variable Volume Flow Systems (See Figs. 19.10 and 19.11)

1. At partial load, the variable volume flow system return temperatures approach the temperature in the secondary medium.
2. Significantly higher pressure differentials occur at part load and must be considered during design unless variable speed pumps are provided.

L. Primary/Secondary/Tertiary Systems (PST Systems; See Figs. 19.12 through 19.21)

1. PST systems decouple system circuits hydraulically, thereby making control, operation, and analysis of large systems less complex.
2. Secondary (tertiary) pumps should always discharge into secondary (tertiary) circuits away from the common piping.
3. *Cross-Over Bridge.* Cross-over bridge is the connection between the primary (secondary) supply main and the primary (secondary) return main. Size cross-over bridge at a pressure drop of 1–4 ft./100 ft.
4. *Common Piping.* Common piping (sometimes called bypass piping) is the length of piping common to both the primary and secondary circuit flow paths and the secondary and tertiary circuit flow paths. Common piping is the interconnection between the primary and secondary circuits and the secondary and tertiary circuits. The common piping is purposely designed to an extremely low or negligible pressure drop and is generally only 6" to 24" long maximum. By designing for an extremely low pressure drop, the common piping ensures hydraulic isolation of the secondary circuit from the primary circuit, and the tertiary circuit from the secondary circuit.

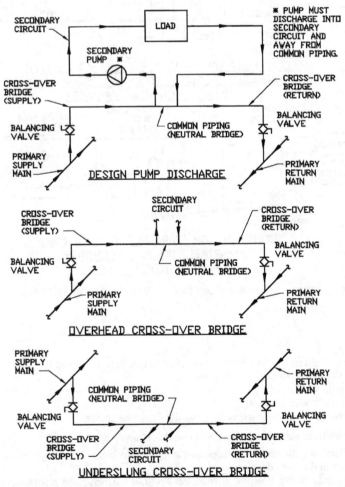

FIGURE 19.12   PRIMARY-SECONDARY TERMINOLOGY.

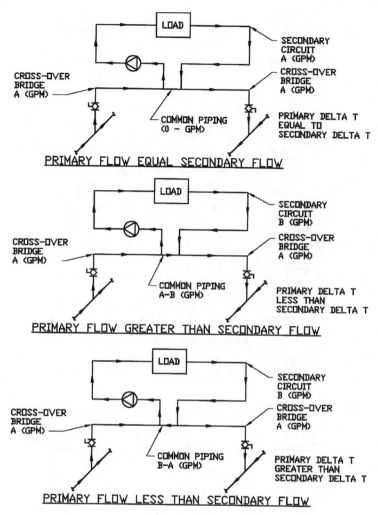

FIGURE 19.13 PRIMARY-SECONDARY FLOW ANALYSIS.

5. Extend common pipe size a minimum of 8 diameters upstream and a minimum of 4 diameters downstream when primary flow rate is considerably less than secondary flow rate (e.g., primary pipe size is smaller than secondary pipe size—use larger pipe size) to prevent any possibility of "jet flow." Common piping (bypass piping) in primary/secondary systems or secondary/tertiary systems should be a minimum of 10 pipe diameters in length and the same size as the larger of the two piping circuits.

6. A one-pipe primary system uses one pipe for supply and return. The secondary circuits are in series. Therefore, this system supplies a different supply water temperature to each secondary circuit, and the secondary circuits must be designed for this temperature change.

7. A two-pipe primary system uses two pipes, one for supply and one for return with a cross-over bridge connecting the two. The secondary circuits are in parallel. Therefore, this system supplies the same supply water temperature to each secondary circuit.

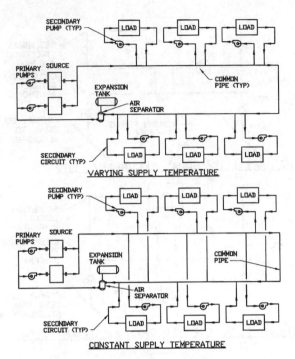

FIGURE 19.14 COUPLED ONE-PIPE PRIMARY-SECONDARY HYDRONIC SYSTEMS.

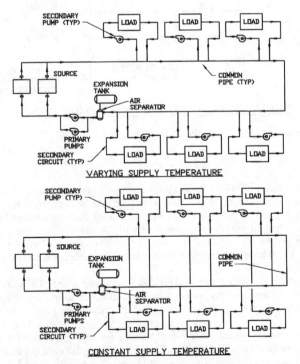

FIGURE 19.15 HEADERED ONE-PIPE PRIMARY-SECONDARY HYDRONIC SYSTEMS.

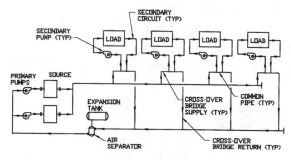

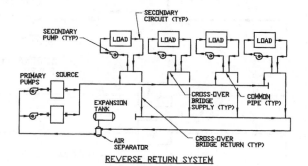

FIGURE 19.16 COUPLED TWO-PIPE PRIMARY-SECONDARY HYDRONIC SYSTEMS.

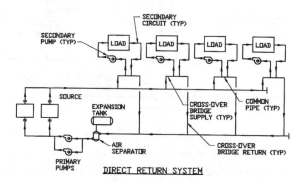

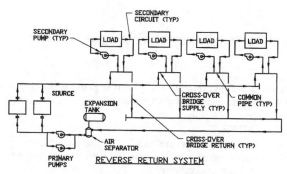

FIGURE 19.17 HEADERED TWO-PIPE PRIMARY-SECONDARY HYDRONIC SYSTEMS.

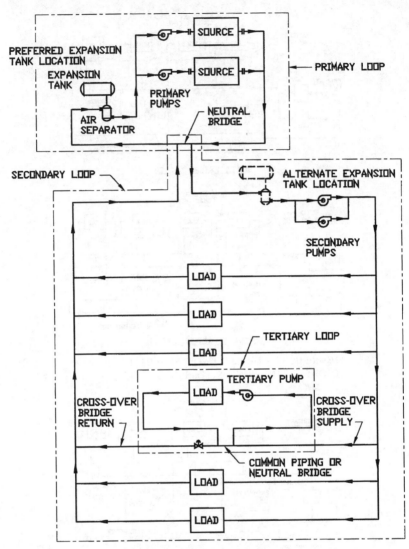

FIGURE 19.18   COUPLED PRIMARY-SECONDARY-TERTIARY HYDRONIC SYSTEMS.

19.06 Hydronic System Design and Piping Installation Guidelines

A. Hydronic systems design principle and goal is to provide the correct water flow at the correct water temperature to the terminal users.

B. Common Design Errors

1. Differential pressure control valves are installed in pump discharge bypasses.
2. Control valves are not selected to provide control with system design pressure differentials at maximum and minimum flows.
3. Control valves are selected with improper pressure drop.

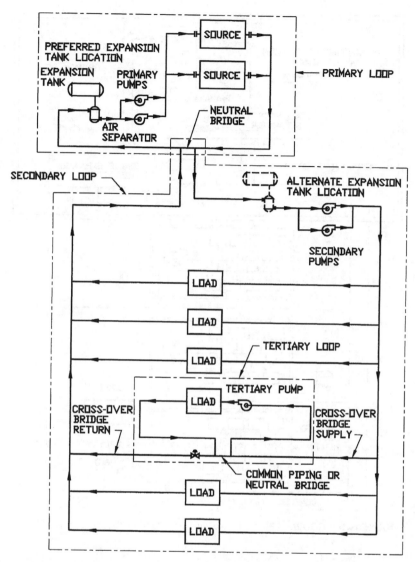

FIGURE 19.19 HEADERED PRIMARY-SECONDARY-TERTIARY HYDRONIC SYSTEMS.

4. Incorrect primary/secondary/tertiary system design.
5. Constant flow secondary or tertiary systems are connected to variable flow primary or secondary systems, respectively.
6. Check valves are not provided in pump discharges when pumps are operating in parallel.
7. Automatic relief valves are oversized, which results in quick, sudden, and sometimes violent system pressure fluctuations.

C. Piping System Arrangements

1. When designing pumping systems for chillers, boilers, and cooling towers, provide either a coupled pumping arrangement (each pump piped directly to each piece of central plant equipment) or provide a headered system (see Fig. 19.22). Hydronic systems should be designed with standby pumps (see Figs. 19.23 and 19.24).

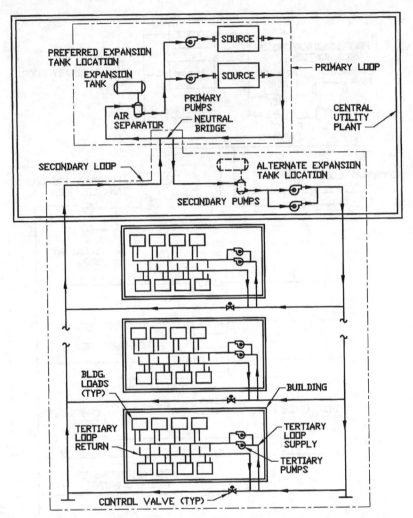

FIGURE 19.20 COUPLED PRIMARY-SECONDARY-TERTIARY HYDRONIC SYSTEMS.

2. Coupled system:
   a. A coupled system should only be used when all the equipment in the system is the same capacity (chillers, boilers, cooling towers, and associated pumps).
3. Headered system:
   a. A headered system is preferred especially when chillers, cooling towers, boilers, and associated pumps are of unequal capacity. Although, the system is easier to design and operate if the equipment is of equal capacity.
   b. When designing a headered system, Griswold valves (flow control devices) must be installed in the supply piping to each piece of equipment to obtain the proper flow through that piece of equipment. In addition to Griswold valves, control valves must be installed to isolate equipment not in service if the system is to be fully automatic. These control valves should be provided with a manual means of opening and closing in case of control system malfunction or failure.
   c. Provide adequate provisions for the expansion and contraction of piping in the boiler, chiller, cooling tower, and pump-headered systems. Provide U-shaped header

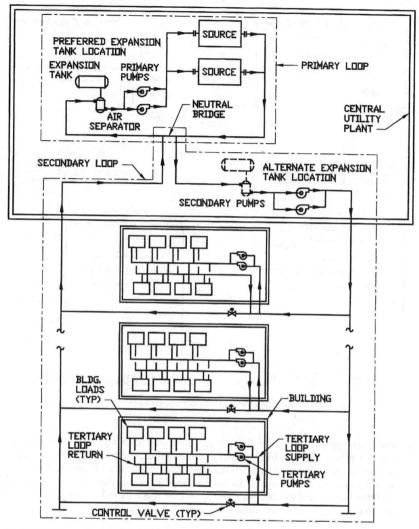

FIGURE 19.21   HEADERED PRIMARY-SECONDARY-TERTIARY HYDRONIC SYSTEMS.

connections for all equipment to accommodate expansion and contraction (first route piping away from the header, then route parallel to the header, and finally route back toward the header; the size of the U-shape will depend on the temperature of the system).

D. The minimum recommended hydronic system pipe size should be 3/4 inch.

E. In general, noise generation in hydronic systems indicates erosion is occurring.

F. Large System Diversities

1. Campus heating: 80 percent.
2. Campus cooling: 65 percent.
3. Constant flow: Load is diversified only; flow is not diversified, resulting in temperature changes.
4. Variable flow: Load and flow are both diversified.

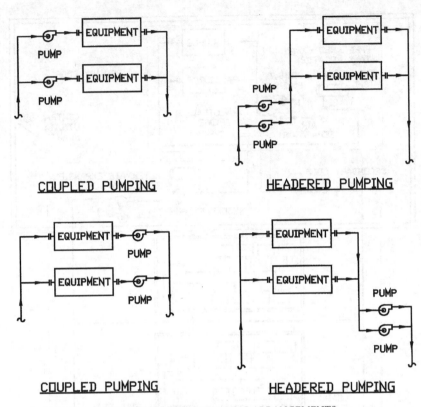

COUPLED PUMPING

HEADERED PUMPING

COUPLED PUMPING

HEADERED PUMPING

FIGURE 19.22   COUPLED-HEADERED PUMPING ARRANGEMENTS.

G. When designing a campus or district type heating or cooling system, the controls at the interface between the central system and the building system should be secured so that access is limited to the personnel responsible for operating the central plant and not accessible to the building operators. Building operators may not fully understand the central plant operation and may unknowingly disrupt the central plant operation with system interface tinkering.

H. Differential pressure control of the system pumps should never be accomplished at the pump. The pressure bypass should be provided at the end of the system or at the end of each of the subsystems regardless of whether the system is a bypass flow system or a variable speed pumping system. Bypass flow need not exceed 20 percent of the pump design flow.

I. Central plant equipment (chillers, boilers, cooling towers, and associated pumps) should be of equal size units; however, the system design may include 1/2-sized units or 1/3-sized units with full-sized equipment. For example, a chiller system may be made up of 1,200-ton, 600-ton, and 400-ton chillers. However, 1/3-sized units have limited application. This permits providing multiple units to achieve the capacity of a single unit and having two or three pumps operate to replace the one larger pump.

J. Pump Discharge Check Valves

1. Pump discharge check valves should be center-guided, spring-loaded, disc-type check valves.

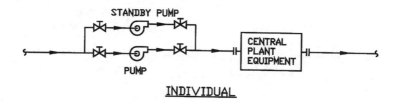

INDIVIDUAL

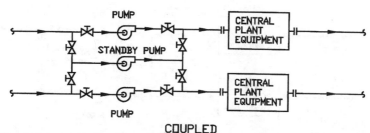

COUPLED

NOTE: STANDBY PUMP MAY SERVE EITHER PIECE OF CENTRAL PLANT EQUIPMENT.

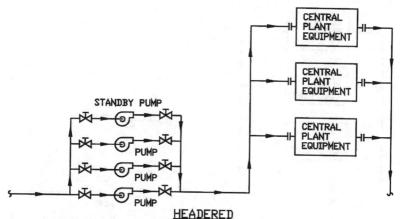

HEADERED

NOTE: STANDBY PUMP MAY SERVE ANY PIECE OF CENTRAL PLANT EQUIPMENT,
PROVIDED ALL EQUIPMENT IS THE SAME CAPACITY.

FIGURE 19.23   STANDBY PUMPS.

2. Pump discharge check valves should be sized so that the check valve is full open at the design flow rate. Generally, this will require the check valve to be one pipe size smaller than the connecting piping.
3. Condenser water system and other open piping system check valves should have globe style bodies to prevent flow reversal and slamming.
4. Installing check valves with 4 to 5 pipe diameters upstream of flow disturbances is recommended by most manufacturers.

K. Install air vents at all high points in water systems. Install drains at all low points in water systems. All automatic air vents, manual air vents, and drains in hydronic systems should be piped to a safe location within 6 inches of the floor, preferably over a floor drain, especially with heating water systems.

L. Thermometers should be installed in both the supply and return piping to all water coils, chillers, boilers, heat exchangers, and other similar equipment. Thermometers

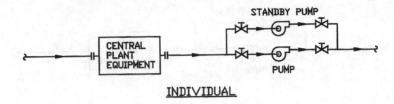

INDIVIDUAL

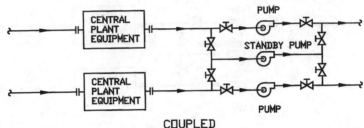

COUPLED

NOTE: STANDBY PUMP MAY SERVE EITHER PIECE OF CENTRAL PLANT EQUIPMENT.

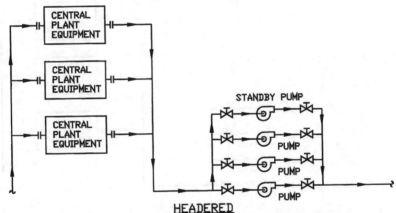

HEADERED

NOTE: STANDBY PUMP MAY SERVE ANY PIECE OF CENTRAL PLANT EQUIPMENT,
PROVIDED ALL EQUIPMENT IS THE SAME CAPACITY.

FIGURE 19.24   STANDBY PUMPS.

should also be installed at each location where major return streams mix at a
location approximately 10 pipe diameters downstream of the mixing point.
Placing thermometers upstream of this point is not required, but often desirable,
because the other return thermometers located upstream will provide the water
temperatures coming into this junction point. Placing thermometers in these
locations will provide assistance in troubleshooting system problems. Liquid-filled-
type thermometers are more accurate than the dial-type thermometers.

M. Select water coils with tube velocities high enough at design flow so that tube
velocities do not end up in the laminar flow region when the flow is reduced
in response to low load conditions. Tube velocities become critical with units
designed for 100 percent outside air at low loads near 32°F. Higher tube velocity
selection results in a higher water pressure drop for the water coil. Sometimes a
trade-off between pressure drop and low load flows must be evaluated.

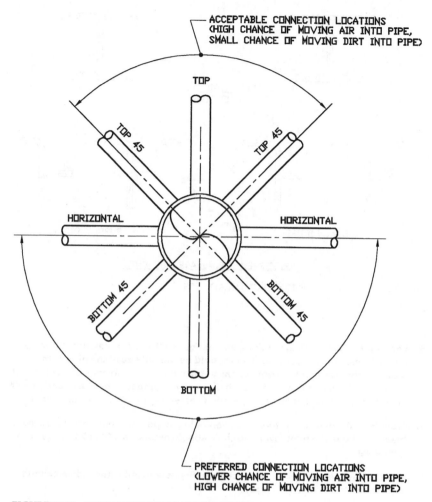

ACCEPTABLE CONNECTION LOCATIONS
(HIGH CHANCE OF MOVING AIR INTO PIPE,
SMALL CHANCE OF MOVING DIRT INTO PIPE)

TOP

TOP 45   TOP 45

HORIZONTAL   HORIZONTAL

BOTTOM 45   BOTTOM 45

BOTTOM

PREFERRED CONNECTION LOCATIONS
(LOWER CHANCE OF MOVING AIR INTO PIPE,
HIGH CHANCE OF MOVING DIRT INTO PIPE)

FIGURE 19.25   HYDRONIC PIPING CONNECTIONS.

N. Install the manual air vent and drain on a coupon rack to relieve pressure from the coupon rack to facilitate removing coupons. Pipe drain to the floor drain.

O. Make piping connections to mains and branches from piping using the following guidelines (see Fig. 19.25):

1. Top of piping: To prevent dirt from entering the main or branch piping.
2. Bottom or side of piping: To prevent air from entering the main or branch piping.

P. Do not use bull head tees (see Fig. 19.26).

Q. Install the manual air vent on a chemical feed tank and also pipe drain to the floor drain.

R. Provide water meters on all makeup water and all blowdown water connections to hydronic systems (heating water, chilled water, condenser water, and steam systems). System water usage is critical in operating the systems, maintaining chemical levels, and troubleshooting the systems. If the project budget permits, these meter readings should be logged and recorded at the building facilities management and control system.

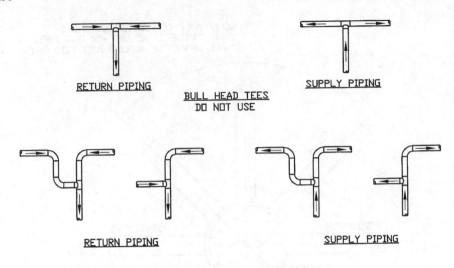

FIGURE 19.26   PIPING DESIGN—BULL HEAD TEES.

S. Locate all valves, strainers, unions, and flanges so they are accessible. All valves (except control valves) and strainers should be the full size of the pipe before reducing the size to make connections to the equipment and controls. Union and/ or flanges should be installed at each piece of equipment, in bypasses and in long piping runs (100 feet or more) to permit disassembly for alteration and repairs.

T. Provide chainwheel operators for all valves in equipment rooms mounted greater than 7'0" above the floor level. The chain should extend to 5'0" to 7'0" above the floor level.

U. All balancing valves should be provided with position indicators and maximum adjustable stops (memory stops).

V. All valves should be installed so the valve remains in service when equipment or piping on the equipment side of the valve is removed.

W. Locate all flow measuring devices in accessible locations with a straight section of pipe upstream (10 pipe diameters) and downstream (5 pipe diameters) of the device or as recommended by the manufacturer.

X. Provide a bypass around water filters and water softeners. Show water filters and water softener feeding hydronic or steam systems on schematic drawings and plans.

Y. Provide vibration isolators for all piping supports connected to, and within 50 feet of, isolated equipment and throughout mechanical equipment rooms, except at base elbow supports and anchor points.

Z. Do not use malleable iron fittings for glycol systems.

AA. Water in a system should be maintained at a pH of approximately 8 to 9. A pH of 7 is neutral; below 7 is acidic; above 7 is alkaline. Closed system water treatment should be 1,600 to 2,000 ppm Borax-Nitrite additive.

BB. Terminal Systems

1. Design for the largest possible system delta T.
2. Better to have terminal coils *slightly* oversized than undersized. Increasing flow rates in terminal coils to twice the design flow rate only increases coil capacity 5 to 16 percent, and tripling the flow rate only increases coil capacity 7 to 22 percent. Grossly oversized terminal unit coils can lead to serious control problems, so care must be taken in properly sizing coils.

CC. Terminal Unit Control Methods

1. Constant supply temperature, variable flow.
2. Variable supply temperature, constant flow.
3. Flow modulation to a minimum value at constant supply temperature. At minimum flow a pump or fan is started to maintain a constant minimum flow at a variable supply temperature.
4. No primary system control, secondary system control is accomplished by blending supply water with return water, or by utilizing face and bypass damper control.

DD. Terminal Unit Design

1. Terminal unit design should be designed for the largest possible system delta T.
2. Terminal unit design should be designed for the closest approach of primary return water temperature and secondary return temperature.
3. Terminals must be selected for full-load and partial-load performance.
4. Select coils with high water velocities at full load, larger pressure drop. This will result in increased performance at partial loads.

EE. Thermal Storage

1. Peak shaving. Constant supply with variable demand.
2. Space heating/cooling. Variable supply with constant-demand waste heat recovery.
3. Variable supply with variable demand.

FF. Provide stop check valves (located closest to the boiler) and isolation valves with a drain between them on both the supply and return connections to all heating water boilers.

19.07 Chilled Water Systems

A. Leaving Water Temperature (LWT): 40–48°F (60°F Maximum)

B. ΔT Range 10–20°F.

C. Chiller Start-up and Shutdown Bypass: When starting a chiller, it takes 5 to 15 minutes from the time the chiller start sequence is initiated until the time the chiller starts to provide chilled water at the design temperature. During this time, the chilled water supply temperature rises above the desired set point. If chilled water temperature is critical and this deviation is unacceptable, the method to correct this problem is to provide the chillers with a bypass that runs from the chiller discharge to the primary pump suction header return. The common pipe only needs to be sized for the flow of one chiller because it is unlikely that more than one chiller will be started at the same time. Chiller system operation with a bypass should be as follows:

1. In the chiller start sequence, the primary chilled water pump is started, the bypass valve is opened, and the supply header valve is closed. When the chilled water supply temperature is reached, as sensed in the bypass, the supply header valve is slowly opened. When the supply header valve is fully opened, the bypass valve is slowly closed.
2. In the chiller stop sequence, the bypass valve is slowly opened. When the bypass valve is fully opened, the supply header valve is slowly closed. When the primary chilled water pump stops, the bypass valve is closed.

D. Large- and campus-chilled water systems should be designed for large delta Ts and for variable flow secondary and tertiary systems.

E. Chilled water pump energy must be accounted for in the chiller capacity because they add heat load to the system (motor out, driven equipment in, see Part 12—Motor Heat Gain).

F. Methods of Maintaining Constant Chilled Water Flow

1. Primary/secondary systems.
2. Bypassing-control.
3. Constant volume flow is only applicable to two chillers in series-flow or single chiller applications.

G. It is best to design chilled water and condenser water systems to pump through the chiller.

H. When combining independent chilled water systems into a central plant . . .

1. Create a central system concept, control scheme, and flow schematics.
2. The system shall only have a single expansion tank connection point sized to handle entire system expansion and contraction.
3. All systems must be altered, if necessary, to be compatible with the central system concept (temperatures, pressures, flow concepts, variable or constant, control concepts).
4. For constant flow and variable flow systems, the secondary chillers are tied into the main chiller plant return main. Chilled water is pumped from the return main through the chiller and back to the return main.
5. District chilled water systems, due to their size, extensiveness, or both, may require that independent plants feed into the supply main at different points. If this is required, design and layout must enable isolating the plant; provide start-up and shutdown bypasses; and provide adequate flow, temperature, pressure, and other control parameter readings and indicators for proper plant operation, as well as other design issues that affect plant operation and optimization.

I. In large systems, it may be beneficial to install a steam-to-water or water-to-water heat exchanger to place an artificial load on the chilled water system to test individual chillers or groups of chillers during plant start-up, after repairs, or for troubleshooting chiller or system problems.

19.08 Low-Temperature Chilled Water Systems (Glycol or Ice Water Systems)

A. Leaving Water Temperature (LWT): 20–40°F (0°F Minimum)

B. ΔT Range 20–40°F

C. The design of low-temperature chilled-water systems is the same as chilled-water systems.

19.09 Heating Water Systems General

A. From a design and practical standpoint, low-temperature heating water systems are often defined as systems with water temperatures of 210°F and less, and high-temperature heating water systems are defined as systems with water temperatures of 211°F and higher.

B. Provide a manual vent on top of a heating water boiler to vent air from the top of the boiler during filling and system operation. Pipe the manual vent discharge to the floor drain.

C. Blowdown separators are not required for hot water boilers, but desirable for maintenance purposes. Install the blowdown separator so the inlet to the separator is at or below the boiler drain to enable the use of the blowdown separator during boiler draining for emergency repairs.

D. Safety: High temperature hydronic systems when operated at higher system temperatures and higher system pressures will result in a lower chance of water hammer and the damaging effects of pipe leaks. These high-temperature heating water systems are also safer than lower-temperature heating water systems because system leaks subcool to temperatures below scalding due to the sudden decrease in pressure and the production of water vapor.

E. Outside air temperature reset of low temperature heating water systems is recommended for energy savings and controllability of terminal units at low load conditions. However, care must be taken with boiler design to prevent thermal shock by low return water temperature or to prevent condensation in the boiler due to low supply water temperature and, therefore, lower combustion stack discharge temperature.

F. Circulating hot water through a boiler that is not operating, in order to keep it hot for standby purposes, creates a natural draft of heated air through the boiler and up the stack, especially in natural draft boilers. Forced draft or induced draft boilers have combustion dampers that close when not firing and therefore reduce, but don't eliminate, this heat loss. Although this heat loss is undesirable for standby boilers, circulating hot water through the boiler is more energy efficient than firing the boiler. Operating a standby boiler may be in violation of air permit regulations in many jurisdictions today.

19.10 Low-Temperature Heating Water Systems

A. Leaving Water Temperature (LWT): 160–200°F (Recommend 180°F, Range: 140–200°F)

B. ΔT Range 20–40°F

C. Low Temperature Water 250°F and Less; 160 psig Maximum

D. The system ΔT is generally limited by the boiler and the maximum temperature difference the boiler can withstand without thermal shock. The following are some common boiler types and the maximum recommended system temperature difference (consult the boiler manufacturer).

1. Steel boilers (fire tube, water tube): 40°F.
2. Cast-iron boilers: 40°F.
3. Modular or copper tube boilers: 100°F (some even higher)

19.11 Medium- and High-Temperature Heating Water Systems

A. Leaving Water Temperature (LWT): 350–450°F

B. ΔT Range 20–100°F

C. Medium Temperature Water 251–350°F, 160 psig Maximum

D. High Temperature Water 351–450°F, 300 psig Maximum

E. The submergence or antiflash margin is the difference between the actual system operating pressure and the vapor pressure of water at the system operating temperature. However, submergence or antiflash margin is often expressed in degrees Fahrenheit—the difference between the temperature corresponding to the vapor pressure equal to the actual system pressure and the system operating temperature.

F. Provide operators on valves on the discharge of the feedwater pumps for medium- and high-temperature systems to provide positive shutoff because the check valves sometimes leak with the large pressure differential. Interlock the valves to open when the pumps operate. Verify that the valve is open with an end switch or a valve positioner.

G. Provide space and racks for spare nitrogen bottles in mechanically pressurized medium- and high-temperature heating water systems.

H. Medium- and High-Temperature Heating Water System Design Principles

1. System pressure must exceed the vapor pressure at the design temperature in all locations in the system. Verify this pressure requirement at the highest location in the system, at the pump suction, and at the control valve when at minimum flow or part load conditions. The greater the elevation difference, above the pressure source (in most cases, the expansion tank), the higher the selected operating temperature should be in the medium- and high-temperature heating water system.
2. Medium- and high-temperature water systems are unforgiving to system design errors in capacity or flow rates.
3. Conversion factors in standard HVAC equations must be adjusted for specific gravity and specific heat at the design temperatures.
4. Thermal expansion and contraction of piping must be considered and are critical in system design.
5. Medium- and high-temperature heating water systems can be transported over essentially unlimited distances.
6. The greater the system delta T, the more economical the system becomes.
7. Use medium- and high-temperature heating water systems when required for process applications because it produces precise temperature control and more uniform surface temperatures in heat transfer devices.
8. The net positive suction head requirements of the medium- and high-temperature system pumps are critical and must be checked for adequate pressure. It is best to locate and design the pumps as follows so cavitation does not occur:
   a. Oversize the pump suction line to reduce resistance.
   b. Locate the pump at a lower level than the expansion tank to take advantage of the static pressure gain.
   c. Elevate the expansion tank above the pumps.
   d. Locate the pumps in the return piping circuit and pump through the boilers, thus reducing the system temperature at the pumps, which reduces the vapor pressure requirements.
9. Either blending fittings or properly designed pipe fittings must be used when blending return water with supply water in large delta T systems or injecting medium- and high-temperature primary supply water into low-temperature secondary circuits. When connecting piping to create a blending tee, the hotter water must always flow downward and the colder water must always flow upward. The blending pipe must remain vertical for a short length equal to a few pipe diameters on either side of the tee. Since turbulence is required for mixing action, it is not desirable to have straight piping for any great distance (a minimum of 10 pipe diameters is adequate).

I. Above approximately 300°F, the bearings and gland seals of a pump must be cooled. Consult factory representatives for all pumps for systems above 250°F to determine specification requirements. Cooling water leaving the pump cooling jacket should not fall below 100°F. The best method for cooling seals is to provide a separate heat exchanger (one at each pump or one for a group of pumps) and circulate the water through the seal chamber. The heat exchanger should be constructed of stainless steel. Another method to cool the seals is to take a side stream flow off of the pump discharge, cool the flow, and inject it into the end face. This is not recommended because the amount of energy wasted is quite substantial.

J. Medium- and high-temperature heating water systems work well for radiant heating systems.

K. Control valves should be placed in the supply to heat exchangers with a check valve in the return. This practice provides a safety shutoff in case of a major leak in the heat exchanger. By placing the control valve in the supply when a leak occurs, the temperature or pressure increases on the secondary side causing the control valve to close while the check valve prevents back flow or pressure from the return. Flashing may occur with the control valve in the supply when a large pressure differential exists or when the system is operated without an antiflash margin. To correct this flashing, control must be split with one control valve in the supply and one control valve in the return.

L. If using medium- or high-temperature heating water systems to produce steam, the steam pressure dictates the delta T and thus the return water temperature.

M. Medium- and High-Temperature Heating Water Systems in Frequent Use

1. Cascade systems with integral expansion space:
   a. Type 1. Feedwater pump piped to steam boiler.
   b. Type 2. Feedwater pump piped to medium- or high-temperature heating water system with steam boiler feedwater provided by medium- and high-temperature heating water system.
2. Flooded generators with external expansion/pressurization provisions.

N. Medium- and High-Temperature Water System Boiler Types

1. Natural circulators, fire tube, and water tube boilers.
2. Controlled (forced) circulation.
3. Combustion (natural and forced), corner tube boilers.

O. Design Requirements

1. Settling chamber to remove any foreign matter, dirt, and debris; oversized header with flanged openings for cleanout.
2. Generator must never be blown down. Blowdown should only be done at the expansion tank or piping system.
3. Boiler safety relief valves should only be tested when water content is cold; otherwise, flashing water-to-steam mixture will erode the valve seat and after opening once or twice the safety relief valves will leak constantly.
4. Boiler safety relief valves must only be considered protection for the boilers. Another safety relief valve must be provided on the expansion tank.
5. Relief valves should be piped to a blowdown tank.

P. Medium- and high-temperature heating water systems may be pressurized by steam systems on the generator discharge or by pump or mechanical means on the suction side of the primary pumps pumping through the boilers.

Q. Steam Pressurized System Characteristics

1. Steam pressurized systems are generally continuously operated with rare shutdowns.

2. A system expansion tank is pressurized with steam and contains a large volume of water at a high temperature, resulting in a considerable ability to absorb load fluctuations.
3. Steam pressurized systems improve the operation of combustion control.
4. A steam pressurized system reduces the need to anticipate load changes.
5. The system is closed and the entry of air or gas is prevented, thus reducing or eliminating corrosion or flow restricting accumulations.
6. Generally, these systems can operate at a lower pressure than pump or mechanically pressurized systems.
7. Steam pressurized systems have a higher first cost.
8. These systems require greater space requirements.
9. The large pressurization tank must be located above and over generators.
10. Pipe discharges into a steam pressurized expansion tank should be vertically upward or should not exceed an angle greater than 45 degrees with respect to the vertical.

R. Mechanically Pressurized System Characteristics

1. Mechanically pressurized systems have flexibility in their expansion tank location.
2. Mechanically pressurized systems should be designed to pump through the generator. Place the expansion and pressurization means at the pump suction inlet.
3. Mechanically pressurized systems are best suited for intermittently operated systems.
4. A submergence or antiflash margin must be provided.
5. A nitrogen supply must be kept on hand. The system cannot operate without nitrogen.
6. Mechanically pressurized systems have a lower first cost.
7. Mechanically pressurized systems require less expansion tank space.
8. Startup and shutdown of these systems is simplified.

S. Pumps in medium- and high-temperature heating water systems should be provided with 1/2 to 3/4 inch bypasses around the check valve and shutoff valves on the pump discharge in order to . . .

1. Refill the pump piping after repairs have been made.
2. Allow for opening the system shutoff valve (often the gate valve), which becomes difficult to open against the pressure differentials experienced.
3. Allow for a slow warming of the pump and pump seals, and for letting sealing surfaces seat properly.

T. Double valves should be installed on both the supply and return side of the equipment for isolation on heating water systems above 250°F with a drain between these valves to visually confirm isolation. The double valving of systems ensures isolation because of the large pressure differentials that occur when the system is opened for repairs. Double valve all of the following.

1. Equipment.
2. Drains.
3. Vents.
4. Gauges.
5. Instrumentation.
6. Double drain and vent valve operation: Fully open the valve closest to the system piping first. Then, open the second valve, modulating the second valve to control flow to the desired discharge rate. Close the second valve first when finished draining or venting. Operating in this fashion keeps the valve closest to the system from being eroded and thus allowing the valve to provide tight shutoff when needed. In addition, this operation allows for the replacement of the second valve with the system in operation since this valve receives most of the wear and tear during operation.

U. Do not use screw fittings because high- and medium-temperature water is very penetrating. Use welded or flanged fittings in lieu of screwed fittings. Do not use union joints.

V. Use of dissimilar metals must be avoided. Use only steel pipe, fittings, valves, flanges, and other devices.

W. Do not use cast-iron or bronze body valves.

X. Use valves with metal-to-metal seats.

Y. Do not use lubricated plug valves.

19.12 Boiler Warming Techniques

A. To provide fully automatic heating water system controls, the controls must look at and evaluate the boiler metal temperature (water temperature) and the refractory temperature prior to starting the primary pumps or enabling the boiler to fire.

B. First, the boiler system design must circulate system water through the boilers to keep the boiler water temperature at system temperature when the boiler is in standby mode, as discussed for boiler warming pump arrangements in the following.

C. Second, the design must look at the water temperature prior to starting the primary pumps to verify the boiler is ready for service.

D. Third, the design must look at refractory temperature to prevent the boiler from going to high fire if the refractory is not at the appropriate temperature. However, the refractory temperature is usually handled by the boiler control package.

E. Boiler warming pumps should be piped to both the system header and the boiler supply piping, thus allowing the boiler to be kept warm (in standby mode) from the system water flow or to warm the boiler when it has been out of service for repairs without the risk of shocking the boiler with system water temperature (see Figs. 19.27 and 19.28).

F. Boiler warming pumps should be selected for 0.1 GPM/BHP (range 0.05 to 0.1 GPM/BHP). At 0.1 GPM/BHP, it takes 45 to 75 minutes to completely exchange the water in the boiler. This flow rate is sufficient to offset the heat loss by radiation and stack losses on boilers when in standby mode of operation. In addition, this flow rate allows the system to keep the boiler warm without firing the boiler, thus allowing for more efficient system operation. For example, it takes 8 to 16 hours to bring a boiler online from a cold start. Therefore, the standby boiler must be kept warm to enable immediate startup of the boiler upon failure of an operating boiler.

G. Heating Water System Warm-Up Procedure

1. Heating water system startup should not exceed a 120°F temperature rise per hour, but boiler or heat exchanger manufacturer limitations should be consulted.
2. It is recommended that no more than a 25°F temperature rise per hour be used when warming heating water systems. Slow warming of the heating water system allows for system piping, supports, hangers, and anchors to keep up with system expansion.
3. Low-temperature heating water systems (250°F and less) should be warmed slowly at a 25°F temperature rise per hour until the system design temperature is reached.
4. Medium- and high-temperature heating water systems (above 250°F) should be warmed slowly at a 25°F temperature rise per hour until a 250°F system temperature is reached. At this temperature, the system should be permitted to settle for at least eight hours or more (preferably overnight). The temperature and pressure mainte-nance time gives the system piping, hangers, supports, and anchors a chance to catch up with the system expansion. After allowing the system to settle, the system can be warmed up to 350°F or the system design temperature in 25°F temperature incre-ments and 25 psig pressure increments, semi-alternating between temperature and pressure increases, and allowing the system to settle for an hour before increasing the temperature or pressure to the next increment. When the system reaches 350°F and the design temperature is above 350°F, the system should be allowed to settle

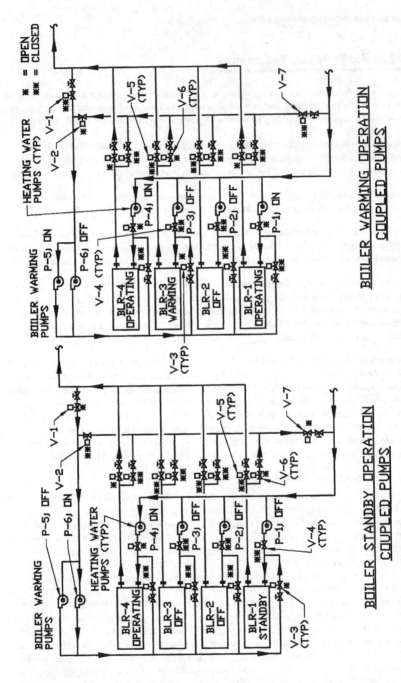

FIGURE 19.27   BOILER STANDBY AND WARMING DIAGRAM—COUPLED PUMPS.

244

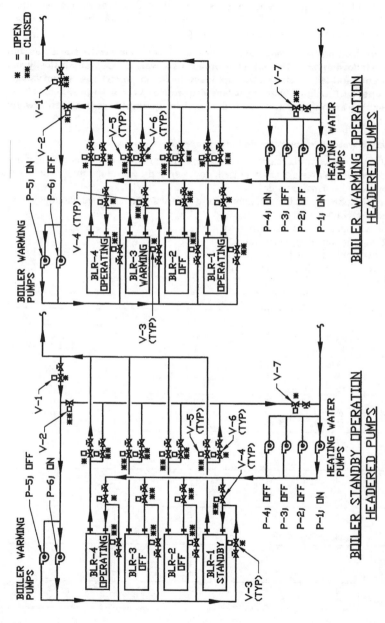

FIGURE 19.28   BOILER STANDBY AND WARMING DIAGRAM—HEADERED PUMPS.

245

for at least eight hours or more (preferably overnight). The temperature and pressure maintenance time gives the system piping, hangers, supports, and anchors a chance to catch up with the system expansion. After allowing the system to settle, the system can be warmed up to 455°F or the system design temperature in 25°F temperature increments and 25 psig pressure increments, semi-alternating between temperature and pressure increases, and allowing the system to settle for an hour before increasing the temperature or pressure to the next increment.

H. **Provide heating water systems with warm-up valves for in-service startup as shown in the following table. This will allow operators to warm these systems slowly and prevent a sudden shock or catastrophic system failure when large system valves are opened. Providing warming valves also reduces wear on large system valves when they are only opened a small amount in an attempt to control system warm-up speed.**

I. **Heating Water System Warming Valve Procedure**

1. First, open the warming return valve slowly to pressurize the equipment without flow.
2. Once the system pressure has stabilized, slowly open the warming supply valve to establish flow and warm the system.
3. When the system pressure and temperature have stabilized, proceed with the following listed items one at a time:
   a. Slowly open the main return valve.
   b. Close the warming return valve.
   c. Slowly open the main supply valve.
   d. Close the warming supply valve.

Bypass and Warming Valves		
	Nominal Pipe Size	
Main Valve Nominal Pipe Size	Series A Warming Valves	Series B Bypass Valves
4	1/2	1
5	3/4	1-1/4
6	3/4	1-1/4
8	3/4	1-1/2
10	1	1-1/2
12	1	2
14	1	2
16	1	3
18	1	3
20	1	3
24	1	4
30	1	4
36	1	6
42	1	6
48	1	8
54	1	8
60	1	10
72	1	10
84	1	12
96	1	12

Notes:

1 Series A valve sizes are utilized in steam service for warming up before the main line is opened, and for balancing pressures where lines are of limited volume.
2 Series B valve sizes are utilized in pipe lines conveying gases or liquids where bypassing may facilitate the operation of the main valve through balancing the pressures on both sides of the disc or discs thereof. The valves in the larger sizes may be of the bolted on type.

19.13 Dual Temperature Water Systems

A. Leaving Cooling Water Temperature: 40–48°F (60°F Maximum)

B. Cooling ΔT Range: 10–20°F

C. Leaving Heating Water Temperature: 160–200°F (Recommend 180°F, Range: 140–200°F)

D. ΔT Range: 20–40°F.

E. Two-pipe switch-over systems provide heating or cooling but not both.

F. Three-pipe systems provide heating and cooling at the same time with a blended return water temperature causing energy waste.

G. Four-Pipe Systems

1. Hydraulically joined at the terminal user (most common with fan coil systems with a single coil). Must design the heating and cooling systems with a common and single expansion tank connected at the generating end. At the terminal units, the heating and cooling supplies should be connected and the heating and cooling returns should be connected.
2. Hydraulically joined at the generator end (most common with condenser water heat recovery systems).
3. Hydraulically joined at both ends.

H. Design of dual temperature water systems is the same as chilled water and heating water systems.

19.14 Condenser Water Systems

A. Entering Water Temperature (EWT): 85°F

B. ΔT Range: 10–20°F

C. Normal ΔT: 10°F

D. Design of condenser water systems is the same as chilled water systems.

E. When using condenser water systems in a waterside economizer operation to produce chilled water, remember to insulate the condenser water piping with the same insulation thickness as the chilled water system.

19.15 Water Source Heat Pump Loop

A. Range: 60–90°F

B. ΔT Range: 10–20°F

19.16 Hydronic System Equation Factors

A. $H = 500 \times GPM \times \Delta T$

B. Substitute the equation factors in the following table for the number 500 in the previous equation for the design water temperatures indicated. Generally, it is acceptable to use 500 for hydronic systems up to 200°F water.

Water Equation Factors		
System Type	System Temperature Range °F	Equation Factor
Low-Temperature (Glycol) Chilled Water	0–40	See Note 2
Chilled Water	40–60	500
Condenser Water Heat Pump Loop	60–110	500
Low-Temperature Heating Water	110–150	490
	151–200	485
	201–250	480
Medium-Temperature Heating Water	251–300	475
	301–350	470
High-Temperature Heating Water	351–400	470
	401–450	470

Notes:
 1 Water equation corrections for temperature, density, and specific heat.
 2 For glycol system equation factors, see Part 20.

C. Water Equation Factor Derivations

1. Standard water conditions:
    a. Temperature: 60°F.
    b. Pressure: 14.7 psia (sea level).
    c. Density: 62.4 lbs./ft.^{3}
2. Water equation examples:
    $H = m \times c_w \times \Delta T$
    Water @ 250°F
    $c_w = 1.02$ Btu/~~Lb H$_2$O~~°F $\times$ 62.4 ~~Lbs.H$_2$O~~/ft$^3 \times 1.0$ ft^3/7.48052 gal.
        $\times 60$ min./h $\times 0.94$ (SG)
      $= 480$ Btu min./h °F gal.
    $H_{250°F} = 480$ Btu min./h °F gal. $\times$ GPM (gal./min.) $\times \Delta T$ (°F)
    $H_{250°F} = 480 \times$ GPM $\times \Delta T$ (°F)
    Water @ 450°F
    $c_w = 1.13$ Btu/~~Lb H$_2$O~~ °F $\times$ 62.4 ~~Lbs.H$_2$O~~/ft$^3 \times 1.0$ ft^3/7.48052 gal.
        $\times 60$ min./h $\times 0.83$ (SG)
      $= 470$ Btu min./h °F gal.
    $H_{450°F} = 470$ Btu min./h °F gal. $\times$ GPM (gal./min.) $\times \Delta T$ (°F)
    $H_{450°F} = 470 \times$ GPM $\times \Delta T$ (°F)

19.17 Hydronic System Design Temperatures and Pressures

A. When designing medium- and high-temperature heating water systems, the appropriate system operating pressure or antiflash margin must be maintained to prevent water from becoming steam and creating water hammer.

B. Antiflash margin is the difference between the actual system operating pressure and the vapor pressure of water at the system operating temperature. However, antiflash margin is often expressed in degrees Fahrenheit—the difference between the temperature corresponding to the vapor pressure equal to the actual system pressure and the system operating temperature.

Hydronic System Design Temperatures and Pressures								
Water Temperature °F	Vapor Pressure psig	System Operating Pressure Antiflash Margin						
		10°F	20°F	30°F	40°F	50°F	60°F	70°F
200	−3.2	−0.6	2.5	6	10	15	21	27
210	−0.6	2.5	6	10	15	21	27	35
212	0.0	3	7	11	16	22	29	36
215	0.9	4	8	13	18	24	31	39
220	2.5	6	10	15	21	27	35	43
225	4.2	8	13	18	24	30	39	48
230	6.1	10	15	21	27	35	43	52
240	10.3	15	21	27	34	43	52	63
250	15.1	21	27	34	43	52	63	75
260	20.7	27	34	43	52	63	75	88
270	27.2	34	43	52	63	75	88	103
275	30.7	39	47	58	69	81	96	111
280	34.5	43	52	63	75	88	103	120
290	42.8	52	63	75	88	103	120	138
300	52.3	63	75	88	103	120	138	159
310	62.9	75	88	103	120	138	159	181
320	74.9	88	103	120	138	159	181	206
325	81.4	96	111	129	148	170	193	219
330	88.3	103	120	138	159	181	206	232
340	103.2	120	138	159	181	206	232	262
350	119.8	138	159	181	206	232	262	294
360	138.2	159	181	206	232	262	294	329
370	158.5	181	206	232	262	294	329	367
375	169.5	193	219	247	277	311	347	387
380	180.9	206	232	262	294	329	367	407
390	205.5	232	262	294	329	367	407	452
400	232.4	262	294	329	367	407	452	500
410	261.8	294	329	367	407	452	500	551
420	293.8	329	367	407	452	500	551	606
425	310.9	347	387	429	475	524	578	635
430	328.6	367	407	452	500	551	606	665
440	366.5	407	452	500	551	606	665	729
450	407.4	452	500	551	606	665	729	797
455	429.1	475	525	578	635	697	762	832

Notes:
1 Safety: High-temperature hydronic systems when operated at higher system temperatures and higher system pressures will result in a lower chance of water hammer and the damaging effects of pipe leaks. These high-temperature heating water systems are also safer than lower-temperature heating water systems because system leaks subcool to temperatures below scalding due to the sudden decrease in pressure and the production of water vapor.
2 The antiflash margin of 40°F minimum is recommended for nitrogen or mechanically pressurized systems.

19.18 Piping Materials

A. 125 Psi (289 ft.) and Less

1. 2" and smaller:
    a. Pipe: black steel pipe, ASTM A53, Schedule 40, Type E or S, Grade B
        Fittings: black malleable iron screw fittings, 150 lbs. ANSI/ASME B16.3
        Joints: pipe threads, general purpose (American) ANSI/ASME B1.20.1.
    b. Pipe: black steel pipe, ASTM A53, Schedule 40, Type E or S, Grade B
        Fittings: cast-iron threaded fittings, 150 lbs. ANSI/ASME B16.4
        Joints: pipe threads, general purpose (American) ANSI/ASME B1.20.1.

   c. Pipe: type "L" copper tubing, ASTM B88, Hard Drawn
     Fittings: wrought copper solder joint fittings, ANSI/ASME B16.22
     Joints: solder joint with 95-5 tin antimony solder, 96-4 tin silver solder, or 94-6
     tin silver solder, ASTM B32.

2. 2-1/2" through 10":
   a. Pipe: black steel pipe, ASTM A53, Schedule 40, Type E or S, Grade B
     Fittings: steel butt-welding fittings ANSI/ASME B16.9
     Joints: welded pipe, ANSI/AWS D1.1 and ANSI/ASME Sec 9.
   b. Pipe: black steel pipe, ASTM A53, Schedule 40, Type E or S, Grade B
     Fittings: factory-grooved end fittings equal to Victaulic full-flow. Tees shall be equal
     to Victaulic Style 20, 25, 27, or 29.
     Joints: Mechanical couplings equal to Victaulic couplings Style 75 or 77 with Grade E
     gaskets, lubricated per the manufacturer's recommendation.

3. 12" and larger:
   a. Pipe: black steel pipe, ASTM A53, 3/8" wall, Type E or S, Grade B
     Fittings: steel butt-welding fittings ANSI/ASME B16.9
     Joints: welded pipe, ANSI/AWS D1.1 and ANSI/ASME Sec 9.
   b. Pipe: black steel pipe, ASTM A53, 3/8" wall, Type E or S, Grade B
     Fittings: Factory-grooved end fittings equal to Victaulic full-flow. Tees shall be equal
     to Victaulic Style 20, 25, 27, or 29.
     Joints: mechanical couplings equal to Victaulic couplings Style 75 or 77 with Grade E
     gaskets, lubricated per manufacturer's recommendation.

4. Mechanical joint manufacturers:
   a. Victaulic.
   b. Anvil Gruvlok.
   c. Grinnell.

B. 126–250 Psig (290–578 ft.)

1. 1-1/2" and smaller:
   a. Pipe: black steel pipe, ASTM A53, Schedule 40, Type E or S, Grade B
     Fittings: forged steel socket-weld, 300 lbs., ANSI B16.11
     Joints: welded pipe, ANSI/AWS D1.1 and ANSI/ASME Sec 9.
   b. Pipe: carbon steel pipe, ASTM A106, Schedule 80, Grade B
     Fittings: forged steel socket-weld, 300 lbs., ANSI B16.11
     Joints: welded pipe, ANSI/AWS D1.1 and ANSI/ASME Sec 9.

2. 2" and larger:
   a. Pipe: black steel pipe, ASTM A53, Schedule 40, Type E or S, Grade B
     Fittings: steel butt-welding fittings, 300 lbs., ANSI/ASME B16.9
     Joints: welded pipe, ANSI/AWS D1.1 and ANSI/ASME Sec 9.
   b. Pipe: carbon steel pipe, ASTM A106, Schedule 80, Grade B
     Fittings: steel butt-welding fittings, 300 lbs., ANSI/ASME B16.9
     Joints: welded pipe, ANSI/AWS D1.1 and ANSI/ASME Sec 9.

19.19 Expansion Tanks and Air Separators

A. Minimum (Fill) Pressure

1. Height of system + 5 to 10 psi or 5–10 psi, whichever is greater.

B. Maximum (System) Pressure

1. 150-lbs. systems: 45–125 psi.
2. 250-lbs. systems: 125–225 psi.

C. System Volume Estimate

1. 12 gal./ton.
2. 35 gal./BHP.

D. Connection Location

1. Suction side of pump(s).
2. Suction side of primary pumps when used in primary/secondary/tertiary systems. An alternate location in primary/secondary/tertiary systems with a single secondary circuit may be the suction side of the secondary pumps.

E. Expansion Tank Design Considerations

1. Solubility of air in water. The amount of air that water can absorb and hold in solution is temperature- and pressure-dependent. As temperature increases, maximum solubility decreases, and as pressure increases, maximum solubility increases. Therefore, expansion tanks are generally connected to the suction side of the pump (the lowest pressure point).
2. Expansion tank sizing. If due to space or structural limitations, the expansion tank must be undersized, the minimum expansion tank size should be capable of handling at least 1/2 of the system expansion volume. With less than this capacity, system startup becomes a tedious and extremely sensitive process. If the expansion tank is under-sized, an automatic drain should be provided and operated by the control system in addition to the manual drain. Size both the manual and automatic drains to enable a quick dump of a waterlogged tank (especially critical with undersized tanks) within the limits of the nitrogen fill speed and system pressure requirements.
3. System volume changes:
   a. System startup and shutdown result in the largest change in system volume.
   b. System volume expansion and contraction must be evaluated at full load and partial load. Variations caused by load changes are described in the following:
      1. In constant flow systems, heating water return temperatures rise and chilled water temperatures drop as load decreases until at no load the return temperature is equal to the supply temperature. Heating systems expand and cooling systems contract at part load.
      2. In variable flow systems, heating water return temperatures drop and chilled water return temperatures rise as load decreases until at no load the return temperature equals the temperature in the secondary medium. Heating systems contract, and cooling systems expand at part load.
4. Expansion tanks are used to accept system volume changes, and a gas cushion (usually air or nitrogen) pressure is maintained by releasing the gas from the tank and readmitting the gas into the tank as the system water expands and contracts, respectively. Expansion tanks are used where constant pressurization in the system must be maintained.
5. Cushion tanks are used in conjunction with expansion tanks and are limited in size. As system water expands, pressure increases in the cushion tank until reaching the relief point, at which time it discharges to a lower-pressure expansion tank. As the system water contracts, pressure decreases in the cushion tank until reaching a low limit, at which time the pump starts and pumps the water from the low pressure expansion tank to the cushion tank, thus increasing the pressure. Cushion tank relief and makeup flow rates are based on the initial expansion of a heating system, or the initial contraction of a cooling system during start-up, because this will be the largest change in system volume for either system.
6. Compression tanks build their own pressure through the thermal expansion of the system contents. Compression tanks are not recommended on medium- or high-temperature heating water systems.
7. When expansion tank level transmitters are provided for building automation control systems, the expansion tank level should be provided from the level transmitter with local readout at the expansion tank, compression tank, or cushion tank. Also provide a sight glass or some other means of visually verifying the level in the tank and the accuracy of transmitter.
8. When expansion tank pressure transmitters are provided for building automation control systems, the expansion tank pressure should be provided from the pressure transmitter with local readout at the expansion tank, compression tank, or cushion tank. Also provide a pressure gauge at the tank to verify the transmitter.

9. Nitrogen relief from the expansion, cushion, or compression tank must be vented to outside (the noise when discharging is quite deafening). The vent can be tied into the vent off of the blowdown separator. Also need to provide nitrogen pressure monitoring and alarms and manual nitrogen relief valves.

10. Expansion tank sizing can be simplified using the following tables and their respective correction factors. These tables can be especially helpful for preliminary sizing.

    a. Low-temperature systems. Tables on pages 253 through 256.

    b. Medium-temperature systems. Tables on pages 257 through 260.

    c. High-temperature systems. Tables on pages 260 through 264.

11. Figure 19.29 is a photograph of an expansion tank in its installed condition.

F. Air Separators

1. Air separators shall be full line size.

2. Figure 19.30 is a photograph of an air separator in its installed condition.

FIGURE 19.29   PHOTOGRAPH OF AN EXPANSION TANK.

FIGURE 19.30   PHOTOGRAPH OF AN AIR SEPARATOR.

EXPANSION TANK SIZING—LOW TEMPERATURE SYSTEMS

Tank Size Expressed as a Percentage of System Volume				
	Expansion Tank Type			
Maximum System Temperature °F	Closed Tank	Open Tank	Diaphragm Tank	
			Tank Volume	Acceptance Volume
100	2.21	1.37	1.32	0.59
110	3.08	1.87	1.83	0.82
120	3.71	2.24	2.21	0.99
130	4.81	2.87	2.86	1.28
140	5.67	3.37	3.37	1.51
150	6.77	3.99	4.03	1.80
160	7.87	4.61	4.68	2.10
170	9.20	5.36	5.48	2.45

(Continued)

EXPANSION TANK SIZING—LOW TEMPERATURE SYSTEMS (*Continued*)

Maximum System Temperature °F	Tank Size Expressed as a Percentage of System Volume			
	Expansion Tank Type			
	Closed Tank	Open Tank	Diaphragm Tank	
			Tank Volume	Acceptance Volume
180	10.53	6.11	6.27	2.81
190	11.87	6.86	7.06	3.16
200	13.20	7.61	7.86	3.52
210	14.77	–	8.79	3.93
220	16.34	–	9.72	4.35
230	17.90	–	10.66	4.77
240	19.71	–	11.73	5.25
250	21.51	–	12.80	5.73

Notes:
1 Table based on initial temperature: 50°F.
2 Table based on initial pressure: 10 psig.
3 Table based on maximum operating pressure: 30 psig.
4. For initial and maximum pressures different than those listed above, multiply the tank size only (not the Acceptance Volume) by correction factors contained in the Low Temperature System Correction Factor tables that follow.

CLOSED EXPANSION TANK SIZING
LOW TEMPERATURE SYSTEM CORRECTION FACTORS

Initial Pressure psig	Pressure Increase—psig Initial Pressure + Pressure Increase = Maximum Operating Pressure									
	5	10	15	20	25	30	35	40	45	50
5	1.76	1.06	0.83	0.71	0.64	0.59	0.56	0.53	0.51	0.50
10	2.66	1.55	1.18	**1.00**	0.89	0.82	0.76	0.72	0.69	0.67
15	3.73	2.14	1.60	1.34	1.18	1.07	0.99	0.94	0.89	0.86
20	4.99	2.81	2.08	1.72	1.50	1.36	1.25	1.17	1.11	1.06
25	6.43	3.57	2.62	2.15	1.86	1.67	1.53	1.43	1.35	1.29
30	8.05	4.43	3.22	2.62	2.26	2.02	1.84	1.71	1.61	1.53
35	9.85	5.37	3.88	3.14	2.69	2.39	2.18	2.02	1.89	1.80
40	11.83	6.41	4.60	3.70	3.16	2.80	2.54	2.35	2.20	2.07
45	13.99	7.54	5.39	4.31	3.66	3.23	2.93	2.70	2.52	2.37
50	16.34	8.75	6.23	4.96	4.21	3.70	3.34	3.07	2.86	2.69
55	18.86	10.06	7.13	5.66	4.78	4.20	3.78	3.46	3.22	3.02
60	21.57	11.46	8.09	6.41	5.40	4.72	4.24	3.88	3.60	3.37
65	24.46	12.95	9.11	7.20	6.05	5.28	4.73	4.32	4.00	3.75
70	27.53	14.53	10.20	8.03	6.73	5.87	5.25	4.78	4.42	4.13
75	30.77	16.20	11.34	8.91	7.45	6.48	5.79	5.27	4.86	4.54
80	34.21	17.96	12.55	9.84	8.21	7.13	6.36	5.78	5.33	4.96
85	37.82	19.81	13.81	10.81	9.01	7.81	6.95	6.31	5.81	5.41
90	41.61	21.75	15.13	11.83	9.84	8.52	7.57	6.86	6.31	5.87
95	45.59	23.79	16.52	12.89	10.71	9.25	8.22	7.44	6.83	6.35
100	49.74	25.91	17.97	13.99	11.61	10.02	8.89	8.04	7.37	6.84

Notes:
1 Table based on initial temperature: 50°F.
2 Table based on initial pressure: 200 psig.
3 Table based on maximum operating pressure: 300 psig.

CLOSED EXPANSION TANK SIZING
LOW TEMPERATURE SYSTEM CORRECTION FACTORS

Initial Pressure psig	Pressure Increase—psig Initial Pressure + Pressure Increase = Maximum Operating Pressure									
	55	60	65	70	75	80	85	90	95	100
5	0.48	0.47	0.47	0.46	0.45	0.44	0.44	0.43	0.43	0.43
10	0.65	0.63	0.62	0.61	0.59	0.59	0.58	0.57	0.56	0.56
15	0.83	0.80	0.78	0.77	0.75	0.74	0.73	0.72	0.71	0.70
20	1.03	0.99	0.96	0.94	0.92	0.90	0.89	0.87	0.86	0.85
25	1.24	1.19	1.16	1.13	1.10	1.08	1.06	1.04	1.02	1.00
30	1.47	1.41	1.37	1.33	1.29	1.26	1.24	1.21	1.19	1.17
35	1.71	1.65	1.59	1.54	1.50	1.46	1.43	1.40	1.37	1.35
40	1.98	1.89	1.82	1.77	1.71	1.67	1.63	1.59	1.56	1.53
45	2.26	2.16	2.07	2.00	1.94	1.89	1.84	1.80	1.76	1.73
50	2.55	2.44	2.34	2.26	2.18	2.12	2.06	2.01	1.97	1.93
55	2.86	2.73	2.62	2.52	2.44	2.36	2.30	2.24	2.19	2.14
60	3.19	3.04	2.91	2.80	2.70	2.62	2.54	2.48	2.42	2.36
65	3.54	3.36	3.21	3.09	2.98	2.88	2.80	2.72	2.65	2.59
70	3.90	3.70	3.53	3.39	3.27	3.16	3.06	2.98	2.90	2.83
75	4.27	4.05	3.87	3.71	3.57	3.45	3.34	3.24	3.16	3.08
80	4.67	4.42	4.21	4.04	3.88	3.75	3.63	3.52	3.43	3.34
85	5.08	4.81	4.58	4.38	4.21	4.06	3.92	3.81	3.70	3.61
90	5.51	5.21	4.95	4.73	4.54	4.38	4.23	4.10	3.99	3.88
95	5.95	5.62	5.34	5.10	4.89	4.71	4.55	4.41	4.28	4.17
100	6.41	6.05	5.74	5.48	5.26	5.06	4.88	4.73	4.59	4.46

Notes:
1 Table based on initial temperature: 50°F.
2 Table based on initial pressure: 10 psig.
3 Table based on maximum operating pressure: 30 psig.

DIAPHRAGM EXPANSION TANK SIZING
LOW TEMPERATURE SYSTEM CORRECTION FACTORS

Initial Pressure psig	Pressure Increase—psig Initial Pressure + Pressure Increase = Maximum Operating Pressure									
	5	10	15	20	25	30	35	40	45	50
5	2.21	1.33	1.04	0.89	0.80	0.74	0.70	0.67	0.64	0.62
10	2.66	1.55	1.18	**1.00**	0.89	0.82	0.76	0.72	0.69	0.67
15	3.11	1.78	1.33	1.11	0.98	0.89	0.83	0.78	0.74	0.71
20	3.55	2.00	1.48	1.22	1.07	0.96	0.89	0.84	0.79	0.76
25	4.00	2.22	1.63	1.34	1.16	1.04	0.95	0.89	0.84	0.80
30	4.45	2.45	1.78	1.45	1.25	1.11	1.02	0.95	0.89	0.85
35	4.89	2.67	1.93	1.56	1.34	1.19	1.08	1.00	0.94	0.89
40	5.34	2.89	2.08	1.67	1.43	1.26	1.15	1.06	0.99	0.94
45	5.79	3.12	2.23	1.78	1.52	1.34	1.21	1.12	1.04	0.98
50	6.24	3.34	2.38	1.89	1.61	1.41	1.27	1.17	1.09	1.03
55	6.68	3.57	2.53	2.01	1.69	1.49	1.34	1.23	1.14	1.07
60	7.13	3.79	2.68	2.12	1.78	1.56	1.40	1.28	1.19	1.12

(Continued)

DIAPHRAGM EXPANSION TANK SIZING
LOW TEMPERATURE SYSTEM CORRECTION FACTORS (*Continued*)

Initial Pressure psig	Pressure Increase—psig Initial Pressure + Pressure Increase = Maximum Operating Pressure									
	5	10	15	20	25	30	35	40	45	50
65	7.58	4.01	2.82	2.23	1.87	1.64	1.47	1.34	1.24	1.16
70	8.03	4.24	2.97	2.34	1.96	1.71	1.53	1.39	1.29	1.21
75	8.47	4.46	3.12	2.45	2.05	1.79	1.59	1.45	1.34	1.25
80	8.92	4.68	3.27	2.57	2.14	1.86	1.66	1.51	1.39	1.29
85	9.37	4.91	3.42	2.68	2.23	1.93	1.72	1.56	1.44	1.34
90	9.82	5.13	3.57	2.79	2.32	2.01	1.79	1.62	1.49	1.38
95	10.26	5.36	3.72	2.90	2.41	2.08	1.85	1.67	1.54	1.43
100	10.71	5.58	3.87	3.01	2.50	2.16	1.91	1.73	1.59	1.47

Notes:
1 Table based on initial temperature: 50°F.
2 Table based on initial pressure: 10 psig.
3 Table based on maximum operating pressure: 30 psig.

DIAPHRAGM EXPANSION TANK SIZING
LOW TEMPERATURE SYSTEM CORRECTION FACTORS

Initial Pressure psig	Pressure Increase—psig Initial Pressure + Pressure Increase = Maximum Operating Pressure									
	55	60	65	70	75	80	85	90	95	100
5	0.61	0.59	0.58	0.57	0.56	0.56	0.55	0.55	0.54	0.54
10	0.65	0.63	0.62	0.61	0.59	0.59	0.58	0.57	0.56	0.56
15	0.69	0.67	0.65	0.64	0.62	0.61	0.60	0.60	0.59	0.58
20	0.73	0.71	0.69	0.67	0.65	0.64	0.63	0.62	0.61	0.60
25	0.77	0.74	0.72	0.70	0.68	0.67	0.66	0.64	0.63	0.63
30	0.81	0.78	0.76	0.73	0.71	0.70	0.68	0.67	0.66	0.65
35	0.85	0.82	0.79	0.77	0.74	0.73	0.71	0.69	0.68	0.67
40	0.89	0.86	0.82	0.80	0.77	0.75	0.74	0.72	0.71	0.69
45	0.93	0.89	0.86	0.83	0.80	0.78	0.76	0.74	0.73	0.71
50	0.97	0.93	0.89	0.86	0.83	0.81	0.79	0.77	0.75	0.74
55	1.01	0.97	0.93	0.89	0.86	0.84	0.81	0.79	0.78	0.76
60	1.06	1.00	0.96	0.92	0.89	0.87	0.84	0.82	0.80	0.78
65	1.10	1.04	1.00	0.96	0.92	0.89	0.87	0.84	0.82	0.80
70	1.14	1.08	1.03	0.99	0.95	0.92	0.89	0.87	0.85	0.83
75	1.18	1.12	1.06	1.02	0.98	0.95	0.92	0.89	0.87	0.85
80	1.22	1.15	1.10	1.05	1.01	0.98	0.95	0.92	0.89	0.87
85	1.26	1.19	1.13	1.08	1.04	1.01	0.97	0.94	0.92	0.89
90	1.30	1.23	1.17	1.12	1.07	1.03	1.00	0.97	0.94	0.92
95	1.34	1.27	1.20	1.15	1.10	1.06	1.02	0.99	0.96	0.94
100	1.38	1.30	1.24	1.18	1.13	1.09	1.05	1.02	0.99	0.96

Notes:
1 Table based on initial temperature: 50°F.
2 Table based on initial pressure: 10 psig.
3 Table based on maximum operating pressure: 30 psig.

EXPANSION TANK SIZING—MEDIUM TEMPERATURE SYSTEMS

Maximum System Temperature °F	Tank Size Expressed as a Percentage of System Volume			
	Expansion Tank Type			
	Closed Tank	Open Tank	Diaphragm Tank	
			Tank Volume	Acceptance Volume
250	263.25	–	18.02	5.73
260	285.30	–	19.53	6.21
270	310.23	–	21.24	6.75
280	335.16	–	22.95	7.29
290	360.08	–	24.65	7.83
300	387.88	–	26.56	8.44
310	415.67	–	28.46	9.04
320	443.47	–	30.36	9.65
330	474.13	–	32.46	10.32
340	504.80	–	34.56	10.98
350	538.33	–	36.86	11.71

Notes:
1 Table based on initial temperature: 50°F.
2 Table based on initial pressure: 200 psig.
3 Table based on maximum operating pressure: 300 psig.
4 For initial and maximum pressures different than those listed above, multiply the tank size only (not the Acceptance Volume) by correction factors contained in the Medium Temperature System Correction Factor tables that follow.

CLOSED EXPANSION TANK SIZING
MEDIUM TEMPERATURE SYSTEM CORRECTION FACTORS

Initial Pressure psig	Pressure Increase—psig Initial Pressure + Pressure Increase = Maximum Operating Pressure									
	10	20	30	40	50	60	70	80	90	**100**
30	0.36	0.21	0.16	0.14	0.13	0.12	0.11	0.10	0.10	0.10
40	0.52	0.30	0.23	0.19	0.17	0.15	0.14	0.14	0.13	0.13
50	072	0.41	0.30	0.25	0.22	0.20	0.18	0.17	0.16	0.16
60	0.94	0.52	0.39	0.32	0.28	0.25	0.23	0.21	0.20	0.19
70	1.19	0.66	0.48	0.39	0.34	0.30	0.28	0.26	0.24	0.23
80	1.47	0.80	0.58	0.47	0.41	0.36	0.33	0.31	0.29	0.27
90	1.78	0.97	0.70	0.56	0.48	0.43	0.39	0.36	0.34	0.32
100	2.12	1.14	0.82	0.66	0.56	0.49	0.45	0.41	0.39	0.36
110	2.49	1.34	0.95	0.76	0.64	0.57	0.51	0.47	0.44	0.41
120	2.88	1.54	1.09	0.87	0.74	0.65	0.58	0.54	0.50	0.47
130	3.31	1.76	1.25	0.99	0.83	0.73	0.66	0.60	0.56	0.52
140	3.77	2.00	1.41	1.11	0.94	0.82	0.73	0.67	0.62	0.58
150	4.26	2.25	1.58	1.25	1.05	0.91	0.82	0.75	0.69	0.65
160	4.78	2.52	1.76	1.39	1.16	1.01	0.90	0.82	0.76	0.71
170	5.32	2.80	1.96	1.54	1.28	1.11	0.99	0.90	0.83	0.78
180	5.90	3.09	2.16	1.69	1.41	1.22	1.09	0.99	0.91	0.85
190	6.50	3.40	2.37	1.85	1.54	1.34	1.19	1.08	0.99	.92

(Continued)

CLOSED EXPANSION TANK SIZING
MEDIUM TEMPERATURE SYSTEM CORRECTION FACTORS (*Continued*)

Initial Pressure psig	Pressure Increase—psig Initial Pressure + Pressure Increase = Maximum Operating Pressure									
	10	20	30	40	50	60	70	80	90	**100**
200	7.14	3.73	2.59	2.02	1.68	1.45	1.29	1.17	1.08	1.00
210	7.81	4.07	2.82	2.20	1.83	1.58	1.40	1.27	1.16	1.08
220	8.50	4.42	3.06	2.39	1.98	1.71	1.51	1.37	1.25	1.16
230	9.22	4.79	3.32	2.58	2.13	1.84	1.63	1.47	1.35	1.25
240	9.98	5.18	3.58	2.78	2.30	1.98	1.75	1.58	1.44	1.34
250	10.76	5.58	3.85	2.98	2.47	2.12	1.87	1.69	1.54	1.43
260	11.57	5.99	4.13	3.20	2.64	2.27	2.00	1.80	1.65	1.52

Notes:
1 Table based on initial temperature: 50°F.
2 Table based on initial pressure: 200 psig.
3 Table based on maximum operating pressure: 300 psig.

CLOSED EXPANSION TANK SIZING
MEDIUM TEMPERATURE SYSTEM CORRECTION FACTORS

Initial Pressure psig	Pressure Increase—psig Initial Pressure + Pressure Increase = Maximum Operating Pressure									
	110	120	130	140	150	160	170	180	190	200
30	0.09	0.09	0.09	0.09	0.09	0.08	0.08	0.08	0.08	0.08
40	0.12	0.12	0.12	0.11	0.11	0.11	0.11	0.11	0.10	0.10
50	0.15	0.15	0.14	0.14	0.14	0.13	0.13	0.13	0.13	0.13
60	0.19	0.18	0.17	0.17	0.17	0.16	0.16	0.16	0.15	0.15
70	0.22	0.21	0.21	0.20	0.20	0.19	0.19	0.18	0.18	0.18
80	0.26	0.25	0.24	0.23	0.23	0.22	0.22	0.21	0.21	0.21
90	0.30	0.29	0.28	0.27	0.26	0.26	0.25	0.25	0.24	0.24
100	0.35	0.33	0.32	0.31	0.30	0.29	0.28	0.28	0.27	0.27
110	0.39	0.38	0.36	0.35	0.34	0.33	0.32	0.31	0.31	0.30
120	0.44	0.42	0.41	0.39	0.38	0.37	0.36	0.35	0.34	0.33
130	0.50	0.47	0.45	0.44	0.42	0.41	0.40	0.39	0.38	0.37
140	0.55	0.52	0.50	0.48	0.47	0.45	0.44	0.43	0.42	0.41
150	0.61	0.58	0.55	0.53	0.51	0.49	0.48	0.47	0.46	0.44
160	0.67	0.63	0.61	0.58	0.56	0.54	0.52	0.51	0.50	0.48
170	0.73	0.69	0.66	0.63	0.61	0.59	0.57	0.55	0.54	0.53
180	0.80	0.76	0.72	0.69	0.66	0.64	0.62	0.60	0.58	0.57
190	0.87	0.82	0.78	0.75	0.72	0.69	0.67	0.65	0.63	0.61
200	0.94	0.89	0.84	0.81	0.77	0.74	0.72	0.70	0.68	0.66
210	1.01	0.96	0.91	0.87	0.83	0.80	0.77	0.75	0.73	0.71
220	1.09	1.03	0.97	0.93	0.89	0.86	0.83	0.80	0.78	0.75
230	1.17	1.10	1.04	1.00	0.95	0.92	0.88	0.85	0.83	0.81
240	1.25	1.18	1.12	1.06	1.02	0.98	0.94	0.91	0.88	0.86
250	1.33	1.26	1.19	1.13	1.08	1.04	1.00	0.97	0.94	0.91
260	1.42	1.34	1.27	1.20	1.15	1.10	1.06	1.03	0.99	0.96

Notes:
1 Table based on initial temperature: 50°F.
2 Table based on initial pressure: 200 psig.
3 Table based on maximum operating pressure: 300 psig.

DIAPHRAGM EXPANSION TANK SIZING
MEDIUM TEMPERATURE SYSTEM CORRECTION FACTORS

Initial Pressure psig	Pressure Increase—psig Initial Pressure + Pressure Increase = Maximum Operating Pressure									
	10	20	30	40	50	60	70	80	90	100
30	1.74	1.03	0.79	0.67	0.60	0.55	0.52	0.50	0.48	0.46
40	2.06	1.19	0.90	0.75	0.67	0.61	0.57	0.54	0.51	0.49
50	2.37	1.35	1.00	0.83	0.73	0.66	0.61	0.57	0.55	0.52
60	2.69	1.50	1.11	0.91	0.79	0.71	0.66	0.61	0.58	0.56
70	3.01	1.66	1.21	0.99	0.86	0.77	0.70	0.65	0.62	0.59
80	3.33	1.82	1.32	1.07	0.92	0.82	0.75	0.69	0.65	0.62
90	3.64	1.98	1.43	1.15	0.98	0.87	0.79	0.73	0.69	0.65
100	3.96	2.14	1.53	1.23	1.05	0.93	0.84	0.77	0.72	0.68
110	4.28	2.30	1.64	1.31	1.11	0.98	0.88	0.81	0.76	0.71
120	4.60	2.46	1.74	1.39	1.17	1.03	0.93	0.85	0.79	0.75
130	4.92	2.62	1.85	1.47	1.24	1.08	0.97	0.89	0.83	0.78
140	5.23	2.78	1.96	1.55	1.30	1.14	1.02	0.93	0.86	0.81
150	5.55	2.93	2.06	1.63	1.36	1.19	1.07	0.97	0.90	0.84
160	5.87	3.09	2.17	1.71	1.43	1.24	1.11	1.01	0.93	0.87
170	6.19	3.25	2.27	1.79	1.49	1.30	1.16	1.05	0.97	0.90
180	6.50	3.41	2.38	1.86	1.56	1.35	1.20	1.09	1.01	0.94
190	6.82	3.57	2.49	1.94	1.62	1.40	1.25	1.13	1.04	0.97
200	7.14	3.73	2.59	2.02	1.68	1.45	1.29	1.17	1.08	**1.00**
210	7.46	3.89	2.70	2.10	1.75	1.51	1.34	1.21	1.11	1.03
220	7.78	4.05	2.80	2.18	1.81	1.56	1.38	1.25	1.15	1.06
230	8.09	4.21	2.91	2.26	1.87	1.61	1.43	1.29	1.18	1.10
240	8.41	4.36	3.02	2.34	1.94	1.67	1.47	1.33	1.22	1.13
250	8.73	4.52	3.12	2.42	2.00	1.72	1.52	1.37	1.25	1.16
260	9.05	4.68	3.23	2.50	2.06	1.77	1.56	1.41	1.29	1.19

Notes:
1 Table based on initial temperature: 50°F.
2 Table based on initial pressure: 200 psig.
3 Table based on maximum operating pressure: 300 psig.

DIAPHRAGM EXPANSION TANK SIZING
MEDIUM TEMPERATURE SYSTEM CORRECTION FACTORS

Initial Pressure psig	Pressure Increase—psig Initial Pressure + Pressure Increase = Maximum Operating Pressure									
	110	120	130	140	150	160	170	180	190	200
30	0.45	0.44	0.43	0.42	0.41	0.41	0.40	0.40	0.39	0.39
40	0.48	0.46	0.45	0.44	0.43	0.43	0.42	0.41	0.41	0.40
50	0.50	0.49	0.48	0.46	0.45	0.45	0.44	0.43	0.43	0.42
60	0.53	0.52	0.50	0.49	0.48	0.47	0.46	0.45	0.44	0.44
70	0.56	0.54	0.52	0.51	0.50	0.49	0.48	0.47	0.46	0.45
80	0.59	0.57	0.55	0.53	0.52	0.51	0.49	0.48	0.48	0.47
90	0.62	0.60	0.57	0.56	0.54	0.53	0.51	0.50	0.49	0.48
100	0.65	0.62	0.60	0.58	0.56	0.55	0.53	0.52	0.51	0.50

(Continued)

DIAPHRAGM EXPANSION TANK SIZING
MEDIUM TEMPERATURE SYSTEM CORRECTION FACTORS (*Continued*)

Initial Pressure psig	Pressure Increase—psig Initial Pressure + Pressure Increase = Maximum Operating Pressure									
	110	120	130	140	150	160	170	180	190	200
110	0.68	0.65	0.62	0.60	0.58	0.57	0.55	0.54	0.53	0.52
120	0.71	0.67	0.65	0.62	0.60	0.59	0.57	0.56	0.54	0.53
130	0.74	0.70	0.67	0.65	0.62	0.61	0.59	0.57	0.56	0.55
140	0.76	0.73	0.70	0.67	0.65	0.63	0.61	0.59	0.58	0.56
150	0.79	0.75	0.72	0.69	0.67	0.64	0.63	0.61	0.59	0.58
160	0.82	0.78	0.74	0.71	0.69	0.66	0.64	0.63	0.61	0.60
170	0.85	0.81	0.77	0.74	0.71	0.68	0.66	0.64	0.63	0.61
180	0.88	0.83	0.79	0.76	0.73	0.70	0.68	0.66	0.64	0.63
190	0.91	0.86	0.82	0.78	0.75	0.72	0.70	0.68	0.66	0.64
200	0.94	0.89	0.84	0.81	0.77	0.74	0.72	0.70	0.68	0.66
210	0.97	0.91	0.87	0.83	0.79	0.76	0.74	0.71	0.69	0.67
220	1.00	0.94	0.89	0.85	0.81	0.78	0.76	0.73	0.71	0.69
230	1.02	0.97	0.92	0.87	0.84	0.80	0.78	0.75	0.73	0.71
240	1.05	0.99	0.94	0.90	0.86	0.82	0.79	0.77	0.74	0.72
250	1.08	1.02	0.96	0.92	0.88	0.84	0.81	0.79	0.76	0.74
260	1.11	1.05	0.99	0.94	0.90	0.86	0.83	0.80	0.78	0.75

Notes:
1 Table based on initial temperature: 50°F.
2 Table based on initial pressure: 200 psig.
3 Table based on maximum operating pressure: 300 psig.

EXPANSION TANK SIZING—HIGH TEMPERATURE SYSTEMS

Tank Sized Expressed as a Percentage of System Volume				
	Expansion Tank Type			
			Diaphragm Tank	
Maximum System Temperature °F	Closed Tank	Open Tank	Tank Volume	Acceptance Volume
350	1,995.03	–	47.71	11.71
360	2,119.30	–	50.68	12.44
370	2,243.58	–	53.65	13.17
380	2,378.48	–	56.88	13.96
390	2,524.02	–	60.36	14.82
400	2,669.56	–	63.84	15.67
410	2,815.10	–	67.32	16.53
420	2,981.90	–	71.31	17.51
430	3,138.07	–	75.04	18.42
440	3,315.51	–	79.29	19.46
450	3,492.95	–	83.53	20.51

Notes:
1 Table based on initial temperature: 50°F.
2 Table based on initial pressure: 600 psig.
3 Table based on maximum operating pressure: 800 psig.
4 For initial and maximum pressures different than those listed above, multiply the tank size (the Acceptance Volume) by correction factors contained in the High Temperature System Correction Factor tables that follow.

CLOSED EXPANSION TANK SIZING
HIGH TEMPERATURE SYSTEM CORRECTION FACTORS

Initial Pressure psig	Pressure Increase—psig Initial Pressure + Pressure Increase = Maximum Operating Pressure									
	20	40	60	80	100	120	140	160	180	200
160	0.68	0.37	0.27	0.22	0.19	0.17	0.16	0.15	0.14	0.13
180	0.83	0.46	0.33	0.27	0.23	0.20	0.19	0.17	0.16	0.15
200	1.01	0.55	0.39	0.32	0.27	0.24	0.22	0.20	0.19	0.18
220	1.19	0.64	0.46	0.37	0.31	0.28	0.25	0.23	0.22	0.20
240	1.40	0.75	0.53	0.43	0.36	0.32	0.29	0.26	0.25	0.23
260	1.62	0.86	0.61	0.49	0.41	0.36	0.32	0.30	0.28	0.26
280	1.85	0.98	0.70	0.55	0.46	0.41	0.37	0.33	0.31	0.29
300	2.10	1.11	0.78	0.62	0.52	0.46	0.41	0.37	0.35	0.32
320	2.37	1.25	0.88	0.69	0.58	0.51	0.45	0.41	0.38	0.36
340	2.65	1.40	0.98	0.77	0.64	0.56	0.50	0.46	0.42	0.39
360	2.95	1.55	1.08	0.85	0.71	0.62	0.55	0.50	0.46	0.43
380	3.27	1.71	1.19	0.94	0.78	0.68	0.60	0.55	0.50	0.47
400	3.60	1.88	1.31	1.02	0.85	0.74	0.66	0.59	0.55	0.51
420	3.95	2.06	1.43	1.12	0.93	0.80	0.71	0.65	0.59	0.55
440	4.31	2.25	1.56	1.21	1.01	0.87	0.77	0.70	0.64	0.59
460	4.69	2.44	1.69	1.31	1.09	0.94	0.83	0.75	0.69	0.64
480	5.08	2.64	1.83	1.42	1.17	1.01	0.90	0.81	0.74	0.69
500	5.50	2.85	1.97	1.53	1.26	1.09	0.96	0.87	0.79	0.73
520	5.92	3.07	2.12	1.64	1.36	1.17	1.03	0.93	0.85	0.78
540	6.37	3.29	2.27	1.76	1.45	1.25	1.10	0.99	0.90	0.84
560	6.82	3.53	2.43	1.88	1.55	1.33	1.17	1.05	0.96	0.89
580	7.30	3.77	2.59	2.00	1.65	1.41	1.25	1.12	1.02	0.94
600	7.79	4.02	2.76	2.13	1.75	1.50	1.32	1.19	1.08	**1.00**
620	8.30	4.28	2.93	2.26	1.86	1.59	1.40	1.26	1.15	1.06
640	8.82	4.54	3.11	2.40	1.97	1.69	1.48	1.33	1.21	1.12
660	9.36	4.81	3.30	2.54	2.09	1.78	1.57	1.41	1.28	1.18
680	9.91	5.10	3.49	2.69	2.20	1.88	1.65	1.48	1.35	1.24
700	10.49	5.39	3.69	2.84	2.33	1.99	1.74	1.56	1.42	1.31

Notes:
1 Table based on initial temperature: 50°F.
2 Table based on initial pressure: 600 psig.
3 Table based on maximum operating pressure: 800 psig.

CLOSED EXPANSION TANK SIZING
HIGH TEMPERATURE SYSTEM CORRECTION FACTORS

Initial Pressure psig	Pressure Increase—psig Initial Pressure + Pressure Increase = Maximum Operating Pressure									
	220	240	260	280	300	320	340	360	380	400
160	0.13	0.12	0.12	0.11	0.11	0.11	0.11	0.10	0.10	0.10
180	0.15	0.14	0.14	0.13	0.13	0.13	0.12	0.12	0.12	0.12
200	0.17	0.16	0.16	0.15	0.15	0.14	0.14	0.14	0.13	0.13
220	0.19	0.19	0.18	0.17	0.17	0.16	0.16	0.15	0.15	0.15
240	0.22	0.21	0.20	0.19	0.19	0.18	0.18	0.17	0.17	0.17
260	0.25	0.24	0.23	0.22	0.21	0.20	0.20	0.19	0.19	0.19
280	0.28	0.26	0.25	0.24	0.23	0.23	0.22	0.21	0.21	0.20
300	0.31	0.29	0.28	0.27	0.26	0.25	0.24	0.24	0.23	0.22
320	0.34	0.32	0.31	0.29	0.28	0.27	0.27	0.26	0.25	0.25
340	0.37	0.35	0.33	0.32	0.31	0.30	0.29	0.28	0.27	0.27
360	0.40	0.38	0.37	0.35	0.34	0.32	0.31	0.31	0.30	0.29
380	0.44	0.42	0.40	0.38	0.37	0.35	0.34	0.33	0.32	0.31
400	0.48	0.45	0.43	0.41	0.39	0.38	0.37	0.36	0.35	0.34
420	0.52	0.49	0.46	0.44	0.43	0.41	0.40	0.38	0.37	0.36
440	0.56	0.53	0.50	0.48	0.46	0.44	0.42	0.41	0.40	0.39
460	0.60	0.56	0.54	0.51	0.49	0.47	0.45	0.44	0.43	0.41
480	0.64	0.60	0.57	0.55	0.52	0.50	0.49	0.47	0.45	0.44
500	0.69	0.65	0.61	0.58	0.56	0.54	0.52	0.50	0.48	0.47
520	0.73	0.69	0.65	0.62	0.59	0.57	0.55	0.53	0.51	0.50
540	0.78	0.73	0.69	0.66	0.63	0.61	0.58	0.56	0.54	0.53
560	0.83	0.78	0.74	0.70	0.67	0.64	0.62	0.60	0.58	0.56
580	0.88	0.83	0.78	0.74	0.71	0.68	0.65	0.63	0.61	0.59
600	0.93	0.87	0.83	0.78	0.75	0.72	0.69	0.66	0.64	0.62
620	0.98	0.92	0.87	0.83	0.79	0.76	0.73	0.70	0.68	0.66
640	1.04	0.97	0.92	0.87	0.83	0.80	0.76	0.74	0.71	0.69
660	1.10	1.03	0.97	0.92	0.88	0.84	0.80	0.77	0.75	0.72
680	1.15	1.08	1.02	0.97	0.92	0.88	0.84	0.81	0.78	0.76
700	1.21	1.14	1.07	1.01	0.87	0.92	0.89	0.85	0.82	0.80

Notes:
 1 Table based on initial temperature: 50°F.
 2 Table based on initial pressure: 600 psig.
 3 Table based on maximum operating pressure: 800 psig.

DIAPHRAGM EXPANSION TANK SIZING
HIGH TEMPERATURE SYSTEM CORRECTION FACTORS

Initial Pressure psig	Pressure Increase—psig Initial Pressure + Pressure Increase = Maximum Operating Pressure									
	20	40	60	80	100	120	140	160	180	200
160	2.39	1.32	0.96	0.78	0.67	0.60	0.55	0.51	0.48	0.46
180	2.64	1.44	1.04	0.84	0.72	0.64	0.59	0.54	0.51	0.48
200	2.88	1.56	1.12	0.90	0.77	0.68	0.62	0.57	0.54	0.51
220	3.13	1.69	1.21	0.97	0.82	0.73	0.66	0.61	0.57	0.53
240	3.37	1.81	1.29	1.03	0.87	0.77	0.69	0.64	0.59	0.56
260	3.62	1.93	1.37	1.09	0.92	0.81	0.73	0.67	0.62	0.58
280	3.86	2.05	1.45	1.15	0.97	0.85	0.76	0.70	0.65	0.61
300	4.11	2.18	1.53	1.21	1.02	0.89	0.80	0.73	0.67	0.63
320	4.35	2.30	1.61	1.27	1.07	0.93	0.83	0.76	0.70	0.66
340	4.60	2.42	1.70	1.33	1.12	0.97	0.87	0.79	0.73	0.68
360	4.84	2.55	1.78	1.40	1.17	1.01	0.90	0.82	0.76	0.71
380	5.09	2.67	1.86	1.46	1.21	1.05	0.94	0.85	0.78	0.73
400	5.34	2.79	1.94	1.52	1.26	1.09	0.97	0.88	0.81	0.75
420	5.58	2.91	2.02	1.58	1.31	1.13	1.01	0.91	0.84	0.78
440	5.83	3.04	2.11	1.64	1.36	1.18	1.04	0.94	0.87	0.80
460	6.07	3.16	2.19	1.70	1.41	1.22	1.08	0.97	0.89	0.83
480	6.32	3.28	2.27	1.76	1.46	1.26	1.11	1.00	0.92	0.85
500	6.56	3.40	2.35	1.82	1.51	1.30	1.15	1.04	0.95	0.88
520	6.81	3.53	2.43	1.89	1.56	1.34	1.18	1.07	0.97	0.90
540	7.05	3.65	2.52	1.95	1.61	1.38	1.22	1.10	1.00	0.93
560	7.30	3.77	2.60	2.01	1.66	1.42	1.25	1.13	1.03	0.95
580	7.55	3.90	2.68	2.07	1.71	1.46	1.29	1.16	1.06	0.98
600	7.79	4.02	2.76	2.13	1.75	1.50	1.32	1.19	1.08	1.00
620	8.04	4.14	2.84	2.19	1.80	1.54	1.36	1.22	1.11	1.02
640	8.28	4.26	2.92	2.25	1.85	1.58	1.39	1.25	1.14	1.05
660	8.53	4.39	3.01	2.32	1.90	1.63	1.43	1.28	1.17	1.07
680	8.77	4.51	3.09	2.38	1.95	1.67	1.46	1.31	1.19	1.10
700	9.02	4.63	3.17	2.44	2.00	1.71	1.50	1.34	1.22	1.12

Notes:
1 Table based on initial temperature: 50°F.
2 Table based on initial pressure: 600 psig.
3 Table based on maximum operating pressure: 800 psig.

DIAPHRAGM EXPANSION TANK SIZING
HIGH TEMPERATURE SYSTEM CORRECTION FACTORS

Initial Pressure psig	Pressure Increase—psig Initial Pressure + Pressure Increase = Maximum Operating Pressure									
	220	240	260	280	300	320	340	360	380	400
160	0.44	0.42	0.41	0.40	0.39	0.38	0.37	0.36	0.36	0.35
180	0.46	0.44	0.43	0.42	0.40	0.39	0.39	0.38	0.37	0.36
200	0.49	0.47	0.45	0.43	0.42	0.41	0.40	0.39	0.38	0.38
220	0.51	0.49	0.47	0.45	0.44	0.43	0.41	0.41	0.40	0.39
240	0.53	0.51	0.49	0.47	0.45	0.44	0.43	0.42	0.41	0.40
260	0.55	0.53	0.50	0.49	0.47	0.46	0.44	0.43	0.42	0.41
280	0.57	0.55	0.52	0.50	0.49	0.47	0.46	0.45	0.44	0.43
300	0.60	0.57	0.54	0.52	0.50	0.49	0.47	0.46	0.45	0.44
320	0.62	0.59	0.56	0.54	0.52	0.50	0.49	0.47	0.46	0.45
340	0.64	0.61	0.58	0.56	0.54	0.52	0.50	0.49	0.47	0.46
360	0.66	0.63	0.60	0.57	0.55	0.53	0.52	0.50	0.49	0.48
380	0.69	0.65	0.62	0.59	0.57	0.55	0.53	0.51	0.50	0.49
400	0.71	0.67	0.64	0.61	0.58	0.56	0.54	0.53	0.51	0.50
420	0.73	0.69	0.66	0.63	0.60	0.58	0.56	0.54	0.53	0.51
440	0.75	0.71	0.67	0.64	0.62	0.59	0.57	0.56	0.54	0.52
460	0.78	0.73	0.69	0.66	0.63	0.61	0.59	0.57	0.55	0.54
480	0.80	0.75	0.71	0.68	0.65	0.63	0.60	0.58	0.57	0.55
500	0.82	0.77	0.73	0.70	0.67	0.64	0.62	0.60	0.58	0.56
520	0.84	0.79	0.75	0.71	0.68	0.66	0.63	0.61	0.59	0.57
540	0.86	0.81	0.77	0.73	0.70	0.67	0.65	0.62	0.60	0.59
560	0.89	0.83	0.79	0.75	0.72	0.69	0.66	0.64	0.62	0.60
580	0.91	0.85	0.81	0.77	0.73	0.70	0.67	0.65	0.63	0.61
600	0.93	0.87	0.73	0.78	0.75	0.72	0.69	0.66	0.64	0.62
620	0.95	0.89	0.84	0.80	0.76	0.73	0.70	0.68	0.66	0.64
640	0.98	0.92	0.86	0.82	0.78	0.75	0.72	0.69	0.67	0.65
660	1.00	0.94	0.88	0.84	0.80	0.76	0.73	0.71	0.68	0.66
680	1.02	0.96	0.90	0.85	0.81	0.78	0.75	0.72	0.69	0.67
700	1.04	0.98	0.92	0.87	0.83	0.79	0.76	0.73	0.71	0.68

Notes:
1 Table based on initial temperature: 50°F.
2 Table based on initial pressure: 600 psig.
3 Table based on maximum operating pressure: 800 psig.

Glycol Piping Systems

20.01 Glycol System Piping

A. Glycol piping is a special type of hydronic piping.

B. Design and sizing of glycol piping systems are identical to chilled water or heating water piping systems, except that the flows are increased to account for the differences in the thermal properties of glycol versus water.

20.02 Glycol System Design Considerations

A. HVAC system glycol applications should use an industrial-grade ethylene glycol (phosphate-based) or propylene glycol (phosphate-based) with corrosion inhibitors without fouling. Specify glycol to have *zero* silicate content.

B. Automobile antifreeze solutions should *not* be used for HVAC systems because they contain silicates to protect aluminum engine parts. These silicates can cause fouling in HVAC systems.

C. Consider having the antifreeze dyed to facilitate leak detection.

D. Glycol systems should be filled with a high-quality water, preferably distilled or deionized (deionized is recommended) water, or filled with prediluted solutions of industrial-grade glycol. Water should have less than 25 ppm of chloride and sulfate, and less than 50 ppm of hard-water ions (Ca++, Mg++). City water is treated with chlorine, which is corrosive.

E. Automatic makeup water systems should be avoided to prevent system contamination or system dilution. A low-level liquid alarm should be used in lieu of an automatic fill line.

F. Systems should be clean with little or no corrosion.

G. Industrial-grade glycol will last up to 20 years in a system if properly maintained.

H. Propylene glycol should be used where low oral toxicity is important or where incidental contact with drinking water is possible.

I. Expansion tank sizing is critical to the design of glycol systems. The design should allow for a glycol level of about two-thirds full during operation. Glycol will expand about 6 percent.

J. Water quality should be analyzed at each site for careful evaluation of the level of corrosion protection required.

K. Foaming of a glycol system is usually caused by air entrainment, improper expansion tank design, contamination by organics (oil, gas) or solids, or improper system operation. Foaming will reduce heat transfer and aggravate cavitation corrosion.

L. A buffering agent should be added to maintain fluid alkalinity, minimize acidic corrosive attack, and counteract fluid degradation. Proper buffering agents will reduce fluid maintenance, extend fluid life, and be less sensitive to contamination.

M. A nonabsorbent bypass filter, of the sock or cartridge variety, should be installed in each glycol system.

N. An annual chemical analysis should be conducted to determine the glycol content, oxidative degradation, foaming agent concentration, inhibitor concentration, buffer concentration, freezing point, and pH, reserve alkalinity.

Ethylene Glycol Characteristics	Propylene Glycol Characteristics
More effective freeze point depression	Less effective freeze point depression
Better heat transfer efficiency	Lower heat transfer efficiency
Lower viscosity	Higher viscosity
Low flammability	Low flammability
Low chemical oxygen demand (more friendly to the environment)	High chemical oxygen demand (less friendly to the environment)
Biodegrades in a reasonable period of time—10–20 days completely	Greater resistance to complete biodegradation—more than 20 days
Noncarcinogenic	Noncarcinogenic
Higher level of acute (short-term) and chronic (long-term) toxicity to humans and animals when taken orally—targets the kidney	Lower level of acute (short-term) and chronic (long-term) toxicity to humans and animals when taken orally
Mild eye irritant	Mild eye irritant
Less irritating to the skin	More irritating to the skin
No adverse reproductive effects in lifetime or three-generation studies	No adverse reproductive effects in lifetime or three-generation studies
At high concentrations during pregnancy will cause birth defects and is toxic to the fetus	At the same concentrations during pregnancy will not cause birth defects
Relatively nontoxic to both sewage microorganisms needed for biodegradation and to aquatic life	Relatively nontoxic to both sewage microorganisms needed for biodegradation and to aquatic life

20.03 Glycol System Equation Factors and Derivations

A. $H = 500 \times GPM \times \Delta T$

B. Substitute the equation factors in the following tables for the number 500 in the preceding equation for the ethylene or propylene glycol indicated.

	Ethylene Glycol				
	Temperature °F				Equation Factor
% Glycol Solution	Freeze Point	Boiling Point	Specific Heat	Specific Gravity (1)	
0	+32	212	1.00	1.000	500
10	+26	214	0.97	1.012	491
20	+16	216	0.94	1.027	483
30	+4	220	0.89	1.040	463
40	−12	222	0.83	1.055	438
50	−34	225	0.78	1.067	416
60	−60	232	0.73	1.079	394
70	<−60	244	0.69	1.091	376
80	−49	258	0.64	1.101	352
90	−20	287	0.60	1.109	333
100	+10	287+	0.55	1.116	307

Note:
1 Specific gravity with respect to water at 60°F.

	Propylene Glycol				
	Temperature °F			Specific Gravity (1)	Equation Factor
% Glycol Solution	Freeze Point	Boiling Point	Specific Heat		
0	+32	212	1.000	1.000	500
10	+26	212	0.980	1.008	494
20	+19	213	0.960	1.017	488
30	+8	216	0.935	1.026	480
40	−7	219	0.895	1.034	463
50	−28	222	0.850	1.041	442
60	<−60	225	0.805	1.046	421
70	<−60	230	0.750	1.048	393
80	<−60	230+	0.690	1.048	362
90	<−60	230+	0.645	1.045	337
100	<−60	230+	0.570	1.040	296

Note:
1 Specific gravity with respect to water at 60°F.

A. Glycol Equation Factor Derivations

1. Standard water conditions:
   a. Temperature: 60°F.
   b. Pressure: 14.7 psia (sea level).
   c. Density: 62.4 lbs./ft.^{3}
2. Water equation examples:
   $H = m \times c_g \times \Delta T$.
   30 percent ethylene glycol.
   $c_g = 0.89$ Btu/~~Lb H$_2$O~~ [AC1]°F $\times$ 62.4 ~~Lbs.H$_2$O/Ft^3~~ $\times$ 1.0 ~~Ft^3~~/7.48052 gal. $\times$
        60 min./h $\times$ 1.040 (SG)
        = 463 Btu min./h °F gal.
   $H_{30\%EG} = 463$ Btu min./h °F gal. $\times$ GPM (gal/min.) $\times$ ΔT (°F).
   $H_{30\%EG} = 463 \times$ GPM $\times$ ΔT (°F).
   50 percent propylene glycol.
   $c_g = 0.85$ Btu/~~Lb H$_2$O~~ °F $\times$ 62.4 ~~Lbs.H$_2$O/Ft^3~~ $\times$ 1.0 ~~Ft^3~~ / 7.48052 gal. $\times$ 60 min./h $\times$
        1.041 (SG)
        = 442 Btu min./h °F gal.
   $H_{50\%PG} = 442$ Btu min./h °F gal. $\times$ GPM (gal/min.) $\times$ ΔT (°F).
   $H_{50\%PG} = 442 \times$ GPM $\times$ ΔT (°F).

Steam Piping Systems

21.01 Steam Piping Systems

A. Steam Pipe Sizing

1. Low-pressure steam systems:
   a. Low-pressure steam: 0–15 psig.
   b. 0.2–3 psi total system pressure drop max.
   c. 1/8–1/2 psi/100 ft.
2. Medium-pressure steam systems:
   a. Medium-pressure steam: 16–100 psig.
   b. 3–10 psi total system pressure drop max.
   c. 1/2–2 psi/100 ft.
3. High-pressure steam systems:
   a. High-pressure steam: 101–300 psig.
   b. 10–60 psi total system pressure drop max.
   c. 2–5 psi/100 ft.
4. Steam velocity:
   a. 15,000 FPM maximum.
   b. 6,000–12,000 FPM recommended.
   c. Low pressure systems: 4,000–6,000 FPM.
   d. Medium pressure systems: 6,000–8,000 FPM.
   e. High pressure systems: 10,000–15,000 FPM.
5. Friction loss estimate:
   a. 2.0 × System Length (ft.) × Friction Rate (ft./100 ft.).
6. Standard steel pipe sizes—1/2", 3/4", 1", 1-1/4", 1-1/2", 2", 2-1/2", 3", 4", 6", 8", 10", 12", 14", 16", 18", 20", 24", 30", 36", 42", 48", 54", 60", 72", 84", 96".
7. Total pressure drop in the steam system should not exceed 20 percent of the total maximum steam pressure at the boiler.
8. Steam condensate liquid to steam volume ratio is 1:1600 at 0 psig.
9. Flash steam: Flash steam is formed when hot steam condensate under pressure is released to a lower pressure; the temperature drops to the boiling point of the lower pressure, causing some of the condensate to evaporate, forming steam. Flash steam occurs whenever steam condensate experiences a drop in pressure and thus produces steam at the lower pressure.
   a. Low-pressure steam systems flash steam is negligible and can generally be ignored.
   b. Medium- and high-pressure steam systems flash steam is important to utilize and consider when sizing condensate piping.
   c. Flash steam recovery requirements:
      1) To utilize flash steam recovery, the condensate must be at a reasonably high pressure (medium- and high-pressure steam systems) and the traps supplying the condensate must be capable of operating with the back pressure of the flash steam system.
      2) There must be a use or demand for the flash steam at the reduced pressure. Demand for steam at the lower pressure should be greater than the supply of flash steam. The demand for steam should occur at the same time as the flash steam supply.
      3) The steam equipment should be in close proximity to the flash steam source to minimize installation and radiation losses and to fully take advantage of the flash steam recovery system. Flash steam recovery systems are especially advantageous when steam is utilized at multiple pressures within the facility and the distribution systems are already in place.
10. Saturated steam:
    a. Saturated steam: Saturated steam is steam that is in equilibrium with the liquid at a given pressure. One pound of steam has a volume of 26.8 cu.ft. at atmospheric pressure (0 psig).

    b. Dry saturated steam: Dry steam is steam which has been completely evaporated and contains no liquid water in the form of mist or small droplets. Steam systems that produce a dry steam supply are superior to systems that produce a wet steam supply.

    c. Wet saturated steam: Wet steam is steam that has not been completely evaporated and contains water in the form of mist or small droplets. Wet steam has a heat content substantially lower than dry steam.

    d. Superheated steam: Superheated steam is dry saturated steam that is heated, which increases the temperature without increasing the system pressure.

11. Steam types:

    a. Plant steam: Steam produced in a conventional boiler system using softened and chemically treated water.

    b. Filtered steam: Plant steam that has been filtered to remove solid particles (no chemical removal).

    c. Clean steam: Steam produced in a clean steam generator using distilled, de-ionized, reverse-osmosis, or ultra-pure water.

    d. Pure steam: Steam produced in a clean steam generator using distilled or de-ionized pyrogen-free water, normally defined uncondensed water for injection.

12. Steam purity versus steam quality:

    a. Steam purity: A qualitative measure of steam contamination caused by dissolved solids, volatiles, or particles in vapor, or by tiny water droplets that may remain in the steam following primary separation in the boiler.

    b. Steam quality: The ratio of the weight of dry steam to the weight of dry saturated steam and entrained water [Example: 0.95 quality refers to 95 parts steam (95 percent) and 5 parts water (5 percent)].

B. Steam System Design and Pipe Installation Guidelines

1. The minimum recommended steam pipe size is 3/4 inch.

2. Locate all valves, strainers, unions, and flanges so they are accessible. All valves (except control valves) and strainers should be the full size of the pipe before reducing the size to make connections to equipment and controls. Union and/or flanges should be installed at each piece of equipment, in bypasses and in long piping runs (100 feet or more), to permit disassembly for alteration and repairs.

3. Provide chainwheel operators for all valves in equipment rooms mounted greater than 7'0" above floor level. The chain should extend to 5'0–7'0" above the floor level.

4. All valves should be installed so the valve remains in service when equipment or piping on the equipment side of the valve is removed.

5. Locate all flow measuring devices in accessible locations with the straight section of the pipe upstream (10 pipe diameters) and downstream (5 pipe diameters) of the device or as recommended by the manufacturer.

6. Provide vibration isolators for all piping supports connected to, and within 50 feet of, isolated equipment, except at base elbow supports and anchor points, throughout mechanical equipment rooms, and for supports of steam mains within 50 feet of boiler or pressure reducing valves.

7. Pitch steam piping downward in the direction of flow 1/4" per 10 ft. (1" per 40 ft.) minimum.

8. Where the length of steam branch lines are less than 8 feet, pitch branch lines downward toward mains 1/2" per foot minimum.

9. Connect all branch lines to the top of steam mains (45 degree preferred, 90 degree acceptable; see Fig. 21.1).

10. Steam piping should be installed with eccentric reducers (flat on the bottom) to prevent accumulation of condensate in the pipe and thus decreasing the risk of water hammer.

11. Drip leg collection points on steam piping should be the same size as the steam piping to prevent steam condensate from passing over the drip leg and thus decreasing the risk of water hammer. The drip leg collection point should be a minimum of 12 inches long including a minimum 6-inch-long dirt leg with the steam trap outlet above the dirt leg.

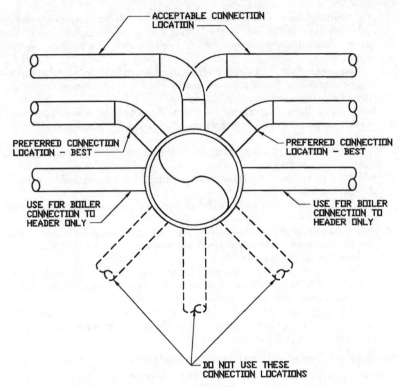

ACCEPTABLE CONNECTION
LOCATION

PREFERRED CONNECTION
LOCATION - BEST

PREFERRED CONNECTION
LOCATION - BEST

USE FOR BOILER
CONNECTION TO
HEADER ONLY

USE FOR BOILER
CONNECTION TO
HEADER ONLY

DO NOT USE THESE
CONNECTION LOCATIONS

FIGURE 21.1   STEAM PIPING CONNECTIONS.

12. Drip legs must be installed at all low points, downfed runouts to all equipment, end of mains, bottom of risers, and ahead of all pressure regulators, control valves, isolation valves, and expansion joints.
13. On straight runs with no natural drainage points, install drip legs at intervals not exceeding 200 feet where the pipe is pitched downward in the direction of steam flow, and a maximum of 100 feet where the pipe is pitched up so that condensate flow is opposite of steam flow.
14. Steam traps used on steam mains and branches shall be a minimum 3/4" size.
15. Control of steam systems with more than 2 million Btuh should be accomplished with two or more control valves (see steam PRVs).
16. Double valves should be installed on the supply side of equipment for isolating steam systems, above 100 psig, with a drain between these valves to visually confirm isolation. The reason for the double valving of systems is to ensure isolation because of the large pressure differentials that occur when the system is opened for repairs. Double valve all of the following:
    a. Equipment.
    b. Drains.
    c. Vents.
    d. Gauges.
    e. Instrumentation.
17. Steam in a steam system should be maintained at a pH of approximately 8 to 9. A pH of 7 is neutral; below 7 is acidic; above 7 is alkaline.
18. Provide a stop check valve (located closest to the boiler) and an isolation valve with a drain between these valves on the steam supply connections to all steam boilers.

19. Provide steam systems with warm-up valves for in-service start-up as shown in the following table. This will allow operators to warm these systems slowly and to prevent a sudden shock or catastrophic system failure when large system valves are opened. Providing warming valves also reduces wear on large system valves when they are only opened a small amount in an attempt to control system warm-up speed.

BYPASS AND WARMING VALVES

Main Valve Nominal Pipe Size	Nominal Pipe Size	
	Series A Warming Valves	Series B Bypass Valves
4	1/2	1
5	3/4	1-1/4
6	3/4	1-1/4
8	3/4	1-1/2
10	1	1-1/2
12	1	2
14	1	2
16	1	3
18	1	3
20	1	3
24	1	4
30	1	4
36	1	6
42	1	6
48	1	8
54	1	8
60	1	10
72	1	10
84	1	12
96	1	12

Notes:

1 Series A valve sizes are utilized in steam service for warming up before the main line is opened, and for balancing pressures where lines are of limited volume.

2 Series B valve sizes are utilized in pipe lines conveying gases or liquids where by-passing may facilitate the operation of the main valve through balancing the pressures on both sides of the disc or discs thereof. The valves in the larger sizes may be of the bolted-on type.

20. Steam system warming valve procedure (see Fig. 21.2):
   a. Slowly open the warming supply valve to establish flow and to warm the system.
   b. Once the system pressure and temperature have stabilized, proceed with the following items, one at a time:
      1) Slowly open the main supply valve.
      2) Close the warming supply valve.
21. Steam system warm-up procedure:
   a. Steam system start-up should not exceed 120°F temperature rise per hour, but the boiler or heat exchanger manufacture limitations should be consulted.
   b. It is recommended that no more than a 25°F temperature rise per hour be used when warming steam systems. Slow warming of the steam system allows for system piping, supports, hangers, and anchors to keep up with system expansion.
   c. Low-pressure steam systems (15 psig and less) should be warmed slowly at 25°F temperature rise per hour until system design pressure is reached.
   d. Medium- and high-pressure steam systems (above 15 psig) should be warmed slowly at 25°F temperature rise per hour until 250°F-15 psig system temperature-pressure is reached. At this temperature-pressure, the system should be permitted to settle for at least eight hours or more (preferably overnight). The temperature-pressure maintenance time gives the system piping, hangers, supports, and anchors a chance to catch up with the system expansion. After allowing the system to settle, the system can be warmed up to 120 psig or the system design pressure in 25 psig pressure increments. Allow the system to settle for an hour before

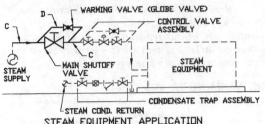

STEAM EQUIPMENT APPLICATION
SERIES A

STEAM SYSTEM
ISOLATION VALVE APPLICATION
SERIES B

NOTES:

1. SERIES A WARMING VALVES COVER STEAM OR MEDIUM/HIGH TEMPERATURE HEATING WATER SERVICE FOR SYSTEM OR EQUIPMENT WARM-UP BEFORE THE MAIN SHUTOFF VALVE TO THE SYSTEM OR DEVICE IS OPENED. WARMING VALVES ARE ALSO USED FOR BALANCING PRESSURES WHERE LINES ARE OF LIMITED VOLUME.
2. SERIES B WARMING VALVES COVER LINES CONVEYING GASES OR LIQUIDS WHERE BYPASSING MAY FACILITATE THE OPERATION OF THE MAIN VALVE BY BALANCING THE PRESSURES ON BOTH SIDES OF THE MAIN VALVE.

MAIN VALVE SIZE (C)	WARMING VALVE SIZE (D)	
	SERIES A WARMING VALVES	SERIES B WARMING VALVES
4″	1/2″	1″
5″, 6″	3/4″	1-1/4″
8″	3/4″	1-1/2″
10″	1″	1-1/2″
12″, 14″	1″	2″
16″, 18″, 20″	1″	3″
24″, 30″	1″	4″
36″, 42″	1″	6″
48″, 54″	1″	8″
60″, 72″	1″	10″
84″, 96″	1″	12″

FIGURE 21.2   STEAM SYSTEM WARMING VALVES.

increasing the pressure to the next increment. When the system reaches 120 psig and the design pressure is above 120 psig, the system should be allowed to settle for at least eight hours or more (preferably overnight). The pressure maintenance time gives the system piping, hangers, supports, and anchors a chance to catch up with the system expansion. After allowing the system to settle, the system can be warmed up to 300 psig or the system design pressure in 25 psig pressure increments; allow the system to settle for an hour before increasing the pressure to the next increment.

C. Low-Pressure Steam Pipe Materials (0–15 psig):

1. 2" and smaller:
   a. Pipe: Black Steel Pipe, *ASTM A53, Schedule 40,* Type E or S, Grade B
      Fittings: Black Cast Iron Screw Fittings, 125 lb., *ANSI/ASME B16.4*
      Joints: Pipe Threads, General Purpose (American) *ANSI/ASME B1.20.1.*
2. 2-1/2" through 10":
   a. Pipe: Black Steel Pipe, *ASTM A53, Schedule 40,* Type E or S, Grade B
      Fittings: Steel Butt-Welding Fittings, 125 lb., *ANSI/ASME B16.9*
      Joints: Welded pipe, *ANSI/AWS D1.1* and *ANSI/ASME Sec. 9.*
3. 12" and larger:
   a. Pipe: Black Steel Pipe, *ASTM A53,* 3/8" wall, Type E or S, Grade B
      Fittings: Steel Butt-Welding Fittings, 125 lb., *ANSI/ASME B16.9*
      Joints: Welded pipe, *ANSI/AWS D1.1* and *ANSI/ASME Sec. 9.*

D. Medium-Pressure Steam Pipe (16–100 psig):

1. 1-1/2" and Smaller:
   a. Pipe: Black Steel Pipe, *ASTM A53, Schedule 40,* Type E or S, Grade B
      Fittings: Forged Steel Socket-Weld, 150 lb., *ANSI B16.11*
      Joints: Welded pipe, *ANSI/AWS D1.1* and *ANSI/ASME Sec. 9.*
   b. Pipe: Carbon Steel Pipe, *ASTM A106, Schedule 40,* Grade B
      Fittings: Forged Steel Socket-Weld, 150 lb., *ANSI B16.11*
      Joints: Welded pipe, *ANSI/AWS D1.1* and *ANSI/ASME Sec. 9.*

2. 2" through 10":
  a. Pipe: Black Steel Pipe, *ASTM A53, Schedule 40,* Type E or S, Grade B
     Fittings: Steel Butt-Welding Fittings, 150 lb., *ANSI/ASME B16.9*
     Joints: Welded pipe, ANSI/AWS D1.1 and ANSI/ASME Sec. 9
  b. Pipe: Carbon Steel Pipe, *ASTM A106, Schedule 40,* Grade B
     Fittings: Steel Butt-Welding Fittings, 150 lb., *ANSI/ASME B16.9*
     Joints: Welded pipe, ANSI/AWS D1.1 and ANSI/ASME Sec. 9.
3. 12" and larger:
  a. Pipe: Black Steel Pipe, *ASTM A53, 3/8" wall,* Type E or S, Grade B
     Fittings: Steel Butt-Welding Fittings, 150 lb., *ANSI/ASME B16.9*
     Joints: Welded pipe, *ANSI/AWS D1.1* and *ANSI/ASME Sec. 9.*
  b. Pipe: Carbon Steel Pipe, *ASTM A106, 3/8" wall,* Grade B
     Fittings: Steel Butt-Welding Fittings, 150 lb., *ANSI/ASME B16.9*
     Joints: Welded pipe, *ANSI/AWS D1.1* and *ANSI/ASME Sec. 9.*

E. High-Pressure Steam Pipe (100–300 psig):

1. 1-1/2" and smaller:
  a. Pipe: Black Steel Pipe, *ASTM A53, Schedule 80,* Type E or S, Grade B
     Fittings: Forged Steel Socket-Weld, 300 lb., *ANSI B16.11*
     Joints: Welded pipe, *ANSI/AWS D1.1* and *ANSI/ASME Sec. 9.*
  b. Pipe: Carbon Steel Pipe, *ASTM A106, Schedule 80,* Grade B
     Fittings: Forged Steel Socket-Weld, 300 lb., *ANSI B16.11*
     Joints: Welded pipe, *ANSI/AWS D1.1* and *ANSI/ASME Sec. 9.*
2. 2" and larger:
  a. Pipe: Black Steel Pipe, *ASTM A53, Schedule 80,* Type E or S, Grade B
     Fittings: Steel Butt-Welding Fittings, 300 lb., *ANSI/ASME B16.9*
     Joints: Welded pipe, *ANSI/AWS D1.1* and *ANSI/ASME Sec. 9.*
  b. Pipe: Carbon Steel Pipe, *ASTM A106, Schedule 80,* Grade B
     Fittings: Steel Butt-Welding Fittings, 300 lb., *ANSI/ASME B16.9*
     Joints: Welded pipe, *ANSI/AWS D1.1* and *ANSI/ASME Sec. 9.*

F. Pipe Testing

1. 1.5 × System Working Pressure.
2. 100 psi minimum.

G. Steam Pressure Reducing Valves (PRV)

 1. PRV types:
  a. Direct acting:
    1) Low cost.
    2) Limited ability to respond to changing load and pressure.
    3) Suitable for systems with low flow requirements.
    4) Suitable for systems with constant loads.
    5) Limited control of downstream pressure.
  b. Pilot-operated:
    1) Close control of downstream pressure over a wide range of upstream pressures.
    2) Suitable for systems with varying loads.
    3) Ability to respond to changing loads and pressures.
    4) Types:
      a) Pressure-operated-pilot.
      b) Temperature-pressure-operated-pilot.
2. Use multiple stage reduction where greater than 100 psig reduction is required or where greater than 50 psig reduction is required to deliver a pressure less than 25 psig operating pressure or when intermediate steam pressure is required.
3. Use multiple PRVs where system steam capacity exceeds 2" PRV size, when normal operation calls for 10 percent of design load for sustained periods, or when there are

two distinct load requirements (i.e., summer/winter). Provide the number of PRVs to suit the project.
   a. If the system capacity for a single PRV exceeds the 2" PRV size but is not larger than the 4" PRV size, use two PRVs with 1/3 and 2/3 capacity split.
   b. If system capacity for a single PRV exceeds the 4" PRV size, use three PRVs with 25 percent, 25 percent, and 50 percent, or 15 percent, 35 percent, and 50 percent capacity split to suit the project.
4. The smallest PRV should be no greater than 1/3 of the system capacity. The maximum size PRV should be 4" (6" when 4" PRV will require more than three valves per stage).
5. The PRV bypass should be two pipe sizes smaller than the largest PRV.
6. Provide 10 pipe diameters from the PRV inlet to the upstream header.
7. Provide 20 pipe diameters from the PRV outlet to the downstream header.
8. Maximum pipe velocity upstream and downstream of PRV:
   a. 8" and smaller: 10,000 FPM.
   b. 10" and larger: 8,000 FPM.
   c. Where low sound levels are required, reduce velocities by 25–50 percent.
   d. If the outlet velocity exceeds the preceding listings, use a noise suppressor.
9. Avoid abrupt changes in pipe size. Use concentric reducers.
10. Limit pipe diameter changes to two pipe sizes per stage of expansion.

H. Safety Relief Valves

1. The safety relief valve must be capable of handling the volume of steam as determined by the high pressure side of the largest PRV, or the bypass, whichever is greater.
2. Use multiple safety relief valves if the capacity of a 4" safety relief valve is exceeded. Each valve must have a separate connection to the pipeline.
3. Safety, relief, and safety relief valve testing is dictated by the insurance underwriter.

I. Steam Systems

1. Residential steam systems are low-pressure steam systems normally with gravity return condensate systems (see Figs. 21.3 through 21.5).
2. Commercial low-pressure steam systems may be provided with either gravity or pumped condensate return systems (see Figs. 21.6 and 21.7).
3. Commercial and industrial medium- and high-pressure steam systems are generally provided with pumped condensate return systems (see Figs. 21.8 and 21.9).

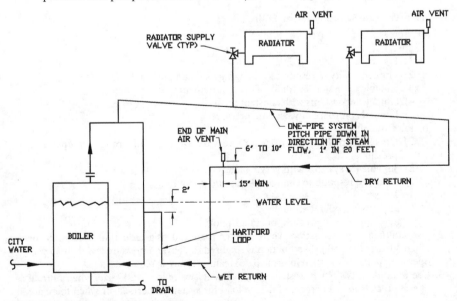

FIGURE 21.3   RESIDENTIAL LP STEAM SYSTEMS 1.

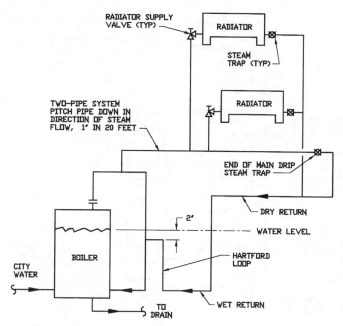

FIGURE 21.4 RESIDENTIAL LP STEAM SYSTEMS 2.

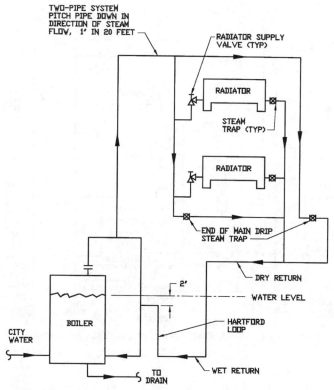

FIGURE 21.5 RESIDENTIAL LP STEAM SYSTEMS 3.

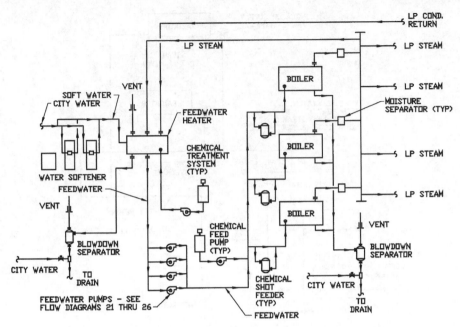

FIGURE 21.6   LOW-PRESSURE STEAM SYSTEMS—GRAVITY RETURN.

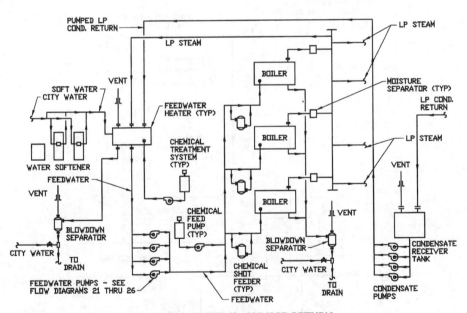

FIGURE 21.7   LOW-PRESSURE STEAM SYSTEMS—PUMPED RETURN.

HIGH PRESSURE STEAM SYSTEM KEYED NOTES:

① BOILER

② DEAERATOR OR FEEDWATER HEATER

③ BLOWDOWN FLASH TANK

④ FEEDWATER PUMPS – SEE FLOW DIAGRAMS 21 THROUGH 26

⑤ CONDENSATE RECEIVER TANK

⑥ CONDENSATE PUMPS

⑦ BLOWDOWN SEPARATOR

⑧ BLOWDOWN HEAT EXCHANGER

⑨ SAMPLE COOLER

⑩ FLASH TANK

⑪ CHEMICAL TREATMENT SYSTEMS

⑫ CHEMICAL FEED PUMPS

⑬ CHEMICAL SHOT FEEDER

⑭ WATER TREATMENT SYSTEM

⑮ MOISTURE SEPARATOR

⑯ PRV STATION

⑰ TEMPERATURE CONTROL

⑱ LEVEL CONTROL

⑲ TOP BLOWDOWN CONTROLLER

⑳ TO DRAIN

㉑ EXHAUST HEAD

㉒ VENT, TERMINATE A MINIMUM OF 7'-6" ABOVE ROOF

㉓ STEAM SYSTEM #1 (LOW PRESSURE STEAM)

㉔ STEAM SYSTEM #2 (MEDIUM PRESSURE STEAM)

㉕ STEAM SYSTEM #3 (HIGH PRESSURE STEAM)

㉖ FLASH STEAM TO LP STEAM SYSTEM

㉗ COMBINED HIGH-PRESSURE & MEDIUM-PRESSURE CONDENSATE RETURNS

㉘ LOW PRESSURE CONDENSATE RETURN

㉙ MEDIUM-PRESSURE CONDENSATE RETURN

㉚ HIGH-PRESSURE CONDENSATE RETURN

㉛ PUMPED CONDENSATE RETURN

㉜ CITY WATER

㉝ TREATED WATER

㉞ HEATED SOFT WATER

㉟ FEEDWATER

FIGURE 21.8 HP STEAM SYSTEM FLOW DIAGRAM NOTES.

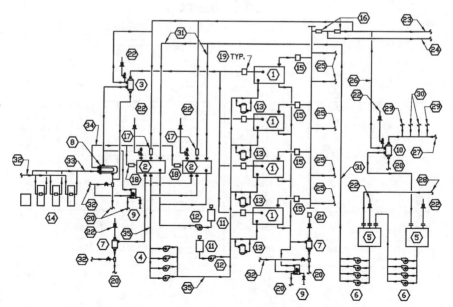

FIGURE 21.9 HIGH-PRESSURE STEAM SYSTEMS—PUMPED RETURN.

21.02 Steam System Design Criteria

STEAM SYSTEM DESIGN CRITERIA

System Type	Initial Steam Pressure psig	Maximum Pressure Drop psig/100 ft.	Maximum Total System Pressure Drop psig	Maximum Velocity FPM
Low Pressure	1	1/8	0.2	4,000
	3	1/8	0.6	4,000
	5	1/4	1.0	6,000
Velocity Range 4,000–6,000 FPM	7	1/4	1.5	6,000
	10	1/2	2.0	6,000
	12	1/2	2.5	6,000
	15	1/2	3	6,000
Medium Pressure	20	1/2	4	8,000
	25	1/2–1	5	8,000
	30	1/2–1	5–6	8,000
Velocity Range 6,000–12,000 FPM	40	1	6–8	10,000
	50	1	8–10	10,000
	60	1	10–12	12,000
	75	1–2	12–15	12,000
	85	1–2	12–15	12,000
	100	1–2	15–20	12,000
High Pressure	120	2	20–24	15,000
	125	2	20–24	15,000
	150	2	24–30	15,000
	175	2	24–30	15,000
Velocity Range 6,000–15,000 FPM	200	2–5	30–40	15,000
	225	2–5	30–40	15,000
	250	2–5	30–50	15,000
	275	2–5	30–50	15,000
	300	2–5	40–60	15,000

21.03 Steam Tables

STEAM TABLES

Steam Pressure psig	Steam Pressure psia	Saturation Temperature °F	Specific Volume cu.ft./lb.	Heat Content (above 32°F) Btu/lb.		
				Sensible	Latent	Total
0	14.7	212.0	26.800	180.2	970.4	1,150.6
1	15.7	215.3	25.212	183.5	968.2	1,151.7
2	16.7	218.5	23.798	186.7	966.2	1,152.9
3	17.7	221.5	22.536	189.7	964.3	1,154.0
4	18.7	224.4	21.407	192.6	962.4	1,155.0
5	19.7	227.1	20.387	195.4	960.6	1,156.0
6	20.7	229.8	19.467	198.1	958.9	1,157.0
7	21.7	232.3	18.626	200.7	957.3	1,158.0
8	22.7	234.8	17.855	203.1	955.7	1,158.8
9	23.7	237.1	17.147	205.5	954.2	1,159.7
10	24.7	239.4	16.496	207.8	952.7	1,160.5
11	25.7	241.6	15.895	210.1	951.2	1,161.3
12	26.7	243.7	15.337	212.2	949.8	1,162.0
13	27.7	245.8	14.817	214.4	948.4	1,162.8
14	28.7	247.8	14.334	216.4	947.1	1,163.5
15	29.7	249.8	13.881	218.3	945.8	1,164.1
16	30.7	251.7	13.458	220.3	944.5	1,164.8
17	31.7	253.5	13.059	222.2	943.3	1,165.5
18	32.7	255.3	12.685	224.0	942.1	1,166.1
19	33.7	257.1	12.332	225.8	940.9	1,166.7

(Continued)

STEAM TABLES (*Continued*)

Steam Pressure psig	Steam Pressure psia	Saturation Temperature °F	Specific Volume cu.ft./lb.	Heat Content (above 32°F) Btu/lb.		
				Sensible	Latent	Total
20	34.7	258.8	11.998	227.5	939.7	1,167.2
21	35.7	260.5	11.684	229.2	938.5	1,167.7
22	36.7	262.1	11.385	230.9	937.4	1,168.3
23	37.7	263.7	11.102	232.5	936.3	1,168.8
24	38.7	265.3	10.833	234.1	935.2	1,169.3
25	39.7	266.8	10.577	235.7	934.1	1,169.8
26	40.7	268.3	10.333	237.3	933.1	1,170.4
27	41.7	269.8	10.101	238.7	932.1	1,170.8
28	42.7	271.2	9.879	240.2	931.1	1,171.3
29	43.7	272.6	9.666	241.7	930.1	1,171.8
30	44.7	274.0	9.463	243.1	929.1	1,172.2
31	45.7	275.4	9.269	244.5	928.2	1,172.7
32	46.7	276.8	9.082	245.9	927.2	1,173.1
33	47.7	278.1	8.904	247.2	926.3	1,173.5
34	48.7	279.4	8.732	248.5	925.4	1,173.9
35	49.7	280.6	8.567	249.9	924.5	1,174.4
36	50.7	281.9	8.408	251.1	923.6	1,174.7
37	51.7	283.1	8.255	252.4	922.7	1,175.1
38	52.7	284.4	8.109	253.7	921.8	1,175.5
39	53.7	285.6	7.966	254.9	921.0	1,175.9
40	54.7	286.7	7.843	256.1	920.1	1,176.2
41	55.7	287.9	7.697	257.3	919.3	1,176.6
42	56.7	289.1	7.570	258.5	918.4	1,176.9
43	57.7	290.2	7.447	259.6	917.6	1,177.2
44	58.7	291.3	7.327	260.8	916.8	1,177.6
45	59.7	292.4	7.212	261.9	916.0	1,177.9
46	60.7	293.5	7.100	263.0	915.2	1,178.2
47	61.7	294.6	6.992	264.2	914.4	1,178.6
48	62.7	295.6	6.887	265.3	913.6	1,178.9
49	63.7	296.7	6.785	266.3	912.9	1,179.2
50	64.7	297.7	6.686	267.4	912.1	1,179.5
51	65.7	298.7	6.591	268.4	911.4	1,179.8
52	66.7	299.7	6.498	269.4	910.6	1,180.0
53	67.7	300.7	6.407	270.5	909.9	1,180.4
54	68.7	301.7	6.391	271.5	909.2	1,180.7
55	69.7	302.7	6.234	272.5	908.5	1,181.0
56	70.7	303.6	6.151	273.5	907.7	1,181.2
57	71.7	304.6	6.070	274.5	907.0	1,181.5
58	72.7	305.5	5.991	275.4	906.3	1,181.7
59	73.7	306.4	5.915	276.4	905.6	1,182.0
60	74.7	307.4	5.840	277.3	905.0	1,182.3
61	75.7	308.3	5.768	278.3	904.3	1,182.6
62	76.7	309.2	5.696	279.2	903.6	1,182.8
63	77.7	310.1	5.627	280.1	902.9	1,183.0
64	78.7	310.9	5.560	281.0	902.3	1,183.3
65	79.7	311.8	5.494	281.9	901.6	1,183.5
66	80.7	312.7	5.430	282.8	901.0	1,183.8
67	81.7	313.5	5.367	283.7	900.3	1,184.0
68	82.7	314.4	5.306	284.6	899.7	1,184.3
69	83.7	315.2	5.246	285.5	899.0	1,184.5
70	84.7	316.0	5.187	286.3	898.4	1,184.7
71	85.7	316.9	5.130	287.2	897.8	1,185.0
72	86.7	317.7	5.075	288.0	897.1	1,185.1
73	87.7	318.5	5.020	288.9	896.5	1,185.4
74	88.7	319.3	4.966	289.7	895.9	1,185.6
75	89.7	320.1	4.914	290.5	895.3	1,185.8
76	90.7	320.9	4.863	291.3	894.7	1,186.0
77	91.7	321.6	4.813	292.1	894.1	1,186.2
78	92.7	322.4	4.764	292.9	893.5	1,186.4
79	93.7	323.2	4.715	293.7	892.9	1,186.6

(*Continued*)

STEAM TABLES (*Continued*)

Steam Pressure psig	Steam Pressure psia	Saturation Temperature °F	Specific Volume cu.ft./lb.	Heat Content (above 32°F) Btu/lb.		
				Sensible	Latent	Total
80	94.7	323.9	4.668	294.5	892.3	1,186.8
81	95.7	324.7	4.623	295.3	891.7	1,187.0
82	96.7	325.4	4.578	296.1	891.1	1,187.2
83	97.7	326.2	4.533	296.9	890.6	1,187.5
84	98.7	326.9	4.489	297.6	890.0	1,187.6
85	99.7	327.6	4.447	298.4	889.4	1,187.8
86	100.7	328.4	4.405	299.1	888.8	1,187.9
87	101.7	329.1	4.364	299.9	888.3	1,188.2
88	102.7	329.8	4.324	300.6	887.7	1,188.3
89	103.7	330.5	4.284	301.4	887.1	1,188.5
90	104.7	331.2	4.245	302.1	886.6	1,188.7
91	105.7	331.9	4.207	302.8	886.1	1,188.9
92	106.7	332.6	4.170	303.5	885.5	1,189.0
93	107.7	333.3	4.133	304.3	885.0	1,189.3
94	108.7	333.9	4.098	305.0	884.4	1,189.4
95	109.7	334.6	4.062	305.7	883.9	1,189.6
96	110.7	335.3	4.048	306.4	883.3	1,189.7
97	111.7	336.0	3.993	307.1	882.8	1,189.9
98	112.7	336.6	3.959	307.8	882.2	1,190.0
99	113.7	337.3	3.926	308.4	881.8	1,190.2
100	114.7	337.9	3.894	309.1	881.2	1,190.3
101	115.7	338.6	3.862	309.8	880.7	1,190.5
102	116.7	339.2	3.830	310.5	880.2	1,190.7
103	117.7	339.9	3.799	311.1	879.7	1,190.8
104	118.7	340.5	3.769	311.8	879.2	1,191.0
105	119.7	341.1	3.739	312.5	878.7	1,191.2
106	120.7	341.7	3.710	313.1	878.1	1,191.2
107	121.7	342.4	3.681	313.8	877.6	1,191.4
108	122.7	343.0	3.652	314.4	877.1	1,191.5
109	123.7	343.6	3.624	315.1	876.6	1,191.7
110	124.7	344.2	3.596	315.7	876.1	1,191.8
111	125.7	344.8	3.569	316.3	875.6	1,191.9
112	126.7	345.4	3.543	317.0	875.1	1,192.1
113	127.7	346.0	3.516	317.6	874.6	1,192.2
114	128.7	346.6	3.490	318.2	874.2	1,192.4
115	129.7	347.2	3.465	318.9	873.7	1,192.6
116	130.7	347.8	3.440	319.5	873.2	1,192.7
117	131.7	348.4	3.415	320.1	872.7	1,192.8
118	132.7	348.9	3.390	320.7	872.2	1,192.9
119	133.7	349.5	3.366	321.3	871.7	1,193.0
120	134.7	350.1	3.342	321.9	871.3	1,193.2
121	135.7	350.7	3.319	322.5	870.8	1,193.3
122	136.7	351.2	3.296	323.1	870.3	1,193.4
123	137.7	351.8	3.273	323.7	869.8	1,193.5
124	138.7	352.4	3.251	324.3	869.4	1,193.7
125	139.7	352.9	3.228	324.9	868.9	1,193.8
126	140.7	353.5	3.206	325.5	868.4	1,193.9
127	141.7	354.0	3.185	326.0	868.0	1,194.0
128	142.7	354.6	3.163	326.6	867.5	1,194.1
129	143.7	355.1	3.142	327.2	867.0	1,194.2
130	144.7	355.7	3.121	327.8	866.6	1,194.4
131	145.7	356.2	3.101	328.4	866.1	1,194.5
132	146.7	356.7	3.081	328.9	865.7	1,194.6
133	147.7	357.3	3.061	329.5	865.2	1,194.7
134	148.7	357.8	3.042	330.0	864.8	1,194.8
135	149.7	358.3	3.022	330.6	864.3	1,194.9
136	150.7	358.8	3.003	331.1	863.9	1,195.0
137	151.7	359.4	2.984	331.7	863.4	1,195.1
138	152.7	359.9	2.965	332.2	863.0	1,195.2
139	153.7	360.4	2.947	332.8	862.5	1,195.3

(*Continued*)

STEAM TABLES (*Continued*)

Steam Pressure psig	Steam Pressure psia	Saturation Temperature °F	Specific Volume cu.ft./lb.	Heat Content (above 32°F) Btu/lb.		
				Sensible	Latent	Total
140	154.7	360.9	2.928	333.3	862.1	1,195.4
141	155.7	361.4	2.910	333.9	861.6	1,195.5
142	156.7	361.9	2.893	334.4	861.2	1,195.6
143	157.7	362.4	2.875	334.9	860.8	1,195.7
144	158.7	362.9	2.858	335.5	860.4	1,195.9
145	159.7	363.4	2.841	336.0	859.9	1,195.9
146	160.7	363.9	2.824	336.5	859.5	1,196.0
147	161.7	364.4	2.807	337.1	859.0	1,196.1
148	162.7	364.9	2.791	337.6	858.6	1,196.2
149	163.7	365.4	2.775	338.1	858.2	1,196.3
150	164.7	365.9	2.759	338.6	857.8	1,196.4
151	165.7	366.4	2.743	339.1	857.3	1,196.4
152	166.7	366.9	2.727	339.7	856.9	1,196.6
153	167.7	367.4	2.712	340.2	856.5	1,196.7
154	168.7	367.9	2.696	340.7	856.1	1,196.8
155	169.7	368.3	2.681	341.2	855.7	1,196.9
156	170.7	368.8	2.666	341.7	855.3	1,197.0
157	171.7	369.3	2.651	342.2	854.8	1,197.0
158	172.7	369.7	2.636	342.7	854.4	1,197.1
159	173.7	370.2	2.621	343.2	854.0	1,197.2
160	174.7	370.7	2.607	343.7	853.6	1,197.3
161	175.7	371.1	2.593	344.2	853.2	1,197.4
162	176.7	371.6	2.579	344.7	852.8	1,197.5
163	177.7	372.1	2.565	345.2	852.4	1,197.6
164	178.7	372.5	2.551	345.7	852.0	1,197.7
165	179.7	373.0	2.537	346.1	851.6	1,197.7
166	180.7	373.4	2.524	346.6	851.2	1,197.8
167	181.7	373.9	2.511	347.1	850.8	1,197.9
168	182.7	374.4	2.498	347.6	850.4	1,198.0
169	183.7	374.8	2.484	348.1	850.0	1,198.1
170	184.7	375.2	2.471	348.5	849.6	1,198.1
171	185.7	375.7	2.459	349.0	849.2	1,198.2
172	186.7	376.1	2.446	349.5	848.8	1,198.3
173	187.7	376.6	2.434	350.0	848.4	1,198.4
174	188.7	377.0	2.421	350.4	848.1	1,198.5
175	189.7	377.4	2.409	350.9	847.7	1,198.6
176	190.7	377.9	2.397	351.4	847.2	1,198.6
177	191.7	378.3	2.385	351.8	846.9	1,198.7
178	192.7	378.8	2.373	352.3	846.5	1,198.8
179	193.7	379.2	2.361	352.8	846.1	1,198.9
180	194.7	379.6	2.349	353.2	845.7	1,198.9
181	195.7	380.0	2.337	353.7	845.3	1,199.0
182	196.7	380.5	2.326	354.1	844.9	1,199.0
183	197.7	380.9	2.315	354.6	844.5	1,199.1
184	198.7	381.3	2.304	355.1	844.1	1,199.2
185	199.7	381.7	2.292	355.5	843.8	1,199.3
186	200.7	382.2	2.281	355.9	843.4	1,199.3
187	201.7	382.6	2.271	356.3	843.1	1,199.4
188	202.7	383.0	2.260	356.8	842.7	1,199.5
189	203.7	383.4	2.249	357.2	842.3	1,199.5
190	204.7	383.8	2.238	357.7	841.9	1,199.6
191	205.7	384.2	2.228	358.1	841.6	1,199.7
192	206.7	384.6	2.218	358.5	841.2	1,199.7
193	207.7	385.0	2.207	359.0	840.8	1,199.8
194	208.7	385.4	2.197	359.4	840.5	1,199.9
195	209.7	385.8	2.187	359.9	840.1	1,200.0
196	210.7	386.3	2.177	360.3	839.7	1,200.0
197	211.7	386.7	2.167	360.7	839.4	1,200.1
198	212.7	387.1	2.158	361.2	838.9	1,200.1
199	213.7	387.5	2.148	361.6	838.6	1,200.2

(*Continued*)

STEAM TABLES (*Continued*)

Steam Pressure psig	Steam Pressure psia	Saturation Temperature °F	Specific Volume cu.ft./lb.	Heat Content (above 32°F) Btu/lb.		
				Sensible	Latent	Total
200	214.7	387.9	2.138	362.1	838.2	1,200.3
201	215.7	388.2	2.128	362.5	837.8	1,200.3
202	216.7	388.6	2.119	362.9	837.5	1,200.4
203	217.7	389.0	2.110	363.3	837.1	1,200.4
204	218.7	389.4	2.100	363.8	836.8	1,200.6
205	219.7	389.8	2.091	364.2	836.4	1,200.6
206	220.7	390.2	2.082	364.6	836.0	1,200.6
207	221.7	390.6	2.073	365.0	835.7	1,200.7
208	222.7	391.0	2.064	365.4	835.3	1,200.7
209	223.7	391.4	2.055	365.8	835.0	1,200.8
210	224.7	391.8	2.046	366.2	834.6	1,200.8
211	225.7	392.1	2.037	366.6	834.2	1,200.8
212	226.7	392.5	2.028	367.0	833.9	1,200.9
213	227.7	392.9	2.020	367.5	833.5	1,201.0
214	228.7	393.3	2.011	367.9	833.2	1,201.1
215	229.7	393.6	2.003	368.3	832.8	1,201.1
216	230.7	394.0	1.994	368.7	832.5	1,201.2
217	231.7	394.4	1.986	369.1	832.1	1,201.2
218	232.7	394.8	1.978	369.5	831.8	1,201.3
219	233.7	395.2	1.970	369.9	831.4	1,201.3
220	234.7	395.5	1.961	370.3	831.1	1,201.4
221	235.7	395.9	1.953	370.7	830.8	1,201.5
222	236.7	396.3	1.945	371.1	830.4	1,201.5
223	237.7	396.6	1.937	371.5	830.1	1,201.6
224	238.7	397.0	1.929	371.9	829.7	1,201.6
225	239.7	397.4	1.921	372.3	829.4	1,201.7
226	240.7	397.7	1.914	372.7	829.0	1,201.7
227	241.7	398.1	1.906	373.0	828.7	1,201.7
228	242.7	398.4	1.898	373.4	828.3	1,201.7
229	243.7	398.8	1.891	373.8	828.0	1,201.8
230	244.7	399.2	1.883	374.2	827.6	1,201.8
231	245.7	399.5	1.876	374.6	827.3	1,201.9
232	246.7	399.9	1.869	375.0	826.9	1,201.9
233	247.7	400.2	1.862	375.3	826.6	1,201.9
234	248.7	400.6	1.854	375.7	826.2	1,201.9
235	249.7	400.9	1.847	376.1	825.9	1,202.0
236	250.7	401.3	1.840	376.5	825.6	1,202.1
237	251.7	401.6	1.833	376.8	825.3	1,202.1
238	252.7	402.0	1.826	377.2	824.9	1,202.1
239	253.7	402.3	1.819	377.6	824.6	1,202.2
240	254.7	402.7	1.812	378.0	824.3	1,202.3
241	255.7	403.0	1.805	378.4	824.0	1,202.4
242	256.7	403.4	1.798	378.7	823.7	1,202.4
243	257.7	403.7	1.791	379.1	823.3	1,202.4
244	258.7	404.1	1.785	379.5	822.9	1,202.4
245	259.7	404.4	1.778	379.9	822.6	1,202.5
246	260.7	404.7	1.771	380.3	822.3	1,202.6
247	261.7	405.1	1.765	380.6	822.0	1,202.6
248	262.7	405.4	1.758	381.0	821.6	1,202.6
249	263.7	405.8	1.752	381.3	821.3	1,202.6
250	264.7	406.1	1.745	381.7	821.0	1,202.7
251	265.7	406.4	1.739	382.1	820.7	1,202.8
252	266.7	406.8	1.733	382.4	820.4	1,202.8
253	267.7	407.1	1.726	382.8	820.0	1,202.8
254	268.7	407.4	1.720	383.2	819.6	1,202.8
255	269.7	407.8	1.714	383.6	819.3	1,202.9
256	270.7	408.1	1.707	383.9	819.0	1,202.9
257	271.7	408.4	1.701	384.3	818.7	1,203.0
258	272.7	408.8	1.695	384.6	818.4	1,203.0
259	273.7	409.1	1.689	385.0	818.0	1,203.0

(*Continued*)

STEAM TABLES (*Continued*)

Steam Pressure psig	Steam Pressure psia	Saturation Temperature °F	Specific Volume cu.ft./lb.	Heat Content (above 32°F) Btu/lb.		
				Sensible	Latent	Total
260	274.7	409.4	1.683	385.3	817.7	1,203.0
261	275.7	409.7	1.677	385.7	817.4	1,203.1
262	276.7	410.1	1.671	386.0	817.1	1,203.1
263	277.7	410.4	1.666	386.4	816.7	1,203.1
264	278.7	410.7	1.660	386.7	816.4	1,203.1
265	279.7	411.1	1.654	387.1	816.1	1,203.2
266	280.7	411.4	1.648	387.5	815.8	1,203.3
267	281.7	411.7	1.642	387.8	815.5	1,203.3
268	282.7	412.0	1.637	388.2	815.2	1,203.4
269	283.7	412.3	1.631	388.5	814.9	1,203.4
270	284.7	412.7	1.625	388.9	814.6	1,203.5
271	285.7	413.0	1.620	389.2	814.3	1,203.5
272	286.7	413.3	1.614	389.5	814.0	1,203.5
273	287.7	413.6	1.609	389.9	813.6	1,203.5
274	288.7	413.9	1.603	390.3	813.3	1,203.6
275	289.7	414.2	1.598	390.6	813.0	1,203.6
276	290.7	414.5	1.593	390.9	812.7	1,203.6
277	291.7	414.9	1.587	391.3	812.3	1,203.6
278	292.7	415.2	1.582	391.6	812.0	1,203.6
279	293.7	415.5	1.577	392.0	811.7	1,203.7
280	294.7	415.8	1.571	392.3	811.4	1,203.7
281	295.7	416.1	1.566	392.6	811.1	1,203.7
282	296.7	416.4	1.561	393.0	810.8	1,203.8
283	297.7	416.7	1.556	393.3	810.5	1,203.8
284	298.7	417.0	1.551	393.7	810.2	1,203.9
285	299.7	417.3	1.546	394.0	809.9	1,203.9
286	300.7	417.6	1.541	394.3	809.6	1,203.9
287	301.7	417.9	1.536	394.7	809.3	1,204.0
288	302.7	418.2	1.531	395.0	809.0	1,204.0
289	303.7	418.5	1.526	395.3	808.7	1,204.0
290	304.7	418.8	1.521	395.7	808.4	1,204.1
291	305.7	419.2	1.516	396.0	808.1	1,204.1
292	306.7	419.5	1.511	396.3	807.8	1,204.1
293	307.7	419.8	1.507	396.6	807.5	1,204.1
294	308.7	420.1	1.502	397.0	807.2	1,204.2
295	309.7	420.4	1.497	397.3	806.9	1,204.2
296	310.7	420.6	1.492	397.6	806.6	1,204.2
297	311.7	420.9	1.488	397.9	806.3	1,204.2
298	312.7	421.2	1.483	398.3	806.0	1,204.3
299	313.7	421.5	1.478	398.6	805.7	1,204.3
300	314.7	421.8	1.474	398.9	805.4	1,204.3
310	324.7	424.7	1.429	402.1	802.4	1,204.5
320	334.7	427.6	1.387	405.3	799.4	1,204.7
330	344.7	430.4	1.347	408.3	796.5	1,204.8
340	354.7	433.1	1.310	411.3	793.7	1,205.0
350	364.7	435.7	1.274	414.3	790.9	1,205.2
360	374.7	438.3	1.240	417.1	788.1	1,205.2
370	384.7	440.9	1.208	420.0	785.4	1,205.4
380	394.7	443.4	1.178	422.8	782.6	1,205.4
390	404.7	445.8	1.149	425.5	780.0	1,205.5
400	414.7	448.2	1.121	428.2	777.4	1,205.6
410	424.7	450.6	1.095	430.8	774.8	1,205.6
420	434.7	452.9	1.069	433.4	772.2	1,205.6
430	444.7	455.2	1.045	436.0	769.6	1,205.6
440	454.7	457.4	1.022	438.6	767.0	1,205.6
450	464.7	459.6	1.000	441.0	764.5	1,205.5
460	474.7	461.8	0.979	443.5	762.0	1,205.5
470	484.7	463.9	0.958	445.9	759.5	1,205.4
480	494.7	466.0	0.939	448.3	757.1	1,205.4
490	504.7	468.1	0.920	450.6	754.7	1,205.3

(*Continued*)

STEAM TABLES (*Continued*)

Steam Pressure psig	Steam Pressure psia	Saturation Temperature °F	Specific Volume cu.ft./lb.	Heat Content (above 32°F) Btu/lb.		
				Sensible	Latent	Total
500	514.7	470.1	0.901	453.0	752.2	1,205.2
510	524.7	472.1	0.884	455.2	749.9	1,205.1
520	534.7	474.1	0.867	457.5	747.5	1,205.0
530	544.7	476.0	0.851	459.8	745.1	1,204.9
540	554.7	478.0	0.835	461.9	742.8	1,204.7
550	564.7	479.9	0.820	464.1	740.5	1,204.6
560	574.7	481.7	0.805	466.3	738.2	1,204.5
570	584.7	483.6	0.791	468.4	735.9	1,204.3
580	594.7	485.4	0.777	470.6	733.6	1,204.2
590	604.7	487.2	0.764	472.7	731.3	1,204.0
600	614.7	488.9	0.751	474.7	729.1	1,203.8
610	624.7	490.7	0.739	476.7	726.9	1,203.6
620	634.7	492.4	0.727	478.7	724.7	1,203.4
630	644.7	494.1	0.715	480.7	722.5	1,203.2
640	654.7	495.8	0.704	482.7	720.3	1,203.0
650	664.7	497.5	0.693	484.7	718.1	1,202.8
660	674.7	499.2	0.682	486.7	715.9	1,202.6
670	684.7	500.8	0.671	488.6	713.8	1,202.4
680	694.7	502.4	0.661	490.5	711.6	1,202.1
690	704.7	504.0	0.651	492.4	709.5	1,201.9
700	714.7	505.5	0.642	494.3	707.3	1,201.6
710	724.7	507.1	0.632	496.2	705.2	1,201.4
720	734.7	508.7	0.623	498.0	703.1	1,201.1
730	744.7	510.2	0.614	499.8	701.0	1,200.8
740	754.7	511.7	0.606	501.6	698.9	1,200.5
750	764.7	513.2	0.597	503.4	696.8	1,200.2
760	774.7	514.7	0.589	505.2	694.8	1,200.0
770	784.7	516.1	0.581	507.0	692.7	1,199.7
780	794.7	517.6	0.573	508.8	690.7	1,199.5
790	804.7	519.0	0.566	510.5	688.6	1,199.1
800	814.7	520.5	0.558	512.3	686.5	1,198.8
810	824.7	521.9	0.551	514.0	684.6	1,198.6
820	834.7	523.3	0.544	515.7	682.6	1,198.3
830	844.7	524.7	0.537	517.4	680.6	1,198.0
840	854.7	526.0	0.530	519.0	678.6	1,197.6
850	864.7	527.4	0.523	520.7	676.6	1,197.3
860	874.7	528.7	0.517	522.4	674.6	1,197.0
870	884.7	530.1	0.511	524.1	672.6	1,196.7
880	894.7	531.4	0.504	525.7	670.6	1,196.3
890	904.7	532.7	0.498	527.4	668.6	1,196.0
900	914.7	534.0	0.492	529.0	666.6	1,195.6
910	924.7	535.3	0.486	530.6	664.7	1,195.3
920	934.7	536.6	0.481	532.2	662.7	1,194.9
930	944.7	537.9	0.475	533.8	660.7	1,194.5
940	954.7	539.1	0.470	535.4	658.7	1,194.1
950	964.7	540.4	0.464	536.9	656.8	1,193.7
960	974.7	541.6	0.459	538.5	654.9	1,193.4
970	984.7	542.9	0.454	540.0	653.0	1,193.0
980	994.7	544.1	0.449	541.6	651.0	1,192.6
990	1,004.7	545.3	0.444	543.1	649.1	1,192.2
1,000	1,014.7	546.5	0.439	544.6	647.2	1,191.8
1,050	1,064.7	552.4	0.416	552.2	637.6	1,189.8
1,100	1,114.7	558.1	0.395	559.5	628.2	1,187.7
1,150	1,164.7	563.6	0.375	566.7	618.9	1,185.6
1,200	1,214.7	568.9	0.357	573.7	609.6	1,183.3
1,250	1,264.7	574.0	0.341	580.6	600.3	1,180.9
1,300	1,314.7	579.0	0.325	587.4	591.1	1,178.5
1,350	1,364.7	583.9	0.311	594.0	581.9	1,175.9
1,400	1,414.7	588.6	0.298	600.5	572.8	1,173.3
1,450	1,464.7	593.2	0.285	607.0	563.6	1,170.6

(*Continued*)

STEAM TABLES (Continued)

Steam Pressure psig	Steam Pressure psia	Saturation Temperature °F	Specific Volume cu.ft./lb.	Heat Content (above 32°F) Btu/lb.		
				Sensible	Latent	Total
1,500	1,514.7	597.7	0.274	613.4	554.5	1,167.9
1,550	1,564.7	602.0	0.263	619.6	545.4	1,165.0
1,600	1,614.7	606.3	0.252	625.8	536.2	1,162.0
1,650	1,664.7	610.4	0.242	632.0	527.1	1,159.1
1,700	1,714.7	614.5	0.233	638.0	517.9	1,155.9
1,750	1,764.7	618.5	0.224	644.1	508.7	1,152.8
1,800	1,814.7	622.3	0.216	650.0	499.4	1,149.4
1,850	1,864.7	626.1	0.208	655.9	490.0	1,145.9
1,900	1,914.7	629.8	0.200	661.8	480.6	1,142.4
1,950	1,964.7	633.5	0.193	667.7	471.2	1,138.9
2,000	2,014.7	637.0	0.187	673.6	461.5	1,135.1
2,050	2,064.7	640.5	0.179	679.4	451.8	1,131.3
2,100	2,114.7	643.9	0.173	685.3	442.1	1,127.2
2,150	2,164.7	647.3	0.167	691.1	432.1	1,123.2
2,200	2,214.7	650.6	0.161	697.0	422.0	1,119.0
2,250	2,264.7	653.8	0.155	702.8	411.7	1,114.5
2,300	2,314.7	657.0	0.150	708.7	401.3	1,110.0
2,350	2,364.7	660.1	0.144	714.6	390.6	1,105.2
2,400	2,414.7	663.2	0.139	720.6	379.7	1,100.3
2,450	2,464.7	666.2	0.134	726.6	368.5	1,095.1
2,500	2,514.7	669.2	0.129	732.7	357.1	1,089.8

21.04 Steam Flow through Orifices

STEAM FLOW THROUGH ORIFICES

Orifice Dia. Inches	Steam Flow lbs./h Steam Pressure psig												
	2	5	10	15	25	50	75	100	125	150	200	250	300
1/32	0.3	0.5	0.6	0.7	0.9	1.5	2.1	2.7	3.3	3.9	5.1	6.3	7.4
1/16	1.3	1.9	2.3	2.8	3.8	6.1	8.5	10.8	13.2	15.6	20.3	25.1	29.8
3/32	2.8	4.2	5.3	6.3	8.5	13.8	19.1	24.4	29.7	35.1	45.7	56.4	67.0
1/8	4.5	7.5	9.4	11.2	15.0	24.5	34.0	43.4	52.9	62.4	81.3	100	119
5/32	7.8	11.7	14.6	17.6	23.5	38.3	53.1	67.9	82.7	97.4	127	156	186
3/16	11.2	16.7	21.0	25.3	33.8	55.1	76.4	97.7	119	140	183	226	268
7/32	15.3	22.9	28.7	34.4	46.0	75.0	104	133	162	191	249	307	365
1/4	20.0	29.8	37.4	45.0	60.1	98.0	136	173	212	250	325	401	477
9/32	25.2	37.8	47.4	56.9	76.1	124	172	220	268	316	412	507	603
5/16	31.2	46.6	58.5	70.3	94.0	153	212	272	331	390	508	627	745
11/32	37.7	56.4	70.7	85.1	114	185	257	329	400	472	615	758	901
3/8	44.9	67.1	84.2	101	135	221	306	391	476	561	732	902	1,073
13/32	52.7	78.8	98.8	119	159	259	359	459	559	659	859	1,059	1,259
7/16	61.1	91.4	115	138	184	300	416	532	648	764	996	1,228	1,460
15/32	70.2	105	131	158	211	344	478	611	744	877	1,144	1,410	1,676
1/2	79.8	119	150	180	241	392	544	695	847	998	1,301	1,604	1,907

Note:
1 Steam leaks and energy wasted: A 1/8" diameter hole in a steam pipe can discharge 62.4 lbs.Stm./h at 150 psig, resulting in 30 tons of coal, 4800 gallons of oil, or 7500 therms of gas to be wasted each year (assuming 8400 hour per year operation).

21.05 Flash Steam

FLASH STEAM

| Steam Press. psig | Flash Steam Flow lbs. Steam/h per 100 lbs. of Steam Condensate/h |

Steam Press. psig	Condensate Pressure psig																
	0	1	3	5	7	10	12	15	20	25	30	40	50	60	75	85	100
0	0.0																
1	0.3	0.0															
3	1.0	0.6	0.0														
5	1.6	1.2	0.6	0.0													
7	2.1	1.8	1.1	0.6	0.0												
10	2.8	2.5	1.9	1.3	0.7	0.0											
12	3.3	3.0	2.3	1.7	1.2	0.5	0.0										
15	3.9	3.6	3.0	2.4	1.8	1.1	0.6	0.0									
20	4.9	4.5	3.9	3.3	2.8	2.1	1.6	1.0	0.0								
25	5.7	5.4	4.8	4.2	3.7	2.9	2.5	1.8	0.9	0.0							
30	6.5	6.2	5.5	5.0	4.4	3.7	3.3	2.6	1.7	0.8	0.0						
40	7.8	7.5	6.9	6.3	5.8	5.1	4.6	4.0	3.0	2.2	1.4	0.0					
50	9.0	8.7	8.1	7.5	7.0	6.3	5.8	5.2	4.2	3.4	2.6	1.2	0.0				
60	10.0	9.7	9.1	8.5	8.0	7.3	6.9	6.2	5.3	4.5	3.7	2.3	1.1	0.0			
75	11.4	11.1	10.5	9.9	9.4	8.7	8.2	7.6	6.7	5.9	5.1	3.7	2.5	1.5	0.0		
85	12.2	11.9	11.3	10.7	10.2	9.5	9.1	8.5	7.5	6.7	6.0	4.6	3.4	2.3	0.9	0.0	
100	13.3	13.0	12.4	11.8	11.3	10.6	10.2	9.6	8.7	7.9	7.1	5.8	4.6	3.5	2.1	1.2	0.0
120	14.6	14.3	13.7	13.2	12.7	12.0	11.5	11.0	10.0	9.2	8.5	7.2	6.0	4.9	3.5	2.6	1.5
125	14.9	14.6	14.0	13.5	13.0	12.3	11.9	11.3	10.4	9.5	8.8	7.5	6.3	5.3	3.8	3.0	1.8
150	16.3	16.0	15.4	14.9	14.4	13.7	13.3	12.7	11.8	11.0	10.3	9.0	7.8	6.8	5.4	4.5	3.3
175	17.6	17.3	16.7	16.2	15.7	15.0	14.6	14.0	13.1	12.3	11.6	10.3	9.2	8.1	6.7	5.9	4.7
200	18.7	18.4	17.9	17.4	16.9	16.2	15.8	15.2	14.3	13.5	12.8	11.5	10.4	9.4	8.0	7.2	6.0
225	19.8	19.5	18.9	18.4	17.9	17.3	16.9	16.3	15.4	14.6	13.9	12.6	11.5	10.5	9.1	8.3	7.2
250	20.8	20.5	19.9	19.4	18.9	18.3	17.8	17.3	16.4	15.6	14.9	13.7	12.5	11.5	10.2	9.4	8.2
275	21.7	21.4	20.8	20.3	19.8	19.2	18.8	18.2	17.4	16.6	15.9	14.6	13.5	12.5	11.2	10.4	9.2
300	22.5	22.2	21.7	21.2	20.7	20.1	19.7	19.1	18.2	17.5	16.8	15.5	14.4	13.4	12.1	11.3	10.2

21.06 Warm-up Loads

LOW-PRESSURE STEAM PIPING WARM-UP LOADS

Pipe Size	Pounds of Steam per 100 Feet of Pipe							
	Steam Pressure psig							
	0	1	3	5	7	10	12	15
1/2"	1	1	2	2	2	2	2	2
3/4"	2	2	2	2	2	2	2	2
1"	3	3	3	3	3	3	4	4
1-1/4"	4	4	4	4	4	5	5	5
1-1/2"	5	5	5	5	5	6	6	6
2"	6	6	7	7	7	7	8	8
2-1/2"	10	10	10	11	11	12	12	13
3"	13	13	14	14	15	15	16	17
4"	18	19	19	20	21	22	23	24
5"	25	25	26	27	28	30	31	32
6"	32	33	34	36	37	39	40	42
8"	48	49	51	54	56	58	60	62
10"	68	70	73	76	79	83	85	89
12"	83	85	89	93	96	101	104	108
14"	92	94	98	103	106	111	115	119
16"	105	108	113	118	122	128	132	137
18"	119	122	127	133	137	144	149	154
20"	132	135	142	148	153	160	166	172

(Continued)

LOW-PRESSURE STEAM PIPING WARM-UP LOADS (*Continued*)

Pipe Size	Pounds of Steam per 100 Feet of Pipe							
	Steam Pressure psig							
	0	1	3	5	7	10	12	15
22"	146	150	157	164	169	177	183	190
24"	159	163	170	178	184	193	199	207
26"	173	177	185	194	200	210	217	225
28"	187	191	200	209	216	226	234	243
30"	200	205	214	224	232	243	251	260
32"	214	219	229	239	247	259	268	278
34"	227	233	243	254	263	275	284	295
36"	241	246	258	269	278	292	301	313
42"	281	288	301	314	325	341	352	366
48"	321	328	343	358	371	389	402	417
54"	361	370	387	404	418	438	453	470
60"	402	411	430	449	465	487	503	523
72"	483	494	517	539	558	585	604	628
84"	564	577	603	629	652	683	706	733
96"	645	660	690	720	745	781	807	838
Corr. Factor	1.50	1.49	1.46	1.44	1.43	1.41	1.40	1.39

Notes:
1 Table based on 70°F ambient temperature, standard weight steel pipe to 250 psig, and extra-strong weight steel pipe above 250 psig.
2 For ambient temperatures of 0°F, multiply table values by correction factor.

MEDIUM-PRESSURE STEAM PIPING WARM-UP LOADS

Pipe Size	Pounds of Steam per 100 Feet of Pipe								
	Steam Pressure psig								
	20	25	30	40	50	60	75	85	100
1/2"	2	2	2	2	2	3	3	3	3
3/4"	3	3	3	3	3	3	4	4	4
1"	4	4	4	5	5	5	5	6	6
1-1/4"	5	6	6	6	7	7	7	8	8
1-1/2"	6	7	7	7	8	8	9	9	10
2"	8	9	9	10	11	11	12	12	13
2-1/2"	13	14	15	16	17	17	19	19	20
3"	18	18	19	21	22	23	24	25	27
4"	25	26	27	29	31	32	35	36	38
5"	34	35	37	40	42	44	47	49	51
6"	44	46	48	51	55	57	61	63	66
8"	66	69	72	77	82	86	92	95	100
10"	94	98	102	110	116	122	130	135	142
12"	115	120	125	134	142	149	159	165	173
14"	126	132	138	148	157	164	175	182	191
16"	145	152	158	170	180	188	201	209	219
18"	163	171	178	191	203	213	227	235	247
20"	182	191	198	213	226	237	252	262	275
22"	201	211	220	236	250	262	279	290	304
24"	219	229	239	257	272	285	304	316	331
26"	238	250	260	279	296	310	331	344	360
28"	257	269	280	301	319	334	356	370	388
30"	275	289	300	323	342	358	382	397	416
32"	294	308	321	344	365	382	408	424	444
34"	312	327	341	366	388	407	434	450	472
36"	331	347	361	388	411	431	459	477	500
42"	386	405	422	453	480	503	536	557	584
48"	441	463	482	517	548	574	612	636	667
54"	497	521	542	583	617	647	690	717	751
60"	552	579	603	648	686	719	767	797	835

(Continued)

MEDIUM-PRESSURE STEAM PIPING WARM-UP LOADS (*Continued*)

Pipe Size	Pounds of Steam per 100 Feet of Pipe								
	Steam Pressure psig								
	20	25	30	40	50	60	75	85	100
72"	664	696	724	778	825	864	921	957	1,003
84"	775	812	846	908	963	1,009	1,075	1,117	1,171
96"	886	929	967	1,039	1,101	1,153	1,230	1,278	1,340
Corr. Factor	1.37	1.36	1.35	1.32	1.31	1.29	1.28	1.27	1.26

Notes:
1 Table based on 70°F ambient temperature, standard weight steel pipe to 250 psig, and extra-strong weight steel pipe above 250 psig.
2 For ambient temperatures of 0°F, multiply table values by the correction factor.

HIGH-PRESSURE STEAM PIPING WARM-UP LOADS

Pipe Size	Pounds of Steam per 100 Feet of Pipe								
	Steam Pressure psig								
	120	125	150	175	200	225	250	275	300
1/2"	3	3	3	4	4	4	4	4	5
3/4"	4	4	4	5	5	5	5	6	7
1"	6	6	7	7	7	8	8	8	11
1-1/4"	8	9	9	9	10	10	11	11	15
1-1/2"	10	10	11	11	12	12	13	13	18
2"	14	14	14	15	16	17	17	18	25
2-1/2"	21	22	23	24	25	26	27	28	39
3"	28	28	30	32	33	34	36	37	52
4"	40	40	43	45	47	49	51	53	75
5"	54	55	58	61	64	66	69	71	104
6"	70	71	75	79	83	86	89	92	144
8"	106	107	113	119	125	129	134	139	218
10"	150	152	161	169	177	184	191	197	275
12"	183	186	197	206	216	225	233	241	329
14"	202	204	217	227	238	247	257	266	362
16"	231	234	248	261	273	284	295	305	416
18"	261	264	280	294	308	320	332	343	470
20"	290	294	312	327	343	356	370	382	523
22"	322	326	345	362	380	394	409	423	578
24"	350	354	375	394	413	429	445	460	631
26"	381	386	409	429	449	467	485	501	384
28"	410	416	440	462	484	503	522	540	739
30"	440	446	472	496	519	540	560	579	794
32"	469	476	504	529	554	576	598	618	844
34"	499	506	536	562	589	612	635	657	900
36"	528	536	567	596	624	648	673	696	955
42"	617	626	663	696	729	757	786	813	1,116
48"	705	714	757	794	832	865	898	928	1,275
54"	794	805	852	895	937	974	1,011	1,045	1,436
60"	883	894	948	995	1,042	1,083	1,124	1,162	1,597
72"	1,060	1,075	1,139	1,195	1,252	1,301	1,350	1,396	1,946
84"	1,238	1,254	1,329	1,395	1,461	1,518	1,576	1,630	2,241
96"	1,415	1,435	1,520	1,595	1,671	1,737	1,803	1,864	2,510
Corr. Factor	1.25	1.25	1.24	1.23	1.22	1.22	1.21	1.21	1.20

Notes:
1 Table based on 70°F ambient temperature, standard weight steel pipe to 250 psig, and extra-strong weight steel pipe above 250 psig.
2 For ambient temperatures of 0°F, multiply table values by the correction factor.

21.07 Steam Operating Loads

LOW-PRESSURE STEAM PIPING OPERATING LOADS

| Pipe Size | Pounds of Steam per Hour per 100 Feet of Pipe | | | | | | | |
| | Steam Pressure psig | | | | | | | |
	0	1	3	5	7	10	12	15
1/2"	2	2	2	2	3	3	3	3
3/4"	3	3	3	3	3	3	3	4
1"	3	3	3	4	4	4	4	4
1-1/4"	4	4	4	4	5	5	5	5
1-1/2"	4	4	5	5	5	6	6	6
2"	5	5	6	6	6	7	7	7
2-1/2"	6	6	7	7	7	8	8	9
3"	7	8	8	8	9	9	10	10
4"	9	9	10	10	11	12	12	13
5"	11	11	12	13	13	14	15	15
6"	13	13	14	15	15	16	17	18
8"	16	17	18	19	19	21	22	23
10"	20	20	21	23	24	25	26	28
12"	23	24	25	26	28	29	31	32
14"	25	26	27	29	30	32	33	35
16"	28	29	31	32	34	36	38	40
18"	31	32	34	36	38	40	42	44
20"	34	35	37	39	41	44	46	48
22"	37	38	41	43	45	48	50	53
24"	40	41	44	47	49	52	54	57
26"	47	48	51	54	57	60	63	66
28"	50	52	55	58	61	65	68	72
30"	54	56	59	62	65	70	73	77
32"	57	59	63	67	70	74	78	82
34"	61	63	67	71	74	79	83	87
36"	65	67	71	75	78	84	87	92
42"	75	78	83	87	92	98	102	107
48"	86	89	94	100	105	112	117	123
54"	97	100	106	112	118	125	131	138
60"	108	111	118	125	131	139	146	153
72"	129	133	141	150	157	167	175	184
84"	151	156	165	175	183	195	204	215
96"	172	178	189	200	209	223	233	245
Corr. Factor	1.70	1.68	1.66	1.64	1.60	1.58	1.57	1.55

Notes:
1 Table based on 70°F ambient temperature, standard weight steel pipe to 250 psig, and extra-strong weight steel pipe above 250 psig.
2 For ambient temperatures of 0°F, multiply the table values by the correction factor.
3 Table values include convection and radiation loads with 80 percent efficient insulation.

MEDIUM-PRESSURE STEAM PIPING OPERATING LOADS

| Pipe Size | Pounds of Steam per Hour per 100 Feet of Pipe | | | | | | | |
| | Steam Pressure psig | | | | | | | |
	20	25	30	40	50	60	75	85	100
1/2"	3	3	4	4	4	5	5	5	6
3/4"	4	4	4	5	5	6	6	6	7
1"	5	5	5	6	6	7	7	8	8
1-1/4"	6	6	6	7	8	8	9	9	10
1-1/2"	6	7	7	8	9	9	10	11	11
2"	8	8	9	10	11	11	12	13	14
2-1/2"	9	10	10	12	12	13	14	15	16
3"	11	12	12	14	15	16	17	18	19
4"	14	15	16	17	18	20	21	23	24

(Continued)

MEDIUM-PRESSURE STEAM PIPING OPERATING LOADS (*Continued*)

	Pounds of Steam per Hour per 100 Feet of Pipe								
	Steam Pressure psig								
Pipe Size	20	25	30	40	50	60	75	85	100
5"	17	18	19	21	22	24	26	27	29
6"	19	21	22	24	26	28	30	32	34
8"	25	26	28	30	33	35	38	40	43
10"	30	32	34	37	40	43	47	49	53
12"	35	37	39	43	47	50	54	57	61
14"	38	40	43	47	51	54	59	62	67
16"	43	45	48	53	57	61	67	70	75
18"	47	50	53	59	64	68	74	78	84
20"	52	56	59	65	70	75	82	86	92
22"	57	60	64	70	76	81	89	94	101
24"	61	65	69	76	83	88	96	102	109
26"	72	77	81	89	97	103	113	110	117
28"	77	82	87	96	104	111	122	129	138
30"	83	88	93	103	112	119	131	138	148
32"	88	94	100	110	119	127	139	147	157
34"	94	100	106	117	127	135	148	156	167
36"	99	106	112	124	134	143	157	166	177
42"	116	124	131	144	157	167	183	193	207
48"	132	141	149	165	179	191	209	221	236
54"	149	159	168	186	201	215	235	248	266
60"	165	177	187	206	224	239	261	276	295
72"	199	212	224	247	268	287	314	331	354
84"	232	247	261	289	313	334	366	386	413
96"	265	283	299	330	358	382	418	442	472
Corr. Factor	1.52	1.51	1.50	1.48	1.47	1.45	1.43	1.42	1.41

Notes:
1 Table based on 70°F ambient temperature, standard weight steel pipe to 250 psig, and extra-strong weight steel pipe above 250 psig.
2 For ambient temperatures of 0°F, multiply the table values by the correction factor.
3 Table values include convection and radiation loads with 80 percent efficient insulation.

HIGH-PRESSURE STEAM PIPING OPERATING LOADS

	Pounds of Steam per Hour per 100 Feet of Pipe								
	Steam Pressure psig								
Pipe Size	120	125	150	175	200	225	250	275	300
1/2"	6	6	7	7	8	8	8	9	9
3/4"	7	7	8	9	9	10	10	11	11
1"	9	9	10	10	11	12	12	13	14
1-1/4"	11	11	12	13	14	14	15	16	17
1-1/2"	12	12	13	14	15	16	17	18	19
2"	15	15	16	18	19	20	21	22	23
2-1/2"	18	18	19	21	22	23	25	26	27
3"	21	21	23	25	26	28	29	31	32
4"	26	27	29	31	33	35	37	38	40
5"	31	32	35	37	40	42	44	46	49
6"	37	38	41	44	46	49	52	54	57
8"	47	48	52	55	59	62	66	69	72
10"	57	58	63	67	72	76	80	84	88
12"	66	68	73	79	84	89	93	98	103
14"	72	74	80	85	91	96	102	107	112
16"	81	83	90	96	103	109	115	121	126
18"	91	92	100	107	115	121	128	134	141
20"	100	102	110	118	126	134	141	148	155
22"	109	111	120	129	138	146	154	161	169
24"	118	120	130	140	149	158	167	175	183
26"	127	129	140	150	161	170	179	188	197

(*Continued*)

MEDIUM-PRESSURE STEAM PIPING OPERATING LOADS (*Continued*)

Pipe Size	Pounds of Steam per Hour per 100 Feet of Pipe								
	Steam Pressure psig								
	120	125	150	175	200	225	250	275	300
28"	149	152	165	177	189	182	192	201	211
30"	160	163	177	190	203	214	226	237	249
32"	170	174	189	202	216	229	241	253	265
34"	181	185	200	215	230	243	256	269	282
36"	192	195	212	228	243	257	271	285	299
42"	224	228	248	265	284	300	317	332	348
48"	256	261	283	303	324	343	362	380	398
54"	287	293	318	341	365	386	407	427	448
60"	319	326	354	379	406	429	452	475	498
72"	383	391	425	455	487	514	543	570	597
84"	447	456	495	531	568	600	633	665	697
96"	511	521	566	607	649	686	724	760	796
Corr. Factor	1.39	1.39	1.39	1.38	1.37	1.37	1.36	1.36	1.35

Notes:
1 Table based on 70°F ambient temperature, standard weight steel pipe to 250 psig, and extra-strong weight steel pipe above 250 psig.
2 For ambient temperatures of 0°F, multiply the table values by the correction factor.
3 Table values include convection and radiation loads with 80 percent efficient insulation.

21.08 Boiling Points of Water

BOILING POINTS OF WATER

Psia	Boiling Point °F	Psia	Boiling Point °F	Psia	Boiling Point °F
0.5	79.6	44	273.1	150	358.5
1	101.7	46	275.8	175	371.8
2	126.0	48	278.5	200	381.9
3	141.4	50	281.0	225	391.9
4	152.9	52	283.5	250	401.0
5	162.2	54	285.9	275	409.5
6	170.0	56	288.3	300	417.4
7	176.8	58	290.5	325	424.8
8	182.8	60	292.7	350	431.8
9	188.3	62	294.9	375	438.4
10	193.2	64	297.0	400	444.7
11	197.7	66	299.0	425	450.7
12	201.9	68	301.0	450	456.4
13	205.9	70	303.0	475	461.9
14	209.6	72	304.9	500	467.1
14.69	212.0	74	306.7	525	472.2
15	213.0	76	308.5	550	477.1
16	216.3	78	310.3	575	481.8
17	219.4	80	312.1	600	486.3
18	222.4	82	313.8	625	490.7
19	225.2	84	315.5	650	495.0
20	228.0	86	317.1	675	499.2
22	233.0	88	318.7	700	503.2
24	237.8	90	320.3	725	507.2
26	242.3	92	321.9	750	511.0
28	246.4	94	323.4	775	514.7
30	250.3	96	324.9	800	518.4
32	254.1	98	326.4	825	521.9
34	257.6	100	327.9	850	525.4
36	261.0	105	331.4	875	528.8
38	264.2	110	334.8	900	532.1
40	267.3	115	338.1	950	538.6
42	270.2	120	341.3	1000	544.8

21.09 Steam Heating Units of Measure

COMPARISON OF COMMON STEAM HEATING UNITS OF MEASURE

MBH (1000 Btuh)	Steam lbs./h	EDR sq.ft.	Boiler hp	Condensate Flow Rate GPM	Cond. Pump Capacity GPM (2)
10	10.6	42	0.3	0.02	0.06
25	26.4	104	0.7	0.05	0.15
50	52.9	208	1.5	0.10	0.30
75	79.3	313	2.2	0.16	0.48
100	105.8	417	2.9	0.21	0.63
200	211.5	833	5.8	0.41	1.23
300	317.3	1,250	8.7	0.62	1.86
400	423.0	1,667	11.6	0.83	2.49
500	528.8	2,083	14.5	1.03	3.09
750	793.1	3,125	21.7	1.55	4.65
1,000	1,058	4,167	29.0	2.07	6.21
1,250	1,322	5,208	36.2	2.58	7.74
1,500	1,418	6,250	43.5	3.10	9.30
1,750	1,851	7,292	50.7	3.62	10.8
2,000	2,115	8,333	58.0	4.13	12.4
2,500	2,644	10,417	72.5	5.17	15.5
3,000	3,173	12,500	87.0	6.20	18.6
4,000	4,230	16,667	115.9	8.27	24.8
5,000	5,288	20,833	144.9	10.3	30.9
7,500	7,931	31,250	217.4	15.5	46.5
10,000	10,575	41,667	289.9	20.7	62.1
15,000	15,862	62,500	434.8	31.0	93.0
20,000	21,150	83,333	579.7	41.3	124
25,000	26,438	104,167	724.6	51.7	155
30,000	31,725	125,000	869.6	62.0	186
35,000	37,014	145,833	1,015	72.3	217
40,000	42,301	166,667	1,159	82.7	248
50,000	52,876	208,333	1,449	103.3	310

Notes:
1 Steam flow rate is based on 15 psig steam with an enthalpy of 945.6 Btu/lb.
2 Condensate pump capacity is equal to three times the condensate flow rate.

21.10 Low-Pressure Steam Pipe Sizing Tables (15 psig and Less)

1 PSIG STEAM PIPING SYSTEMS—STEEL PIPE

	Steam Flow lbs./h								
	Pressure Drop psig/100 ft.			Velocity FPM (mph)					
Pipe Size	0.125	0.25	0.5	2,000 (23)	4,000 (45)	6,000 (68)	8,000 (91)	10,000 (114)	12,000 (136)
1/2	4	6	9		Pressure these				
3/4	10	14	20	18					
1	20	28	40	29		drop	governs		
1-1/4	44	62	87	49		pipe	sizes		
1-1/2	68	96	135	67	135			with	
2	137	194	274	111	222				
2-1/2	226	320	452	158	317				
3	414	585	822	245	489	734			
4	874	1,236	1,748	421	842	1,263	1,685		
5	1,608	2,274	3,217	659	1,318	1,978	2,637		
6	2,654	3,753	5,308	956	1,912	2,867	3,823	4,779	
8	5,525	7,813		1,655	3,310	4,965	6,620	8,275	9,930

(Continued)

1 PSIG STEAM PIPING SYSTEMS—STEEL PIPE (Continued)

Pipe Size	Pressure Drop psig/100 ft.			Steam Flow lbs./h					
				Velocity FPM (mph)					
	0.125	0.25	0.5	2,000 (23)	4,000 (45)	6,000 (68)	8,000 (91)	10,000 (114)	12,000 (136)
10	10,082	14,258		2,609	5,218	7,826	10,435	13,044	15,653
12	16,181			3,742	7,483	11,225	14,967	18,708	22,450
14	20,959			4,562	9,123	13,685	18,247	22,809	27,370
16	30,212			6,043	12,086	18,128	24,171	30,214	36,257
18	41,576			7,732	15,463	23,195	30,927	38,659	46,390
20	55,192			9,629	19,257	28,886	38,514	48,143	57,771
22				11,733	23,466	35,200	46,933	58,666	70,399
24				14,046	28,092	42,137	56,183	70,229	84,275
26				16,566	33,132	49,698	66,265	82,831	99,397
28			with sizes	19,294	38,589	57,883	77,178	96,472	115,767
30				22,231	44,461	66,692	88,922	111,153	133,384
32		governs pipe		25,375	50,749	76,124	101,498	126,873	152,248
34	Velocity these			28,726	57,453	86,179	114,906	143,632	172,359
36				32,286	64,572	96,859	129,145	161,431	193,717
42				44,213	88,425	132,638	176,851	221,064	265,276
48				58,010	116,020	174,030	232,040	290,050	348,060
54				73,678	147,356	221,034	294,712	368,390	442,069
60				91,217	182,434	273,651	364,868	456,085	547,302
72				131,907	263,815	395,722	527,629	659,537	791,444
84				180,081	360,162	540,243	720,324	900,404	1,080,485
96				235,738	471,475	707,213	942,951	1,178,689	1,414,426

Notes:
1 Maximum recommended pressure drop/velocity: 0.125 psig/100 ft./4,000 FPM.
2 Table based on Standard Weight Steel Pipe using steam equations in Part 3.

3 PSIG STEAM PIPING SYSTEMS—STEEL PIPE

Pipe Size	Pressure Drop psig/100 ft.			Steam Flow lbs./h					
				Velocity FPM (mph)					
	0.125	0.25	0.5	2,000 (23)	4,000 (45)	6,000 (68)	8,000 (91)	10,000 (114)	12,000 (136)
1/2	5	6	9	20	Pressure these				
3/4	10	15	21						
1	21	30	42	32		drop			
1-1/4	46	65	92	55		pipe			
1-1/2	72	101	143	75			governs sizes	with	
2	145	205	290	124	248				
2-1/2	239	338	478	177	354				
3	437	619	870	274	547	821			
4	924	1,307	1,849	471	942	1,413			
5	1,701	2,405	3,402	737	1,475	2,212	2,949		
6	2,807	3,969	5,614	1,069	2,138	3,207	4,276	5,345	
8	5,843	8,263		1,851	3,702	5,553	7,404	9,255	11,106

(Continued)

3 PSIG STEAM PIPING SYSTEMS—STEEL PIPE (Continued)

				Steam Flow lbs./h					
	Pressure Drop psig/100 ft.			Velocity FPM (mph)					
Pipe Size	0.125	0.25	0.5	2,000 (23)	4,000 (45)	6,000 (68)	8,000 (91)	10,000 (114)	12,000 (136)
10	10,662	15,078		2,918	5,835	8,753	11,670	14,588	17,506
12	17,112	24,200		4,185	8,369	12,554	16,738	20,923	25,108
14	22,165			5,102	10,204	15,305	20,407	25,509	30,611
16	31,951			6,758	13,516	20,275	27,033	33,791	40,549
18	43,968			8,647	17,294	25,941	34,588	43,235	51,883
20	58,368			10,768	21,537	32,305	43,074	53,842	64,611
22	75,290			13,122	26,245	39,367	52,489	65,611	78,734
24				15,709	31,417	47,126	62,834	78,543	94,252
26				18,527	37,055	55,582	74,110	92,637	111,164
28			with sizes	21,579	43,157	64,736	86,315	107,893	129,472
30				24,862	49,725	74,587	99,450	124,312	149,175
32		governs		28,379	56,757	85,136	113,515	141,893	170,272
34		pipe		32,127	64,255	96,382	128,509	160,637	192,764
36				36,109	72,217	108,326	144,434	180,543	216,651
42	Velocity			49,447	98,894	148,341	197,788	247,235	296,682
48	these			64,878	129,755	194,633	259,511	324,388	389,266
54				82,401	164,801	247,202	329,603	412,003	494,404
60				102,016	204,032	306,048	408,064	510,080	612,096
72				147,524	295,047	442,571	590,094	737,618	885,141
84				201,400	402,801	604,201	805,601	1,007,001	1,208,402
96				263,646	527,292	790,939	1,054,585	1,318,231	1,581,877

Notes:
1 Maximum recommended pressure drop/velocity: 0.125 psig/100 ft./4,000 FPM.
2 Table based on Standard Weight Steel Pipe using steam equations in Part 3.

5 PSIG STEAM PIPING SYSTEMS—STEEL PIPE

				Steam Flow lbs./h					
	Pressure Drop psig/100 ft.			Velocity FPM (mph)					
Pipe Size	0.125	0.25	0.5	2,000 (23)	4,000 (45)	6,000 (68)	8,000 (91)	10,000 (114)	12,000 (136)
1/2	5	7	10		Pressure these				
3/4	11	15	22	22					
1	22	31	44	35		drop			
1-1/4	48	69	97	61		pipe			
1-1/2	75	106	150	83			governs sizes	with	
2	153	216	305	137	275				
2-1/2	251	355	503	196	392				
3	460	651	914	302	605	907			
4	972	1,375	1,944	521	1,042	1,563			
5	1,789	2,529	3,577	815	1,631	2,446	3,261		
6	2,952	4,174	5,903	1,182	2,364	3,546	4,728		
8	6,144	8,689		2,047	4,094	6,141	8,188	10,235	12,282
10	11,212	15,856		3,226	6,453	9,679	12,906	16,132	19,359
12	17,995	25,449		4,628	9,255	13,883	18,510	23,138	27,765
14	23,309	32,964		5,642	11,284	16,926	22,567	28,209	33,851
16	33,599			7,474	14,947	22,421	29,894	37,368	44,842
18	46,237			9,562	19,125	28,687	38,250	47,812	57,375
20	61,380			11,908	23,817	35,725	47,633	59,542	71,450
22	79,175			14,511	29,023	43,534	58,045	72,557	87,068
24	99,764			17,371	34,743	52,114	69,486	86,857	104,229
26				20,489	40,977	61,466	81,955	102,443	122,932
28				23,863	47,726	71,589	95,452	119,314	143,177
30				27,494	54,988	82,483	109,977	137,471	164,965
32				31,383	62,765	94,148	125,531	156,913	188,296

(Continued)

5 PSIG STEAM PIPING SYSTEMS—STEEL PIPE (Continued)

Pipe Size	Pressure Drop psig/100 ft.			Steam Flow lbs./h					
				Velocity FPM (mph)					
	0.125	0.25	0.5	2,000 (23)	4,000 (45)	6,000 (68)	8,000 (91)	10,000 (114)	12,000 (136)
34				35,528	71,056	106,585	142,113	177,641	213,169
36				39,931	79,862	119,792	159,723	199,654	239,585
42				54,681	109,362	164,044	218,725	273,406	328,087
48	Velocity these	governs pipe	with sizes	71,745	143,491	215,236	286,981	358,727	430,472
54				91,123	182,247	273,370	364,493	455,616	546,740
60				112,815	225,630	338,445	451,260	564,075	676,890
72				163,140	326,280	489,419	652,559	815,699	978,839
84				222,720	445,439	668,159	890,879	1,113,598	1,336,318
96				291,555	583,109	874,664	1,166,219	1,457,774	1,749,328

Notes:
1 Maximum recommended pressure drop/velocity: 0.25 psig/100 ft./6,000 FPM.
2 Table based on Standard Weight Steel Pipe using steam equations in Part 3.

7 PSIG STEAM PIPING SYSTEMS—STEEL PIPE

Pipe Size	Pressure Drop psig/100 ft.			Steam Flow lbs./h					
				Velocity FPM (mph)					
	0.125	0.25	0.5	2,000 (23)	4,000 (45)	6,000 (68)	8,000 (91)	10,000 (114)	12,000 (136)
1/2	5	7	10		Pressure these				
3/4	11	16	23						
1	23	33	46	39		drop			
1-1/4	51	72	101	67		pipe			
1-1/2	79	111	157	91			governs sizes	with	
2	160	226	319	150	300				
2-1/2	263	372	526	214	429				
3	481	680	956	331	662				
4	1,016	1,438	2,033	570	1,139	1,709			
5	1,870	2,645	3,741	892	1,783	2,675	3,567		
6	3,087	4,365	6,174	1,293	2,586	3,879	5,171		
8	6,426	9,087	12,851	2,239	4,477	6,716	8,955	11,194	
10	11,726	16,583		3,529	7,057	10,586	14,115	17,644	21,172
12	18,819	26,614		5,061	10,122	15,183	20,244	25,306	30,367
14	24,376	34,473		6,170	12,341	18,511	24,682	30,852	37,023
16	35,138			8,174	16,348	24,521	32,695	40,869	49,043
18	48,354			10,458	20,917	31,375	41,833	52,292	62,750
20	64,191			13,024	26,048	39,072	52,096	65,120	78,144
22	82,801			15,871	31,742	47,613	63,483	79,354	95,225
24	104,332			18,999	37,998	56,997	75,996	94,995	113,993
26	128,924			22,408	44,816	67,224	89,633	112,041	134,449
28			with sizes	26,099	52,197	78,296	104,394	130,493	156,591
30				30,070	60,140	90,210	120,280	150,350	180,421
32		governs pipe		34,323	68,646	102,968	137,291	171,614	205,937
34				38,857	77,713	116,570	155,427	194,284	233,140
36				43,672	87,344	131,015	174,687	218,359	262,031
42	Velocity these			59,804	119,608	179,412	239,216	299,020	358,824
48				78,467	156,934	235,401	313,868	392,335	470,801
54				99,660	199,321	298,981	398,641	498,301	597,962
60				123,384	246,768	370,153	493,537	616,921	740,305
72				178,424	356,847	535,271	713,695	892,119	1,070,542
84				243,585	487,171	730,756	974,342	1,217,927	1,461,513
96				318,869	637,739	956,608	1,275,478	1,594,347	1,913,217

Notes:
1 Maximum recommended pressure drop/velocity: 0.25 psig/100 ft./6,000 FPM.
2 Table based on Standard Weight Steel Pipe using steam equations in Part 3.

10 PSIG STEAM PIPING SYSTEMS—STEEL PIPE

	Steam Flow lbs./h								
	Pressure Drop psig/100 ft.			Velocity FPM (mph)					
Pipe Size	0.25	0.5	1	2,000 (23)	4,000 (45)	6,000 (68)	8,000 (91)	10,000 (114)	12,000 (136)
1/2	8	11	15	15	Pressure				
3/4	17	24	34	27	these				
1	35	49	69	44		drop	governs		
1-1/4	76	108	152	76	151	pipe	sizes		
1-1/2	118	167	236	103	206			with	
2	240	339	479	169	339				
2-1/2	395	558	790	242	484	725			
3	723	1,016	1,445	373	747	1,120			
4	1,527	2,160	3,054	643	1,286	1,929	2,572		
5	2,810	3,974	5,620	1,006	2,013	3,019	4,025	5,031	
6	4,637	6,558		1,459	2,918	4,377	5,836	7,295	8,754
8	9,654	13,652		2,526	5,053	7,579	10,105	12,632	15,158
10	17,616	24,912		3,982	7,964	11,946	15,929	19,911	23,893
12	28,273			5,711	11,423	17,134	22,846	28,557	34,268
14	36,621			6,963	13,927	20,890	27,853	34,816	41,780
16	52,789			9,224	18,448	27,672	36,896	46,120	55,344
18				11,802	23,604	35,406	47,208	59,011	70,813
20				14,697	29,395	44,092	58,790	73,487	88,185
22				17,910	35,820	53,730	71,641	89,551	107,461
24				21,440	42,880	64,320	85,760	107,201	128,641
26				25,287	50,575	75,862	101,150	126,437	151,724
28			with	29,452	58,904	88,356	117,808	147,260	176,712
30			sizes	33,934	67,868	101,802	135,735	169,669	203,603
32		governs		38,733	77,466	116,199	154,932	193,665	232,398
34	Velocity	pipe		43,849	87,699	131,548	175,398	219,247	263,097
36	these			49,283	98,567	147,850	197,133	246,416	295,700
42				67,488	134,977	202,465	269,954	337,442	404,930
48				88,549	177,098	265,648	354,197	442,746	531,295
54				112,466	224,932	337,397	449,863	562,329	674,795
60				139,238	278,476	417,714	556,952	696,190	835,428
72				201,350	402,699	604,049	805,399	1,006,749	1,208,098
84				274,884	549,768	824,653	1,099,537	1,374,421	1,649,305
96				359,841	719,683	1,079,524	1,439,366	1,799,207	2,159,049

Notes:
1 Maximum recommended pressure drop/velocity: 0.5 psig/100 ft./6,000 FPM.
2 Table based on Standard Weight Steel Pipe using steam equations in Part 3.

12 PSIG STEAM PIPING SYSTEMS—STEEL PIPE

	Steam Flow lbs./h								
Pipe Size	Pressure Drop psig/100 ft.			Velocity FPM (mph)					
	0.25	0.5	1	2,000 (23)	4,000 (45)	6,000 (68)	8,000 (91)	10,000 (114)	12,000 (136)
1/2	8	11	16		Pressure				
3/4	18	25	36	29	these				
1	36	51	72	47		drop	governs		
1-1/4	79	112	158	81		pipe	sizes		
1-1/2	123	173	245	111	221			with	
2	249	352	497	182	365				
2-1/2	410	579	819	260	520	780			
3	750	1,054	1,499	402	803	1,205			
4	1,584	2,240	3,168	692	1,383	2,075	2,767		
5	2,915	4,122	5,830	1,083	2,165	3,248	4,331	5,413	
6	4,810	6,803		1,570	3,139	4,709	6,279	7,849	9,418
8	10,013	14,161		2,718	5,436	8,154	10,873	13,591	16,309
10	18,272			4,284	8,569	12,853	17,138	21,422	25,706
12	29,326			6,145	12,290	18,435	24,580	30,725	36,870
14	37,986			7,492	14,984	22,475	29,967	37,459	44,951

(Continued)

12 PSIG STEAM PIPING SYSTEMS—STEEL PIPE (*Continued*)

Pipe Size	Pressure Drop psig/100 ft.			Steam Flow lbs./h					
				Velocity FPM (mph)					
	0.25	0.5	1	2,000 (23)	4,000 (45)	6,000 (68)	8,000 (91)	10,000 (114)	12,000 (136)
16	54,755			9,924	19,848	29,773	39,697	49,621	59,545
18	75,351			12,698	25,396	38,094	50,792	63,490	76,188
20				15,813	31,626	47,439	63,252	79,066	94,879
22				19,270	38,539	57,809	77,079	96,348	115,618
24				23,068	46,135	69,203	92,270	115,338	138,406
26				27,207	54,414	81,621	108,828	136,034	163,241
28				31,688	63,375	95,063	126,750	158,438	190,126
30				36,510	73,019	109,529	146,039	182,548	219,058
32		governs	with	41,673	83,346	125,019	166,693	208,366	250,039
34	Velocity	pipe	sizes	47,178	94,356	141,534	188,712	235,890	283,068
36	these			53,024	103,048	159,073	212,097	265,121	318,145
42				72,611	145,223	217,834	290,445	363,056	435,668
48				95,271	190,542	285,812	381,093	476,354	571,625
54				121,003	242,006	363,008	484,011	605,014	726,017
60				149,807	299,615	449,422	599,229	749,036	898,844
72				216,634	433,267	649,901	866,535	1,083,168	1,299,802
84				295,750	591,500	887,250	1,183,000	1,478,750	1,774,500
96				387,156	774,312	1,161,469	1,548,625	1,935,781	2,322,937

Notes:
1 Maximum recommended pressure drop/velocity: 0.5 psig/100 ft./6,000 FPM.
2 Table based on Standard Weight Steel Pipe using steam equations in Part 3.

15 PSIG STEAM PIPING SYSTEMS—STEEL PIPE

Pipe Size	Pressure Drop psig/100 ft.			Steam Flow lbs./h					
				Velocity FPM (mph)					
	0.25	0.5	1	2,000 (23)	4,000 (45)	6,000 (68)	8,000 (91)	10,000 (114)	12,000 (136)
1/2	8	12	16		Pressure				
3/4	19	26	37	32	these				
1	38	53	75	52		drop	governs		
1-1/4	83	117	166	90		pipe	sizes		
1-1/2	129	182	258	122	244			with	
2	261	370	523	201	403				
2-1/2	430	609	861	287	575	862			
3	788	1,107	1,575	444	887	1,331			
4	1,665	2,354	3,329	764	1,528	2,291	3,055		
5	3,063	4,332	6,126	1,196	2,391	3,587	4,782	5,978	
6	5,055	7,149	10,110	1,733	3,467	5,200	6,934	8,667	
8	10,522	14,881		3,002	6,003	9,005	12,006	15,008	18,010
10	19,201	27,155		4,731	9,463	14,194	18,925	23,656	28,388
12	30,817			6,786	13,572	20,358	27,143	33,929	40,715
14	39,918			8,273	16,546	24,820	33,093	41,366	49,639
16	57,540			10,959	21,918	32,878	43,837	54,796	65,755
18	79,183			14,022	28,045	42,067	56,089	70,112	84,134
20				17,462	34,925	52,387	69,849	87,312	104,774
22			with	21,279	42,559	63,838	85,118	106,397	127,676
24			sizes	25,473	50,947	76,420	101,894	127,367	152,840
26		governs		30,044	60,089	90,133	120,178	150,222	180,267
28		pipe		34,992	69,985	104,977	139,970	174,962	209,955
30	Velocity			40,317	80,635	120,952	161,270	201,587	241,905
32	these			46,019	92,039	138,058	184,078	230,097	276,117
34				52,098	104,197	156,295	208,394	260,492	312,590
36				58,554	117,109	175,663	234,218	292,772	351,326
42				80,184	160,368	240,553	320,737	400,921	481,105
48				105,207	210,414	315,621	420,828	526,035	631,242
54				133,623	267,245	400,868	534,491	668,114	801,736
60				165,431	330,863	496,294	661,725	827,157	992,588
72				239,227	478,455	717,682	956,909	1,196,137	1,435,364
84				326,595	653,190	979,785	1,306,380	1,632,975	1,959,570
96				427,534	855,069	1,282,603	1,710,138	2,137,672	2,565,207

Notes:
1 Maximum recommended pressure drop/velocity: 0.5 psig/100 ft./6,000 FPM.
2 Table based on Standard Weight Steel Pipe using steam equations in Part 3.

21.11 Medium-Pressure Steam Pipe Sizing Tables (20–100 psig)

20 PSIG STEAM PIPING SYSTEMS—STEEL PIPE

Pipe Size	Steam Flow lbs./h								
	Pressure Drop psig/100 ft.			Velocity FPM (mph)					
	0.25	0.5	1	4,000 (45)	6,000 (68)	8,000 (91)	10,000 (114)	12,000 (136)	15,000 (170)
1/2	9	13	18						
3/4	20	29	40						
1	41	57	81		Pressure	drop			
1-1/4	89	126	178		these	pipe	governs		
1-1/2	139	196	277				sizes	with	
2	281	397	562	466					
2-1/2	463	655	926	665					
3	847	1,191	1,695	1,026	1,540				
4	1,790	2,532	3,581	1,767	2,651	3,535			
5	3,295	4,659	6,589	2,766	4,150	5,533			
6	5,437	7,689	10,874	4,011	6,016	8,022	10,027		
8	11,318	16,006	22,636	6,945	10,418	13,891	17,364	20,836	
10	20,653	29,208		10,948	16,421	21,895	27,369	32,843	41,054
12	33,148	46,878		15,702	23,553	31,403	39,254	47,105	58,881
14	42,936	60,720		19,143	28,715	38,286	47,858	57,430	71,787
16	61,891	87,527		25,358	38,038	50,717	63,396	76,075	95,094
18	85,170	120,449		32,446	48,669	64,892	81,115	97,338	121,673
20	113,063			40,406	60,609	80,812	101,015	121,218	151,522
22	145,843			49,238	73,857	98,476	123,095	147,714	184,643
24	183,768			58,943	88,414	117,885	147,357	176,828	221,035
26	227,082			69,519	104,279	139,039	173,799	208,558	260,698
28	276,022		with	80,969	121,453	161,937	202,422	242,906	303,632
30	330,813		sizes	93,290	139,935	186,580	233,225	279,870	349,838
32	397,670			106,484	159,726	212,968	266,210	319,451	399,314
34		governs		120,550	180,825	241,100	301,375	361,650	452,062
36		pipe		135,488	203,232	270,977	338,721	406,465	508,081
42				185,537	278,306	371,075	463,844	556,612	695,765
48	Velocity			243,437	365,156	486,875	608,593	730,312	912,890
54	these			309,188	463,782	618,376	772,970	927,564	1,159,456
60				382,790	574,185	765,580	956,974	1,148,369	1,435,462
72				553,546	830,318	1,107,091	1,383,864	1,660,637	2,075,796
84				755,705	1,133,557	1,511,409	1,889,262	2,267,114	1,833,893
96				989,267	1,483,901	1,978,534	2,473,168	2,967,802	3,709,752

Notes:
1 Maximum recommended pressure drop/velocity: 0.5 psig/100 ft./8,000 FPM.
2 Table based on Standard Weight Steel Pipe using steam equations in Part 3.

25 PSIG STEAM PIPING SYSTEMS—STEEL PIPE

Pipe Size	Steam Flow lbs./h								
	Pressure Drop psig/100 ft.			Velocity FPM (mph)					
	0.25	0.5	1	4,000 (45)	6,000 (68)	8,000 (91)	10,000 (114)	12,000 (136)	15,000 (170)
1/2	9	13	19						
3/4	21	30	43						
1	43	61	86		Pressure				
1-1/4	95	134	190		these	drop	governs		
1-1/2	148	209	295			pipe	sizes	with	
2	299	423	599	529					
2-1/2	493	697	986	754					
3	902	1,269	1,805	1,164	1,747				
4	1,907	2,697	3,814	2,005	3,008				
5	3,509	4,963	7,018	3,138	4,708	6,277			
6	5,791	8,190	11,582	4,550	6,825	9,100	11,376		
8	12,055	17,048	24,110	7,879	11,819	15,759	19,698	23,638	
10	21,998	31,110	43,996	12,420	18,629	24,839	31,049	37,259	
12	35,306	49,930		17,813	26,719	35,626	44,532	53,438	66,798
14	45,731	64,674		21,717	32,576	43,434	54,293	65,151	81,439

(Continued)

25 PSIG STEAM PIPING SYSTEMS—STEEL PIPE (*Continued*)

Pipe Size	Pressure Drop psig/100 ft.			Steam Flow lbs./h					
				Velocity FPM (mph)					
	0.25	0.5	1	4,000 (45)	6,000 (68)	8,000 (91)	10,000 (114)	12,000 (136)	15,000 (170)
16	65,920	93,225		28,768	43,152	52,536	71,920	86,304	107,880
18	90,715	128,291		36,809	55,213	73,617	92,021	110,426	138,032
20	120,424	170,306		45,839	68,758	91,677	114,597	137,516	171,895
22	155,339			55,858	83,788	111,717	139,646	167,575	209,469
24	195,732			66,868	100,302	133,735	167,169	200,603	250,754
26	241,867			78,867	118,300	157,733	197,167	236,600	295,750
28	293,993			91,855	137,783	183,710	229,638	275,565	344,457
30	352,351			105,833	158,750	211,667	264,583	317,500	396,875
32	417,171		with	120,801	181,201	241,602	302,002	362,403	453,004
34	488,677		sizes	136,758	205,137	273,517	341,896	410,275	512,844
36	567,084	governs		153,705	230,558	307,410	384,263	461,116	576,395
42		pipe		210,484	315,725	420,967	526,209	631,451	789,314
48				276,168	414,253	552,337	690,421	828,505	1,035,632
54				350,760	526,140	701,519	876,899	1,052,279	1,315,349
60	Velocity			434,257	651,386	868,515	1,085,643	1,302,772	1,628,465
72	these			627,972	941,958	1,255,944	1,569,930	1,882,916	2,354,984
84				857,312	1,285,968	1,714,624	2,143,280	2,571,936	3,214,920
96				1,122,278	1,683,417	2,244,556	2,805,695	3,366,834	4,208,542

Notes:
1 Maximum recommended pressure drop/velocity: 0.5 psig/100 ft./8,000 FPM.
2 Table based on Standard Weight Steel Pipe using steam equations in Part 3.

30 PSIG STEAM PIPING SYSTEMS—STEEL PIPE

Pipe Size	Pressure Drop psig/100 ft.			Steam Flow lbs./h					
				Velocity FPM (mph)					
	0.25	0.5	1	4,000 (45)	6,000 (68)	8,000 (91)	10,000 (114)	12,000 (136)	15,000 (170)
1/2	10	14	20						
3/4	23	32	45						
1	46	65	91	Pressure					
1-1/4	101	142	201	these					
1-1/2	156	221	312		drop	governs			
2	317	448	633	591		pipe	sizes		
2-1/2	521	737	1,043	843				with	
3	954	1,342	1,909	1,302					
4	2,017	2,852	4,034	2,243	3,364				
5	3,711	5,249	7,423	3,510	5,266	7,021			
6	6,125	8,662	12,249	5,090	7,634	10,179			
8	12,749	18,030	25,499	8,813	13,220	17,626	22,033		
10	23,265	32,902	46,530	13,891	20,837	27,783	34,729	41,674	
12	37,340	52,806	74,679	19,924	29,886	39,848	49,810	59,772	
14	48,365	68,399		24,291	36,436	48,582	60,727	72,873	91,091
16	69,717	98,595		32,178	48,266	64,355	80,444	96,533	120,666
18	95,940	135,680		41,171	61,757	82,342	102,928	123,513	154,391
20	127,361	180,116		51,271	76,907	102,543	128,178	153,814	192,268
22	164,286	232,336		62,479	93,718	124,957	156,197	187,436	234,295
24	207,006			74,793	112,189	149,586	186,982	224,378	280,473
26	255,799			88,214	132,321	176,428	220,534	264,641	330,802
28	310,927			102,742	154,113	205,483	256,854	308,225	385,281
30	372,647			118,376	177,565	236,753	295,941	355,129	443,912
32	441,200		with	135,118	202,677	270,236	337,795	405,354	506,693
34	516,825		sizes	152,967	229,450	305,933	382,417	458,900	573,625
36	599,748			171,922	257,883	343,844	429,805	515,766	644,708
42		governs		235,430	353,145	470,860	588,575	706,290	882,862
48		pipe		308,900	463,349	617,799	772,249	926,699	1,158,373
54				392,331	588,497	784,662	980,828	1,176,994	1,471,242
60	Velocity			485,725	728,587	971,450	1,214,312	1,457,175	1,821,468
72	these			702,398	1,053,597	1,404,796	1,755,995	2,107,194	2,633,993
84				958,919	1,438,379	1,917,839	2,397,398	2,876,758	3,595,948
96				1,255,289	1,882,933	2,510,577	3,138,222	3,765,866	4,707,332

Notes:
1 Maximum recommended pressure drop/velocity: 0.5 psig/100 ft./8,000 FPM.
2 Table based on Standard Weight Steel Pipe using steam equations in Part 3.

40 PSIG STEAM PIPING SYSTEMS—STEEL PIPE

Pipe Size	Pressure Drop psig/100 ft.			Steam Flow lbs./h — Velocity FPM (mph)					
	0.5	1	2	4,000 (45)	6,000 (68)	8,000 (91)	10,000 (114)	12,000 (136)	15,000 (170)
1/2	16	22	31						
3/4	35	50	71						
1	71	100	142	Pressure	drop				
1-1/4	156	221	312	these	pipe		governs	with	
1-1/2	242	343	485	433			sizes		
2	492	695	984	713					
2-1/2	810	1,145	1,620	1,017	1,526				
3	1,473	2,097	2,965	1,571	2,356				
4	3,133	4,430	6,265	2,705	4,058	5,410			
5	5,764	8,152	11,529	4,234	6,352	8,469	10,586		
6	9,513	13,453	19,026	6,139	9,209	12,278	15,348	18,418	
8	19,802	28,005	39,605	10,631	15,946	21,261	26,577	31,892	
10	36,136	51,103		16,757	25,135	33,513	41,891	50,270	62,837
12	57,996	82,019		24,033	36,050	48,066	60,083	72,100	90,124
14	75,122	106,239		29,301	43,951	58,602	73,252	87,903	109,878
16	108,286			38,814	58,221	77,628	97,035	116,442	145,553
18	149,016			49,662	74,493	99,325	124,156	148,987	186,234
20	197,819			61,846	92,769	123,692	154,615	185,537	231,922
22	255,172		with	75,364	113,047	150,729	188,411	226,093	282,617
24	321,526		sizes	90,218	135,327	180,437	225,546	270,655	338,319
26	397,311	governs		106,407	159,611	212,815	266,018	319,222	399,028
28		pipe		123,932	185,897	247,863	309,829	371,795	464,743
30				142,791	214,186	285,582	356,977	428,373	535,466
32	Velocity			162,985	244,478	325,971	407,464	488,956	611,195
34	these			184,515	276,773	369,030	461,288	553,546	691,932
36				207,380	311,070	414,760	518,450	622,140	777,675
42				283,986	425,979	567,972	709,965	851,958	1,064,947
48				372,608	558,912	745,216	931,521	1,117,825	1,397,281
54				473,247	709,871	946,494	1,183,118	1,419,742	1,774,677
60				585,903	878,854	1,171,805	1,464,757	1,757,708	2,197,135
72				847,264	1,270,895	1,694,527	2,118,159	2,541,791	3,177,239
84				1,156,691	1,735,036	2,313,382	2,891,727	3,470,073	4,337,591
96				1,514,184	2,271,277	3,028,369	3,785,461	4,542,553	5,678,192

Notes:
1 Maximum recommended pressure drop/velocity: 1.0 psig/100 ft./10,000 FPM.
2 Table based on Standard Weight Steel Pipe using steam equations in Part 3.

50 PSIG STEAM PIPING SYSTEMS—STEEL PIPE

Pipe Size	Pressure Drop psig/100 ft.			Steam Flow lbs./h — Velocity FPM (mph)					
	0.5	1	2	4,000 (45)	6,000 (68)	8,000 (91)	10,000 (114)	12,000 (136)	15,000 (170)
1/2	17	24	34						
3/4	38	54	76						
1	77	109	154	Pressure	drop				
1-1/4	169	239	338	these	pipe		governs	with	
1-1/2	263	371	525	508			sizes		
2	533	753	1,065	837					
2-1/2	877	1,241	1,755	1,194					
3	1,569	2,271	3,212	1,843	2,765				
4	3,393	4,799	6,786	3,174	4,761	6,348			
5	6,244	8,830	12,488	4,968	7,453	9,937	12,421		
6	10,304	14,573	20,609	7,203	10,805	14,407	18,008		
8	21,450	30,335	42,900	12,473	18,710	24,947	31,183	37,420	

(Continued)

50 PSIG STEAM PIPING SYSTEMS—STEEL PIPE (*Continued*)

Pipe Size	Pressure Drop psig/100 ft.			Steam Flow lbs./h Velocity FPM (mph)					
	0.5	1	2	4,000 (45)	6,000 (68)	8,000 (91)	10,000 (114)	12,000 (136)	15,000 (170)
10	39,142	55,355		19,661	29,492	39,322	49,153	58,983	73,729
12	62,822	88,844		28,199	42,298	56,398	70,497	84,597	105,746
14	81,373	115,078		34,380	51,570	68,759	85,949	103,139	128,924
16	117,296	165,882		45,542	68,313	91,084	113,854	136,625	170,782
18	161,415			58,270	87,406	116,541	145,676	174,811	218,514
20	214,279			72,566	108,849	145,132	181,414	217,697	272,122
22	276,404			88,428	132,641	176,855	221,069	265,283	331,604
24	348,279			105,856	158,784	211,712	264,640	317,568	396,961
26	430,370			124,851	187,277	249,703	312,128	374,554	468,192
28	523,121		with sizes	145,413	218,120	290,826	363,533	436,239	545,299
30	626,961			167,541	251,312	335,083	418,853	502,624	628,280
32				191,236	286,854	382,473	478,091	573,709	717,136
34		governs pipe		216,498	324,747	432,996	541,245	649,493	811,867
36				243,326	364,989	486,652	608,315	729,978	912,472
42				333,210	499,815	666,420	833,025	999,630	1,249,538
48	Velocity these			437,194	655,790	874,387	1,092,984	1,311,581	1,639,476
54				555,277	832,915	1,110,553	1,388,192	1,665,830	2,082,288
60				687,459	1,031,189	1,374,918	1,718,648	2,062,378	2,577,972
72				994,123	1,491,184	1,988,245	2,485,307	2,982,368	3,727,960
84				1,357,184	2,035,776	2,714,368	3,392,960	4,071,552	5,089,440
96				1,776,643	2,664,965	3,553,286	4,441,608	5,329,929	6,662,412

Notes:
1 Maximum recommended pressure drop/velocity: 1.0 psig/100 ft./10,000 FPM.
2 Table based on Standard Weight Steel Pipe using steam equations in Part 3.

60 PSIG STEAM PIPING SYSTEMS—STEEL PIPE

Pipe Size	Pressure Drop psig/100 ft.			Steam Flow lbs./h Velocity FPM (mph)					
	0.5	1	2	4,000 (45)	6,000 (68)	8,000 (91)	10,000 (114)	12,000 (136)	15,000 (170)
1/2	18	25	36						
3/4	41	58	82						
1	82	116	164		Pressure these				
1-1/4	181	256	362			drop pipe	governs sizes	with	
1-1/2	281	397	562						
2	570	806	1,140	957					
2-1/2	938	1,327	1,877	1,366					
3	1,707	2,429	3,436	2,109	3,164				
4	3,630	5,133	7,260	3,632	5,449				
5	6,680	9,446	13,359	5,686	8,529	11,371			
6	11,023	15,589	22,046	8,243	12,365	16,487	20,608		
8	22,946	32,451	45,893	14,274	21,412	28,549	35,686	42,823	
10	41,873	59,217	83,745	22,500	33,750	45,000	56,249	67,499	
12	67,204	95,041		32,270	48,406	64,541	80,676	96,811	121,014
14	87,049	123,106		39,344	59,015	78,687	98,359	118,031	147,539
16	125,479	177,454		52,117	78,176	104,235	130,293	156,352	195,440
18	172,676	244,200		66,684	100,026	133,368	166,710	200,052	250,064
20	229,227	324,176		83,043	124,565	166,086	207,608	249,129	311,412
22	295,686			101,195	151,793	202,391	252,988	303,586	379,482
24	372,575			121,140	181,710	242,280	302,851	363,421	454,276
26	460,392			142,878	214,317	285,756	357,195	428,634	535,792
28	559,614			166,408	249,613	332,817	416,021	499,225	624,032
30	670,697			191,732	287,598	383,464	479,329	575,195	718,994
32	794,082		with sizes	218,848	328,272	437,696	547,120	656,544	820,680
34		governs pipe		247,757	371,635	495,514	619,392	743,271	929,088
36				278,459	417,688	556,917	696,146	835,376	1,044,220
42				381,321	571,981	762,641	953,302	1,143,962	1,429,952
48	Velocity these			500,318	750,477	1,000,636	1,250,795	1,500,954	1,876,192
54				635,450	953,176	1,270,901	1,588,626	1,906,351	2,382,939
60				786,718	1,180,077	1,573,436	1,966,795	2,360,154	2,950,193
72				1,137,659	1,706,489	2,275,318	2,844,148	3,412,977	4,266,221
84				1,553,141	2,329,711	3,106,282	3,882,852	4,659,423	5,824,279
96				2,033,164	3,049,746	4,066,328	5,082,909	6,099,491	7,624,364

Notes:
1 Maximum recommended pressure drop/velocity: 1.0 psig/100 ft./12,000 FPM.
2 Table based on Standard Weight Steel Pipe using steam equations in Part 3.

75 PSIG STEAM PIPING SYSTEMS—STEEL PIPE

Pipe Size	Pressure Drop psig/100 ft.			Velocity FPM (mph)					
	0.5	1	2	4,000 (45)	6,000 (68)	8,000 (91)	10,000 (114)	12,000 (136)	15,000 (170)
1/2	20	28	39						
3/4	45	63	89						
1	90	127	179		Pressure				
1-1/4	197	279	394		these	drop	governs	with	
1-1/2	306	433	612			pipe	sizes		
2	621	879	1,243	1,138					
2-1/2	1,023	1,447	2,046	1,624					
3	1,862	2,649	3,746	2,507					
4	3,957	5,597	7,915	4,318	6,477				
5	7,283	10,299	14,565	6,758	10,138	13,517			
6	12,018	16,997	24,036	9,799	14,698	19,597			
8	25,018	35,380	50,035	16,967	25,451	33,935	42,419		
10	45,652	64,562	91,304	26,745	40,117	53,489	66,862	80,234	
12	73,270	103,620		38,359	57,538	76,718	95,897	115,077	143,846
14	94,906	134,218		46,766	70,150	93,533	116,916	140,299	175,374
16	136,805	193,471		61,950	92,925	123,900	154,876	185,851	232,313
18	188,261	266,242		79,265	118,897	158,530	198,162	237,795	297,244
20	249,917	353,436		98,711	148,066	197,422	246,777	296,132	370,165
22	322,374			120,288	180,431	240,575	300,719	360,863	451,079
24	406,203			143,996	215,993	287,991	359,989	431,987	539,983
26	501,947			169,834	254,752	339,669	424,586	509,503	636,879
28	610,125			197,804	296,707	395,609	494,511	593,413	741,767
30	731,235		with	227,905	341,858	455,811	569,764	683,716	854,646
32	865,756		sizes	260,138	390,206	520,275	650,344	780,413	975,516
34	1,014,152	governs		294,501	441,751	589,001	736,252	883,502	1,104,378
36	1,176,871	pipe		330,995	496,492	661,990	827,487	992,985	1,241,231
42				453,264	679,896	906,527	1,133,159	1,359,791	1,699,739
48				594,712	892,068	1,189,424	1,486,780	1,784,136	2,230,170
54	Velocity			755,340	1,133,009	1,510,679	1,888,349	2,266,019	2,832,524
60	these			935,147	1,402,720	1,870,293	2,337,867	2,805,440	3,506,800
72				1,352,299	2,028,449	2,704,598	3,380,748	4,056,898	5,071,122
84				1,846,169	2,769,254	3,692,339	4,615,423	5,538,508	6,923,135
96				2,416,757	3,625,136	4,833,514	6,041,893	7,250,271	9,062,839

Notes:
1. Maximum recommended pressure drop/velocity: 1.0 psig/100 ft./12,000 FPM.
2. Table based on Standard Weight Steel Pipe using steam equations in Part 3.

85 PSIG STEAM PIPING SYSTEMS—STEEL PIPE

Pipe Size	Pressure Drop psig/100 ft.			Velocity FPM (mph)					
	0.5	1	2	4,000 (45)	6,000 (68)	8,000 (91)	10,000 (114)	12,000 (136)	15,000 (170)
1/2	21	29	41						
3/4	47	66	94						
1	94	133	188		Pressure				
1-1/4	207	293	415		these	drop	governs	with	
1-1/2	322	455	644			pipe	sizes		
2	653	924	1,306	1,258					
2-1/2	1,076	1,521	2,151	1,794					
3	1,957	2,784	3,938	2,771					
4	4,160	5,883	8,320	4,771	7,157				
5	7,655	10,826	15,311	7,468	11,202	14,936			
6	12,633	17,866	25,267	10,828	16,241	21,655			
8	26,298	37,192	52,597	18,749	28,124	37,499	46,873		
10	47,989	67,867	95,979	29,553	44,330	59,107	73,883	88,660	
12	77,021	108,925	154,043	42,387	63,580	84,774	105,967	127,161	
14	99,765	141,089		51,678	77,516	103,355	129,194	155,033	193,791

(Continued)

85 PSIG STEAM PIPING SYSTEMS—STEEL PIPE (*Continued*)

Pipe Size	Pressure Drop psig/100 ft.			Steam Flow lbs./h Velocity FPM (mph)					
	0.5	1	2	4,000 (45)	6,000 (68)	8,000 (91)	10,000 (114)	12,000 (136)	15,000 (170)
16	143,808	203,376		68,456	102,684	136,911	171,139	205,367	256,709
18	197,899	279,872		87,589	131,383	175,178	218,972	262,766	328,458
20	262,712	371,531		109,077	163,615	218,153	272,692	327,230	409,037
22	338,879	479,247		132,919	199,379	265,839	332,298	398,758	498,447
24	426,999			159,117	238,675	318,234	397,792	477,350	596,688
26	527,645			187,669	281,504	375,338	469,173	563,007	703,759
28	641,361			218,576	327,865	437,153	546,441	655,729	819,661
30	768,671			251,838	377,758	503,677	629,596	755,515	944,394
32	910,079			287,455	431,183	574,910	718,638	862,366	1,077,957
34	1,066,072			325,427	488,140	650,854	813,567	976,281	1,220,351
36	1,237,121			365,753	548,630	731,507	914,384	1,097,260	1,371,575
42	1,845,105			500,862	751,293	1,001,724	1,252,155	1,502,586	1,878,232
48		governs pipe	with sizes	657,164	985,746	1,314,328	1,642,910	1,971,492	2,464,365
54				834,660	1,251,989	1,669,319	2,086,649	2,503,979	3,129,973
60	Velocity these			1,033,349	1,550,023	2,066,697	2,583,372	3,100,046	3,875,057
72				1,494,307	2,241,461	2,988,614	3,735,768	4,482,922	5,603,652
84				2,040,040	3,060,060	4,080,080	5,100,099	6,120,119	7,650,149
96				2,670,546	4,005,820	5,341,093	6,676,366	8,011,639	10,014,549

Notes:
1 Maximum recommended pressure drop/velocity: 1.0 psig/100 ft./12,000 FPM.
2 Table based on Standard Weight Steel Pipe using steam equations in Part 3.

100 PSIG STEAM PIPING SYSTEMS—STEEL PIPE

Pipe Size	Pressure Drop psig/100 ft.			Steam Flow lbs./h Velocity FPM (mph)					
	0.5	1	2	4,000 (45)	6,000 (68)	8,000 (91)	10,000 (114)	12,000 (136)	15,000 (170)
1/2	22	31	44						
3/4	50	71	100						
1	101	142	201	Pressure these					
1-1/4	222	313	443						
1-1/2	344	486	688			drop pipe	governs sizes		
2	698	987	1,396						
2-1/2	1,149	1,625	2,299	2,049				with	
3	2,091	2,975	4,208	3,164					
4	4,446	6,287	8,891	5,449	8,173				
5	8,181	11,569	16,362	8,529	12,793				
6	13,501	19,093	27,001	12,365	18,548	24,730			
8	28,104	39,744	56,207	21,412	32,117	42,823	53,529		
10	51,283	72,526	102,567	33,750	50,624	67,499	84,374	101,249	
12	82,308	116,402	164,617	48,406	72,608	96,811	121,014	145,217	
14	106,613	150,773	213,226	59,015	88,523	118,031	147,539	177,046	
16	153,680	217,336		78,176	117,264	156,352	195,440	234,528	293,160
18	211,483	299,083		100,026	150,039	200,052	250,064	300,077	375,097
20	280,745	397,033		124,565	186,847	249,129	311,412	373,694	467,118
22	362,139	512,142		151,793	227,689	303,586	379,482	455,379	569,223
24	456,309	645,318		181,710	272,565	363,421	454,276	545,131	681,414
26	563,863	797,422		214,317	321,475	428,634	535,792	642,951	803,688
28	685,384			249,613	374,419	499,225	624,032	748,838	936,048
30	821,433			287,598	431,397	575,195	718,994	862,793	1,078,491
32	972,548		with sizes	328,272	492,408	656,544	820,680	984,816	1,231,020
34	1,139,248			371,635	557,453	743,271	929,088	1,114,906	1,393,632
36	1,322,038			417,688	626,532	835,376	1,044,220	1,253,064	1,566,330
42	1,971,754	governs pipe		571,981	857,971	1,143,962	1,429,952	1,715,943	2,144,929
48	2,783,057			750,477	1,125,715	1,500,954	1,876,192	2,251,430	2,814,288
54				953,176	1,429,763	1,906,351	2,382,939	2,859,527	3,574,408
60				1,180,077	1,770,116	2,360,154	2,950,193	3,540,231	4,425,289
72	Velocity these			1,706,489	2,559,733	3,412,977	4,266,221	5,119,466	6,399,332
84				2,329,711	3,494,567	4,659,423	5,824,279	6,989,134	8,736,418
96				3,049,746	4,574,618	6,099,491	7,624,364	9,149,237	11,436,546

Notes:
1 Maximum recommended pressure drop/velocity: 1.0 psig/100 ft./12,000 FPM.
2 Table based on Standard Weight Steel Pipe using steam equations in Part 3.

Refer to the online resource for Section 21.12 High-Pressure Steam Pipe Sizing Tables (120-300 psig). www.mheducation.com/HVACequations

Steam Condensate Piping Systems

22.01 Steam Condensate Piping

A. Steam Condensate Pipe Sizing

1. Steam condensate pipe sizing criteria limits:
   a. Pressure drop: 1/16–1.0 psig/100 ft.
   b. Velocity–liquid systems: 150 ft./min. max.
   c. Velocity–vapor systems: 5,000 ft./min. max.
2. Recommended steam condensate pipe sizing criteria:
   a. Low-pressure systems:
      1) Pressure drop: 1/8–1/4 psig/100 ft.
      2) Velocity–vapor systems: 5,000 ft. per minute.
   b. Medium-pressure systems:
      1) Pressure drop: 1/8–1/4 psig/100 ft.
      2) Velocity–vapor systems: 5,000 ft. per minute.
   c. High-pressure systems:
      1) Pressure drop: 1/4–1/2 psig/100 ft.
      2) Velocity–vapor systems: 5,000 ft. per minute.
3. *Wet Returns.* Return pipes contain only liquid, no vapor. Wet condensate returns connect to the boiler below the waterline so the piping is always flooded.
4. *Dry Returns.* Return pipes contain saturated liquid and saturated vapor (most common). Dry condensate returns connect to the boiler above the waterline so the piping is not flooded and must be pitched in the direction of flow. Dry condensate returns often carry steam, air, and condensate.
5. *Open Returns.* The return system is vented to the atmosphere and condensate lines are essentially at atmospheric pressure (gravity flow lines).
6. *Closed Returns.* The return system is not vented to the atmosphere.
7. Steam traps and steam condensate piping should be selected to discharge at four times the condensate rating of air handling heating coils and three times the condensate rating of all other equipment for system startup.
8. Steam condensate liquid to steam volume ratio is 1:1600 at 0 psig.
9. *Flash Steam.* Flash steam is formed when hot steam condensate under pressure is released to a lower pressure; the temperature drops to the boiling point of the lower pressure, causing some of the condensate to evaporate forming steam. Flash steam occurs whenever steam condensate experiences a drop in pressure and thus produces steam at the lower pressure.
   a. Low-pressure steam systems' flash steam is negligible and can be generally be ignored.
   b. Medium- and high-pressure steam systems' flash steam is important to utilize and consider when sizing condensate piping.
   c. Flash steam recovery requirements:
      1) To utilize flash steam recovery, the condensate must be at a reasonably high pressure (medium- and high-pressure steam systems) and the traps supplying the condensate must be capable of operating with the back pressure of the flash steam system.
      2) There must be a use or demand for the flash steam at the reduced pressure. Demand for steam at the lower pressure should be greater than the supply of flash steam. The demand for steam should occur at the same time as the flash steam supply.
      3) The steam equipment should be in close proximity to the flash steam source to minimize installation and radiation losses and to fully take advantage of the flash steam recovery system. Flash steam recovery systems are especially advantageous when steam is utilized at multiple pressures within the facility and the distribution systems are already in place.

B. Steam Condensate System Design and Pipe Installation Guidelines

1. The minimum recommended steam condensate pipe size is 3/4 in.
2. Locate all valves, strainers, unions, and flanges so they are accessible. All valves (except control valves) and strainers should be the full size of the pipe before reducing the size to make connections to equipment and controls. Union and/or flanges should be installed at each piece of equipment, in bypasses and in long piping runs (100 ft. or more), to permit disassembly for alteration and repairs.
3. Provide chainwheel operators for all valves in equipment rooms mounted greater than 7'0" above floor level. The chain should extend to 5'0"–7'0" above the floor level.
4. All valves should be installed so the valve remains in service when equipment or piping on the equipment side of the valve is removed.
5. Locate all flow measuring devices in accessible locations with a straight section of pipe upstream (10 pipe diameters) and downstream (5 pipe diameters) of the device, or as recommended by the manufacturer.
6. Provide vibration isolators for all piping supports connected to, and within 50 ft. of, isolated equipment, except at base elbow supports and anchor points, throughout mechanical equipment rooms, and for supports of steam mains within 50 ft. of the boiler or pressure reducing valves.
7. Drip leg collection points on steam piping should be the same size as the steam piping to prevent steam condensate from passing over the drip leg and thus decreasing the risk of water hammer. The drip leg collection point should be a minimum of 12 in. long, including a minimum 6-in.-long dirt leg with the steam trap outlet above the dirt leg.
8. Pitch all steam return lines downward in the direction of condensate flow 1/2" per 10 ft. (1" per 20 ft.) minimum.
9. Drip legs must be installed at all low points, downfed runouts to all equipment, at the end of mains, the bottom of risers, and ahead of all pressure regulators, control valves, isolation valves, and expansion joints.
10. On straight runs with no natural drainage points, install drip legs at intervals not exceeding 200 ft. where the pipe is pitched downward in the direction of steam flow, and a maximum of 100 ft. where the pipe is pitched up so that condensate flow is opposite of steam flow.
11. Steam traps used on steam mains and branches shall be at minimum 3/4" size.
12. When elevating steam condensate to an overhead return main, it requires 1 psi to elevate condensate 2 ft. Try to avoid elevating condensate.
13. Steam condensate in a steam system should be maintained at a pH of approximately 8–9. A pH of 7 is neutral; below 7 is acidic; above 7 is alkaline.

C. Low-Pressure Steam Condensate Pipe Materials (0–15 psig)

1. 2" and smaller:
   - a. Pipe: black steel pipe, *ASTM A53, Schedule 80,* Type E or S, Grade B.
     Fittings: black cast iron screw fittings, 250 lbs., *ANSI/ASME B16.4.*
     Joints: pipe threads, general purpose (American) *ANSI/ASME B1.20.1.*
2. 2-1/2" and larger:
   - a. Pipe: black steel pipe, *ASTM A53, Schedule 80,* Type E or S, Grade B.
     Fittings: steel butt-welding fittings, 250 lbs., *ANSI/ASME B16.9.*
     Joints: welded pipe, ANSI/AWS D1.1 and ANSI/ASME Sec. 9.

D. Medium-Pressure Steam Condensate Pipe Materials (16–100 psig)

1. 2" and smaller:
   - a. Pipe: black steel pipe, *ASTM A53, Schedule 80,* Type E or S, Grade B.
     Fittings: black cast iron screw fittings, 250 lbs., *ANSI/ASME B16.4.*
     Joints: pipe threads, general purpose (American) *ANSI/ASME B1.20.1.*
2. 2-1/2" and larger:
   - a. Pipe: black steel pipe, *ASTM A53, Schedule 80,* Type E or S, Grade B.
     Fittings: steel butt-welding fittings, 250 lbs., *ANSI/ASME B16.9.*
     Joints: welded pipe, ANSI/AWS D1.1 and ANSI/ASME Sec. 9.

E. High-Pressure Steam Condensate Pipe Materials (100–300 psig)

1. 1-1/2" and smaller:
    a. Pipe: black steel pipe, *ASTM A53, Schedule 80,* Type E or S, Grade B.
       Fittings: forged steel socket-weld, 300 lbs., *ANSI B16.11.*
       Joints: welded pipe, ANSI/AWS D1.1 and ANSI/ASME Sec. 9.
    b. Pipe: carbon steel pipe, *ASTM A106, Schedule 80,* Grade B.
       Fittings: forged steel socket-weld, 300 lbs., *ANSI B16.11.*
       Joints: welded pipe, ANSI/AWS D1.1 and ANSI/ASME Sec. 9.
2. 2" and larger:
    a. Pipe: black steel pipe, *ASTM A53, Schedule 80,* Type E or S, Grade B.
       Fittings: steel butt-welding fittings, 300 lbs., *ANSI/ASME B16.9.*
       Joints: welded pipe, ANSI/AWS D1.1 and ANSI/ASME Sec. 9.
    b. Pipe: carbon steel pipe, *ASTM A106, Schedule 80,* Grade B.
       Fittings: steel butt-welding fittings, 300 lbs., *ANSI/ASME B16.9.*
       Joints: welded pipe, ANSI/AWS D1.1 and ANSI/ASME Sec. 9.

F. Pipe Testing

1. 1.5 × system working pressure.
2. 100 psi minimum.

G. Steam Traps

1. Steam trap types:
    a. A steam trap is a self-actuated valve that closes in the presence of steam and opens in the presence of steam condensate or noncondensable gases.
    b. Thermostatic traps: React to differences in temperature between steam and cooled condensate. Condensate must be subcooled for the trap to operate properly. Thermostatic traps work best in drip and tracing services and where steam temperature and pressure are constant and predictable.
       1) Liquid expansion thermostatic trap.
       2) Balanced pressure thermostatic trap:
           a) Balanced pressure traps change their actuation temperature automatically with changes in steam pressure. Balanced pressure traps are used in applications where system pressure varies.
           b) During startup and operation, this trap discharges air and other noncondensables very well. This trap is often used as a standalone air vent in steam systems.
           c) The balanced pressure trap will cause condensate to back up in the system.
       3) Bimetal thermostatic trap:
           a) Bimetal traps are rugged and resist damage from steam system events such as water hammer, freezing, superheated steam, and vibration.
           b) Bimetal traps cannot compensate for steam system pressure changes.
           c) Bimetal traps have a slow response time to changing process pressure and temperature conditions.
       4) Bellows thermostatic trap.
       5) Capsule thermostatic trap.
    c. Mechanical traps: Operate according to the difference in density between steam and condensate (buoyancy operated).
       1) Float and thermostatic (F&T) traps:
           a) Process or modulating applications—will work in almost any application—heat exchangers, coils, humidifiers, etc.
           b) The simplest type of mechanical trap.
           c) The F&T trap is the only trap that provides continuous, immediate, and modulating condensate discharge.
           d) A thermostat valve opens when cold or when below saturation (steam) temperature in order to allow air to bleed out during system startup and operation. The valve closes when the system reaches steam temperature.

    2) Inverted bucket traps:

        a) Work best in applications with constant load and constant pressure—drips.

        b) When the inverted bucket is filled with steam, it rises and closes the discharge valve preventing the discharge of steam. When the inverted bucket is filled with condensate, it drops, opening the valve and discharging the condensate.

        c) Inverted bucket traps are poor at removing air and other noncondensable gases.

  d. Kinetic traps: Rely on the difference in flow characteristics of steam and condensate and the pressure created by flash steam.

    1) Thermodynamic traps:

        a) Thermodynamic traps work best in drip and tracing services.

        b) Thermodynamic traps can remove air and other noncondensables during startup only if the system pressures are increased slowly. Because of this, thermodynamic traps often require a separate air vent.

        c) These traps snap open and snap shut and the sound can be annoying if used in noise-sensitive areas.

        d) The thermodynamic trap is rugged because it has only one moving part and is resistant to water hammer, superheated steam, freezing, and vibration.

    2) Impulse or piston traps.

    3) Orifice traps.

2. Steam trap selection:

  a. HVAC equipment steam traps should be selected to discharge three to four times the condensate rating of the equipment for system startup.

  b. Boiler header steam traps should be selected to discharge three to five times the condensate carryover rating of the boilers (typically 10 percent).

  c. Steam main piping steam traps should be selected to discharge two to three times the condensate generated during the start-up mode caused by radiation losses.

  d. Steam branch piping steam traps should be selected to discharge three times the condensate generated during the startup mode caused by radiation losses.

  e. Use float and thermostatic (F&T) traps for all steam-supplied equipment.

    1) Thermostatic traps may be used for steam radiators, steam finned tube, and other noncritical equipment, in lieu of F&T traps.

    2) A combination of an inverted bucket trap and an F&T trap, in parallel with an F&T trap installed above an inverted bucket trap, may be used in lieu of F&T traps.

  f. Use inverted bucket traps for all pipeline drips.

3. Steam trap functions:

  a. Steam traps allow condensate to flow from the heat exchanger or other device to minimize fouling, prevent damage, and to allow the heat transfer process to continue.

  b. Steam traps prevent steam escape from the heat exchanger or other devices.

  c. Steam traps vent air or other noncondensable gases to prevent corrosion and allow heat transfer.

4. Common steam trap problems:

  a. Steam leakage: Like all valves, the steam trap seat is subject to damage, corrosion, and/or erosion. When the trap seat is damaged, the valve will not seal; thus, the steam trap will leak live steam.

  b. Air binding: Air, carbon dioxide, hydrogen, and other noncondensable gases trapped in a steam system will reduce heat transfer and can defeat steam trap operation.

  c. Insufficient pressure difference: Steam traps rely on a positive pressure difference between the upstream steam pressure and the downstream condensate pressure to discharge condensate. When this is not maintained, the discharge of condensate is impeded.

    1) Overloading of the condensate return system is one cause: too much back pressure.

    2) Steam pressure that is too low is another cause.

d. Dirt: Steam condensate often contains dirt, particles of scale and corrosion, and other impurities from the system that can erode and damage the steam traps. Strainers should always be placed upstream of the steam traps to extend life.

e. Freezing: Freezing is normally only a problem when the steam system is shut down or idles, and liquid condensate remains in the trap.

f. Noise: Thermodynamic traps are generally the only trap that produces noise when they operate. All other traps operate relatively quietly.

g. Maintenance: Steam traps, as with all valves, must be maintained. Most steam traps can be maintained inline without removing the body from the connecting piping.

5. Steam trap characteristics are given in the following table.

STEAM TRAP COMPARISON

	Steam Trap Type		
Characteristic	Inverted Bucket	Float & Thermostatic	Liquid Expansion Thermostatic
Method of Operation	Intermittent, condensate drainage is continuous; discharge is intermittent	Continuous	Intermittent
No Load	Small dribble	No action	No action
Light Load	Intermittent	Usually continuous but may cycle at high pressures	Continuous; usually dribble action
Normal Load	Intermittent	Usually continuous but may cycle at high pressures	May blast at high pressures
Full or Overload	Continuous	Continuous	Continuous
Energy Conservation	Excellent	Good	Fair
Resistance to Wear	Excellent	Good	Fair
Corrosion Resistance	Excellent	Good	Good
Resistance to Hydraulic Shock	Excellent	Poor	Poor
Vent Air and CO_2 at Steam Temperature	Yes	No	No
Capability to Vent Air at Very Low Pressure (1/4 psig)	Poor	Excellent	Good
Capability to Handle Startup Air Loads	Fair	Excellent	Excellent
Operation Against Back Pressure	Excellent	Excellent	Excellent
Resistance to Damage from Freezing; Cast Iron Trap Not Recommended	Good	Poor	Good
Capability to Purge System	Excellent	Fair	Good
Performance on Very Light Loads	Excellent	Excellent	Excellent
Responsiveness to Slugs of Condensate	Immediate	Immediate	Delayed
Capability to Handle Dirt	Excellent	Poor	Fair
Comparative Physical Size	Large	Large	Small
Capability to Handle Flash Steam	Fair	Poor	Poor
Usual Mechanical Failure Mode	Open	Closed with air vent open	Open or closed depending on design
Subcooling	No	No	Yes
Venting	Fair	Excellent	Excellent
Seat Pressure Rating	Yes	Yes	N/a

(Continued)

STEAM TRAP COMPARISON (*Continued*)

Characteristic	Steam Trap Type		
	Inverted Bucket	Float & Thermostatic	Liquid Expansion Thermostatic
Advantages	Rugged	Continuous condensate discharge	Utilizes sensible heat of condensate
	Tolerates water hammer without damage	Handles rapid pressure changes	Allows discharge of noncondensables at startup to the set point temperature
		High noncondensable capacity	Not affected by superheated steam, water hammer, or vibration
			Resists freezing
Disadvantages	Discharges noncondensables slowly (additional air vent required)	Float can be damaged by water hammer	Element subject to corrosion damage
	Level of condensate can freeze, damaging the trap body	Level of condensate in chamber can freeze, damaging float and body	Condensate backs up into the drain line and/or process
	Must have water seal to operate; subject to losing prime	Some thermostatic air vent designs are susceptible to corrosion	
	Pressure fluctuations and superheated steam can cause loss of the water seal		
Recommended Services	Continuous operation where noncondensable venting is not critical and rugged construction is important	Heat exchangers with high and variable heat transfer rates	Ideal for tracing used for freeze protection
		When condensate pump is required	Freeze protection—water and condensate lines and traps
		Batch processes that require frequent startup of an air-filled system	Noncritical temperature control of heated tanks

STEAM TRAP COMPARISON

Characteristic	Steam Trap Type		
	Balanced Pressure Thermostatic	Bimetal Thermostatic	Thermodynamic
Method of Operation	Intermittent	Intermittent	Intermittent
No Load	No action	No action	No action
Light Load	Continuous; usually dribble action	Continuous; usually dribble action	Intermittent
Normal Load	May blast at high pressures	May blast at high pressures	Intermittent
Full or Overload	Continuous	Continuous	Continuous
Energy Conservation	Fair	Fair	Poor
Resistance to Wear	Fair	Fair	Poor
Corrosion Resistance	Good	Good	Excellent
Resistance to Hydraulic Shock	Good	Good	Excellent
Vent Air and CO_2 at Steam Temperature	No	No	No
Capability to Vent Air at Very Low Pressure (1/4 psig)	Good	Good	Not recommended for low-pressure applications
Capability to Handle Startup Air Loads	Excellent	Excellent	Poor

(*Continued*)

TEAM TRAP COMPARISON (*Continued*)

Characteristic	Steam Trap Type		
	Balanced Pressure Thermostatic	Bimetal Thermostatic	Thermodynamic
Operation Against Back Pressure	Excellent	Excellent	Poor
Resistance to Damage from Freezing; Cast Iron Trap Not Recommended	Good	Good	Good
Capability to Purge System	Good	Good	Excellent
Performance on Very Light Loads	Excellent	Excellent	Poor
Responsiveness to Slugs of Condensate	Delayed	Delayed	Delayed
Capability to Handle Dirt	Fair	Fair	Poor
Comparative Physical Size	Small	Small	Small
Capability to Handle Flash Steam	Poor	Poor	Poor
Usual Mechanical Failure Mode	Open or closed depending on design	Open or closed depending on design	Open, dirt can cause to fail closed
Subcooling	Yes	Yes	No
Venting	Excellent	Excellent	Fair
Seat Pressure Rating	N/a	N/a	N/a
Advantages	Small and lightweight	Small and lightweight	Rugged, withstands corrosion, water hammer, high pressure, and superheated steam
	Maximum discharge of noncondensables at startup	Maximum discharge of noncondensables at startup	Handles wide pressure range
	Unlikely to freeze	Unlikely to freeze and unlikely to be damaged if it does freeze	Compact and simple
		Rugged; withstands corrosion, water hammer, high pressure, and superheated steam	Audible operations warn when repair is needed
Disadvantages	Some types of damage by water hammer, corrosion, and superheated steam	Responds slowly to load and pressure changes	Poor operation with very low-pressure steam or high back pressure
	Condensate backs up into the drain line and/or process	More condensate backup than balance pressure thermostatic trap	Requires slow pressure buildup to remove air at startup to prevent air binding
		Back pressure changes operating characteristics	Noisy operation
Recommended Services	Batch processing requiring rapid discharge of noncondensables at startup	Drip legs on constant-pressure steam mains	Steam main drips, tracers
	Drip legs on steam mains and tracing	Installations subject to ambient conditions below freezing	Constant-pressure, constant-load applications
	Installations subject to ambient conditions below freezing		Installations subject to ambient conditions below freezing

6. Steam trap inspection:
a. Method #1 is shown in the following table:

Trap Failure Rate	Steam Trap Inspection Frequency
Over 10%	Every 2 months
5–10%	Every 3 months
Less than 5%	Every 6 months

b. Method #2 is shown in the following table:

System Pressure	Steam Trap Inspection Frequency
0–30 psig	Annually
30–100 psig	Semi-annually
100–250 psig	Quarterly or monthly
Over 250 psig	Monthly or weekly

22.02 Steam Condensate System Design Criteria

STEAM CONDENSATE SYSTEM DESIGN CRITERIA

System Type	Initial Steam Pressure psig	Maximum System Back Pressure psig	Maximum Pressure Drop psig/100 ft.	Maximum Velocity FPM
Low Pressure	1	0	1/8	5,000
	3	0	1/8	5,000
	5	0	1/8	5,000
	7	0	1/8	5,000
	10	3	1/4	5,000
	12	4	1/4	5,000
	15	5	1/4	5,000
Medium Pressure	20	6	1/4	5,000
	25	8	1/4	5,000
	30	10	1/4	5,000
	40	13	1/4	5,000
	50	16	1/4	5,000
	60	20	1/4	5,000
	75	25	1/4	5,000
	85	28	1/4	5,000
	100	33	1/4	5,000
High Pressure	120	40	1/4	5,000
	125	41	1/4	5,000
	150	50	1/4	5,000
	175	58	1/4	5,000
	200	66	1/2	5,000
	225	75	1/2	5,000
	250	83	1/2	5,000
	275	91	1/2	5,000
	300	100	1/2	5,000

22.03 Low-Pressure Steam Condensate System Pipe Sizing Tables (15 psig and Less)

1 PSIG STEAM CONDENSATE PIPING SYSTEMS—STEEL PIPE

Pipe Size	Pressure Drop psig/100 ft.			Velocity FPM (mph)			
	0.125	0.25	0.5	2,000 (23)	3,000 (34)	4,000 (45)	5,000 (57)
	Steam Condensate Flow lbs./h 0 psig Back Pressure						
1/2	843	1,192	1,686		Pressure	drop	
3/4	2,067	2,923	4,134	3,954	with these	pipe	governs
1	4,329	6,122	8,658	6,577			sizes
1-1/4	9,965	14,093	19,930	11,729	17,594		
1-1/2	15,758	22,285	31,515	16,158	24,237		
2	32,660	46,189	65,321	27,000	40,500	54,000	

(Continued)

1 PSIG STEAM CONDENSATE PIPING SYSTEMS—STEEL PIPE (Continued)

Pipe Size	Pressure Drop psig/100 ft.			Velocity FPM (mph)			
	0.125	0.25	0.5	2,000 (23)	3,000 (34)	4,000 (45)	5,000 (57)
2-1/2	54,310	76,806		38,753	58,130	77,507	96,883
3	100,865	142,645		60,396	90,594	120,792	150,989
4	216,701			105,124	157,686	210,247	262,809
5	405,300			166,358	249,536	332,715	415,894
6				238,345	357,518	476,690	595,863
8				417,533	626,299	835,065	1,043,831
10				682,684	1,024,026	1,365,369	1,706,711
12				991,486	1,487,228	1,982,971	2,478,714
14				1,213,661	1,820,491	2,427,322	3,034,152
16				1,615,821	2,423,731	3,231,642	4,039,552
18				2,075,432	3,113,148	4,150,864	5,188,580
20				2,592,495	3,888,742	5,184,990	6,481,237
22				3,167,009	4,750,513	6,334,018	7,917,522
24				3,798,974	5,698,462	7,597,949	9,497,436
26				4,488,391	6,732,587	8,976,782	11,220,978
28				5,235,260	7,852,889	10,470,519	13,088,149
30				6,039,579	9,059,369	12,079,159	15,098,948
32	Velocity	governs	with	6,901,350	10,352,026	13,802,701	17,253,376
34	these	pipe	sizes	7,820,573	11,730,859	15,641,146	19,551,432
36				8,797,247	13,195,870	17,594,494	21,993,117
42				12,071,977	18,107,966	24,143,954	30,179,943
48				15,863,770	23,795,655	31,727,540	39,659,425
54				20,172,626	30,258,938	40,345,251	50,431,564
60				24,998,544	37,497,816	49,997,088	62,496,360
72				36,201,568	54,302,353	72,403,137	90,503,921
84				49,472,844	74,209,266	98,945,687	123,682,109
96				64,812,370	97,218,554	129,624,739	162,030,924

Notes:
1 Maximum recommended pressure drop/velocity: 0.125 psig/100 ft./5,000 FPM.
2 Table based on heavy weight steel pipe using steam equations in Part 3.

3 PSIG STEAM CONDENSATE PIPING SYSTEMS—STEEL PIPE

Pipe Size	Pressure Drop psig/100 ft.			Velocity FPM (mph)			
	0.125	0.25	0.5	2,000 (23)	3,000 (34)	4,000 (45)	5,000 (57)
	Steam Condensate Flow lbs./h 0 psig Back Pressure						
1/2	293	414	586		Pressure	drop	
3/4	718	1,015	1,436	1,373	with these	pipe	
1	1,504	2,127	3,008	2,285			governs
1-1/4	3,462	4,895	6,923	4,074	6,112		sizes
1-1/2	5,474	7,741	10,947	5,613	8,419		
2	11,345	16,045	22,690	9,379	14,069	18,758	
2-1/2	18,866	26,680		13,462	20,193	26,923	33,654
3	35,037	49,550		20,980	31,469	41,959	52,449
4	75,275			36,517	54,775	73,033	91,292
5	140,788			57,787	86,681	115,575	144,468
6				82,794	124,190	165,587	206,984
8				145,038	217,556	290,075	362,594
10				237,143	355,714	474,286	592,857
12				344,411	516,616	688,822	861,027
14				421,588	632,381	843,175	1,053,969
16				561,285	841,928	1,122,570	1,403,213
18				720,940	1,081,409	1,441,879	1,802,349
20			with	900,551	1,350,826	1,801,102	2,251,377
22			sizes	1,100,119	1,650,178	2,200,238	2,750,297
24		governs		1,319,644	1,979,466	2,639,287	3,299,109
26	Velocity	pipe		1,559,125	2,338,688	3,118,251	3,897,813
28	these			1,818,564	2,727,846	3,637,128	4,546,410
30				2,097,959	3,146,939	4,195,918	5,244,898
32				2,397,311	3,595,967	4,794,622	5,993,278
34				2,716,620	4,074,930	5,433,240	6,791,550
36				3,055,886	4,583,829	6,111,771	7,639,714
42				4,193,424	6,290,135	8,386,847	10,483,559
48				5,510,573	8,265,859	11,021,145	13,776,432
54				7,007,333	10,511,000	14,014,666	17,518,333
60				8,683,705	13,025,557	17,367,409	21,709,262

(Continued)

3 PSIG STEAM CONDENSATE PIPING SYSTEMS—STEEL PIPE (Continued)

Pipe Size	Pressure Drop psig/100 ft.			Velocity FPM (mph)			
	0.125	0.25	0.5	2,000 (23)	3,000 (34)	4,000 (45)	5,000 (57)
72				12,575,282	18,862,922	25,150,563	31,438,204
84				17,185,304	25,777,955	34,370,607	42,963,259
96				22,513,770	33,770,656	45,027,541	56,284,426
	Steam Condensate Flow lbs./h 1 psig Back Pressure						
1/2	461	653	923		Pressure	drop	
3/4	1,132	1,601	2,264	2,232	with these	pipe	
1	2,370	3,352	4,741	3,713			governs
1-1/4	5,456	7,716	10,912	6,621	9,932		sizes
1-1/2	8,628	12,201	17,255	9,121	13,682		
2	17,882	25,289	35,764	15,242	22,862	30,483	
2-1/2	29,736	42,053		21,876	32,814	43,753	54,691
3	55,226	78,101		34,093	51,140	68,187	85,234
4	118,648			59,342	89,014	118,685	148,356
5	221,910			93,909	140,863	187,818	234,772
6				134,546	201,819	269,092	336,365
8				235,697	353,546	471,394	589,243
10				385,375	578,063	770,751	963,438
12				559,694	839,541	1,119,388	1,399,234
14				685,112	1,027,668	1,370,224	1,712,779
16				912,131	1,368,197	1,824,262	2,280,328
18				1,171,582	1,757,373	2,343,163	2,928,954
20				1,463,464	2,195,195	2,926,927	3,658,659
22			with	1,787,777	2,681,665	3,575,554	4,469,442
24			sizes	2,144,521	3,216,782	4,289,043	5,361,304
26		governs		2,533,697	3,800,546	5,067,395	6,334,243
28		pipe		2,955,305	4,432,957	5,910,609	7,388,262
30	Velocity			3,409,343	5,114,015	6,818,686	8,523,358
32	these			3,895,813	5,843,720	7,791,626	9,739,533
34				4,414,714	6,622,071	8,829,429	11,036,786
36				4,966,047	7,449,070	9,932,094	12,415,117
42				6,814,632	10,221,949	13,629,265	17,036,581
48				8,955,100	13,432,650	17,910,200	22,387,750
54				11,387,450	17,081,174	22,774,899	28,468,624
60				14,111,681	21,167,521	28,223,362	35,279,202
72				20,435,790	30,653,684	40,871,579	51,089,474
84				27,927,426	41,891,139	55,854,852	69,818,565
96				36,586,590	54,879,885	73,173,180	91,466,476

Notes:
1 Maximum recommended pressure drop/velocity: 0.125 psig/100 ft./5,000 FPM.
2 Table based on heavy weight steel pipe using steam equations in Part 3.

5 PSIG STEAM CONDENSATE PIPING SYSTEMS—STEEL PIPE

Pipe Size	Pressure Drop psig/100 ft.			Velocity FPM (mph)			
	0.125	0.25	0.5	2,000 (23)	3,000 (34)	4,000 (45)	5,000 (57)
	Steam Condensate Flow lbs./h 0 psig Back Pressure						
1/2	183	259	366		Pressure	drop	
3/4	449	635	898	858	with these	pipe	
1	940	1,329	1,880	1,428			governs
1-1/4	2,163	3,060	4,327	2,546	3,820		sizes
1-1/2	3,421	4,838	6,842	3,508	5,262		
2	7,091	10,028	14,182	5,862	8,793	11,724	
2-1/2	11,791	16,675		8,414	12,620	16,827	21,034
3	21,898	30,969		13,112	19,668	26,224	32,781
4	47,047			22,823	34,234	45,646	57,057
5	87,993			36,117	54,176	72,234	90,293
6				51,746	77,619	103,492	129,365
8				90,649	135,973	181,297	226,621

(Continued)

5 PSIG STEAM CONDENSATE PIPING SYSTEMS—STEEL PIPE (*Continued*)

Pipe Size	Pressure Drop psig/100 ft.			Velocity FPM (mph)			
	0.125	0.25	0.5	2,000 (23)	3,000 (34)	4,000 (45)	5,000 (57)
10				148,214	222,322	296,429	370,536
12				215,257	322,885	430,513	538,142
14				263,492	395,238	526,984	658,730
16				350,803	526,205	701,606	877,008
18				450,587	675,881	901,174	1,126,468
20				562,844	844,266	1,125,689	1,407,111
22				687,574	1,031,361	1,375,149	1,718,936
24				824,777	1,237,166	1,649,555	2,061,943
26				974,453	1,461,680	1,948,907	2,436,133
28	Velocity	governs	with	1,136,602	1,704,904	2,273,205	2,841,506
30	these	pipe	sizes	1,311,224	1,966,837	2,622,449	3,278,061
32				1,498,319	2,247,479	2,996,639	3,745,799
34				1,697,888	2,546,831	3,395,775	4,244,719
36				1,909,929	2,864,893	3,819,857	4,774,821
42				2,620,890	3,931,335	5,241,780	6,552,224
48				3,444,108	5,166,162	6,888,216	8,610,270
54				4,379,583	6,569,375	8,759,166	10,948,958
60				5,427,315	8,140,973	10,854,631	13,568,289
72				7,859,551	11,789,327	15,719,102	19,648,878
84				10,740,815	16,111,222	21,481,629	26,852,037
96				14,071,107	21,106,660	28,142,213	35,177,766
Steam Condensate Flow lbs./h 1 psig Back Pressure							
1/2	240	340	481				
3/4	590	834	1,179	1,163	Pressure	drop	
1	1,235	1,746	2,470	1,934	with these	pipe	governs
1-1/4	2,843	4,020	5,685	3,450	5,175		sizes
1-1/2	4,495	6,357	8,990	4,752	7,128		
2	9,317	13,176	18,634	7,941	11,911	15,882	
2-1/2	15,493	21,910		11,398	17,097	22,795	28,494
3	28,773	40,691		17,763	26,644	35,526	44,407
4	61,817			30,918	46,377	61,836	77,295
5	115,617			48,927	73,391	97,855	122,318
6				70,100	105,149	140,199	175,249
8				122,800	184,200	245,600	307,001
10				200,784	301,176	401,568	501,960
12				291,605	437,408	583,210	729,013
14				356,949	535,423	713,898	892,372
16				475,228	712,842	950,456	1,188,070
18				610,404	915,606	1,220,808	1,526,010
20				762,477	1,143,715	1,524,954	1,906,192
22			with	931,447	1,397,170	1,862,894	2,328,617
24			sizes	1,117,314	1,675,971	2,234,627	2,793,284
26		governs		1,320,078	1,980,116	2,640,155	3,300,194
28	Velocity	pipe		1,539,739	2,309,608	3,079,477	3,849,346
30	these			1,776,296	2,664,445	3,552,593	4,440,741
32				2,029,751	3,044,627	4,059,503	5,074,378
34				2,300,103	3,450,155	4,600,207	5,750,258
36				2,587,352	3,881,028	5,174,704	6,468,380
42				3,550,481	5,325,721	7,100,962	8,876,202
48				4,665,682	6,998,524	9,331,365	11,664,206
54				5,932,957	8,899,435	11,865,914	14,832,392
60				7,352,304	11,028,457	14,704,609	18,380,761
72				10,647,218	15,970,827	21,294,436	26,618,045
84				14,550,424	21,825,635	29,100,847	36,376,059
96				19,061,921	28,592,881	38,123,842	47,654,802

Notes:
1 Maximum recommended pressure drop/velocity: 0.125 psig/100 ft./5,000 FPM.
2 Table based on heavy weight steel pipe using steam equations in Part 3.

7 PSIG STEAM CONDENSATE PIPING SYSTEMS—STEEL PIPE

Pipe Size	Pressure Drop psig/100 ft.			Velocity FPM (mph)			
	0.125	0.25	0.5	2,000 (23)	3,000 (34)	4,000 (45)	5,000 (57)
Steam Condensate Flow lbs./h 0 psig Back Pressure							
1/2	136	192	271		Pressure	drop	
3/4	333	471	666	636	with these	pipe	
1	697	986	1,394	1,059			governs
1-1/4	1,604	2,269	3,208	1,888	2,832		sizes
1-1/2	2,537	3,587	5,073	2,601	3,902		
2	5,258	7,435	10,515	4,346	6,520	8,693	
2-1/2	8,743	12,364		6,238	9,358	12,477	15,596
3	16,237	22,962		9,722	14,583	19,444	24,306
4	34,884			16,922	25,384	33,845	42,306
5	65,243			26,780	40,169	53,559	66,949
6				38,368	57,552	76,736	95,919
8				67,213	100,819	134,425	168,031
10				109,896	164,843	219,791	274,739
12				159,605	239,408	319,210	399,013
14				195,370	293,055	390,740	488,425
16				260,108	390,162	520,215	650,269
18				334,094	501,141	668,188	835,235
20				417,328	625,993	834,657	1,043,321
22			with	509,811	764,717	1,019,622	1,274,528
24			sizes	611,542	917,313	1,223,084	1,528,856
26		governs		722,522	1,083,782	1,445,043	1,806,304
28	Velocity	pipe		842,749	1,264,124	1,685,498	2,106,873
30	these			972,225	1,458,337	1,944,450	2,430,562
32				1,110,949	1,666,424	2,221,898	2,777,373
34				1,258,921	1,888,382	2,517,843	3,147,304
36				1,416,142	2,124,213	2,832,284	3,540,355
42				1,943,294	2,914,941	3,886,588	4,858,235
48				2,553,680	3,830,520	5,107,360	6,384,200
54				3,247,301	4,870,951	6,494,601	8,118,252
60				4,024,156	6,036,234	8,048,312	10,060,390
72				5,827,570	8,741,354	11,655,139	14,568,924
84				7,963,921	11,945,882	15,927,842	19,909,803
96				10,433,211	15,649,816	20,866,421	26,083,027
Steam Condensate Flow lbs./h 1 psig Back Pressure							
1/2	166	235	333		Pressure	drop	
3/4	408	577	816	805	with these	pipe	
1	854	1,208	1,709	1,338			governs
1-1/4	1,967	2,781	3,933	2,387	3,580		sizes
1-1/2	3,110	4,398	6,220	3,288	4,932		
2	6,446	9,116	12,892	5,494	8,241	10,988	
2-1/2	10,719	15,159		7,886	11,828	15,771	19,714
3	19,907	28,153		12,289	18,434	24,579	30,724
4	42,769			21,391	32,086	42,782	53,477
5	79,991			33,851	50,776	67,702	84,627
6				48,499	72,749	96,998	121,248
8				84,961	127,441	169,921	212,402
10				138,914	208,372	277,829	347,286
12				201,750	302,625	403,500	504,375
14				246,959	370,438	493,918	617,397
16			with	328,791	493,187	657,583	821,979
18			sizes	422,314	633,471	844,629	1,055,786
20		governs		527,528	791,291	1,055,055	1,318,819
22		pipe		644,431	966,647	1,288,862	1,611,078
24	Velocity			773,025	1,159,538	1,546,050	1,932,563
26	these			913,310	1,369,964	1,826,619	2,283,274
28				1,065,284	1,597,926	2,130,568	2,663,211
30				1,228,949	1,843,424	2,457,899	3,072,373
32				1,404,305	2,106,457	2,808,609	3,510,762
34				1,591,351	2,387,026	3,182,701	3,978,376
36				1,790,087	2,685,130	3,580,173	4,475,217
42				2,456,437	3,684,656	4,912,875	6,141,093

(Continued)

[7]8888888888888888888888888888888888

7 PSIG STEAM CONDENSATE PIPING SYSTEMS—STEEL PIPE (*Continued*)

Pipe Size	Pressure Drop psig/100 ft.			Velocity FPM (mph)			
	0.125	0.25	0.5	2,000 (23)	3,000 (34)	4,000 (45)	5,000 (57)
48				3,228,001	4,842,002	6,456,002	8,070,003
54				4,104,778	6,157,167	8,209,557	10,261,946
60				5,086,769	7,630,153	10,173,537	12,716,922
72				7,366,389	11,049,584	14,732,779	18,415,973
84				10,066,863	15,100,294	20,133,726	25,167,157
96				13,188,190	19,782,284	26,376,379	32,970,474

Notes:
1 Maximum recommended pressure drop/velocity: 0.125 psig/100 ft./5,000 FPM.
2 Table based on heavy weight steel pipe using steam equations in Part 3.

10 PSIG STEAM CONDENSATE PIPING SYSTEMS—STEEL PIPE

Pipe Size	Pressure Drop psig/100 ft.			Velocity FPM (mph)			
	0.125	0.25	0.5	2,000 (23)	3,000 (34)	4,000 (45)	5,000 (57)
Steam Condensate Flow lbs./h 0 psig Back Pressure							
1/2	101	143	202	473	Pressure with these	drop pipe	governs sizes
3/4	247	350	494				
1	518	732	1,035	786			
1-1/4	1,191	1,685	2,383	1,402	2,104		
1-1/2	1,884	2,664	3,768	1,932	2,898		
2	3,905	5,523	7,810	3,228	4,842	6,457	
2-1/2	6,494	9,183		4,634	6,950	9,267	11,584
3	12,060	17,055		7,221	10,832	14,442	18,053
4	25,910			12,569	18,854	25,138	31,423
5	48,460			19,891	29,836	39,781	49,726
6				28,498	42,747	56,996	71,244
8				49,922	74,884	99,845	124,806
10				81,625	122,438	163,251	204,063
12				118,547	177,821	237,094	296,368
14				145,112	217,667	290,223	362,779
16				193,196	289,794	386,392	482,990
18				248,149	372,224	496,299	620,374
20				309,972	464,958	619,944	774,931
22		with sizes		378,664	567,996	757,328	946,660
24				454,225	681,338	908,450	1,135,563
26		governs pipe		536,655	804,983	1,073,311	1,341,639
28	Velocity these			625,955	938,932	1,251,910	1,564,887
30				722,124	1,083,185	1,444,247	1,805,309
32				825,161	1,237,742	1,650,323	2,062,904
34				935,068	1,402,603	1,870,137	2,337,671
36				1,051,845	1,577,767	2,103,689	2,629,612
42				1,443,389	2,165,083	2,886,777	3,608,471
48				1,896,755	2,845,133	3,793,510	4,741,888
54				2,411,944	3,617,917	4,823,889	6,029,861
60				2,988,956	4,483,435	5,977,913	7,472,391
72				4,328,448	6,492,673	8,656,897	10,821,121
84				5,915,231	8,872,847	11,830,463	14,788,078
96				7,749,305	11,623,958	15,498,610	19,373,263
Steam Condensate Flow lbs./h 1 psig Back Pressure							
1/2	118	167	235	569	Pressure with these	drop pipe	governs sizes
3/4	289	408	578				
1	605	855	1,210	947			
1-1/4	1,392	1,969	2,784	1,689	2,534		
1-1/2	2,201	3,113	4,403	2,327	3,491		
2	4,563	6,452	9,125	3,889	5,833	7,778	
2-1/2	7,587	10,730		5,582	8,372	11,163	13,954
3	14,091	19,927		8,699	13,048	17,397	21,747
4	30,272			15,141	22,711	30,282	37,852

(Continued)

10 PSIG STEAM CONDENSATE PIPING SYSTEMS—STEEL PIPE (*Continued*)

Pipe Size	Pressure Drop psig/100 ft.			Velocity FPM (mph)			
	0.125	0.25	0.5	2,000 (23)	3,000 (34)	4,000 (45)	5,000 (57)
5	56,619			23,960	35,940	47,921	59,901
6				34,329	51,493	68,657	85,821
8				60,137	90,205	120,273	150,342
10				98,326	147,489	196,652	245,816
12				142,803	214,204	285,605	357,006
14				174,802	262,203	349,604	437,005
16				232,725	349,087	465,450	581,812
18				298,922	448,383	597,844	747,305
20				373,394	560,091	746,788	933,485
22				456,141	684,211	912,281	1,140,352
24				547,162	820,743	1,094,324	1,367,905
26		governs	with	646,458	969,687	1,292,916	1,616,144
28		pipe	sizes	754,028	1,131,043	1,508,057	1,885,071
30	Velocity			869,874	1,304,810	1,739,747	2,174,684
32	these			993,993	1,490,990	1,987,987	2,484,984
34				1,126,388	1,689,582	2,252,776	2,815,970
36				1,267,057	1,900,586	2,534,114	3,167,643
42				1,738,713	2,608,069	3,477,426	4,346,782
48				2,284,840	3,427,261	4,569,681	5,712,101
54				2,905,440	4,358,160	5,810,880	7,263,599
60				3,600,511	5,400,767	7,201,022	9,001,278
72				5,214,070	7,821,105	10,428,140	13,035,174
84				7,125,516	10,688,274	14,251,032	17,813,790
96				9,334,850	14,002,275	18,669,700	23,337,125
Steam Condensate Flow lbs./h 3 psig Back Pressure							
1/2	167	236	333		Pressure	drop	
3/4	408	578	817		with these	pipe	
1	855	1,210	1,711	1,417			governs
1-1/4	1,969	2,784	3,938	2,527	3,791		sizes
1-1/2	3,113	4,403	6,227	3,481	5,222		
2	6,453	9,126	12,906	5,817	8,726	11,635	
2-1/2	10,730	15,175		8,350	12,524	16,699	20,874
3	19,928	28,183		13,013	19,519	26,025	32,531
4	42,814			22,649	33,974	45,299	56,623
5	80,076			35,842	53,764	71,685	89,606
6				51,352	77,029	102,705	128,381
8				89,959	134,939	179,918	224,898
10				147,087	220,631	294,174	367,718
12				213,620	320,429	427,239	534,049
14				261,488	392,232	522,976	653,721
16				348,135	522,203	696,270	870,338
18				447,160	670,740	894,321	1,117,901
20				558,564	837,845	1,117,127	1,396,409
22			with	682,345	1,023,517	1,364,690	1,705,862
24			sizes	818,504	1,227,757	1,637,009	2,046,261
26		governs		967,042	1,450,563	1,934,084	2,417,605
28		pipe		1,127,958	1,691,937	2,255,916	2,819,895
30	Velocity			1,301,252	1,951,878	2,602,504	3,253,130
32	these			1,486,924	2,230,386	2,973,848	3,717,310
34				1,684,974	2,527,461	3,369,949	4,212,436
36				1,895,403	2,843,104	3,790,805	4,738,507
42				2,600,957	3,901,435	5,201,913	6,502,392
48				3,417,914	5,126,871	6,835,828	8,544,785
54				4,346,274	6,519,411	8,692,549	10,865,686
60				5,386,038	8,079,057	10,772,076	13,465,095
72				7,799,775	11,699,663	15,599,551	19,499,438
84				10,659,126	15,988,688	21,318,251	26,647,814
96				13,964,089	20,946,133	27,928,178	34,910,222

Notes:
1 Maximum recommended pressure drop/velocity: 0.25 psig/100 ft./5,000 FPM.
2 Table based on heavy weight steel pipe using steam equations in Part 3.

12 PSIG STEAM CONDENSATE PIPING SYSTEMS—STEEL PIPE

Pipe Size	Pressure Drop psig/100 ft.			Velocity FPM (mph)			
	0.125	0.25	0.5	2,000 (23)	3,000 (34)	4,000 (45)	5,000 (57)
Steam Condensate Flow lbs./h 0 psig Back Pressure							
1/2	87	123	174	408	Pressure with these	drop pipe	governs sizes
3/4	213	301	426				
1	446	631	893	678			
1-1/4	1,028	1,453	2,055	1,210	1,814		
1-1/2	1,625	2,298	3,250	1,666	2,499		
2	3,368	4,763	6,736	2,784	4,177	5,569	
2-1/2	5,601	7,921	with sizes	3,996	5,995	7,993	9,991
3	10,402	14,710		6,228	9,342	12,457	15,571
4	22,347			10,841	16,261	21,682	27,102
5	41,797			17,156	25,733	34,311	42,889
6		governs pipe		24,579	36,869	49,159	61,448
8				43,058	64,587	86,116	107,645
10	Velocity these			70,402	105,603	140,804	176,005
12				102,247	153,370	204,494	255,617
14				125,159	187,738	250,318	312,897
16				166,632	249,947	333,263	416,579
18				214,029	321,043	428,058	535,072
20				267,351	401,027	534,702	668,378
22				326,598	489,897	653,196	816,494
24				391,769	587,654	783,538	979,423
26				462,865	694,298	925,731	1,157,163
28				539,886	809,829	1,079,772	1,349,715
30				622,832	934,247	1,245,663	1,557,079
32				711,702	1,067,553	1,423,404	1,779,254
34				806,497	1,209,745	1,612,993	2,016,241
36				907,216	1,360,824	1,814,432	2,268,040
42				1,244,923	1,867,384	2,489,845	3,112,307
48				1,635,951	2,453,927	3,271,903	4,089,878
54				2,080,302	3,120,453	4,160,604	5,200,755
60				2,577,975	3,866,962	5,155,950	6,444,937
72				3,733,287	5,599,930	7,466,573	9,333,217
84				5,101,887	7,652,831	10,203,774	12,754,718
96				6,683,776	10,025,663	13,367,551	16,709,439
Steam Condensate Flow lbs./h 1 psig Back Pressure							
1/2	100	141	199	482	Pressure with these	drop pipe	governs sizes
3/4	245	346	489				
1	512	724	1,024	802			
1-1/4	1,179	1,667	2,357	1,430	2,146		
1-1/2	1,864	2,636	3,728	1,970	2,956		
2	3,863	5,463	7,726	3,293	4,939	6,585	
2-1/2	6,424	9,085	with sizes	4,726	7,089	9,452	11,815
3	11,930	16,872		7,365	11,048	14,730	18,413
4	25,631			12,820	19,229	25,639	32,049
5	47,939			20,287	30,430	40,574	50,717
6		governs pipe		29,066	43,598	58,131	72,664
8				50,917	76,376	101,834	127,293
10	Velocity these			83,252	124,878	166,504	208,130
12				120,909	181,364	241,819	302,274
14				148,003	222,005	296,006	370,008
16				197,046	295,569	394,091	492,614
18				253,094	379,641	506,189	632,736
20				316,149	474,223	632,298	790,372
22				386,210	579,314	772,419	965,524
24				463,276	694,915	926,553	1,158,191
26				547,349	821,024	1,094,699	1,368,373
28				638,428	957,642	1,276,856	1,596,070
30				736,513	1,104,770	1,473,026	1,841,283
32				841,604	1,262,406	1,683,208	2,104,011

(Continued)

12 PSIG STEAM CONDENSATE PIPING SYSTEMS—STEEL PIPE (Continued)

Pipe Size	Pressure Drop psig/100 ft.			Velocity FPM (mph)			
	0.125	0.25	0.5	2,000 (23)	3,000 (34)	4,000 (45)	5,000 (57)
34				953,701	1,430,552	1,907,403	2,384,253
36				1,072,805	1,609,207	2,145,609	2,682,011
42				1,472,151	2,208,226	2,944,301	3,680,376
48				1,934,551	2,901,827	3,869,102	4,836,378
54				2,460,007	3,690,010	4,920,013	6,150,016
60				3,048,516	4,572,775	6,097,033	7,621,291
72				4,414,700	6,622,050	8,829,400	11,036,750
84				6,033,102	9,049,654	12,066,205	15,082,756
96				7,903,723	11,855,585	15,807,447	19,759,308
Steam Condensate Flow lbs./h 3 psig Back Pressure							
1/2	134	189	268		Pressure	drop	
3/4	329	465	657		with these	pipe	
1	688	973	1,376	1,140			governs
1-1/4	1,584	2,240	3,168	2,033	3,049		sizes
1-1/2	2,504	3,542	5,009	2,801	4,201		
2	5,191	7,341	10,382	4,680	7,020	9,359	
2-1/2	8,632	12,207		6,717	10,075	13,434	16,792
3	16,031	22,671		10,468	15,702	20,936	26,170
4	34,442			18,220	27,330	36,440	45,550
5	64,417			28,833	43,250	57,666	72,083
6				41,310	61,965	82,620	103,275
8				72,367	108,551	144,734	180,918
10				118,323	177,485	236,647	295,809
12				171,845	257,768	343,690	429,613
14				210,353	315,529	420,705	525,882
16				280,055	420,083	560,111	700,139
18				359,716	539,573	719,431	899,289
20				449,333	674,000	898,667	1,123,333
22			with	548,909	823,363	1,097,817	1,372,272
24			sizes	658,441	987,662	1,316,883	1,646,103
26		governs		777,932	1,166,898	1,555,863	1,944,829
28		pipe		907,380	1,361,069	1,814,759	2,268,449
30	Velocity			1,046,785	1,570,177	2,093,570	2,616,962
32	these			1,196,148	1,794,222	2,392,296	2,990,369
34				1,355,468	2,033,202	2,710,936	3,388,670
36				1,524,746	2,287,119	3,049,492	3,811,865
42				2,092,325	3,138,488	4,184,650	5,230,813
48				2,749,522	4,124,283	5,499,044	6,873,804
54				3,496,336	5,244,504	6,992,672	8,740,841
60				4,332,768	6,499,153	8,665,537	10,931,921
72				6,274,486	9,411,729	12,548,972	15,686,215
84				8,574,674	12,862,012	17,149,349	21,436,686
96				11,233,334	16,850,001	22,466,667	28,083,334

Notes:
1 Maximum recommended pressure drop/velocity: 0.25 psig/100 ft./5,000 FPM.
2 Table based on heavy weight steel pipe using steam equations in Part 3.

15 PSIG STEAM CONDENSATE PIPING SYSTEMS—STEEL PIPE

Pipe Size	Pressure Drop psig/100 ft.			Velocity FPM (mph)			
	0.125	0.25	0.5	2,000 (23)	3,000 (34)	4,000 (45)	5,000 (57)
Steam Condensate Flow lbs./h 0 psig Back Pressure							
1/2	73	103	146		Pressure	drop	
3/4	179	253	358	342	with these	pipe	
1	375	530	750	570			governs
1-1/4	863	1,221	1,726	1,016	1,524		sizes
1-1/2	1,365	1,930	2,730	1,400	2,099		
2	2,829	4,001	5,658	2,339	3,508	4,677	

(Continued)

15 PSIG STEAM CONDENSATE PIPING SYSTEMS—STEEL PIPE (*Continued*)

Pipe Size	Pressure Drop psig/100 ft.			Velocity FPM (mph)			
	0.125	0.25	0.5	2,000 (23)	3,000 (34)	4,000 (45)	5,000 (57)
2-1/2	4,704	6,652		3,357	5,035	6,713	8,391
3	8,736	12,355		5,231	7,847	10,462	13,078
4	18,769			9,105	13,658	18,210	22,763
5	35,105			14,409	21,613	28,818	36,022
6				20,644	30,966	41,288	51,610
8				36,164	54,246	72,328	90,411
10				59,130	88,695	118,260	147,825
12				85,877	128,815	171,753	214,692
14				105,120	157,680	210,240	262,801
16				139,953	209,929	279,906	349,882
18				179,762	269,643	359,524	449,405
20				224,547	336,820	449,094	561,367
22			with	274,308	411,462	548,616	685,770
24			sizes	329,045	493,568	658,090	822,613
26		governs		388,758	583,137	777,517	971,896
28		pipe		453,448	680,172	906,895	1,133,619
30	Velocity			523,113	784,670	1,046,226	1,307,783
32	these			597,755	896,632	1,195,510	1,494,387
34				677,372	1,016,059	1,354,745	1,693,431
36				761,966	1,142,949	1,523,933	1,904,916
42				1,045,604	1,568,406	2,091,209	2,614,011
48				1,374,027	2,061,041	2,748,055	3,435,068
54				1,747,235	2,620,853	3,494,471	4,368,088
60				2,165,228	3,247,842	4,330,456	5,413,071
72				3,135,569	4,703,353	6,271,138	7,838,922
84				4,285,049	6,427,574	8,570,099	10,712,624
96				5,613,670	8,420,505	11,227,340	14,034,175
Steam Condensate Flow lbs./h 1 psig Back Pressure							
1/2	82	116	164		Pressure	drop	
3/4	202	285	403	398	with these	pipe	
1	422	597	845	661			governs
1-1/4	972	1,375	1,944	1,180	1,769		sizes
1-1/2	1,537	2,174	3,074	1,625	2,438		
2	3,186	4,506	6,372	2,715	4,073	5,431	
2-1/2	5,298	7,492		3,897	5,846	7,795	9,744
3	9,839	13,915		6,074	9,111	12,148	15,185
4	21,139			10,572	15,859	21,145	26,431
5	39,536			16,731	25,096	33,462	41,827
6				23,971	35,956	47,942	59,927
8				41,992	62,988	83,984	104,980
10				68,659	102,988	137,318	171,647
12				99,716	149,573	199,431	249,289
14				122,060	183,090	244,120	305,150
16				162,506	243,759	325,012	406,265
18				208,730	313,095	417,460	521,825
20				260,732	391,098	521,464	651,830
22			with	318,512	477,768	637,024	796,280
24			sizes	382,070	573,105	764,140	955,175
26		governs		451,406	677,109	902,812	1,128,515
28		pipe		526,520	789,780	1,053,040	1,316,299
30	Velocity			607,412	911,118	1,214,823	1,518,529
32	these			694,082	1,041,122	1,388,163	1,735,204
34				786,530	1,179,794	1,573,059	1,966,324
36				884,755	1,327,133	1,769,511	2,211,889
42				1,214,101	1,821,152	2,428,202	3,035,253
48				1,595,449	2,393,173	3,190,898	3,988,622
54				2,028,798	3,043,198	4,057,597	5,071,996
60				2,514,150	3,771,225	5,028,300	6,285,375
72				3,640,859	5,461,289	7,281,718	9,102,148
84				4,975,576	7,463,364	9,951,152	12,438,940
96				6,518,301	9,777,451	13,036,601	16,295,751

(*Continued*)

15 PSIG STEAM CONDENSATE PIPING SYSTEMS—STEEL PIPE (*Continued*)

Pipe Size	Pressure Drop psig/100 ft.			Velocity FPM (mph)			
	0.125	0.25	0.5	2,000 (23)	3,000 (34)	4,000 (45)	5,000 (57)
Steam Condensate Flow lbs./h 3 psig Back Pressure							
1/2	105	149	211		Pressure	drop	
3/4	258	366	517		with these	pipe	
1	541	766	1,083	897			governs
1-1/4	1,246	1,762	2,492	1,599	2,399		sizes
1-1/2	1,970	2,786	3,941	2,203	3,305		
2	4,084	5,775	8,168	3,682	5,522	7,363	
2-1/2	6,791	9,604		5,284	7,926	10,568	13,210
3	12,612	17,836		8,235	12,353	16,470	20,588
4	27,096			14,334	21,501	28,668	35,835
5	50,677			22,683	34,025	45,367	56,709
6				32,499	48,749	64,999	81,248
8				56,932	85,398	113,864	142,330
10				93,087	139,630	186,173	232,716
12				135,193	202,789	270,386	337,982
14				165,487	248,231	330,975	413,718
16				220,323	330,485	440,647	550,808
18				282,993	424,490	565,986	707,483
20				353,497	530,245	706,993	883,741
22			with	431,834	647,751	863,667	1,079,584
24			sizes	518,005	777,007	1,036,009	1,295,011
26		governs		612,009	918,014	1,224,018	1,530,023
28	Velocity	pipe		713,848	1,070,771	1,427,695	1,784,619
30	these			823,520	1,235,279	1,647,039	2,058,799
32				941,025	1,411,538	1,882,051	2,352,563
34				1,066,365	1,599,547	2,132,730	2,665,912
36				1,199,538	1,799,307	2,399,076	2,998,845
42				1,646,060	2,469,090	3,292,120	4,115,150
48				2,163,085	3,244,628	4,326,171	5,407,713
54				2,750,614	4,125,921	5,501,228	6,876,535
60				3,408,646	5,112,970	6,817,293	8,521,616
72				4,936,221	7,404,332	9,872,443	12,340,554
84				6,745,810	10,118,715	13,491,620	16,864,526
96				8,837,413	13,256,119	17,674,826	22,093,532
Steam Condensate Flow lbs./h 5 psig Back Pressure							
1/2	138	195	276		Pressure	drop	
3/4	338	478	676		with these	pipe	
1	708	1,001	1,416	1,233			governs
1-1/4	1,630	2,305	3,260	2,200			sizes
1-1/2	2,577	3,645	5,154	3,030	4,545		
2	5,342	7,554	10,683	5,063	7,595	10,126	
2-1/2	8,883	12,562	17,765	7,267	10,901	14,534	
3	16,497	23,330		11,326	16,988	22,651	28,314
4	35,442			19,713	29,569	39,426	49,282
5	66,288			31,196	46,793	62,391	77,989
6	107,770			44,695	67,042	89,390	111,737
8				78,296	117,445	156,593	195,741
10				128,018	192,027	256,036	320,045
12				185,925	278,888	371,850	464,813
14				227,588	341,382	455,176	568,970
16			with	303,002	454,502	606,003	757,504
18		governs	sizes	389,189	583,783	778,377	972,972
20	Velocity	pipe		486,149	729,224	972,298	1,215,373
22	these			593,883	890,825	1,187,766	1,484,708
24				712,390	1,068,586	1,424,781	1,780,976
26				841,671	1,262,507	1,683,342	2,104,178
28				981,725	1,472,588	1,963,450	2,454,313
30				1,132,553	1,698,829	2,265,105	2,831,382
32				1,294,154	1,941,230	2,588,307	3,235,384

(*Continued*)

15 PSIG STEAM CONDENSATE PIPING SYSTEMS—STEEL PIPE (*Continued*)

Pipe Size	Pressure Drop psig/100 ft.			Velocity FPM (mph)			
	0.125	0.25	0.5	2,000 (23)	3,000 (34)	4,000 (45)	5,000 (57)
34				1,466,528	2,199,792	2,933,056	3,666,319
36				1,649,675	2,474,513	3,299,351	4,124,189
42				2,263,759	3,395,638	4,527,517	5,659,397
48				2,974,802	4,462,204	5,949,605	7,437,006
54				3,782,807	5,674,210	7,565,613	9,457,017
60				4,687,772	7,031,657	9,375,543	11,719,429
72				6,788,583	10,182,874	13,577,165	16,971,457
84				9,277,236	13,915,854	18,554,472	23,193,090
96				12,153,731	18,230,597	24,307,462	30,384,328

Notes:
1. Maximum recommended pressure drop/velocity: 0.25 psig/100 ft./5,000 FPM.
2. Table based on heavy weight steel pipe using steam equations in Part 3.

22.04 Medium-Pressure Steam Condensate System Pipe Sizing Tables (20–100 psig)

20 PSIG STEAM CONDENSATE PIPING SYSTEMS—STEEL PIPE

Pipe Size	Pressure Drop psig/100 ft.			Velocity FPM (mph)			
	0.125	0.25	0.5	2,000 (23)	3,000 (34)	4,000 (45)	5,000 (57)
	Steam Condensate Flow lbs./h 0 psig Back Pressure						
1/2	59	83	118		Pressure	drop	
3/4	144	204	288	276	with these	pipe	
1	302	427	604	459			governs
1-1/4	695	983	1,390	818	1,227		sizes
1-1/2	1,099	1,555	2,199	1,127	1,691		
2	2,279	3,222	4,557	1,884	2,826	3,767	
2-1/2	3,789	5,359		2,704	4,056	5,407	6,759
3	7,037	9,952		4,214	6,320	8,427	10,534
4	15,119			7,334	11,001	14,668	18,336
5	28,277			11,606	17,410	23,213	29,016
6				16,629	24,943	33,257	41,572
8				29,130	43,695	58,260	72,825
10				47,629	71,444	95,258	119,073
12				69,173	103,760	138,347	172,934
14				84,674	127,011	169,348	211,685
16				112,732	169,098	225,463	281,829
18				144,798	217,196	289,595	361,994
20				180,872	271,308	361,743	452,179
22		with		220,954	331,431	441,908	552,385
24		sizes		265,045	397,567	530,089	662,612
26		governs		313,144	469,715	626,287	782,859
28	Velocity	pipe		365,251	547,876	730,501	913,127
30	these			421,366	632,049	842,732	1,053,415
32				481,490	722,234	962,979	1,203,724
34				545,621	818,432	1,091,243	1,364,053
36				613,761	920,642	1,227,523	1,534,404
42				842,231	1,263,346	1,684,462	2,105,577
48				1,106,775	1,660,162	2,213,549	2,766,937
54				1,407,392	2,111,089	2,814,785	3,518,481
60				1,744,084	2,616,127	3,488,169	4,360,211
72				2,525,691	3,788,536	5,051,382	6,314,227
84				3,451,594	5,177,391	6,903,187	8,628,984
96				4,521,793	6,782,690	9,043,586	11,304,483

(Continued)

20 PSIG STEAM CONDENSATE PIPING SYSTEMS—STEEL PIPE (*Continued*)

Pipe Size	Pressure Drop psig/100 ft.			Velocity FPM (mph)			
	0.125	0.25	0.5	2,000 (23)	3,000 (34)	4,000 (45)	5,000 (57)
Steam Condensate Flow lbs./h 5 psig Back Pressure							
1/2	98	139	197		Pressure	drop	
3/4	241	341	482		with these	pipe	
1	505	714	1,010	880			governs
1-1/4	1,163	1,644	2,325	1,569			sizes
1-1/2	1,839	2,600	3,677	2,162	3,242		
2	3,811	5,389	7,622	3,612	5,418	7,224	
2-1/2	6,337	8,962	12,674	5,184	7,776	10,369	
3	11,769	16,644		8,080	12,119	16,159	20,199
4	25,284			14,063	21,095	28,126	35,158
5	47,290			22,255	33,382	44,510	55,637
6	76,883			31,885	47,828	63,770	79,713
8				55,856	83,785	111,713	139,641
10				91,328	136,991	182,655	228,319
12				132,638	198,957	265,276	331,595
14				162,360	243,540	324,720	405,900
16				216,160	324,240	432,320	540,400
18				277,646	416,468	555,291	694,114
20				346,817	520,225	693,634	867,042
22				423,674	635,510	847,347	1,059,184
24				508,216	762,324	1,016,432	1,270,541
26		governs	with	600,445	900,667	1,200,889	1,501,111
28		pipe	sizes	700,358	1,050,538	1,400,717	1,750,896
30	Velocity			807,958	1,211,937	1,615,916	2,019,895
32	these			923,243	1,384,865	1,846,487	2,308,109
34				1,046,215	1,569,322	2,092,429	2,615,536
36				1,176,871	1,765,307	2,353,742	2,942,178
42				1,614,956	2,422,433	3,229,911	4,037,389
48				2,122,211	3,183,317	4,244,422	5,305,528
54				2,698,638	4,047,957	5,397,276	6,746,595
60				3,344,236	5,016,354	6,688,472	8,360,589
72				4,842,945	7,264,418	9,685,891	12,107,363
84				6,618,340	9,927,509	13,236,679	16,545,849
96				8,670,419	13,005,628	17,340,838	21,676,047
Steam Condensate Flow lbs./h 7 psig Back Pressure							
1/2	123	174	246		Pressure	drop	
3/4	301	426	602		with these	pipe	
1	631	892	1,262	1,150			governs
1-1/4	1,452	2,053	2,904	2,050			sizes
1-1/2	2,296	3,247	4,592	2,824	4,236		
2	4,759	6,730	9,518	4,719	7,079	9,438	
2-1/2	7,913	11,191	15,827	6,773	10,160	13,547	
3	14,697	20,785		10,556	15,834	21,112	26,390
4	31,575	44,654		18,374	27,560	36,747	45,934
5	59,055			29,076	43,614	58,152	72,690
6	96,011			41,658	62,487	83,316	104,145
8				72,976	109,464	145,953	182,441
10				119,320	178,979	238,639	298,299
12				173,292	259,938	346,584	433,230
14				212,124	318,185	424,247	530,309
16				282,413	423,620	564,826	706,033
18		governs	with	362,744	544,116	725,488	906,860
20	Velocity	pipe	sizes	453,116	679,674	906,232	1,132,790
22	these			553,530	830,294	1,107,059	1,383,824
24				663,985	995,977	1,327,969	1,659,961
26				784,481	1,176,721	1,568,962	1,961,202
28				915,018	1,372,528	1,830,037	2,287,546
30				1,055,597	1,583,396	2,111,195	2,638,993
32				1,206,218	1,809,327	2,412,435	3,015,544

(*Continued*)

20 PSIG STEAM CONDENSATE PIPING SYSTEMS—STEEL PIPE (*Continued*)

Pipe Size	Pressure Drop psig/100 ft.			Velocity FPM (mph)			
	0.125	0.25	0.5	2,000 (23)	3,000 (34)	4,000 (45)	5,000 (57)
34				1,366,879	2,050,319	2,733,759	3,417,198
36				1,537,582	2,306,374	3,075,165	3,843,956
42				2,109,940	3,164,909	4,219,879	5,274,849
48				2,772,669	4,159,003	5,545,338	6,931,672
54				3,525,771	5,288,656	7,051,541	8,814,426
60				4,369,244	6,553,866	8,738,489	10,923,111
72				6,327,308	9,490,963	12,654,617	15,818,271
84				8,646,861	12,970,292	17,293,722	21,617,153
96				11,327,903	16,991,854	22,655,806	28,319,757

Notes:
1 Maximum recommended pressure drop/velocity: 0.25 psig/100 ft./5,000 FPM.
2 Table based on heavy weight steel pipe using steam equations in Part 3.

25 PSIG STEAM CONDENSATE PIPING SYSTEMS—STEEL PIPE

Pipe Size	Pressure Drop psig/100 ft.			Velocity FPM (mph)			
	0.125	0.25	0.5	2,000 (23)	3,000 (34)	4,000 (45)	5,000 (57)
	Steam Condensate Flow lbs./h 0 psig Back Pressure						
1/2	50	71	100		Pressure	drop	
3/4	123	174	246	235	with these	pipe	governs
1	257	364	515	391			sizes
1-1/4	593	838	1,185	697	1,046		
1-1/2	937	1,325	1,874	961	1,441		
2	1,942	2,746	3,884	1,605	2,408	3,211	
2-1/2	3,229	4,567		2,304	3,456	4,609	5,761
3	5,997	8,482		3,591	5,387	7,182	8,978
4	12,885			6,251	9,376	12,501	15,626
5	24,099			9,892	14,837	19,783	24,729
6				14,172	21,258	28,344	35,430
8				24,826	37,239	49,653	62,066
10				40,592	60,888	81,184	101,480
12				58,953	88,430	117,906	147,383
14				72,164	108,245	144,327	180,409
16				96,076	144,114	192,152	240,190
18				123,404	185,106	246,808	308,510
20				154,148	231,223	308,297	385,371
22			with	188,309	282,463	376,617	470,772
24			sizes	225,885	338,827	451,770	564,712
26		governs		266,877	400,316	533,755	667,193
28		pipe		311,286	466,929	622,571	778,214
30	Velocity			359,110	538,665	718,220	897,775
32	these			410,351	615,526	820,701	1,025,876
34				465,007	697,511	930,014	1,162,518
36				523,080	784,619	1,046,159	1,307,699
42				717,793	1,076,690	1,435,586	1,794,483
48				943,251	1,414,877	1,886,502	2,358,128
54				1,199,453	1,799,180	2,398,907	2,998,634
60				1,486,400	2,229,600	2,972,800	3,716,000
72				2,152,526	3,228,789	4,305,051	5,381,314
84				2,941,629	4,412,443	5,883,257	7,354,071
96				3,853,708	5,780,563	7,707,417	9,634,271
	Steam Condensate Flow lbs./h 5 psig Back Pressure						
1/2	78	111	157		Pressure	drop	
3/4	192	272	384	701	with these	pipe	governs
1	402	569	805				sizes
1-1/4	926	1,310	1,852	1,250			
1-1/2	1,464	2,071	2,929	1,722	2,583		
2	3,035	4,293	6,071	2,877	4,316	5,754	

(*Continued*)

25 PSIG STEAM CONDENSATE PIPING SYSTEMS—STEEL PIPE (*Continued*)

Pipe Size	Pressure Drop psig/100 ft.			Velocity FPM (mph)			
	0.125	0.25	0.5	2,000 (23)	3,000 (34)	4,000 (45)	5,000 (57)
2-1/2	5,047	7,138	10,095	4,129	6,194	8,259	16,089
3	9,374	13,257		6,436	9,653	12,871	28,004
4	20,140			11,202	16,803	22,403	
5	37,668			17,727	26,590	35,453	44,316
6				25,397	38,096	50,795	63,493
8				44,491	66,737	88,982	111,228
10				72,745	109,117	145,490	181,862
12				105,650	158,475	211,300	264,124
14				129,324	193,986	258,648	323,310
16				172,177	258,266	344,354	430,443
18				221,152	331,728	442,304	552,880
20				276,249	414,373	552,497	690,621
22				337,467	506,201	674,934	843,668
24				404,807	607,211	809,615	1,012,019
26		governs	with	478,270	717,405	956,539	1,195,674
28	Velocity	pipe	sizes	557,854	836,781	1,115,708	1,394,634
30	these			643,560	965,340	1,287,119	1,608,899
32				735,387	1,103,081	1,470,775	1,838,469
34				833,337	1,250,006	1,666,674	2,083,343
36				937,409	1,406,113	1,874,817	2,343,522
42				1,286,354	1,929,531	2,572,708	3,215,885
48				1,690,396	2,535,595	3,380,793	4,225,991
54				2,149,535	3,224,303	4,299,071	5,373,838
60				2,663,771	3,995,656	5,327,542	6,659,427
72				3,857,532	5,786,298	7,715,064	9,643,830
84				5,271,680	7,907,520	10,543,360	13,179,200
96				6,906,214	10,359,322	13,812,429	17,265,536
Steam Condensate Flow lbs./h 7 psig Back Pressure							
1/2	94	133	188		Pressure	drop	
3/4	231	326	461		with these	pipe	
1	483	683	966	880			governs
1-1/4	1,112	1,572	2,224	1,570			sizes
1-1/2	1,758	2,486	3,516	2,162	3,244		
2	3,644	5,153	7,288	3,613	5,420	7,227	
2-1/2	6,059	8,569	12,119	5,186	7,780	10,373	
3	11,254	15,915		8,083	12,124	16,166	20,207
4	24,178	34,192		14,069	21,103	28,138	35,172
5	45,220			22,264	33,396	44,528	55,660
6	73,517			31,898	47,847	63,796	79,745
8				55,879	83,819	111,758	139,698
10				91,365	137,047	182,729	228,412
12				132,692	199,038	265,384	331,730
14				162,426	243,639	324,852	406,065
16				216,248	324,372	432,495	540,619
18				277,758	416,637	555,516	694,396
20				346,957	520,436	693,915	867,394
22				423,846	635,768	847,691	1,059,614
24				508,422	762,634	1,016,845	1,271,056
26		governs	with	600,688	901,032	1,201,376	1,501,720
28	Velocity	pipe	sizes	700,643	1,050,964	1,401,285	1,751,607
30	these			808,286	1,212,429	1,616,572	2,020,715
32				923,618	1,385,427	1,847,236	2,309,045
34				1,046,639	1,569,959	2,093,278	2,616,598
36				1,177,349	1,766,023	2,354,698	2,943,372
42				1,615,611	2,423,416	3,231,222	4,039,027
48				2,123,072	3,184,608	4,246,144	5,307,681
54				2,699,733	4,049,599	5,399,466	6,749,332
60				3,345,593	5,018,389	6,691,186	8,363,982
72				4,844,910	7,267,366	9,689,821	12,112,276
84				6,621,025	9,931,538	13,242,050	16,552,563
96				8,673,937	13,010,905	17,347,874	21,684,842

Notes:
1 Maximum recommended pressure drop/velocity: 0.25 psig/100 ft./5,000 FPM.
2 Table based on heavy weight steel pipe using steam equations in Part 3.

30 PSIG STEAM CONDENSATE PIPING SYSTEMS—STEEL PIPE

| Pipe Size | Pressure Drop psig/100 ft. | | | Velocity FPM (mph) | | | |
	0.125	0.25	0.5	2,000 (23)	3,000 (34)	4,000 (45)	5,000 (57)
	Steam Condensate Flow lbs./h 0 psig Back Pressure						
1/2	44	63	88		Pressure	drop	
3/4	108	153	217	207	with these	pipe	
1	227	321	454	345			governs
1-1/4	523	739	1,046	615	923		sizes
1-1/2	827	1,169	1,653	848	1,272		
2	1,714	2,423	3,427	1,417	2,125	2,833	
2-1/2	2,849	4,030		2,033	3,050	4,066	5,083
3	5,292	7,484		3,169	4,753	6,337	7,922
4	11,369			5,515	8,273	11,030	13,788
5	21,264			8,728	13,092	17,456	21,820
6				12,505	18,757	25,009	31,261
8				21,906	32,858	43,811	54,764
10				35,817	53,725	71,633	89,541
12				52,018	78,026	104,035	130,044
14				63,674	95,511	127,348	159,184
16				84,773	127,159	169,546	211,932
18				108,886	163,329	217,772	272,215
20				136,013	204,020	272,026	340,033
22			with	166,155	249,232	332,309	415,387
24			sizes	199,310	298,965	398,621	498,276
26		governs		235,480	353,220	470,960	588,700
28		pipe		274,664	411,996	549,328	686,660
30	Velocity			316,862	475,293	633,724	792,155
32	these			362,074	543,111	724,148	905,185
34				410,300	615,450	820,601	1,025,751
36				461,541	692,311	923,082	1,153,852
42				633,347	950,020	1,266,694	1,583,367
48				832,280	1,248,421	1,664,561	2,080,701
54				1,058,341	1,587,512	2,116,682	2,645,853
60				1,311,529	1,967,294	2,623,059	3,278,823
72				1,899,287	2,848,931	3,798,575	4,748,218
84				2,595,555	3,893,332	5,191,109	6,488,886
96				3,400,331	5,100,496	6,800,662	8,500,827
	Steam Condensate Flow lbs./h 5 psig Back Pressure						
1/2	66	94	132		Pressure	drop	
3/4	162	230	325		with these	pipe	
1	340	481	680	592			governs
1-1/4	782	1,107	1,565	1,056			sizes
1-1/2	1,237	1,750	2,475	1,455	2,182		
2	2,564	3,627	5,129	2,431	3,646	4,861	
2-1/2	4,264	6,031	8,529	3,489	5,233	6,978	
3	7,920	11,200		5,437	8,156	10,874	13,593
4	17,015			9,464	14,196	18,928	23,660
5	31,824			14,977	22,465	29,953	37,441
6	51,739			21,457	32,186	42,915	53,643
8				37,589	56,383	75,178	93,972
10				61,459	92,189	122,919	153,649
12				89,260	133,889	178,519	223,149
14				109,261	163,892	218,523	273,153
16		governs	with	145,466	218,199	290,932	363,665
18		pipe	sizes	186,843	280,265	373,686	467,108
20	Velocity			233,392	350,089	466,785	583,481
22	these			285,114	427,671	570,227	712,784
24				342,007	513,011	684,014	855,018
26				404,073	606,109	808,145	1,010,182
28				471,310	706,966	942,621	1,178,276
30				543,720	815,580	1,087,441	1,359,301
32				621,302	931,953	1,242,604	1,553,256

(Continued)

30 PSIG STEAM CONDENSATE PIPING SYSTEMS—STEEL PIPE (*Continued*)

Pipe Size	Pressure Drop psig/100 ft.			Velocity FPM (mph)			
	0.125	0.25	0.5	2,000 (23)	3,000 (34)	4,000 (45)	5,000 (57)
34				704,056	1,056,084	1,408,113	1,760,141
36				791,983	1,187,974	1,583,965	1,979,956
42				1,086,794	1,630,191	2,173,588	2,716,985
48				1,428,155	2,142,232	2,856,309	3,570,387
54				1,816,064	2,724,097	3,632,129	4,540,161
60				2,250,523	3,375,785	4,501,047	5,626,309
72				3,259,089	4,888,633	6,518,178	8,147,722
84				4,453,851	6,680,777	8,907,702	11,134,628
96				5,834,810	8,752,215	11,669,620	14,587,025
Steam Condensate Flow lbs./h 7 psig Back Pressure							
1/2	78	110	155		Pressure	drop	
3/4	190	269	381		with these	pipe	
1	399	564	797	727			governs
1-1/4	918	1,298	1,836	1,296			sizes
1-1/2	1,451	2,052	2,903	1,785	2,678		
2	3,008	4,254	6,016	2,983	4,474	5,966	
2-1/2	5,002	7,074	10,004	4,281	6,422	8,562	
3	9,290	13,137		6,672	10,008	13,344	16,680
4	19,958	28,225		11,613	17,420	23,227	29,034
5	37,328			18,378	27,567	36,756	45,945
6	60,686			26,331	39,496	52,662	65,827
8				46,127	69,190	92,253	115,316
10				75,419	113,128	150,838	188,547
12				109,534	164,300	219,067	273,834
14				134,078	201,117	268,156	335,195
16				178,506	267,760	357,013	446,266
18				229,282	343,922	458,563	573,204
20				286,404	429,605	572,807	716,009
22				349,873	524,809	699,745	874,681
24				419,688	629,533	839,377	1,049,221
26		governs	with	495,851	743,777	991,702	1,239,628
28		pipe	sizes	578,361	867,541	1,156,721	1,445,902
30	Velocity these			667,217	1,000,826	1,334,434	1,668,043
32				762,421	1,143,631	1,524,841	1,906,052
34				863,971	1,295,956	1,727,942	2,159,927
36				971,868	1,457,802	1,943,736	2,429,670
42				1,333,641	2,000,462	2,667,282	3,334,103
48				1,752,536	2,628,804	3,505,072	4,381,340
54				2,228,553	3,342,830	4,457,106	5,571,383
60				2,761,692	4,142,538	5,523,384	6,904,230
72				3,999,336	5,999,005	7,998,673	9,998,341
84				5,465,469	8,198,203	10,930,938	13,663,672
96				7,160,089	10,740,134	14,320,179	17,900,224
Steam Condensate Flow lbs./h 10 psig Back Pressure							
1/2	99	139	197		Pressure		
3/4	242	342	484		with these		
1	506	716	1,013	981		drop	governs
1-1/4	1,166	1,649	2,331	1,749		pipe	sizes
1-1/2	1,843	2,607	3,687	2,409	3,614		
2	3,821	5,403	7,641	4,026	6,039		
2-1/2	6,353	8,985	12,707	5,778	8,668	11,557	
3	11,800	16,687		9,005	13,508	18,011	22,514
4	25,350	35,851		15,675	23,512	31,350	39,187
5	47,413			24,805	37,208	49,610	62,013
6	77,083			35,539	53,309	71,078	88,848
8				62,257	93,386	124,515	155,644
10				101,794	152,690	203,587	254,484
12				147,838	221,758	295,677	369,596
14				180,966	271,450	361,933	452,416

(*Continued*)

STEAM TRAP COMPARISON (*Continued*)

Pipe Size	Pressure Drop psig/100 ft.			Velocity FPM (mph)			
	0.125	0.25	0.5	2,000 (23)	3,000 (34)	4,000 (45)	5,000 (57)
16				240,932	361,398	481,863	602,329
18				309,463	464,195	618,927	773,658
20				386,562	579,842	773,123	966,404
22				472,226	708,339	944,452	1,180,565
24				566,457	849,686	1,132,914	1,416,143
26				669,255	1,003,882	1,338,510	1,673,137
28				780,619	1,170,928	1,561,237	1,951,547
30				900,549	1,350,824	1,801,098	2,251,373
32	Velocity	governs	with	1,029,046	1,543,569	2,058,092	2,572,615
34	these	pipe	sizes	1,166,109	1,749,164	2,332,219	2,915,274
36				1,311,739	1,967,609	2,623,479	3,279,348
42				1,800,028	2,700,041	3,600,055	4,500,069
48				2,365,414	3,548,121	4,730,828	5,913,535
54				3,007,899	4,511,848	6,015,797	7,519,746
60				3,727,481	5,591,222	7,454,963	9,318,703
72				5,397,941	8,096,912	10,795,882	13,494,853
84				7,376,794	11,065,190	14,753,587	18,441,984
96				9,664,039	14,496,058	19,328,077	24,160,097

Notes:
1 Maximum recommended pressure drop/velocity: 0.25 psig/100 ft./5,000 FPM.
2 Table based on heavy weight steel pipe using steam equations in Part 3.

40 PSIG STEAM CONDENSATE PIPING SYSTEMS—STEEL PIPE

Pipe Size	Pressure Drop psig/100 ft.			Velocity FPM (mph)			
	0.125	0.25	0.5	2,000 (23)	3,000 (34)	4,000 (45)	5,000 (57)
	Steam Condensate Flow lbs./h 0 psig Back Pressure						
1/2	37	52	73	172	Pressure	drop	
3/4	90	127	180		with these	pipe	
1	188	266	376	286			governs
1-1/4	433	613	867	510	765		sizes
1-1/2	685	969	1,370	703	1,054		
2	1,420	2,008	2,840	1,174	1,761	2,348	
2-1/2	2,361	3,339		1,685	2,527	3,370	4,212
3	4,385	6,202		2,626	3,939	5,252	6,565
4	9,422			4,571	6,856	9,141	11,426
5	17,622			7,233	10,849	14,466	18,082
6				10,363	15,544	20,726	25,907
8				18,154	27,230	36,307	45,384
10				29,682	44,523	59,364	74,205
12				43,108	64,662	86,216	107,770
14				52,768	79,152	105,536	131,920
16				70,253	105,380	140,506	175,633
18				90,236	135,354	180,472	225,590
20			with	112,717	169,076	225,434	281,793
22			sizes	137,696	206,544	275,392	344,240
24		governs		165,173	247,759	330,346	412,932
26	Velocity	pipe		195,147	292,721	390,295	487,869
28	these			227,620	341,430	455,240	569,050
30				262,590	393,886	525,181	656,476
32				300,059	450,088	600,117	750,147
34				340,025	510,037	680,050	850,062
36				382,489	573,733	764,978	956,222
42				524,869	787,303	1,049,737	1,312,171
48				689,729	1,034,594	1,379,458	1,724,323
54				877,071	1,315,606	1,754,141	2,192,677
60				1,086,893	1,630,340	2,173,786	2,717,233
72				1,573,981	2,360,972	3,147,962	3,934,953
84				2,150,993	3,226,490	4,301,986	5,377,483
96				2,817,929	4,226,894	5,635,858	7,044,823

(*Continued*)

40 PSIG STEAM CONDENSATE PIPING SYSTEMS—STEEL PIPE (*Continued*)

Pipe Size	Pressure Drop psig/100 ft.			Velocity FPM (mph)			
	0.125	0.25	0.5	2,000 (23)	3,000 (34)	4,000 (45)	5,000 (57)
Steam Condensate Flow lbs./h 5 psig Back Pressure							
1/2	52	74	104		Pressure	drop	
3/4	128	180	255		with these	pipe	
1	267	378	534	465			governs
1-1/4	615	870	1,230	830			sizes
1-1/2	972	1,375	1,945	1,143	1,715		
2	2,015	2,850	4,031	1,910	2,865	3,820	
2-1/2	3,351	4,739	6,702	2,742	4,112	5,483	
3	6,224	8,802		4,273	6,409	8,545	10,682
4	13,371			7,437	11,156	14,874	18,593
5	25,008			11,769	17,654	23,538	29,423
6	40,658			16,862	25,293	33,724	42,155
8				29,539	44,308	59,077	73,846
10				48,297	72,445	96,594	120,742
12				70,143	105,215	140,286	175,358
14				85,861	128,791	171,722	214,652
16				114,312	171,468	228,624	285,780
18				146,827	220,241	293,655	367,068
20				183,407	275,111	366,814	458,518
22				224,051	336,077	448,103	560,129
24				268,760	403,140	537,520	671,900
26		governs	with	317,533	476,300	635,067	793,833
28	Velocity	pipe	sizes	370,371	555,556	740,742	925,927
30	these			427,273	640,909	854,546	1,068,182
32				488,239	732,359	976,478	1,220,598
34				553,270	829,905	1,106,540	1,383,175
36				622,365	933,548	1,244,730	1,555,913
42				854,037	1,281,056	1,708,075	2,135,094
48				1,122,290	1,683,434	2,244,579	2,805,724
54				1,427,121	2,140,682	2,854,243	3,567,804
60				1,768,533	2,652,800	3,537,066	4,421,333
72				2,561,096	3,841,644	5,122,193	6,402,741
84				3,499,979	5,249,968	6,999,957	8,749,946
96				4,585,180	6,877,770	9,170,361	11,462,951
Steam Condensate Flow lbs./h 7 psig Back Pressure							
1/2	59	84	119		Pressure	drop	
3/4	146	206	291		with these	pipe	
1	305	432	610	556			governs
1-1/4	702	993	1,405	992			sizes
1-1/2	1,111	1,571	2,221	1,366	2,049		
2	2,302	3,256	4,604	2,283	3,424	4,566	
2-1/2	3,828	5,414	7,656	3,277	4,915	6,553	
3	7,110	10,055		5,106	7,660	10,213	12,766
4	15,275	21,602		8,888	13,332	17,777	22,221
5	28,568			14,066	21,098	28,131	35,164
6	46,446			20,152	30,228	40,304	50,381
8				35,303	52,954	70,605	88,257
10				57,721	86,582	115,443	144,303
12				83,831	125,746	167,661	209,577
14		governs	with	102,616	153,924	205,232	256,539
16	Velocity	pipe	sizes	136,619	204,928	273,237	341,547
18	these			175,479	263,219	350,958	438,698
20				219,197	328,795	438,394	547,992
22				267,772	401,659	535,545	669,431
24				321,206	481,808	642,411	803,014
26				379,496	569,244	758,992	948,740

(*Continued*)

40 PSIG STEAM CONDENSATE PIPING SYSTEMS—STEEL PIPE (*Continued*)

Pipe Size	Pressure Drop psig/100 ft.			Velocity FPM (mph)			
	0.125	0.25	0.5	2,000 (23)	3,000 (34)	4,000 (45)	5,000 (57)
28				442,644	663,966	885,289	1,106,611
30				510,650	765,975	1,021,300	1,276,625
32				583,513	875,270	1,167,026	1,458,783
34				661,234	991,851	1,322,468	1,653,085
36				743,812	1,115,719	1,487,625	1,859,531
42				1,020,693	1,531,039	2,041,386	2,551,732
48				1,341,291	2,011,937	2,682,582	3,353,228
54				1,705,607	2,558,411	3,411,215	4,264,018
60				2,113,642	3,170,462	4,227,283	5,284,104
72				3,060,864	4,591,296	6,121,728	7,652,160
84				4,182,958	6,274,437	8,365,916	10,457,395
96				5,479,924	8,219,886	10,959,848	13,699,810
Steam Condensate Flow lbs./h 10 psig Back Pressure							
1/2	72	102	144	Pressure			
3/4	177	250	353	with these		drop	governs
1	370	523	740	717		pipe	sizes
1-1/4	852	1,205	1,704	1,278			
1-1/2	1,347	1,905	2,694	1,761	2,641		
2	2,792	3,949	5,585	2,942	4,414		
2-1/2	4,643	6,567	9,287	4,223	6,335	8,446	
3	8,624	12,196		6,582	9,872	13,163	16,454
4	18,527	26,202		11,456	17,184	22,912	28,640
5	34,652			18,129	27,193	36,258	45,322
6	56,336			25,974	38,961	51,948	64,934
8				45,501	68,251	91,002	113,752
10				74,396	111,594	148,792	185,989
12				108,048	162,071	216,095	270,119
14				132,259	198,389	264,518	330,648
16				176,085	264,127	352,169	440,212
18				226,171	339,256	452,342	565,427
20				282,518	423,777	565,036	706,295
22				345,126	517,689	690,252	862,815
24				413,995	620,992	827,989	1,034,987
26		governs	with	489,124	733,686	978,248	1,222,810
28		pipe	sizes	570,514	855,771	1,141,029	1,426,286
30	Velocity			658,165	987,248	1,316,331	1,645,413
32	these			752,077	1,128,116	1,504,154	1,880,193
34				852,250	1,278,375	1,704,500	2,130,624
36				958,683	1,438,025	1,917,366	2,396,708
42				1,315,548	1,973,322	2,631,096	3,288,870
48				1,728,760	2,593,140	3,457,520	4,321,900
54				2,198,319	3,297,479	4,396,638	5,495,798
60				2,724,225	4,086,338	5,448,451	6,810,564
72				3,945,079	5,917,619	7,890,158	9,862,698
84				5,391,321	8,086,982	10,782,642	13,478,303
96				7,062,952	10,594,427	14,125,903	17,657,379
Steam Condensate Flow lbs./h 12 psig Back Pressure							
1/2	82	116	164	Pressure			
3/4	201	284	402	with these		drop	governs
1	421	595	842			pipe	sizes
1-1/4	969	1,371	1,938	1,508			
1-1/2	1,533	2,167	3,065	2,077			
2	3,177	4,492	6,353	3,471	5,207		
2-1/2	5,282	7,470	10,564	4,982	7,474	9,965	
3	9,810	13,873		7,765	11,647	15,530	19,412
4	21,076	29,806		13,515	20,273	27,031	33,788
5	39,419			21,388	32,082	42,776	53,470
6	64,086			30,643	45,965	61,286	76,608
8				53,680	80,521	107,361	134,201

(*Continued*)

40 PSIG STEAM CONDENSATE PIPING SYSTEMS—STEEL PIPE (Continued)

Pipe Size	Pressure Drop psig/100 ft.			Velocity FPM (mph)			
	0.125	0.25	0.5	2,000 (23)	3,000 (34)	4,000 (45)	5,000 (57)
10				87,770	131,655	175,540	219,424
12				127,471	191,207	254,942	318,678
14				156,035	234,053	312,070	390,088
16				207,739	311,609	415,478	519,348
18				266,829	400,244	533,659	667,074
20				333,306	499,959	666,612	833,265
22				407,169	610,753	814,338	1,017,922
24				488,418	732,627	976,836	1,221,045
26				577,053	865,580	1,154,106	1,442,633
28	Velocity	governs	with	673,075	1,009,612	1,346,150	1,682,687
30	these	pipe	sizes	776,483	1,164,724	1,552,966	1,941,207
32				887,277	1,330,916	1,774,554	2,218,193
34				1,005,458	1,508,186	2,010,915	2,513,644
36				1,131,024	1,696,537	2,262,049	2,827,561
42				1,552,042	2,328,064	3,104,085	3,880,106
48				2,039,537	3,059,305	4,079,074	5,098,842
54				2,593,508	3,890,262	5,187,016	6,483,770
60				3,213,956	4,820,934	6,427,911	8,034,889
72				4,654,281	6,981,421	9,308,561	11,635,701
84				6,360,512	9,540,767	12,721,023	15,901,279
96				8,332,649	12,498,973	16,665,297	20,831,622

Notes:
1 Maximum recommended pressure drop/velocity: 0.25 psig/100 ft./5,000 FPM.
2 Table based on heavy weight steel pipe using steam equations in Part 3.

50 PSIG STEAM CONDENSATE PIPING SYSTEMS—STEEL PIPE

Pipe Size	Pressure Drop psig/100 ft.			Velocity FPM (mph)			
	0.125	0.25	0.5	2,000 (23)	3,000 (34)	4,000 (45)	5,000 (57)
	Steam Condensate Flow lbs./h 0 psig Back Pressure						
1/2	32	45	64	150	Pressure	drop	governs
3/4	78	111	156		with these	pipe	sizes
1	164	232	328	249			
1-1/4	377	533	754	444	666		
1-1/2	596	843	1,193	611	917		
2	1,236	1,748	2,472	1,022	1,533	2,044	
2-1/2	2,055	2,907		1,467	2,200	2,933	3,666
3	3,817	5,398		2,286	3,428	4,571	5,714
4	8,201			3,978	5,967	7,957	9,946
5	15,338			6,296	9,443	12,591	15,739
6				9,020	13,530	18,040	22,550
8				15,801	23,702	31,602	39,503
10				25,836	38,753	51,671	64,589
12				37,522	56,283	75,044	93,805
14				45,930	68,895	91,860	114,825
16			with	61,149	91,724	122,298	152,873
18			sizes	78,543	117,814	157,085	196,357
20		governs		98,110	147,166	196,221	245,276
22	Velocity	pipe		119,852	179,779	239,705	299,631
24	these			143,769	215,653	287,537	359,421
26				169,859	254,788	339,718	424,647
28				198,123	297,185	396,247	495,308
30				228,562	342,843	457,124	571,405
32				261,175	391,762	522,350	652,937
34				295,962	443,943	591,924	739,905
36				332,923	499,385	665,847	832,308
42				456,852	685,279	913,705	1,142,131
48				600,349	900,524	1,200,698	1,500,873
54				763,414	1,145,120	1,526,827	1,908,534
60				946,046	1,419,069	1,892,092	2,365,115
72				1,370,013	2,055,020	2,740,027	3,425,034
84				1,872,252	2,808,378	3,744,504	4,680,630
96				2,452,762	3,679,143	4,905,523	6,131,904

(Continued)

50 PSIG STEAM CONDENSATE PIPING SYSTEMS—STEEL PIPE (*Continued*)

Pipe	Pressure Drop psig/100 ft.			Velocity FPM (mph)			
Size	0.125	0.25	0.5	2,000 (23)	3,000 (34)	4,000 (45)	5,000 (57)
	Steam Condensate Flow lbs./h 5 psig Back Pressure						
1/2	44	62	88		Pressure	drop	
3/4	108	152	215		with these	pipe	
1	225	318	450	392			governs
1-1/4	518	733	1,037	700			sizes
1-1/2	820	1,159	1,639	964	1,446		
2	1,699	2,403	3,398	1,610	2,416	3,221	
2-1/2	2,825	3,995	5,650	2,311	3,467	4,623	
3	5,247	7,420		3,602	5,403	7,204	9,005
4	11,273			6,270	9,405	12,540	15,675
5	21,083			9,922	14,883	19,844	24,805
6	34,277			14,215	21,323	28,431	35,539
8				24,903	37,354	49,805	62,257
10				40,717	61,075	81,434	101,792
12				59,135	88,702	118,269	147,836
14				72,386	108,578	144,771	180,964
16				96,371	144,557	192,743	240,928
18				123,784	185,675	247,567	309,459
20				154,622	231,934	309,245	386,556
22				188,888	283,332	377,776	472,220
24				226,580	339,870	453,159	566,449
26		governs	with	267,698	401,547	535,396	669,245
28		pipe	sizes	312,243	468,365	624,486	780,608
30	Velocity			360,215	540,322	720,429	900,537
32	these			411,613	617,419	823,225	1,029,032
34				466,437	699,656	932,875	1,166,093
36				524,688	787,033	1,049,377	1,311,721
42				720,001	1,080,002	1,440,002	1,800,003
48				946,152	1,419,229	1,892,305	2,365,381
54				1,203,143	1,804,714	2,406,285	3,007,857
60				1,490,972	2,236,458	2,981,944	3,727,429
72				2,159,146	3,238,720	4,318,293	5,397,866
84				2,950,676	4,426,015	5,901,353	7,376,691
96				3,865,562	5,798,343	7,731,123	9,663,904
	Steam Condensate Flow lbs./h 7 psig Back Pressure						
1/2	49	70	99		Pressure	drop	
3/4	121	171	242		with these	pipe	
1	253	358	507	462			governs
1-1/4	583	825	1,167	824			sizes
1-1/2	923	1,305	1,845	1,135	1,702		
2	1,912	2,704	3,824	1,896	2,844	3,792	
2-1/2	3,180	4,497	6,359	2,722	4,082	5,443	
3	5,905	8,351		4,241	6,362	8,483	10,603
4	12,687	17,942		7,382	11,074	14,765	18,456
5	23,728			11,683	17,524	23,365	29,207
6	38,577			16,738	25,107	33,476	41,845
8				29,322	43,983	58,644	73,305
10		governs	with	47,942	71,914	95,885	119,856
12		pipe	sizes	69,628	104,443	139,257	174,071
14	Velocity			85,231	127,847	170,462	213,078
16	these			113,473	170,210	226,947	283,683
18				145,750	218,625	291,500	364,375
20				182,062	273,092	364,123	455,154

(*Continued*)

50 PSIG STEAM CONDENSATE PIPING SYSTEMS—STEEL PIPE (*Continued*)

Pipe Size	Pressure Drop psig/100 ft.			Velocity FPM (mph)			
	0.125	0.25	0.5	2,000 (23)	3,000 (34)	4,000 (45)	5,000 (57)
22				222,408	333,612	444,815	556,019
24				266,788	400,183	533,577	666,971
26				315,204	472,806	630,407	788,009
28				367,654	551,480	735,307	919,134
30				424,138	636,207	848,276	1,060,345
32				484,657	726,986	969,314	1,211,643
34				549,211	823,816	1,098,422	1,373,027
36				617,799	926,699	1,235,598	1,544,498
42				847,772	1,271,658	1,695,544	2,119,430
48				1,114,056	1,671,084	2,228,112	2,785,140
54				1,416,651	2,124,977	2,833,303	3,541,629
60				1,755,558	2,633,338	3,511,117	4,388,896
72				2,542,307	3,813,460	5,084,614	6,355,767
84				3,474,301	5,211,452	6,948,602	8,685,753
96				4,551,541	6,827,312	9,103,082	11,378,853
Steam Condensate Flow lbs./h 10 psig Back Pressure							
1/2	58	83	117		Pressure with these	drop pipe	governs sizes
3/4	143	203	286				
1	300	424	600	581			
1-1/4	690	976	1,381	1,036			
1-1/2	1,092	1,544	2,184	1,427	2,140		
2	2,263	3,200	4,526	2,384	3,577		
2-1/2	3,763	5,322	7,526	3,422	5,134	6,845	
3	6,989	9,883		5,334	8,001	10,668	13,334
4	15,015	21,234		9,284	13,926	18,568	23,210
5	28,082			14,692	22,038	29,383	36,729
6	45,655			21,049	31,574	42,098	52,623
8				36,874	55,311	73,748	92,185
10				60,291	90,436	120,581	150,726
12				87,562	131,343	175,124	218,905
14				107,183	160,775	214,366	267,958
16				142,699	214,049	285,399	356,749
18				183,290	274,934	366,579	458,224
20				228,953	343,430	457,907	572,383
22				279,691	419,536	559,382	699,227
24				335,502	503,254	671,005	838,756
26		governs pipe	with sizes	396,387	594,581	792,775	990,969
28	Velocity these			462,346	693,520	924,693	1,155,866
30				533,379	800,068	1,066,758	1,333,447
32				609,485	914,228	1,218,971	1,523,713
34				690,666	1,035,998	1,381,331	1,726,664
36				776,919	1,165,379	1,553,839	1,942,299
42				1,066,124	1,599,186	2,132,247	2,665,309
48				1,400,992	2,101,488	2,801,984	3,502,480
54				1,781,524	2,672,286	3,563,048	4,453,810
60				2,207,720	3,311,579	4,415,439	5,519,299
72				3,197,103	4,795,654	6,394,205	7,992,757
84				4,369,141	6,553,712	8,738,282	10,922,853
96				5,723,835	8,585,752	11,447,670	14,309,587
Steam Condensate Flow lbs./h 12 psig Back Pressure							
1/2	65	92	130		Pressure with these	drop pipe	governs sizes
3/4	160	226	320				
1	335	474	670				
1-1/4	771	1,090	1,542	1,199			
1-1/2	1,219	1,724	2,438	1,652			
2	2,526	3,573	5,052	2,761	4,141		
2-1/2	4,201	5,941	8,402	3,962	5,944	7,925	
3	7,802	11,033		6,175	9,263	12,351	15,438
4	16,762	23,704		10,749	16,123	21,497	26,871

(*Continued*)

50 PSIG STEAM CONDENSATE PIPING SYSTEMS—STEEL PIPE (*Continued*)

Pipe Size	Pressure Drop psig/100 ft.			Velocity FPM (mph)			
	0.125	0.25	0.5	2,000 (23)	3,000 (34)	4,000 (45)	5,000 (57)
5	31,349			17,010	25,514	34,019	42,524
6	50,967			24,370	36,555	48,740	60,925
8				42,691	64,037	85,383	106,729
10				69,802	104,704	139,605	174,506
12				101,376	152,065	202,753	253,441
14				124,093	186,140	248,186	310,233
16				165,213	247,819	330,426	413,032
18				212,207	318,310	424,413	530,517
20				265,075	397,612	530,150	662,687
22				323,817	485,726	647,634	809,543
24				388,434	582,651	776,868	971,084
26		governs pipe	with sizes	458,925	688,387	917,849	1,147,311
28	Velocity these			535,290	802,934	1,070,579	1,338,224
30				617,529	926,293	1,235,058	1,543,822
32				705,642	1,058,464	1,411,285	1,764,106
34				799,630	1,199,445	1,599,260	1,999,075
36				899,492	1,349,238	1,798,984	2,248,730
42				1,234,324	1,851,485	2,468,647	3,085,809
48				1,622,023	2,433,035	3,244,046	4,055,058
54				2,062,591	3,093,886	4,125,181	5,156,477
60				2,556,026	3,834,040	5,112,053	6,390,066
72				3,701,502	5,552,253	7,403,004	9,253,755
84				5,058,450	7,587,675	10,116,901	12,646,126
96				6,626,871	9,940,306	13,253,742	16,567,177

Steam Condensate Flow lbs./h 15 psig Back Pressure

Pipe Size	0.125	0.25	0.5	2,000	3,000	4,000	5,000
1/2	77	108	153		Pressure with these	drop pipe	governs sizes
3/4	188	266	376				
1	394	557	788				
1-1/4	907	1,283	1,814	1,483			
1-1/2	1,434	2,028	2,869	2,044			
2	2,973	4,204	5,946	3,415	5,122		
2-1/2	4,943	6,991	9,887	4,901	7,352	9,802	
3	9,181	12,984	18,362	7,638	11,458	15,277	
4	19,724	27,894		13,295	19,943	26,590	33,238
5	36,890	52,171		21,040	31,559	42,079	52,599
6	59,976			30,144	45,216	60,288	75,360
8	126,990			52,806	79,209	105,613	132,016
10				86,341	129,511	172,681	215,851
12				125,395	188,093	250,791	313,488
14				153,494	230,242	306,989	383,736
16				204,356	306,535	408,713	510,891
18				262,484	393,727	524,969	656,211
20				327,878	491,818	655,757	819,696
22				400,539	600,808	801,077	1,001,346
24				480,465	720,697	960,929	1,201,161
26			with sizes	567,657	851,485	1,135,313	1,419,142
28	Velocity these	governs pipe		662,115	993,172	1,324,229	1,655,287
30				763,839	1,145,758	1,527,678	1,909,597
32				872,829	1,309,243	1,745,658	2,182,072
34				989,085	1,483,627	1,978,170	2,472,712
36				1,112,607	1,668,910	2,225,214	2,781,517
42				1,526,769	2,290,154	3,053,539	3,816,923
48				2,006,326	3,009,488	4,012,651	5,015,814
54				2,551,276	3,826,914	5,102,552	6,378,190
60				3,161,620	4,742,431	6,323,241	7,904,051
72				4,578,491	6,867,737	9,156,983	11,446,228
84				6,256,938	9,385,408	12,513,877	15,642,346
96				8,196,962	12,295,443	16,393,924	20,492,404

Notes:
1 Maximum recommended pressure drop/velocity: 0.25 psig/100 ft./5,000 FPM.
2 Table based on heavy weight steel pipe using steam equations in Part 3.

60 PSIG STEAM CONDENSATE PIPING SYSTEMS—STEEL PIPE

Pipe Size	Pressure Drop psig/100 ft.			Velocity FPM (mph)			
	0.125	0.25	0.5	2,000 (23)	3,000 (34)	4,000 (45)	5,000 (57)
Steam Condensate Flow lbs./h 0 psig Back Pressure							
1/2	29	41	57		Pressure	drop	
3/4	70	99	141	134	with these	pipe	
1	147	208	294	224			governs
1-1/4	339	479	677	399	598		sizes
1-1/2	536	757	1,071	549	824		
2	1,110	1,570	2,220	918	1,376	1,835	
2-1/2	1,846	2,610		1,317	1,976	2,634	3,293
3	3,428	4,848		2,053	3,079	4,105	5,131
4	7,365			3,573	5,359	7,145	8,932
5	13,774			5,654	8,481	11,308	14,134
6				8,100	12,150	16,201	20,251
8				14,190	21,285	28,380	35,475
10				23,201	34,802	46,403	58,004
12				33,696	50,544	67,392	84,241
14				41,247	61,870	82,494	103,117
16				54,915	82,372	109,829	137,287
18				70,535	105,802	141,070	176,337
20				88,107	132,161	176,215	220,269
22			with	107,633	161,449	215,265	269,082
24			sizes	129,110	193,666	258,221	322,776
26		governs		152,541	228,811	305,081	381,351
28		pipe		177,923	266,885	355,847	444,808
30	Velocity			205,259	307,888	410,517	513,147
32	these			234,546	351,820	469,093	586,366
34				265,787	398,680	531,573	664,467
36				298,980	448,469	597,959	747,449
42				410,273	615,410	820,546	1,025,683
48				539,139	808,709	1,078,279	1,347,849
54				685,578	1,028,368	1,371,157	1,713,946
60				849,590	1,274,385	1,699,180	2,123,975
72				1,230,331	1,845,497	2,460,663	3,075,828
84				1,681,363	2,522,045	3,362,727	4,203,408
96				2,202,686	3,304,029	4,405,372	5,506,715
Steam Condensate Flow lbs./h 5 psig Back Pressure							
1/2	39	55	77		Pressure	drop	
3/4	95	134	189		with these	pipe	
1	198	280	396	345			governs
1-1/4	456	644	911	615			sizes
1-1/2	721	1,019	1,441	847	1,271		
2	1,494	2,112	2,987	1,416	2,124	2,831	
2-1/2	2,484	3,512	4,967	2,032	3,048	4,064	
3	4,613	6,523		3,167	4,750	6,333	7,917
4	9,910			5,512	8,268	11,024	13,780
5	18,535			8,723	13,084	17,445	21,806
6	30,133			12,497	18,746	24,994	31,243
8				21,892	32,839	43,785	54,731
10				35,795	53,693	71,590	89,488
12				51,986	77,980	103,973	129,966
14				63,636	95,454	127,271	159,089
16				84,722	127,083	169,444	211,805
18		governs	with	108,821	163,231	217,642	272,052
20	Velocity	pipe	sizes	135,932	203,898	271,864	339,830
22	these			166,055	249,083	332,110	415,138
24				199,191	298,786	398,382	497,977
26				235,339	353,009	470,678	588,348
28				274,499	411,749	548,999	686,249
30				316,672	475,008	633,344	791,681
32				361,857	542,786	723,715	904,643

(Continued)

60 PSIG STEAM CONDENSATE PIPING SYSTEMS—STEEL PIPE (*Continued*)

Pipe Size	Pressure Drop psig/100 ft.			Velocity FPM (mph)			
	0.125	0.25	0.5	2,000 (23)	3,000 (34)	4,000 (45)	5,000 (57)
34				410,055	615,082	820,110	1,025,137
36				461,265	691,897	922,529	1,153,161
42				632,968	949,452	1,265,936	1,582,420
48				831,782	1,247,674	1,663,565	2,079,456
54				1,057,708	1,586,562	2,115,416	2,644,270
60				1,310,744	1,966,117	2,621,489	3,276,861
72				1,898,151	2,847,226	3,796,301	4,745,377
84				2,594,001	3,891,002	5,188,002	6,485,003
96				3,398,296	5,097,444	6,796,592	8,495,740
Steam Condensate Flow lbs./h 7 psig Back Pressure							
1/2	43	61	86		Pressure	drop	
3/4	105	149	211		with these	pipe	
1	221	312	441	402			governs
1-1/4	508	718	1,016	717			sizes
1-1/2	803	1,136	1,607	988	1,482		
2	1,665	2,355	3,330	1,651	2,477	3,302	
2-1/2	2,769	3,915	5,537	2,370	3,555	4,740	
3	5,142	7,272		3,693	5,540	7,386	9,233
4	11,047	15,623		6,428	9,642	12,857	16,071
5	20,662			10,173	15,259	20,346	25,432
6	33,591			14,575	21,862	29,150	36,437
8				25,532	38,298	51,064	63,830
10				41,746	62,619	83,493	104,366
12				60,630	90,944	121,259	151,574
14				74,216	111,323	148,431	185,539
16				98,808	148,212	197,615	247,019
18				126,913	190,370	253,826	317,283
20				158,531	237,797	317,063	396,329
22				193,663	290,495	387,326	484,158
24				232,308	348,462	464,616	580,770
26		governs	with	274,466	411,699	548,932	686,165
28	Velocity	pipe	sizes	320,137	480,205	640,274	800,342
30	these			369,321	553,982	738,643	923,303
32				422,019	633,028	844,037	1,055,047
34				478,229	717,344	956,459	1,195,573
36				537,953	806,930	1,075,906	1,344,883
42				738,203	1,107,305	1,476,407	1,845,508
48				970,072	1,455,108	1,940,144	2,425,180
54				1,233,559	1,850,339	2,467,119	3,083,898
60				1,528,665	2,292,998	3,057,330	3,821,663
72				2,213,732	3,320,598	4,427,464	5,534,330
84				3,025,273	4,537,909	6,050,545	7,563,181
96				3,963,287	5,944,931	7,926,574	9,908,218
Steam Condensate Flow lbs./h 10 psig Back Pressure							
1/2	50	71	100		Pressure		
3/4	123	174	246		with these		
1	257	364	514	498		drop	governs
1-1/4	592	837	1,184	888		pipe	sizes
1-1/2	936	1,324	1,873	1,224	1,836		
2	1,941	2,744	3,881	2,045	3,067		
2-1/2	3,227	4,564	6,454	2,935	4,402	5,870	
3	5,993	8,476		4,574	6,861	9,148	11,435
4	12,876	18,209		7,961	11,942	15,923	19,904
5	24,082			12,599	18,898	25,198	31,497
6	39,152			18,051	27,076	36,102	45,127
8				31,621	47,432	63,243	79,053

(*Continued*)

60 PSIG STEAM CONDENSATE PIPING SYSTEMS—STEEL PIPE (Continued)

Pipe Size	Pressure Drop psig/100 ft.			Velocity FPM (mph)			
	0.125	0.25	0.5	2,000 (23)	3,000 (34)	4,000 (45)	5,000 (57)
10				51,702	77,554	103,405	129,256
12				75,089	112,634	150,178	187,723
14				91,915	137,873	183,831	229,788
16				122,373	183,559	244,745	305,931
18				157,181	235,771	314,361	392,952
20				196,340	294,510	392,680	490,850
22				239,850	359,775	479,700	599,625
24				287,711	431,567	575,423	719,278
26				339,924	509,885	679,847	849,809
28	Velocity	governs	with	396,487	594,730	792,974	991,217
30	these	pipe	sizes	457,401	686,102	914,802	1,143,503
32				522,667	784,000	1,045,333	1,306,666
34				592,283	888,424	1,184,566	1,480,707
36				666,250	999,375	1,332,501	1,665,626
42				914,259	1,371,388	1,828,517	2,285,647
48				1,201,426	1,802,139	2,402,852	3,003,565
54				1,527,753	2,291,629	3,055,506	3,819,382
60				1,893,239	2,839,858	3,786,477	4,733,097
72				2,741,688	4,112,532	5,483,376	6,854,220
84				3,746,774	5,620,161	7,493,549	9,366,936
96				4,908,497	7,362,746	9,816,995	12,271,243
Steam Condensate Flow lbs./h 12 psig Back Pressure							
1/2	55	78	111		Pressure		
3/4	136	192	271		with these		
1	284	402	568			drop	governs
1-1/4	654	924	1,307	1,017		pipe	sizes
1-1/2	1,033	1,462	2,067	1,401			
2	2,142	3,029	4,284	2,341	3,511		
2-1/2	3,562	5,037	7,124	3,360	5,040	6,720	
3	6,615	9,356		5,236	7,854	10,472	13,090
4	14,213	20,100		9,114	13,671	18,228	22,785
5	26,582			14,423	21,634	28,846	36,057
6	43,216			20,664	30,996	41,328	51,660
8				36,199	54,299	72,398	90,498
10				59,187	88,781	118,375	147,968
12				85,960	128,940	171,919	214,899
14				105,222	157,833	210,444	263,055
16				140,088	210,132	280,177	350,221
18				179,936	269,903	359,871	449,839
20				224,764	337,146	449,528	561,910
22				274,573	411,860	549,146	686,433
24				329,363	494,045	658,726	823,408
26		governs	with	389,134	583,701	778,268	972,836
28	Velocity	pipe	sizes	453,886	680,829	907,772	1,134,715
30	these			523,619	785,428	1,047,238	1,309,047
32				598,333	897,499	1,196,666	1,495,832
34				678,027	1,017,041	1,356,055	1,695,069
36				762,703	1,144,055	1,525,406	1,906,758
42				1,046,615	1,569,923	2,093,231	2,616,538
48				1,375,356	2,063,034	2,750,712	3,438,390
54				1,748,925	2,623,387	3,497,850	4,372,312
60				2,167,322	3,250,983	4,334,644	5,418,305
72				3,138,601	4,707,901	6,277,202	7,846,502
84				4,289,193	6,433,789	8,578,386	10,722,982
96				5,619,098	8,428,647	11,238,196	14,047,745

(Continued)

60 PSIG STEAM CONDENSATE PIPING SYSTEMS—STEEL PIPE (*Continued*)

Pipe	Pressure Drop psig/100 ft.			Velocity FPM (mph)			
Size	0.125	0.25	0.5	2,000 (23)	3,000 (34)	4,000 (45)	5,000 (57)
Steam Condensate Flow lbs./h 15 psig Back Pressure							
1/2	64	90	128		Pressure		
3/4	157	221	313		with these		
1	328	464	656			drop	governs
1-1/4	755	1,067	1,510	1,235		pipe	sizes
1-1/2	1,194	1,688	2,387	1,701			
2	2,474	3,499	4,948	2,842	4,263		
2-1/2	4,114	5,818	8,228	4,079	6,118	8,158	
3	7,640	10,805	15,281	6,357	9,535	12,713	
4	16,415	23,214		11,064	16,596	22,129	27,661
5	30,700	43,417		17,509	26,264	35,018	43,773
6	49,912			25,086	37,629	50,172	62,715
8	105,681			43,946	65,918	87,891	109,864
10				71,853	107,779	143,706	179,632
12				104,354	156,532	208,709	260,886
14				127,739	191,608	255,477	319,346
16				170,066	255,099	340,132	425,165
18				218,440	327,661	436,881	546,101
20				272,862	409,292	545,723	682,154
22				333,330	499,994	666,659	833,324
24				399,844	599,766	799,688	999,611
26			with	472,406	708,609	944,812	1,181,014
28			sizes	551,014	826,521	1,102,028	1,377,535
30	Velocity	governs		635,669	953,504	1,271,338	1,589,173
32	these	pipe		726,371	1,089,557	1,452,742	1,815,928
34				823,120	1,234,680	1,646,240	2,057,800
36				925,915	1,388,873	1,851,831	2,314,788
42				1,270,583	1,905,874	2,541,165	3,176,456
48				1,669,671	2,504,506	3,339,342	4,174,177
54				2,123,181	3,184,771	4,246,361	5,307,951
60				2,631,111	3,946,667	5,262,222	6,577,778
72				3,810,236	5,715,354	7,620,472	9,525,590
84				5,207,045	7,810,568	10,414,091	13,017,613
96				6,821,539	10,232,309	13,643,079	17,053,848
Steam Condensate Flow lbs./h 20 psig Back Pressure							
1/2	81	114	162		Pressure		
3/4	198	280	396		with these		
1	415	587	830			drop	governs
1-1/4	956	1,352	1,911	1,681		pipe	sizes
1-1/2	1,511	2,137	3,022	2,316			
2	3,132	4,430	6,265	3,870	5,805		
2-1/2	5,209	7,366	10,417	5,555	8,332		
3	9,673	13,680	19,347	8,657	12,985	17,314	
4	20,782	29,391		15,068	22,602	30,136	37,669
5	38,870	54,970		23,845	35,767	47,689	59,612
6	63,194			34,163	51,244	68,326	85,407
8	133,803			59,847	89,770	119,693	149,616
10				97,852	146,778	195,704	244,630
12				142,114	213,170	284,227	355,284
14			with	173,959	260,938	347,918	434,897
16			sizes	231,602	347,403	463,204	579,005
18	Velocity	governs		297,480	446,220	594,960	743,700
20	these	pipe		371,592	557,389	743,185	928,981
22				453,940	680,910	907,880	1,134,849
24				544,522	816,783	1,089,044	1,361,305
26				643,339	965,008	1,286,677	1,608,347

(*Continued*)

60 PSIG STEAM CONDENSATE PIPING SYSTEMS—STEEL PIPE (*Continued*)

Pipe Size	Pressure Drop psig/100 ft.			Velocity FPM (mph)			
	0.125	0.25	0.5	2,000 (23)	3,000 (34)	4,000 (45)	5,000 (57)
28				750,390	1,125,585	1,500,781	1,875,976
30				865,677	1,298,515	1,731,353	2,164,191
32				989,198	1,483,796	1,978,395	2,472,994
34				1,120,953	1,681,430	2,241,907	2,802,383
36				1,260,944	1,891,416	2,521,888	3,152,360
42				1,730,324	2,595,486	3,460,648	4,325,809
48				2,273,816	3,410,724	4,547,633	5,684,541
54				2,891,421	4,337,132	5,782,843	7,228,554
60				3,583,139	5,374,709	7,166,278	8,957,848
72				5,188,913	7,783,369	10,377,825	12,972,282
84				7,091,137	10,636,705	14,182,273	17,727,841
96				9,289,811	13,934,716	18,579,622	23,224,527

Notes:
1 Maximum recommended pressure drop/velocity: 0.25 psig/100 ft./5,000 FPM.
2 Table based on heavy weight steel pipe using steam equations in Part 3.

75 PSIG STEAM CONDENSATE PIPING SYSTEMS—STEEL PIPE

Pipe Size	Pressure Drop psig/100 ft.			Velocity FPM (mph)			
	0.125	0.25	0.5	2,000 (23)	3,000 (34)	4,000 (45)	5,000 (57)
	Steam Condensate Flow lbs./h 0 psig Back Pressure						
1/2	25	36	50				
3/4	62	87	124	118	Pressure	drop	
1	130	183	259	197	with these	pipe	
1-1/4	298	422	596	351	526		governs
1-1/2	471	667	943	483	725		sizes
2	977	1,382	1,954	808	1,212	1,616	
2-1/2	1,625	2,298		1,159	1,739	2,319	2,899
3	3,018	4,268		1,807	2,710	3,614	4,517
4	6,483			3,145	4,718	6,290	7,863
5	12,126			4,977	7,466	9,954	12,443
6				7,131	10,696	14,262	17,827
8				12,492	18,738	24,984	31,230
10				20,425	30,637	40,850	51,062
12				29,664	44,496	59,327	74,159
14				36,311	54,466	72,622	90,777
16				48,343	72,514	96,686	120,857
18				62,094	93,140	124,187	155,234
20				77,563	116,345	155,127	193,908
22			with	94,752	142,128	189,504	236,880
24			sizes	113,659	170,489	227,319	284,148
26		governs		134,286	201,428	268,571	335,714
28		pipe		156,631	234,946	313,261	391,577
30	Velocity			180,695	271,042	361,389	451,736
32	these			206,477	309,716	412,955	516,193
34				233,979	350,969	467,958	584,948
36				263,200	394,799	526,399	657,999
42				361,174	541,761	722,349	902,936
48				474,619	711,928	949,237	1,186,547
54				603,533	905,299	1,207,066	1,508,832
60				747,917	1,121,875	1,495,833	1,869,791
72				1,083,093	1,624,640	2,166,186	2,707,733
84				1,480,149	2,220,223	2,960,297	3,700,371
96				1,939,083	2,908,624	3,878,165	4,847,707
	Steam Condensate Flow lbs./h 5 psig Back Pressure						
1/2	33	47	66		Pressure	drop	
3/4	81	115	163		with these	pipe	
1	170	241	341	297			governs
1-1/4	392	555	785	530			sizes
1-1/2	621	878	1,241	730	1,094		
2	1,286	1,819	2,573	1,219	1,829	2,438	

(Continued)

75 PSIG STEAM CONDENSATE PIPING SYSTEMS—STEEL PIPE (*Continued*)

	Pressure Drop psig/100 ft.			Velocity FPM (mph)			
Pipe Size	0.125	0.25	0.5	2,000 (23)	3,000 (34)	4,000 (45)	5,000 (57)
2-1/2	2,139	3,025	4,278	1,750	2,625	3,500	6,818
3	3,972	5,618		2,727	4,091	5,454	11,867
4	8,534			4,747	7,120	9,494	
5	15,962			7,512	11,268	15,024	18,780
6	25,951			10,762	16,144	21,525	26,906
8				18,854	28,281	37,707	47,134
10				30,827	46,240	61,653	77,067
12				44,771	67,156	89,541	111,927
14				54,803	82,204	109,606	137,007
16				72,963	109,444	145,925	182,406
18				93,716	140,574	187,433	234,291
20				117,064	175,596	234,129	292,661
22				143,007	214,510	286,013	357,516
24				171,543	257,315	343,086	428,858
26		governs	with	202,674	304,011	405,347	506,684
28		pipe	sizes	236,399	354,598	472,797	590,997
30	Velocity			272,718	409,077	545,435	681,794
32	these			311,631	467,447	623,262	779,078
34				353,139	529,708	706,277	882,847
36				397,240	595,861	794,481	993,101
42				545,111	817,667	1,090,222	1,362,778
48				716,330	1,074,495	1,432,660	1,790,825
54				910,897	1,366,345	1,821,793	2,277,242
60				1,128,811	1,693,217	2,257,623	2,822,029
72				1,634,685	2,452,027	3,269,370	4,086,712
84		°		2,233,951	3,350,926	4,467,901	5,584,876
96				2,926,608	4,389,912	5,853,216	7,316,521
Steam Condensate Flow lbs./h 7 psig Back Pressure							
1/2	37	52	73		Pressure	drop	
3/4	90	127	180		with these	pipe	
1	188	266	377	343			governs
1-1/4	433	613	867	612			sizes
1-1/2	685	969	1,370	843	1,264		
2	1,420	2,009	2,840	1,408	2,113	2,817	
2-1/2	2,362	3,340	4,723	2,021	3,032	4,043	
3	4,386	6,203		3,150	4,726	6,301	7,876
4	9,423	13,327		5,483	8,225	10,967	13,709
5	17,625			8,677	13,016	17,355	21,694
6	28,654			12,432	18,649	24,865	31,081
8				21,779	32,669	43,558	54,448
10				35,610	53,415	71,220	89,025
12				51,717	77,576	103,435	129,293
14				63,306	94,960	126,613	158,266
16				84,284	126,425	168,567	210,709
18		governs	with	108,258	162,386	216,515	270,644
20		pipe	sizes	135,228	202,843	270,457	338,071
22	Velocity			165,196	247,794	330,392	412,990
24	these			198,160	297,240	396,320	495,400
26				234,121	351,182	468,242	585,303
28				273,079	409,618	546,158	682,697
30				315,034	472,550	630,067	787,584
32				359,985	539,977	719,970	899,962
34				407,933	611,899	815,866	1,019,832
36				458,878	688,316	917,755	1,147,194
42				629,692	944,539	1,259,385	1,574,231
48				827,478	1,241,217	1,654,956	2,068,695
54				1,052,234	1,578,352	2,104,469	2,630,586
60				1,303,962	1,955,942	2,607,923	3,259,904
72				1,888,328	2,832,492	3,776,656	4,720,820
84				2,580,578	3,870,867	5,161,155	6,451,444
96				3,380,710	5,071,066	6,761,421	8,451,776

(*Continued*)

75 PSIG STEAM CONDENSATE PIPING SYSTEMS—STEEL PIPE (*Continued*)

Pipe Size	Pressure Drop psig/100 ft.			Velocity FPM (mph)			
	0.125	0.25	0.5	2,000 (23)	3,000 (34)	4,000 (45)	5,000 (57)
Steam Condensate Flow lbs./h 10 psig Back Pressure							
1/2	42	60	84		Pressure with these		
3/4	103	146	206				
1	216	306	432	419		drop	governs
1-1/4	498	704	995	747		pipe	sizes
1-1/2	787	1,113	1,574	1,028	1,543		
2	1,631	2,306	3,262	1,718	2,578		
2-1/2	2,712	3,835	5,424	2,466	3,700	4,933	
3	5,037	7,123		3,844	5,766	7,688	9,610
4	10,821	15,303		6,691	10,036	13,381	16,727
5	20,238			10,588	15,882	21,176	26,470
6	32,902			15,170	22,755	30,339	37,924
8				26,574	39,861	53,148	66,436
10				43,450	65,175	86,900	108,625
12				63,104	94,656	126,208	157,760
14				77,244	115,867	154,489	193,111
16				102,840	154,260	205,681	257,101
18				132,093	198,139	264,185	330,231
20				165,001	247,502	330,003	412,504
22				201,567	302,350	403,134	503,917
24				241,789	362,683	483,578	604,472
26		governs	with	285,667	428,501	571,335	714,168
28		pipe	sizes	333,202	499,804	666,405	833,006
30	Velocity			384,394	576,591	768,788	960,985
32	these			439,242	658,863	878,484	1,098,105
34				497,747	746,620	995,494	1,244,367
36				559,908	839,862	1,119,816	1,399,770
42				768,331	1,152,497	1,536,662	1,920,828
48				1,009,663	1,514,494	2,019,326	2,524,157
54				1,283,903	1,925,855	2,567,807	3,209,759
60				1,591,053	2,386,580	3,182,106	3,977,633
72				2,304,079	3,456,118	4,608,158	5,760,197
84				3,148,740	4,723,110	6,297,480	7,871,850
96				4,125,037	6,187,556	8,250,074	10,312,593
Steam Condensate Flow lbs./h 12 psig Back Pressure							
1/2	46	65	92		Pressure with these		
3/4	113	159	225				
1	236	334	472			drop	governs
1-1/4	543	768	1,087	845		pipe	sizes
1-1/2	859	1,215	1,719	1,165			
2	1,781	2,519	3,562	1,946	2,919		
2-1/2	2,961	4,188	5,923	2,793	4,190	5,587	
3	5,500	7,778		4,353	6,530	8,707	10,884
4	11,817	16,711		7,578	11,366	15,155	18,944
5	22,101			11,991	17,987	23,983	29,979
6	35,931			17,180	25,771	34,361	42,951
8				30,097	45,145	60,193	75,242
10				49,209	73,814	98,419	123,023
12				71,468	107,203	142,937	178,671
14				87,483	131,225	174,967	218,708

(*Continued*)

75 PSIG STEAM CONDENSATE PIPING SYSTEMS—STEEL PIPE (Continued)

Pipe Size	Pressure Drop psig/100 ft.			Velocity FPM (mph)			
	0.125	0.25	0.5	2,000 (23)	3,000 (34)	4,000 (45)	5,000 (57)
16				116,472	174,708	232,944	291,180
18				149,602	224,403	299,203	374,004
20				186,873	280,309	373,745	467,182
22				228,285	342,427	456,570	570,712
24				273,838	410,758	547,677	684,596
26				323,533	485,300	647,066	808,833
28				377,369	566,053	754,738	943,422
30				435,346	653,019	870,692	1,088,365
32	Velocity these	governs pipe	with sizes	497,464	746,197	994,929	1,243,661
34				563,724	845,586	1,127,448	1,409,310
36				634,125	951,187	1,268,250	1,585,312
42				870,174	1,305,262	1,740,349	2,175,436
48				1,143,495	1,715,243	2,286,990	2,858,738
54				1,454,087	2,181,130	2,908,174	3,635,217
60				1,801,950	2,702,924	3,603,899	4,504,874
72				2,609,488	3,914,232	5,218,976	6,523,720
84				3,566,111	5,349,166	7,132,221	8,915,276
96				4,671,817	7,007,726	9,343,634	11,679,543
Steam Condensate Flow lbs./h 15 psig Back Pressure							
1/2	52	74	104				
3/4	128	181	256	Pressure with these			
1	268	379	536			drop pipe	governs sizes
1-1/4	617	872	1,234	1,009			
1-1/2	975	1,379	1,951	1,390			
2	2,022	2,859	4,043	2,322	3,483		
2-1/2	3,362	4,754	6,723	3,333	5,000	6,666	
3	6,243	8,830	12,487	5,195	7,792	10,389	
4	13,414	18,970		9,041	13,562	18,083	22,604
5	25,088	35,479		14,308	21,462	28,616	35,770
6	40,787			20,500	30,749	40,999	51,249
8	86,360			35,911	53,867	71,822	89,778
10				58,716	88,075	117,433	146,791
12				85,276	127,914	170,552	213,189
14				104,385	156,577	208,769	260,962
16				138,974	208,460	277,947	347,434
18				178,504	267,756	357,008	446,260
20				222,976	334,463	445,951	557,439
22				272,388	408,583	544,777	680,971
24				326,743	490,114	653,485	816,856
26			with sizes	386,038	579,057	772,076	965,095
28				450,275	675,412	900,549	1,125,687
30	Velocity these	governs pipe		519,453	779,179	1,038,905	1,298,632
32				593,572	890,358	1,187,144	. 1,483,930
34				672,633	1,008,949	1,345,265	1,681,581
36				756,634	1,134,952	1,513,269	1,891,586
42				1,038,288	1,557,432	2,076,575	2,595,719
48				1,364,413	2,046,619	2,728,825	3,411,031
54				1,735,009	2,602,514	3,470,018	4,337,523
60				2,150,077	3,225,116	4,300,154	5,375,193
72				3,113,628	4,670,442	6,227,256	7,784,069
84				4,255,065	6,382,597	8,510,130	10,637,662
96				5,574,388	8,361,582	11,148,776	13,935,970
Steam Condensate Flow lbs./h 20 psig Back Pressure							
1/2	64	90	128				
3/4	157	222	313	Pressure with these			
1	328	464	656			drop pipe	governs sizes
1-1/4	755	1,068	1,511	1,329			
1-1/2	1,195	1,689	2,389	1,831			
2	2,476	3,502	4,952	3,059	4,589		
2-1/2	4,117	5,823	8,234	4,391	6,586		
3	7,647	10,814	15,293	6,843	10,264	13,686	
4	16,428	23,233		11,911	17,866	23,821	29,777

(Continued)

75 PSIG STEAM CONDENSATE PIPING SYSTEMS—STEEL PIPE (*Continued*)

Pipe Size	Pressure Drop psig/100 ft.			Velocity FPM (mph)			
	0.125	0.25	0.5	2,000 (23)	3,000 (34)	4,000 (45)	5,000 (57)
5	30,726	43,453		18,849	28,273	37,697	47,122
6	49,953			27,005	40,508	54,010	67,513
8	105,768			47,307	70,961	94,615	118,268
10				77,350	116,024	154,699	193,374
12				112,337	168,506	224,675	280,843
14				137,510	206,265	275,021	343,776
16				183,076	274,614	366,152	457,690
18				235,151	352,726	470,301	587,877
20				293,735	440,602	587,470	734,337
22				358,829	538,243	717,657	897,071
24				430,432	645,647	860,863	1,076,079
26			with	508,544	762,816	1,017,088	1,271,360
28			sizes	593,166	889,748	1,186,331	1,482,914
30	Velocity	governs		684,297	1,026,445	1,368,593	1,710,742
32	these	pipe		781,937	1,172,906	1,563,874	1,954,843
34				886,087	1,329,130	1,772,174	2,215,217
36				996,746	1,495,119	1,993,492	2,491,865
42				1,367,780	2,051,670	2,735,559	3,419,449
48				1,797,398	2,696,096	3,594,795	4,493,494
54				2,285,600	3,428,400	4,571,200	5,714,000
60				2,832,386	4,248,579	5,664,773	7,080,966
72				4,101,712	6,152,568	8,203,424	10,254,280
84				5,605,375	8,408,062	11,210,749	14,013,436
96				7,343,374	11,015,061	14,686,749	18,358,436
Steam Condensate Flow lbs./h 25 psig Back Pressure							
1/2	78	110	156		Pressure		
3/4	191	270	381		with these		
1	399	565	799			drop	governs
1-1/4	919	1,300	1,839	1,723		pipe	sizes
1-1/2	1,454	2,056	2,908	2,373			
2	3,014	4,262	6,027	3,966	5,948		
2-1/2	5,011	7,087	10,022	5,692	8,538		
3	9,307	13,162	18,614	8,871	13,306	17,741	
4	19,995	28,277		15,440	23,160	30,880	38,600
5	37,397	52,888		24,434	36,651	48,868	61,084
6	60,799	85,983		35,007	52,510	70,014	87,517
8	128,734			61,325	91,988	122,650	153,313
10	246,797			100,269	150,404	200,539	250,673
12				145,625	218,437	291,249	364,061
14				178,257	267,385	356,513	445,642
16				237,324	355,986	474,648	593,310
18				304,829	457,244	609,659	762,073
20				380,773	571,160	761,546	951,933
22				465,155	697,732	930,310	1,162,887
24				557,975	836,962	1,115,950	1,394,937
26			with	659,233	988,850	1,318,466	1,648,083
28			sizes	768,929	1,153,394	1,537,859	1,922,324
30		governs		887,064	1,330,596	1,774,128	2,217,660
32	Velocity	pipe		1,013,637	1,520,455	2,027,273	2,534,092
34	these			1,148,648	1,722,971	2,297,295	2,871,619
36				1,292,097	1,938,145	2,584,193	3,230,242
42				1,773,073	2,659,610	3,546,146	4,432,683
48				2,329,993	3,494,990	4,659,986	5,824,983
54				2,962,857	4,444,285	5,925,714	7,407,142
60				3,671,664	5,507,496	7,343,329	9,179,161
72				5,317,110	7,975,665	10,634,220	13,292,775
84				7,266,330	10,899,495	14,532,660	18,165,825
96				9,519,325	14,278,988	19,038,650	23,798,313

Notes:
1 Maximum recommended pressure drop/velocity: 0.25 psig/100 ft./5,000 FPM.
2 Table based on heavy weight steel pipe using steam equations in Part 3.

85 PSIG STEAM CONDENSATE PIPING SYSTEMS—STEEL PIPE

Pipe Size	Pressure Drop psig/100 ft.			Velocity FPM (mph)			
	0.125	0.25	0.5	2,000 (23)	3,000 (34)	4,000 (45)	5,000 (57)
Steam Condensate Flow lbs./h 0 psig Back Pressure							
1/2	24	33	47	110	Pressure	drop	
3/4	58	82	115		with these	pipe	governs
1	121	171	242	184			sizes
1-1/4	278	393	556	327	491		
1-1/2	440	622	880	451	677		
2	912	1,290	1,824	754	1,131	1,508	
2-1/2	1,516	2,144		1,082	1,623	2,164	2,705
3	2,816	3,982		1,686	2,529	3,372	4,215
4	6,050			2,935	4,402	5,870	7,337
5	11,315			4,644	6,967	9,289	11,611
6				6,654	9,981	13,309	16,636
8				11,657	17,486	23,314	29,143
10				19,060	28,590	38,119	47,649
12				27,681	41,522	55,362	69,203
14				33,884	50,826	67,768	84,710
16				45,112	67,668	90,223	112,779
18				57,944	86,915	115,887	144,859
20				72,379	108,569	144,759	180,948
22			with	88,419	132,629	176,838	221,048
24			sizes	106,063	159,094	212,125	265,157
26		governs		125,310	187,966	250,621	313,276
28		pipe		146,162	219,243	292,324	365,405
30	Velocity			168,618	252,927	337,235	421,544
32	these			192,677	289,016	385,355	481,693
34				218,341	327,511	436,682	545,852
36				245,608	368,413	491,217	614,021
42				337,035	505,552	674,070	842,587
48				442,897	664,346	885,794	1,107,243
54				563,195	844,793	1,126,390	1,407,988
60				697,929	1,046,893	1,395,858	1,744,822
72				1,010,704	1,516,056	2,021,407	2,526,759
84				1,381,222	2,071,832	2,762,443	3,453,054
96				1,809,482	2,714,224	3,618,965	4,523,706
Steam Condensate Flow lbs./h 5 psig Back Pressure							
1/2	31	43	61		Pressure	drop	
3/4	75	106	150		with these	pipe	governs
1	157	223	315	274			sizes
1-1/4	362	512	725	489			
1-1/2	573	810	1,146	674	1,010		
2	1,188	1,680	2,375	1,126	1,689	2,251	
2-1/2	1,975	2,793	3,950	1,616	2,424	3,231	6,295
3	3,668	5,187		2,518	3,777	5,036	10,957
4	7,880			4,383	6,574	8,766	
5	14,738			6,936	10,404	13,871	17,339
6	23,960			9,937	14,906	19,874	24,843
8				17,408	26,111	34,815	43,519
10				28,462	42,693	56,925	71,156
12				41,337	62,005	82,674	103,342
14				50,600	75,899	101,199	126,499
16		governs	with	67,366	101,050	134,733	168,416
18		pipe	sizes	86,528	129,793	173,057	216,321
20	Velocity			108,086	162,128	216,171	270,214
22	these			132,038	198,057	264,076	330,095
24				158,386	237,579	316,772	395,965
26				187,129	280,693	374,258	467,822
28				218,267	327,401	436,534	545,668
30				251,801	377,701	503,601	629,501
32				287,729	431,594	575,459	719,323

(Continued)

85 PSIG STEAM CONDENSATE PIPING SYSTEMS—STEEL PIPE (*Continued*)

Pipe Size	Pressure Drop psig/100 ft.			Velocity FPM (mph)			
	0.125	0.25	0.5	2,000 (23)	3,000 (34)	4,000 (45)	5,000 (57)
34				326,053	489,080	652,107	815,133
36				366,772	550,159	733,545	916,931
42				503,302	754,953	1,006,603	1,258,254
48				661,388	992,082	1,322,776	1,653,470
54				841,032	1,261,548	1,682,064	2,102,579
60				1,042,233	1,563,349	2,084,465	2,605,582
72				1,509,306	2,263,959	3,018,612	3,773,266
84				2,062,609	3,093,913	4,125,217	5,156,522
96				2,702,140	4,053,210	5,404,280	6,755,351
	Steam Condensate Flow lbs./h 7 psig Back Pressure						
1/2	34	48	67		Pressure	drop	
3/4	83	117	165		with these	pipe	
1	173	245	346	315			governs
1-1/4	398	563	797	562			sizes
1-1/2	630	891	1,260	775	1,162		
2	1,305	1,846	2,611	1,294	1,942	2,589	
2-1/2	2,171	3,070	4,341	1,858	2,787	3,716	
3	4,031	5,701		2,896	4,343	5,791	7,239
4	8,661	12,249		5,040	7,560	10,080	12,600
5	16,199			7,976	11,964	15,952	19,940
6	26,337			11,427	17,141	22,854	28,568
8				20,018	30,027	40,036	50,045
10				32,730	49,096	65,461	81,826
12				47,536	71,303	95,071	118,839
14				58,187	87,281	116,375	145,469
16				77,468	116,203	154,937	193,671
18				99,504	149,256	199,008	248,760
20				124,294	186,441	248,588	310,735
22				151,838	227,757	303,676	379,596
24				182,137	273,206	364,274	455,343
26		governs	with	215,190	322,785	430,380	537,976
28	Velocity	pipe	sizes	250,998	376,497	501,996	627,495
30	these			289,560	434,340	579,120	723,900
32				330,876	496,315	661,753	827,191
34				374,947	562,421	749,895	937,369
36				421,773	632,659	843,546	1,054,432
42				578,776	868,163	1,157,551	1,446,939
48				760,568	1,140,853	1,521,137	1,901,421
54				967,151	1,450,726	1,934,302	2,417,877
60				1,198,524	1,797,785	2,397,047	2,996,309
72				1,735,638	2,603,457	3,471,277	4,339,096
84				2,371,913	3,557,869	4,743,826	5,929,782
96				3,107,347	4,661,020	6,214,694	7,768,367
	Steam Condensate Flow lbs./h 10 psig Back Pressure						
1/2	38	54	77		Pressure		
3/4	94	133	188		with these		
1	197	279	395	382		drop	governs
1-1/4	454	642	908	681		pipe	sizes
1-1/2	718	1,016	1,436	939	1,408		
2	1,489	2,105	2,977	1,569	2,353		
2-1/2	2,475	3,501	4,951	2,251	3,377	4,503	
3	4,597	6,502		3,509	5,263	7,018	8,772
4	9,877	13,968		6,107	9,161	12,215	15,268
5	18,473			9,665	14,497	19,329	24,162
6	30,033			13,847	20,770	27,694	34,617
8		governs	with	24,257	36,386	48,514	60,643
10		pipe	sizes	39,661	59,492	79,323	99,153
12	Velocity			57,601	86,402	115,203	144,004
14	these			70,509	105,764	141,018	176,273

(*Continued*)

85 PSIG STEAM CONDENSATE PIPING SYSTEMS—STEEL PIPE (Continued)

	Pressure Drop psig/100 ft.			Velocity FPM (mph)			
Pipe Size	0.125	0.25	0.5	2,000 (23)	3,000 (34)	4,000 (45)	5,000 (57)
16				93,873	140,809	187,746	234,682
18				120,575	180,862	241,149	301,436
20				150,614	225,921	301,228	376,535
22				183,991	275,986	367,982	459,977
24				220,706	331,059	441,411	551,764
26				260,758	391,137	521,516	651,895
28				304,148	456,223	608,297	760,371
30				350,876	526,314	701,752	877,191
32				400,942	601,413	801,884	1,002,355
34				454,345	681,518	908,690	1,135,863
36				511,086	766,629	1,022,172	1,277,715
42				701,335	1,052,003	1,402,671	1,753,338
48				921,624	1,382,436	1,843,248	2,304,059
54				1,171,952	1,757,927	2,343,903	2,929,879
60				1,452,319	2,178,478	2,904,638	3,630,797
72				2,103,171	3,154,757	4,206,343	5,257,928
84				2,874,181	4,311,272	5,748,362	7,185,453
96				3,765,348	5,648,023	7,530,697	9,413,371
Steam Condensate Flow lbs./h 12 psig Back Pressure							
1/2	42	59	83		Pressure		
3/4	102	145	205		with these		
1	214	303	429			drop	governs
1-1/4	494	698	987	768		pipe	sizes
1-1/2	781	1,104	1,561	1,058			
2	1,618	2,288	3,235	1,768	2,652		
2-1/2	2,690	3,804	5,380	2,537	3,806	5,075	
3	4,996	7,065		3,954	5,932	7,909	9,886
4	10,734	15,180		6,883	10,325	13,766	17,208
5	20,075			10,892	16,339	21,785	27,231
6	32,638			15,606	23,409	31,212	39,015
8				27,338	41,008	54,677	68,346
10				44,699	67,049	89,399	111,749
12				64,919	97,378	129,837	162,296
14				79,466	119,199	158,931	198,664
16				105,798	158,696	211,595	264,494
18				135,891	203,837	271,782	339,728
20		governs	with	169,746	254,619	339,493	424,366
22		pipe	sizes	207,363	311,045	414,726	518,408
24	Velocity			248,742	373,113	497,484	621,855
26	these			293,882	440,823	587,764	734,705
28				342,784	514,176	685,568	856,960
30				395,448	593,172	790,896	988,619
32				451,873	677,810	903,746	1,129,683
34				512,060	768,090	1,024,120	1,280,150
36				576,009	864,013	1,152,018	1,440,022
42				790,425	1,185,638	1,580,851	1,976,063
48				1,038,697	1,558,045	2,077,394	2,596,742
54				1,320,824	1,981,236	2,641,647	3,302,059
60				1,636,806	2,455,209	3,273,611	4,092,014
72				2,370,335	3,555,503	4,740,671	5,925,839
84				3,239,286	4,858,929	6,478,572	8,098,215
96				4,243,658	6,365,486	8,487,315	10,609,144
Steam Condensate Flow lbs./h 15 psig Back Pressure							
1/2	47	67	94		Pressure		
3/4	115	163	231		with these		
1	242	342	483			drop	governs
1-1/4	556	786	1,112	909		pipe	sizes
1-1/2	879	1,243	1,758	1,253			
2	1,822	2,577	3,645	2,093	3,140		
2-1/2	3,030	4,285	6,060	3,004	4,507	6,009	
3	5,628	7,959	11,255	4,682	7,023	9,364	
4	12,091	17,099		8,150	12,225	16,300	20,374

(Continued)

85 PSIG STEAM CONDENSATE PIPING SYSTEMS—STEEL PIPE (Continued)

Pipe Size	Pressure Drop psig/100 ft.			Velocity FPM (mph)			
	0.125	0.25	0.5	2,000 (23)	3,000 (34)	4,000 (45)	5,000 (57)
5	22,613	31,980		12,897	19,345	25,794	32,242
6	36,764			18,478	27,717	36,956	46,195
8	77,843			32,369	48,554	64,739	80,923
10				52,925	79,388	105,851	132,313
12				76,865	115,298	153,731	192,163
14				94,090	141,134	188,179	235,224
16				125,267	187,901	250,534	313,168
18				160,899	241,348	321,797	402,247
20				200,984	301,476	401,968	502,460
22				245,524	368,285	491,047	613,809
24				294,517	441,775	589,034	736,292
26			with	347,964	521,946	695,929	869,911
28			sizes	405,866	608,798	811,731	1,014,664
30	Velocity	governs		468,221	702,331	936,442	1,170,552
32	these	pipe		535,030	802,545	1,070,060	1,337,575
34				606,293	909,440	1,212,586	1,515,733
36				682,010	1,023,015	1,364,020	1,705,025
42				935,885	1,403,827	1,871,770	2,339,712
48				1,229,845	1,844,768	2,459,690	3,074,613
54				1,563,891	2,345,836	3,127,782	3,909,727
60				1,938,022	2,907,033	3,876,044	4,845,055
72				2,806,541	4,209,811	5,613,082	7,016,352
84				3,835,402	5,753,103	7,670,803	9,588,504
96				5,024,605	7,536,907	10,049,209	12,561,511
Steam Condensate Flow lbs./h 20 psig Back Pressure							
1/2	57	80	114		Pressure		
3/4	139	197	278		with these		
1	292	412	583			drop	governs
1-1/4	671	949	1,343	1,181		pipe	sizes
1-1/2	1,061	1,501	2,123	1,627			
2	2,200	3,111	4,400	2,718	4,077		
2-1/2	3,658	5,174	7,317	3,902	5,852		
3	6,795	9,609	13,589	6,080	9,121	12,161	
4	14,598	20,644		10,584	15,875	21,167	26,459
5	27,302	38,611		16,748	25,123	33,497	41,871
6	44,387			23,996	35,994	47,992	59,990
8	93,983			42,036	63,054	84,072	105,090
10				68,731	103,096	137,462	171,827
12				99,820	149,730	199,640	249,551
14				122,188	183,282	244,377	305,471
16				162,677	244,015	325,353	406,692
18				208,949	313,424	417,898	522,373
20				261,006	391,509	522,011	652,514
22				318,846	478,269	637,693	797,116
24				382,471	573,706	764,942	956,177
26			with	451,880	677,819	903,759	1,129,699
28			sizes	527,072	790,609	1,054,145	1,317,681
30	Velocity	governs		608,049	912,074	1,216,099	1,520,123
32	these	pipe		694,810	1,042,215	1,389,620	1,737,025
34				787,355	1,181,033	1,574,710	1,968,388
36				885,684	1,328,526	1,771,368	2,214,210
42				1,215,376	1,823,063	2,430,751	3,038,439
48				1,597,123	2,395,685	3,194,247	3,992,809
54				2,030,928	3,046,392	4,061,856	5,077,320
60				2,516,789	3,775,183	5,033,578	6,291,972
72				3,644,681	5,467,021	7,289,361	9,111,701
84				4,980,798	7,471,197	9,961,597	12,451,996
96				6,525,142	9,787,713	13,050,284	16,312,855

(Continued)

85 PSIG STEAM CONDENSATE PIPING SYSTEMS—STEEL PIPE (Continued)

Pipe Size	Pressure Drop psig/100 ft.			Velocity FPM (mph)			
	0.125	0.25	0.5	2,000 (23)	3,000 (34)	4,000 (45)	5,000 (57)
Steam Condensate Flow lbs./h 25 psig Back Pressure							
1/2	68	96	136	Pressure			
3/4	167	236	333	with these			
1	349	494	698			drop	governs
1-1/4	804	1,136	1,607	1,506		pipe	sizes
1-1/2	1,271	1,797	2,542	2,074			
2	2,634	3,725	5,268	3,466	5,199		
2-1/2	4,380	6,194	8,760	4,975	7,462		
3	8,134	11,504	16,268	7,753	11,629	15,506	
4	17,476	24,714		13,495	20,242	26,989	33,737
5	32,685	46,224		21,355	32,033	42,710	53,388
6	53,139	75,150		30,596	45,894	61,192	76,491
8	112,514			53,598	80,398	107,197	133,996
10	215,701			87,636	131,454	175,271	219,089
12				127,276	190,915	254,553	318,191
14				155,797	233,695	311,594	389,492
16				207,422	311,133	414,844	518,555
18				266,422	399,633	532,844	666,055
20				332,797	499,195	665,594	831,992
22				406,547	609,820	813,094	1,016,367
24				487,672	731,508	975,344	1,219,179
26			with	576,172	864,258	1,152,344	1,440,429
28		governs	sizes	672,047	1,008,070	1,344,094	1,680,117
30		pipe		775,297	1,162,945	1,550,593	1,938,242
32	Velocity			885,922	1,328,883	1,771,843	2,214,804
34	these			1,003,922	1,505,883	2,007,843	2,509,804
36				1,129,297	1,693,945	2,258,593	2,823,242
42				1,549,672	2,324,507	3,099,343	3,874,179
48				2,036,422	3,054,632	4,072,843	5,091,054
54				2,589,546	3,884,320	5,179,093	6,473,866
60				3,209,046	4,813,569	6,418,093	8,022,616
72				4,647,171	6,970,757	9,294,342	11,617,928
84				6,350,796	9,526,194	12,701,592	15,876,989
96				8,319,920	12,479,881	16,639,841	20,799,801

Notes:
1 Maximum recommended pressure drop/velocity: 0.25 psig/100 ft./5,000 FPM.
2 Table based on heavy weight steel pipe using steam equations in Part 3.

100 PSIG STEAM CONDENSATE PIPING SYSTEMS—STEEL PIPE

Pipe Size	Pressure Drop psig/100 ft.			Velocity FPM (mph)			
	0.125	0.25	0.5	2,000 (23)	3,000 (34)	4,000 (45)	5,000 (57)
Steam Condensate Flow lbs./h 0 psig Back Pressure							
1/2	22	31	43		Pressure	drop	
3/4	53	75	106	101	with these	pipe	
1	111	157	222	168			governs
1-1/4	255	361	510	300	450		sizes
1-1/2	403	571	807	414	621		
2	836	1,182	1,672	691	1,037	1,382	
2-1/2	1,390	1,966		992	1,488	1,984	2,480
3	2,582	3,652		1,546	2,319	3,092	3,866
4	5,548			2,691	4,037	5,383	6,728
5	10,376			4,259	6,388	8,518	10,647
6				6,102	9,153	12,204	15,255
8				10,689	16,034	21,379	26,723
10				17,478	26,216	34,955	43,694
12				25,383	38,075	50,767	63,458
14				31,071	46,607	62,142	77,678

(Continued)

100 PSIG STEAM CONDENSATE PIPING SYSTEMS—STEEL PIPE (*Continued*)

Pipe Size	Pressure Drop psig/100 ft.			Velocity FPM (mph)			
	0.125	0.25	0.5	2,000 (23)	3,000 (34)	4,000 (45)	5,000 (57)
16				41,367	62,051	82,734	103,418
18				53,134	79,700	106,267	132,834
20				66,371	99,557	132,742	165,928
22				81,079	121,619	162,159	202,698
24				97,258	145,888	194,517	243,146
26				114,908	172,363	229,817	287,271
28				134,029	201,044	268,058	335,073
30				154,621	231,931	309,241	386,552
32	Velocity	governs	with	176,683	265,025	353,366	441,708
34	these	pipe	sizes	200,216	300,325	400,433	500,541
36				225,220	337,831	450,441	563,051
42				309,058	463,586	618,115	772,644
48				406,132	609,198	812,264	1,015,331
54				516,444	774,666	1,032,889	1,291,111
60				639,994	959,991	1,279,988	1,599,984
72				926,805	1,390,208	1,853,610	2,317,013
84				1,266,566	1,899,849	2,533,132	3,166,416
96				1,659,277	2,488,916	3,318,554	4,148,193
Steam Condensate Flow lbs./h 5 psig Back Pressure							
1/2	28	39	56		Pressure	drop	
3/4	68	96	136		with these	pipe	
1	143	202	285	248			governs
1-1/4	328	464	657	443			sizes
1-1/2	519	734	1,038	610	915		
2	1,076	1,522	2,152	1,020	1,530	2,039	
2-1/2	1,789	2,530	3,578	1,464	2,195	2,927	
3	3,323	4,699		2,281	3,422	4,562	5,703
4	7,138			3,970	5,956	7,941	9,926
5	13,351			6,283	9,425	12,566	15,708
6	21,706			9,002	13,503	18,004	22,505
8				15,769	23,654	31,539	39,424
10				25,784	38,676	51,568	64,459
12				37,447	56,170	74,893	93,617
14				45,838	68,757	91,676	114,595
16				61,027	91,540	122,053	152,567
18				78,385	117,578	156,771	195,964
20				97,914	146,871	195,828	244,785
22				119,612	179,419	239,225	299,031
24				143,481	215,221	286,961	358,701
26		governs	with	169,519	254,278	339,037	423,797
28		pipe	sizes	197,727	296,590	395,453	494,316
30	Velocity			228,104	342,156	456,209	570,261
32	these			260,652	390,978	521,304	651,630
34				295,369	443,054	590,739	738,423
36				332,257	498,385	664,513	830,641
42				455,937	683,906	911,875	1,139,843
48				599,147	898,720	1,198,293	1,497,867
54				761,885	1,142,827	1,523,769	1,904,711
60				944,151	1,416,226	1,888,302	2,360,377
72				1,367,270	2,050,904	2,734,539	3,418,174
84				1,868,502	2,802,753	3,737,004	4,671,256
96				2,447,849	3,671,774	4,895,698	6,119,623
Steam Condensate Flow lbs./h 7 psig Back Pressure							
1/2	30	43	61		Pressure	drop	
3/4	74	105	149		with these	pipe	
1	156	221	312	284			governs
1-1/4	359	508	718	507			sizes
1-1/2	568	803	1,135	698	1,047		
2	1,177	1,664	2,353	1,167	1,750	2,333	
2-1/2	1,956	2,767	3,913	1,675	2,512	3,349	
3	3,634	5,139		2,610	3,915	5,220	6,524
4	7,806	11,040		4,543	6,814	9,085	11,356

(*Continued*)

100 PSIG STEAM CONDENSATE PIPING SYSTEMS—STEEL PIPE (*Continued*)

Pipe Size	Pressure Drop psig/100 ft.			Velocity FPM (mph)			
	0.125	0.25	0.5	2,000 (23)	3,000 (34)	4,000 (45)	5,000 (57)
5	14,600			7,189	10,783	14,377	17,971
6	23,737			10,299	15,449	20,598	25,748
8				18,042	27,063	36,084	45,105
10				29,500	44,249	58,999	73,749
12				42,843	64,265	85,687	107,108
14				52,444	78,666	104,888	131,110
16				69,822	104,733	139,643	174,554
18				89,682	134,523	179,364	224,205
20				112,025	168,038	224,050	280,063
22				136,851	205,276	273,701	342,126
24				164,159	246,238	328,317	410,396
26		governs	with	193,949	290,924	387,898	484,873
28	Velocity	pipe	sizes	226,222	339,333	452,445	565,556
30	these			260,978	391,467	521,956	652,445
32				298,216	447,324	596,432	745,540
34				337,937	506,905	675,874	844,842
36				380,140	570,210	760,281	950,351
42				521,646	782,468	1,043,291	1,304,114
48				685,494	1,028,241	1,370,988	1,713,734
54				871,685	1,307,527	1,743,370	2,179,212
60				1,080,219	1,620,329	2,160,438	2,700,548
72				1,564,316	2,346,474	3,128,632	3,910,790
84				2,137,785	3,206,677	4,275,570	5,344,462
96				2,800,625	4,200,938	5,601,251	7,001,564
Steam Condensate Flow lbs./h 10 psig Back Pressure							
1/2	34	49	69		Pressure		
3/4	84	119	169		with these		
1	176	250	353	342		drop	governs
1-1/4	406	574	812	609		pipe	sizes
1-1/2	642	908	1,285	840	1,259		
2	1,331	1,883	2,663	1,403	2,104		
2-1/2	2,214	3,131	4,428	2,014	3,020	4,027	
3	4,112	5,815		3,138	4,707	6,276	7,845
4	8,834	12,493		5,462	8,193	10,924	13,655
5	16,522			8,644	12,966	17,288	21,610
6	26,861			12,384	18,576	24,769	30,961
8				21,695	32,542	43,390	54,237
10				35,472	53,208	70,944	88,680
12				51,517	77,276	103,034	128,793
14				63,061	94,592	126,123	157,653
16				83,957	125,936	167,915	209,894
18				107,839	161,758	215,677	269,597
20				134,705	202,058	269,410	336,763
22				164,557	246,835	329,113	411,391
24				197,393	296,090	394,787	493,483
26		governs	with	233,215	349,823	466,430	583,038
28	Velocity	pipe	sizes	272,022	408,033	544,044	680,055
30	these			313,814	470,721	627,629	784,536
32				358,592	537,887	717,183	896,479
34				406,354	609,531	812,708	1,015,885
36				457,102	685,652	914,203	1,142,754
42				627,255	940,883	1,254,511	1,568,139
48				824,276	1,236,413	1,648,551	2,060,689
54				1,048,162	1,572,243	2,096,324	2,620,405
60				1,298,915	1,948,372	2,597,830	3,247,287
72				1,881,020	2,821,530	3,762,040	4,702,550
84				2,570,590	3,855,886	5,141,181	6,426,476
96				3,367,626	5,051,440	6,735,253	8,419,066

(*Continued*)

100 PSIG STEAM CONDENSATE PIPING SYSTEMS—STEEL PIPE (*Continued*)

Pipe Size	Pressure Drop psig/100 ft.			Velocity FPM (mph)			
	0.125	0.25	0.5	2,000 (23)	3,000 (34)	4,000 (45)	5,000 (57)
Steam Condensate Flow lbs./h 12 psig Back Pressure							
1/2	37	53	74	Pressure with these		drop pipe	governs sizes
3/4	91	129	182				
1	191	270	382				
1-1/4	439	621	878	683			
1-1/2	694	982	1,389	941			
2	1,439	2,035	2,878	1,573	2,359		
2-1/2	2,393	3,384	4,786	2,257	3,386	4,514	
3	4,444	6,285		3,518	5,277	7,036	8,795
4	9,548	13,503		6,123	9,185	12,246	15,308
5	17,859			9,690	14,534	19,379	24,224
6	29,034			13,883	20,824	27,765	34,707
8				24,320	36,479	48,639	60,799
10				39,764	59,645	79,527	99,409
12				57,750	86,625	115,500	144,375
14				70,691	106,036	141,382	176,727
16				94,115	141,173	188,230	235,288
18				120,886	181,328	241,771	302,214
20				151,002	226,504	302,005	377,506
22				184,466	276,698	368,931	461,164
24				221,275	331,912	442,550	553,187
26		governs pipe	with sizes	261,431	392,146	522,861	653,577
28	Velocity these			304,933	457,399	609,866	762,332
30				351,781	527,672	703,562	879,453
32				401,976	602,964	803,952	1,004,940
34				455,517	683,275	911,034	1,138,792
36				512,404	768,606	1,024,808	1,281,011
42				703,144	1,054,716	1,406,288	1,757,860
48				924,001	1,386,001	1,848,001	2,310,002
54				1,174,974	1,762,461	2,349,948	2,937,436
60				1,456,065	2,184,097	2,912,129	3,640,161
72				2,108,596	3,162,893	4,217,191	5,271,489
84				2,881,594	4,322,391	5,763,188	7,203,985
96				3,775,060	5,662,589	7,550,119	9,437,649
Steam Condensate Flow lbs./h 15 psig Back Pressure							
1/2	41	59	83	Pressure with these		drop pipe	governs sizes
3/4	102	144	203				
1	213	301	426				
1-1/4	490	694	981	802			
1-1/2	776	1,097	1,551	1,105			
2	1,608	2,273	3,215	1,847	2,770		
2-1/2	2,673	3,780	5,346	2,650	3,975	5,301	
3	4,964	7,021	9,929	4,130	6,196	8,261	
4	10,666	15,084		7,189	10,784	14,379	17,973
5	19,948	28,211		11,377	17,066	22,754	28,443
6	32,432			16,300	24,451	32,601	40,751
8	68,670			28,555	42,832	57,110	71,387
10				46,689	70,033	93,377	116,721
12				67,807	101,711	135,615	169,518
14				83,002	124,503	166,004	207,505
16			with sizes	110,505	165,758	221,011	276,264
18				141,938	212,907	283,876	354,845
20	Velocity these	governs pipe		177,300	265,950	354,600	443,250
22				216,591	324,886	433,182	541,477
24				259,811	389,716	519,621	649,527
26				306,960	460,440	613,919	767,399
28				358,038	537,057	716,076	895,094
30				413,045	619,567	826,090	1,032,612
32				471,981	707,972	943,962	1,179,953

(*Continued*)

100 PSIG STEAM CONDENSATE PIPING SYSTEMS—STEEL PIPE (*Continued*)

Pipe Size	Pressure Drop psig/100 ft.			Velocity FPM (mph)			
	0.125	0.25	0.5	2,000 (23)	3,000 (34)	4,000 (45)	5,000 (57)
34				534,847	802,270	1,069,693	1,337,116
36				601,641	902,462	1,203,282	1,504,103
42				825,599	1,238,398	1,651,198	2,063,997
48				1,084,918	1,627,378	2,169,837	2,712,296
54				1,379,600	2,069,400	2,759,199	3,448,999
60				1,709,643	2,564,464	3,419,285	4,274,107
72				2,475,814	3,713,721	4,951,628	6,189,535
84				3,383,433	5,075,149	6,766,865	8,458,581
96				4,432,498	6,648,747	8,864,996	11,081,245
Steam Condensate Flow lbs./h 20 psig Back Pressure							
1/2	49	70	99		Pressure		
3/4	121	171	242		with these		
1	253	358	507			drop	governs
1-1/4	583	825	1,166	1,026		pipe	sizes
1-1/2	922	1,304	1,845	1,413			
2	1,912	2,703	3,823	2,362	3,543		
2-1/2	3,179	4,495	6,357	3,390	5,085		
3	5,904	8,349	11,807	5,283	7,925	10,566	
4	12,683	17,937		9,196	13,794	18,392	22,989
5	23,722	33,548		14,552	21,828	29,105	36,381
6	38,567			20,849	31,274	41,699	52,124
8	81,659			36,524	54,786	73,048	91,310
10				59,718	89,578	119,437	149,296
12				86,731	130,097	173,462	216,828
14				106,166	159,249	212,332	265,415
16				141,345	212,018	282,691	353,363
18				181,550	272,325	363,100	453,875
20				226,781	340,171	453,561	566,952
22				277,037	415,555	554,074	692,592
24				332,318	498,478	664,637	830,796
26			with	392,626	588,939	785,252	981,565
28	Velocity	governs	sizes	457,959	686,938	915,918	1,144,897
30	these	pipe		528,317	792,476	1,056,635	1,320,793
32				603,701	905,552	1,207,403	1,509,254
34				684,111	1,026,167	1,368,222	1,710,278
36				769,547	1,154,320	1,539,093	1,923,867
42				1,056,006	1,584,010	2,112,013	2,640,016
48				1,387,697	2,081,545	2,775,393	3,469,242
54				1,764,618	2,646,926	3,529,235	4,411,544
60				2,186,769	3,280,153	4,373,538	5,466,922
72				3,166,763	4,750,144	6,333,526	7,916,907
84				4,327,679	6,491,518	8,655,358	10,819,197
96				5,669,517	8,504,275	11,339,034	14,173,792
Steam Condensate Flow lbs./h 25 psig Back Pressure							
1/2	58	82	116		Pressure		
3/4	142	201	285		with these		
1	298	422	596			drop	governs
1-1/4	686	971	1,373	1,286		pipe	sizes
1-1/2	1,086	1,535	2,171	1,772			
2	2,250	3,182	4,500	2,961	4,441		
2-1/2	3,741	5,291	7,483	4,250	6,374		
3	6,948	9,827	13,897	6,623	9,934	13,246	
4	14,928	21,112		11,527	17,291	23,055	28,819
5	27,920	39,486		18,242	27,363	36,484	45,605
6	45,392	64,195		26,136	39,204	52,272	65,340
8	96,112			45,785	68,677	91,570	114,462
10	184,257			74,860	112,291	149,721	187,151
12				108,722	163,084	217,445	271,806
14				133,085	199,628	266,171	332,713

(*Continued*)

100 PSIG STEAM CONDENSATE PIPING SYSTEMS—STEEL PIPE (*Continued*)

Pipe Size	Pressure Drop psig/100 ft.			Velocity FPM (mph)			
	0.125	0.25	0.5	2,000 (23)	3,000 (34)	4,000 (45)	5,000 (57)
16				177,185	265,777	354,369	442,961
18				227,584	341,376	455,168	568,959
20				284,283	426,424	568,566	710,707
22				347,282	520,923	694,564	868,205
24				416,581	624,871	833,161	1,041,452
26				492,179	738,269	984,359	1,230,449
28				574,078	861,117	1,148,156	1,435,195
30				662,277	993,415	1,324,553	1,655,692
32	Velocity	governs	with	756,775	1,135,163	1,513,550	1,891,938
34	these	pipe	sizes	857,573	1,286,360	1,715,147	2,143,934
36				964,672	1,447,008	1,929,343	2,411,679
42				1,323,766	1,985,649	2,647,532	3,309,414
48				1,739,559	2,609,338	3,479,118	4,348,897
54				2,212,051	3,318,077	4,424,102	5,530,128
60				2,741,243	4,111,864	5,482,485	6,853,106
72				3,969,722	5,954,584	7,939,445	9,924,306
84				5,424,999	8,137,498	10,849,997	13,562,496
96				7,107,071	10,660,606	14,214,142	17,767,677
	Steam Condensate Flow lbs./h 30 psig Back Pressure						
1/2	68	96	136				
3/4	167	236	333				
1	349	493	698		Pressure	drop	
1-1/4	803	1,135	1,606	1,590	with these	pipe	governs
1-1/2	1,269	1,795	2,539	2,191			sizes
2	2,631	3,721	5,262	3,661			
2-1/2	4,375	6,188	8,751	5,254	7,881		
3	8,126	11,492	16,252	8,188	12,282		
4	17,458	24,689	34,916	14,252	21,379	28,505	
5	32,652	46,177		22,554	33,832	45,109	56,386
6	53,085	75,073		32,314	48,471	64,628	80,786
8	112,400			56,608	84,912	113,216	141,520
10	215,482			92,557	138,835	185,113	231,392
12				134,423	201,635	268,846	336,058
14				164,545	246,818	329,090	411,363
16				219,069	328,603	438,138	547,672
18				281,382	422,073	562,764	703,455
20				351,484	527,226	702,968	878,710
22				429,375	644,063	858,750	1,073,438
24				515,055	772,583	1,030,111	1,287,639
26			with	608,525	912,787	1,217,050	1,521,312
28		governs	sizes	709,783	1,064,675	1,419,567	1,774,459
30		pipe		818,831	1,228,247	1,637,662	2,047,078
32	Velocity			935,668	1,403,502	1,871,336	2,339,170
34	these			1,060,294	1,590,441	2,120,587	2,650,734
36				1,192,709	1,789,063	2,385,418	2,981,772
42				1,636,689	2,455,033	3,273,377	4,091,721
48				2,150,770	3,226,155	4,301,541	5,376,926
54				2,734,954	4,102,431	5,469,908	6,837,386
60				3,389,240	5,083,860	6,778,480	8,473,100
72				4,908,118	7,362,177	9,816,236	12,270,296
84				6,707,405	10,061,107	13,414,809	16,768,512
96				8,787,099	13,180,649	17,574,198	21,967,748

Notes:
1 Maximum recommended pressure drop/velocity: 0.25 psig/100 ft./5,000 FPM.
2 Table based on heavy weight steel pipe using steam equations in Part 3.

Refer to the online resource for Section 22.05 High-Pressure Steam Condensate System Pipe Sizing Tables (120-300 psig). www.mheducation.com/HVACequations

AC Condensate Piping

23.01 Air Conditioning (AC) Condensate Piping

A. AC Condensate Flow

1. Range:	0.02–0.08 GPM/ton.
2. Average:	0.04 GPM/ton.
3. Unitary packaged AC equipment:	0.006 GPM/ton.
4. Air handling units (100 percent outside air):	0.100 GPM/1000 CFM.
5. Air handling units (50 percent outdoor air):	0.065 GPM/1000 CFM.
6. Air handling units (25 percent outdoor air):	0.048 GPM/1000 CFM.
7. Air handling units (15 percent outdoor air):	0.041 GPM/1000 CFM.
8. Air handling units (0 percent outdoor air):	0.030 GPM/1000 CFM.

B. AC Condensate Pipe Sizing

1. Minimum pipe sizes are provided in the following table.
2. Pipe size shall not be smaller than the drain pan outlet. The minimum size below grade and below ground floor shall be 2-1/2" (4" Allegheny Co., PA). The drain shall have a slope of not less than 1/8" per foot.
3. Some localities require AC condensate to be discharged to storm sewers. Some require AC condensate to be discharged to sanitary sewers, while some permit AC condensate to be discharged to either storm or sanitary sewers. Verify pipe sizing and discharge requirements with local authorities and codes.

AC Tons	Minimum Drain Size
0–20	1"
21–40	1-1/4"
41–60	1-1/2"
61–100	2"
101–250	3"
251 and larger	4"

Refrigerant Piping Systems

24.01 Refrigerant Systems and Piping

A. Refrigeration System Design Considerations

1. Refrigeration load and system size:
   a. Conduction heat gains, sensible.
   b. Radiation heat gains, sensible.
   c. Convection/infiltration heat gains, sensible and latent.
   d. Internal heat gains, lights, people, equipment.
   e. Product load, sensible and latent.
2. Part load performance, minimum versus maximum load.
3. Piping layout and design:
   a. Ensure proper refrigerant flow to feed evaporators.
   b. Size piping to limit excessive pressure drop and temperature rise and to minimize first cost.
   c. Ensure proper lubricating oil flow to compressors and protect compressors for loss of lubricating oil flow.
   d. To prevent liquid (oil or refrigerant) from entering the compressors.
   e. Maintain a clean and dry system.
   f. To prevent refrigeration system leaks.
4. Refrigerant type selection and refrigerant limitations.
5. System operation, partial year or year round regardless of ambient conditions.
6. Load variations during short time periods.
7. Evaporator frost control.
8. Oil management under varying load conditions.
9. Heat exchange method.
10. Secondary coolant selection.
11. Installed cost, operating costs, maintenance costs, system efficiency, and system simplicity.
12. Safe operation for building inhabitants.
13. Operating pressure and pressure ratios; single stage versus two stage versus multistaged.
14. Special electrical requirements.
15. Refrigerant system capacity estimate:
    a. Packaged systems: 2.0 lbs. refrigerant per ton.
    b. Split systems: 3.0 lbs. refrigerant per ton.

B. Refrigerant Pipe Design Criteria

1. Halocarbon refrigerants:
   a. Liquid lines (condensers to receivers)—100 FPM or less.
   b. Liquid lines (receivers to evaporator)—300 FPM or less.
   c. Compressor suction line—900 to 4,000 FPM.
   d. Compressor discharge line—2,000 to 3,500 FPM.
   e. Defrost gas supply lines—1,000 to 2,000 FPM.
   f. Condensate drop legs—150 FPM or less.
   g. Condensate mains—100 FPM or less.
   h. Pressure loss due to refrigerant liquid risers is 0.5 psi per foot of lift.
   i. Liquid lines should be sized to produce a pressure drop due to friction that corresponds to a 1°F to 2°F change in saturation temperature or less.
   j. Discharge and suction lines should be sized to produce a pressure drop due to friction that corresponds to a 2°F change in saturation temperature or less.
   k. Pump suction pipe sizing should be 2.5 fps maximum. Oversizing of pump suction piping should be limited to one pipe size.
2. Standard steel pipe sizes: 1/2", 3/4", 1", 1-1/4", 1-1/2", 2", 2-1/2", 3", 4", 6", 8", 10", 12", 14", 16", 18", 20".
3. Standard copper pipe sizes: 3/8", 1/2", 5/8", 3/4", 7/8", 1", 1-1/8", 1-1/4", 1-3/8", 1-1/2", 1-5/8", 2", 2-1/8", 2-1/2", 2-5/8", 3", 3-1/8", 3-5/8", 4", 4-1/8", 6", 8", 10", 12".

4. Ammonia refrigerant:
   a. Liquid lines should be sized for 2.0 psi/100 ft. of equivalent pipe length or less. Liquid lines should be sized for a 3:1, 4:1, or 5:1 overfeed ratio (4:1 recommended).
   b. Suction lines should be sized for 0.25, 0.5, or 1.0°F/100 ft. of equivalent pipe length.
   c. Discharge lines should be sized for 1.0°F/100 ft. of equivalent pipe length.
   d. Pump suction pipe sizing should be 3.0 fps maximum. Oversizing of pump suction piping should be limited to one pipe size.
   e. Cooling water flow rate: 0.1 GPM/ton.

C. Halocarbon Refrigerant Pipe Materials

1. Pipe: Type "L (ACR)" copper tubing, *ASTM B280*, hard drawn.
   Fittings: Wrought copper solder joint fittings, *ANSI/ASME B16.22*.
   Joints: Classification BAg-1 (silver) AWS A5.8 Brazed-Silver Alloy brazing. Brazing shall be conducted using a brazing flux. Do not use an acid flux.

D. Ammonia Refrigerant Pipe Materials

1. Liquid lines:
   a. 1-1/2" and smaller: Schedule 80 minimum.
   b. 2" to 6":   Schedule 40 minimum.
   c. 8" and larger: Schedule 30 minimum.
2. Suction, discharge, and vapor lines:
   a. 1-1/2" and smaller: Schedule 80 minimum.
   b. 2" to 6":   Schedule 40 minimum.
   c. 8" and larger: Schedule 30 minimum.
3. Fittings:
   a. Couplings, elbows, tees, and unions for threaded piping systems must be constructed of forged steel with a pressure rating of 300 psi.
   b. Welding fittings must match the weight of the pipe.
   c. Low-pressure side piping, vessels, and flanges should be designed for 150 psi.
   d. High-pressure side piping, vessels, and flanges should be designed for 250 psi if the system is water or evaporative cooled, and 300 psi if the system is air cooled.
4. Joints:
   a. 1-1/4" pipe and smaller may be threaded, although welded systems are superior.
   b. 1-1/2" pipe and larger must be welded.
5. Recommended low pressure side piping requirements:
   a. 1-1/4" and smaller:
      Pipe: Black steel pipe, *ASTM A53, Schedule 80*, Type E or S, Grade B or carbon steel pipe, *ASTM A106, Schedule 80*, Type S, Grade B.
      Fittings: Forged steel threaded fittings, 300 lbs.
      Joints: Pipe threads, general purpose (American) *ANSI/ASME B1.20.1*.
      OR
      Pipe: Black steel pipe, *ASTM A53, Schedule 80*, Type E or S, Grade B or carbon steel pipe, *ASTM A106, Schedule 80*, Type S, Grade B.
      Fittings: Forged steel socket weld, 150 lbs. *ANSI B16.11*.
      Joints: Welded pipe, *ANSI/AWS D1.1* and *ANSI/ASME Sec. 9*.
   b. 1-1/2":
      Pipe: Black steel pipe, *ASTM A53, Schedule 80*, Type E or S, Grade B or carbon steel pipe, *ASTM A106, Schedule 80*, Type S, Grade B.
      Fittings: Forged steel socket weld, 150 lbs. *ANSI B16.11*.
      Joints: Welded pipe, *ANSI/AWS D1.1* and *ANSI/ASME Sec. 9*.
   c. 2" and larger:
      Pipe: Black steel pipe, *ASTM A53, Schedule 40*, Type E or S, Grade B or carbon steel pipe, *ASTM A106, Schedule 40*, Type S, Grade B.
      Fittings: Steel butt-welding fittings, 150 lbs., *ANSI/ASME B16.9*.
      Joints: Welded pipe, *ANSI/AWS D1.1* and *ANSI/ASME Sec. 9*.

6. Recommended high pressure side piping requirements:
   a. 1-1/4" and smaller:
      Pipe: Black steel pipe, *ASTM A53, Schedule 80*, Type E or S, Grade B or carbon
                 steel pipe, *ASTM A106, Schedule 80*, Type S, Grade B.
      Fittings: Forged steel threaded fittings, 300 lbs.
      Joints: Pipe threads, general purpose (American) *ANSI/ASME B1.20.1*.
      OR
      Pipe: Black steel pipe, *ASTM A53, Schedule 80*, Type E or S, Grade B or carbon
                 steel pipe, *ASTM A106, Schedule 80*, Type S, Grade B.
      Fittings: Forged steel socket weld, 300 lbs. *ANSI B16.11*.
      Joints: Welded pipe, *ANSI/AWS D1.1* and *ANSI/ASME Sec. 9*.
   b. 1-1/2":
      Pipe: Black steel pipe, *ASTM A53, Schedule 80*, Type E or S, Grade B or carbon
                 steel pipe, *ASTM A106, Schedule 80*, Type S, Grade B.
      Fittings: Forged steel socket weld, 300 lbs. *ANSI B16.11*.
      Joints: Welded pipe, *ANSI/AWS D1.1* and *ANSI/ASME Sec. 9*.
   c. 2" and larger:
      Pipe: Black steel pipe, *ASTM A53, Schedule 40*, Type E or S, Grade B or carbon
                 steel pipe, *ASTM A106, Schedule 40*, Type S, Grade B.
      Fittings: Steel butt-welding fittings, 300 lbs., *ANSI/ASME B16.9*.
      Joints: Welded pipe, *ANSI/AWS D1.1* and *ANSI/ASME Sec. 9*.

E. Refrigerant Piping Installation

1. Slope piping 1 percent in direction of oil return.
2. Install horizontal hot gas discharge piping with 1/2" per 10 feet downward slope away from the compressor.
3. Install horizontal suction lines with 1/2" per 10 feet downward slope to the compressor, with no long traps or dead ends that may cause oil to separate from the suction gas and return to the compressor in damaging slugs.
4. Liquid lines may be installed level.
5. Provide line size liquid indicators in the main liquid line leaving the condenser or receiver. Install moisture-liquid indicators in liquid lines between filter dryers and thermostatic expansion valves and in the liquid line to receiver.
6. Provide a line size strainer upstream of each automatic valve. Provide a shutoff valve on each side of the strainer.
7. Provide permanent filter dryers in low temperature systems and systems using hermetic compressors.
8. Provide replaceable cartridge filter dryers with three-valve bypass assembly for solenoid valves that are adjacent to receivers.
9. Provide refrigerant charging valve connections in the liquid line between the receiver shutoff valve and expansion valve.
10. Normally, only refrigerant suction lines are insulated, but liquid lines should be insulated where condensation will become a problem, and hot gas lines should be insulated where personal injury from contact may pose a problem.
11. Refrigerant lines should be installed a minimum of 7'6" above the floor.

F. Refrigerant Properties

1. Halocarbon refrigerants absorb 40-80 Btuh/lb., and ammonia absorbs 500–600 Btuh/lb.
2. Ammonia refrigeration systems require smaller piping than halocarbon refrigeration systems for the same pressure drop and capacity.
3. Human or living tissue contact with many refrigerants in their liquid state can cause instant freezing, frostbite, solvent defatting or dehydration, and/or caustic or acid burns.
4. Leak detectors are essential for all halocarbon refrigerants because they are generally heavier than air, are odorless, and can cause suffocation due to oxygen depravation. Ammonia is lighter than air and has a distinctive and unmistakable odor.

5. Ammonia properties:
   a. Refrigerant grade ammonia:
      1) 99.98 percent ammonia minimum.
      2) 0.015 percent water maximum.
      3) 3 ppm oil maximum.
      4) 0.2 mL/g noncondensable gases.
   b. Agricultural grade ammonia:
      1) 99.5 percent ammonia minimum.
      2) 0.5 percent water maximum.
      3) 0.2 percent water minimum.
      4) 5 ppm oil maximum.
   c. Ammonia limitations are shown in the following table.

Concentration of Ammonia in the Air	Limitations/Symptoms
4 ppm	Detectable by human sense of smell.
25 ppm	Maximum ACGIH Permissible Exposure Limit (PEL). Maximum European Government Limit.
30–35 ppm	Uncomfortable—breathing support desired or required. Common level around ammonia print machines. Maximum recommended exposure 15 minutes (ACGIH).
50 ppm	Maximum OSHA and NIOSH Permissible Exposure Limit (PEL).
100 ppm	Noticeable irritation to the eyes, throat, and mucous membranes.
400 ppm	Mucous membranes may be destroyed with prolonged contact with ammonia. No serious health threat with infrequent and less-than-one-hour exposures.
500 ppm	Immediate Danger to Life and Health (IDLH) Limit.
700 ppm	Significant eye irritation.
1,700 ppm	Convulsive coughing occurs. Fatal after short exposures of less than one half hour.
2,500 ppm	Exposure in as short a time as 30 minutes is dangerous. Affects show up several days later—pulmonary edema (water in the lungs).
5,000 ppm and above	Immediate hazard to life due to suffocation. Full face respiratory protection is required, including eyes. Causes respiratory spasm, strangulation, and asphyxia—no exposure permissible.
15,000 ppm and above	Full body protection required. Ammonia reacts with body perspiration to form a caustic solution that attacks the skin causing burns and blisters.
160,000–270,000 ppm	Flammable in air at 68°F.
15.5% by volume	Lower Flammability Limit (LFL); also referred to Lower Explosive Limit (LEL)

6. Refrigerant physical properties are shown in the following table.

Refrigerant Physical Properties								
Refrigerant		ASHRAE Std. 15 Group No.	Molecular Mass	Boiling Point at 14.7 Psia °F	Freezing Point °F	Critical		
No.	Name					Temp. °F	Press. psia	Volume ft.³/lb.
R-11	—	A1	137.38	74.87	−168.0	388.4	639.5	0.0289
R-12	—	A1	120.93	−21.62	−252.0	233.6	596.9	0.0287
R-13	—	A1	104.47	−114.60	−294.0	83.9	561.0	0.0277
R-13B1	—	A1	148.93	−71.95	−270.0	152.6	575.0	0.0215
R-14	—	A1	88.01	−198.30	−299.0	−50.2	543.0	0.0256
R-22	—	A1	86.48	−41.36	−256.0	204.8	721.9	0.0305
R-40	—	B2	50.49	−11.60	−144.0	289.6	968.7	0.0454
R-113	—	A1	187.39	117.63	−31.0	417.4	498.9	0.0278
R-114	—	A1	170.94	38.80	−137.0	294.3	473.0	0.0275
R-115	—	A1	154.48	−38.40	−159.0	175.9	457.6	0.0261
R-123	—	B1	152.93	82.17	−160.9	362.8	532.9	—
R-134a	—	A1	102.03	−15.08	−141.9	214.0	589.8	0.0290
R-142b	—	A2	100.50	14.40	−204.0	278.8	598.0	0.0368
R-152a	—	A2	66.05	−13.00	−178.6	236.3	652.0	0.0439

(*Continued*)

Refrigerant Physical Properties								
Refrigerant		ASHRAE		Boiling Point		Critical		
No.	Name	Std. 15 Group No.	Molecular Mass	at 14.7 Psia °F	Freezing Point °F	Temp. °F	Press. psia	Volume ft.³/lb.
R-170	Ethane	A3	30.07	−127.85	−297.0	90.0	709.8	0.0830
R-290	Propane	A3	44.10	−43.73	−305.8	206.3	617.4	0.0728
R-C318	—	A1	200.04	21.50	−42.5	239.6	403.6	0.0258
R-410A	—	A1	72.58	−60.84	—	161.83	714.50	0.0328
R-500	—	A1	99.31	−28.30	−254.0	221.9	641.9	0.0323
R-502	—	A1	111.63	−49.80	—	179.9	591.0	0.0286
R-503	—	A1	87.50	−127.60	—	67.1	607.0	0.0326
R-600	Butane	A3	58.13	31.10	−217.3	305.6	550.7	0.0702
R-600a	Isobutane	A3	58.13	10.89	−255.5	275.0	529.1	0.0725
R-611	—	B2	60.05	89.20	−146.0	417.2	870.0	0.0459
R-717	Ammonia	B2	17.03	−28.00	−107.9	271.4	1657.0	0.0680
R-744	Carbon dioxide	A1	44.01	−109.20	−69.9	87.9	1070.0	0.0342
R-764	Sulfur dioxide	B1	64.07	14.00	−103.9	315.5	1143.0	0.0306
R-1150	Ethylene	A3	28.05	−154.7	−272.0	48.8	742.2	0.0700
R-1270	Propylene	A3	42.09	−53.86	−301.0	197.2	670.3	0.0720

	Energy Absorption Rate Btu/lb.				
Refrigerant Type	40°F	20°F	0°F	−20°F	−40°F
R-11	80.863	82.507	84.126	85.732	87.335
R-12	64.649	66.953	69.098	71.116	73.038
R-22	86.503	90.344	93.891	97.193	100.296
R-123	76.787	78.078	79.167	80.162	81.340
R-134a	84.011	87.589	90.925	94.063	97.050
R-502	61.687	65.069	68.101	70.795	73.162
R-717 Ammonia	535.936	552.858	568.692	583.540	597.482

Air Handling Units

25.01 Air Handling Units, Air Conditioning Units, Heat Pumps

A. Definitions

1. *Air Handling Units (AHUs)*. AHUs contain fans, filters, coils, and other items but do not contain refrigeration compressors.
2. *Air Conditioning Units (ACUs)*. ACUs are AHUs that contain refrigeration compressors.
3. *Heat Pumps*. Heat pumps are ACUs with refrigeration systems capable of providing heat to the space as well as cooling.

B. Air Handling Unit Types

1. Packaged AHUs (central station AHUs):
   a. 800–50,000 CFM.
   b. 0–9" SP.
   c. 1/4–100 hp.
2. Factory-fabricated AHUs (custom AHUs):
   a. 1,000–125,000 CFM.
   b. 0–13" SP.
   c. 1/4–500 hp.
   d. Shipping limiting factor; two to three times more expensive than packaged AHUs.
3. Field-fabricated AHUs:
   a. 10,000–804,000 CFM.
   b. 0–14" SP.
   c. 2–2500 hp.
   d. Fan size limiting factor.

C. Packaged Equipment, All Spaces

1. 300–500 CFM/ton @ 20°F ΔT.
2. 400 CFM/ton @ 20°F ΔT (typical).

D. Water Source Heat Pumps

1. Water heat rejection:
   a. 2.0–3.0 GPM/ton @ 15–10°F ΔT.
   b. 3.0 GPM/ton @ 10°F ΔT recommended.
2. 85–95°F Condenser water temperature.
3. 60–90°F Heat pump water loop temperatures:
   a. Winter design: 60°F.
   b. Summer design: 90°F.
4. Cooling tower, evap. cooler sizing:
   a. 1.4 × Block Cooling Load.
5. Supplemental heater sizing:
   b. 0.75 × Block Heating Load.

E. Geothermal Source Heat Pumps

1. Efficiencies:
   a. Average: 3.5–4.7 COP; 12–16 EER.
   b. High: 5.3–5.9 COP; 18–20 EER.
2. Vertical wells used for heat transfer are the most common system type in lieu of horizontal heat transfer sites.
3. Length of heat exchanger pipe required:
   a. Range: 130 ft./ton–175 ft./ton.
   b. Average: 150 ft./ton.
4. 50–110°F Heat pump water loop temperatures.
5. If the system is sized to meet cooling requirements, supplemental heat will not be required.

6. If the system is sized to meet heating requirements, a supplemental cooling tower will be required.
7. Pipe spacing:
   a. Commercial: 15 ft. × 15 ft. center to center grid.
   b. Residential: 10 ft. × 10 ft. center to center grid.

F. Air Handling Unit Fans

1. 1/2°F temperature rise for each 1" S.P. from fan heat.
2. See Part 26, Fans for more information.
3. A return air system with more than a 1/2" pressure drop should have a return air fan. A return air fan is also required if you intend to use an economizer and still maintain the space under a neutral or negative pressure.

G. Economizers

1. Water side economizers take advantage of low condenser water temperature to either precool entering air, assist in mechanical cooling, or to provide total system cooling.
2. Air side economizers take advantage of cool outdoor air to either assist in mechanical cooling or to provide total system cooling.
   a. Dry bulb.
   b. Enthalpy—required by energy conservation codes.

H. System Types

1. VAV systems:
   a. Fans selected for 100 percent block airflow.
   b. Normal operation 60–80 percent block airflow.
   c. Minimum airflow 30–50 percent block airflow.
2. Constant volume reheat systems:
   a. Fans selected and operated at a 100 percent sum of peak airflow.
   b. Constant volume systems are generally not permitted by energy conservation codes. If employed, a supply temperature reset must be employed.
3. Hybrid VAV/constant volume reheat systems:
   a. Fans selected for 100 percent block airflow for VAV spaces, plus 100 percent sum of peak airflow for constant volume spaces.
   b. Normal operation 60–80 percent of the system design airflow.
   c. Minimum airflow 30–50 percent of the system design airflow.
4. Dual duct systems:
   a. Cold deck designed for 100 percent of the sum of peak airflow.
   b. Hot deck designed for 75–90 percent of the sum of peak airflow.
   c. Fans selected and operated at 100 percent of the sum of peak airflow.
   d. Dual duct systems are generally not permitted by energy conservation codes. If employed, a cold deck and hot deck supply temperature reset must be employed.
5. Dual duct VAV systems:
   a. Cold deck designed for 100 percent of block airflow.
   b. Hot deck designed for 75–90 percent of block airflow.
   c. Fans selected for 100 percent block airflow.
   d. Normal operation 60–80 percent block airflow.
   e. Minimum airflow 30–50 percent block airflow.
6. Single zone and multizone systems:
   a. Cold deck designed for 100 percent of the sum of peak airflow.
   b. Hot deck designed for 75–90 percent of the sum of peak airflow.
   c. Fans selected and operated at 100 percent of the sum of peak airflow.

I. Clearance Requirements

1. Minimum recommended clearance around air handling units and similar equipment is 24 inches on the nonservice side and 36 inches on the service side. Maintain minimum clearance for coil pull as recommended by the equipment manufacturer; this is generally

equal to the width of the air handling unit. Maintain minimum clearance as required to open access and control doors on air handling units for service, maintenance, and inspection.

2. Mechanical room locations and placement must take into account how large air handling units and similar equipment can be moved into and out of the building during the initial installation and after construction for maintenance and repair and/or replacement.

25.02 Coils

A. General

1. Field-erected and factory-assembled air handling unit coils should be arranged for removal from the upstream side without dismantling supports. Provide galvanized structural steel supports for all coils (except the lowest coil) in banks over two coils high to permit the independent removal of any coil.
2. When air handling units are used to supply makeup air (100 percent OA) for smoke control/smoke management systems, water coil freeze up must be considered. Some possible solutions are listed in the following:
   a. Provide preheat coil in AHU to heat the air from the outside design temperature to 45–50°F.
   b. Provide control of the system to open all water coil control valves serving smoke control/smoke management systems to full open and circulate water through the coils.
   c. Elect not to provide freeze protection with owner concurrence in the event a fire or other emergency occurs on a cold day. Also, many emergency situations are fairly short in duration. A follow-up letter should also be written.
3. Select water coils with tube velocities high enough at design flow so that tube velocities do not end up in the laminar flow region when the flow is reduced in response low-load conditions. Tube velocities become critical with units designed for 100 percent outside air at low loads near 32°F. Higher tube velocity selection results in a higher water pressure drop for the water coil. Sometimes a trade-off between pressure drop and low load flows must be evaluated.
4. It is best to use water coils with same end connections to reduce flow imbalances caused by differences in velocity head.
5. In horizontal water coil headers, supply water flow should be downward, while return water flow should be upward for proper air venting.
6. Water coil flow patterns:
   a. Multiple path, parallel flow, grid type coil.
   b. Series flow, serpentine coil.
   c. Series and parallel flow.

B. Air Handling Unit Coil Designations

1. Preheat coils normally heat air to a desired setpoint level (quite often the setpoint is the cooling coil discharge temperature plus or minus 5°F). This setpoint temperature may or may not be adequate for maintaining space temperature for human comfort; however, it is generally adequate to prevent freezing and also for equipment room heating. Preheat coils may be hot water, steam, or electric type coils, or they may be direct-fired or indirect-fired gas heaters. Preheat coil (water or steam type) freeze protection methods are listed in the following:
   a. Preheat pumps (primary/secondary system).
   b. Internal face and bypass coils.
   c. Integral face and bypass dampers.
   d. Preheat coils are required whenever the design mixed air temperature is below 40°F or when 100 percent outside air units have an outside design temperature below 40°F.

2. Cooling coils provide both the sensible and latent cooling required to maintain temperature and humidity levels. Cooling coils are either chilled water or DX (refrigerant) type coils.
3. Heating coils are designed to maintain space temperatures acceptable for human comfort or process requirements. Heating coils should not operate when cooling coils are operating. Automatic temperature control interlocks should be established to prevent simultaneous operation. Heating coils may be hot water, steam, or electric type coils, or they may be direct-fired or indirect-fired gas heaters.
4. Reheat coils will often operate in conjunction with cooling coils to maintain a temperature and/or relative humidity acceptable for human comfort or process requirements. Reheat coils may be hot water, steam, or electric type coils.

C. Water and Steam Coils

1. Preheat:
   a. Concurrent air/water or steam flow.
   b. Freeze protection:
      1) Preheat pumps (primary/secondary system). See Fig. 25.1.
      2) Face and bypass dampers—internal. See Fig. 25.2.
      3) Integral face and bypass (IFB) coils. See Fig. 25.3.

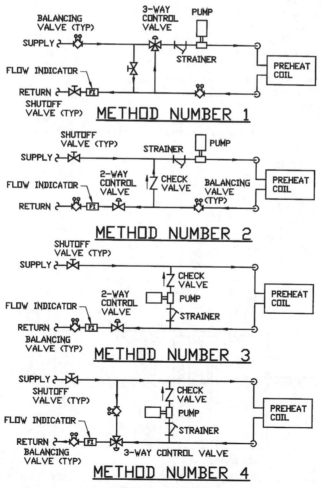

FIGURE 25.1   PREHEAT COIL PIPING DIAGRAMS.

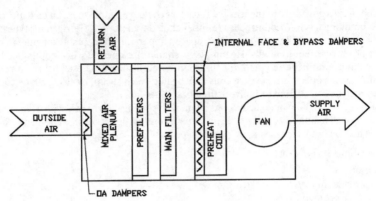

FIGURE 25.2 AIR HANDLING UNITS W/INTERNAL FACE AND BYPASS DAMPERS (PREHEAT COIL FREEZE PROTECTION).

2. Cooling, heating, reheat:
   a. Counter air/water or steam flow.
3. Cooling coil face velocity:
   a. 450–550 fpm range.
   b. 500 fpm recommended.
   c. 450 fpm preferred.
4. Preheat, heating, and reheat coil face velocity:
   a. 500–900 fpm range.
   b. 600–700 fpm recommended.
   c. 600 fpm preferred.
   d. Use a preheat coil whenever the mixed air temperature (outside air and return air) is below 40°F.

D. Refrigerant Coils

1. Cooling:
   a. Counter air/refrigerant flow.
   b. Cooling coil face velocity:
      1) 450–550 fpm range.
      2) 500 fpm recommended.
      3) 450 fpm preferred.

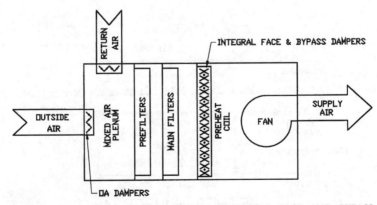

FIGURE 25.3 AIR HANDLING UNITS W/INTEGRAL FACE AND BYPASS DAMPERS (PREHEAT COIL FREEZE PROTECTION).

E. Weight and Volume of Water in Standard Water Coils

1. Weight of water in the tubes:

$$W_{WT} = 0.966 \text{ lbs./row sq.ft.} \times \text{No. of Rows} \times \text{Face Area of Coil}$$

2. Total weight of water in coil:

$$W_{WC} = W_{WT} + W_{WH}$$

3. Total weight of water coils:

$$W_T = W_C + W_{WC}$$

4. Volume of water in coil:

$$V = W_{WC} \times 0.12$$

where

W_{WT} = water weight in the tubes (pounds)
W_{WH} = water weight in the headers/U-bends, from table (pounds)
W_{WC} = water weight in the coil (pounds)
W_C = dry coil weight (pounds)
W_T = total weight of the coil (pounds)
V = volume of the coil (gallons)

	Weight of Water in Coil Headers and U-Bends						
	Number of Rows						
Finned Width	1	2	3	4	5	6	8
6"	0.75	1.75	—	—	—	—	—
9"	1.00	2.75	—	—	—	—	—
12"	1.50	3.26	3.84	4.04	4.75	4.94	7.61
18"	2.75	3.94	4.82	5.07	6.21	8.70	13.10
24"	3.85	5.28	6.50	6.86	8.37	11.61	17.60
30"	4.72	8.66	10.12	10.50	12.48	16.52	24.00
33"	5.21	9.50	11.09	11.58	13.54	17.99	26.10
36"	—	16.34	19.58	22.82	26.06	29.30	32.55
42"	—	18.95	22.73	26.51	30.29	34.07	37.85
48"	—	21.55	25.88	30.20	34.52	38.84	43.16

F. Coil Pressure Drop

1. Air pressure drop (water, steam, refrigerant coils) is given in the following table:
   a. Cooling coils:
      1) Range: 0.5–1.0" WC.
      2) Recommended schedule value: 0.75" WC.
   b. Dehumidification/heat recovery coils:
      1) Range: 1.0–1.5" WC.
      2) Recommended schedule value: 1.25" WC.
   c. Heating coils:
      1) Range: 0.1–0.25" WC.
      2) Recommended schedule value: 0.15" WC.

Number of Rows	Face Velocity (fpm)						
	450	500	550	600	700	800	900
1	0.05–0.15	0.05–0.18	0.08–0.20	0.08–0.25	0.12–0.30	0.15–0.40	0.17–0.50
2	0.10–0.35	0.11–0.50	0.15–0.50	0.16–0.60	0.20–0.80	0.25–0.90	0.32–0.90
4	0.20–0.70	0.22–0.90	0.28–1.00	0.33–1.20	0.40–1.50	0.50–1.80	0.65–1.70
6	0.30–1.10	0.35–1.30	0.45–1.50	0.50–1.70	0.65–2.30	0.75–2.80	1.00–2.70
8	0.40–1.50	0.45–1.75	0.60–2.00	0.60–2.40	0.85–3.00	1.00–3.70	1.30–3.70
10	0.50–1.75	0.60–2.25	0.70–2.50	0.80–3.00	1.10–3.80	1.30–4.50	1.70–4.50

Notes:
1 Lower pressure drop is for 70 fins/ft.
2 Higher pressure drop is for 170 fins/ft.
3 Pressure drops in in. W.G.

2. Water pressure drop is given in the following table:
   a. Cooling coils:
      1) Range: 10–20 ft. H_2O.
      2) Recommended schedule value: 15 ft. H_2O.
   b. Dehumidification/heat recovery coils:
      1) Range: 10–20 ft. H_2O.
      2) Recommended schedule value: 15 ft. H_2O.
   c. Heating coils:
      1) Range: 1–5 ft. H_2O.
      2) Recommended schedule value: 2.5 ft. H_2O.

Finned Width	Finned Length											
	12	24	36	48	60	72	84	96	108	120	132	144
12	0.11	0.13	0.14	0.15	0.16	0.17	0.18	0.19	0.20	0.21	0.22	0.23
	8.77	10.1	11.6	13.1	14.6	16.2	17.7	19.2	20.7	22.2	23.7	25.2
18	0.07	0.09	0.10	0.11	0.12	0.13	0.14	0.15	0.16	0.17	0.18	0.19
	6.31	7.65	9.16	10.7	12.2	13.7	15.2	16.7	18.2	19.7	21.2	22.3
24	0.09	0.11	0.12	0.13	0.14	0.15	0.16	0.17	0.18	0.19	0.20	0.21
	8.21	9.55	11.1	12.6	14.1	15.6	17.1	18.6	20.1	21.7	23.2	24.7
30	0.12	0.14	0.15	0.16	0.17	0.18	0.19	0.20	0.21	0.22	0.23	0.24
	10.3	11.6	13.2	14.7	16.2	17.7	19.2	20.7	22.2	23.7	25.3	26.8
33	0.15	0.17	0.18	0.19	0.20	0.21	0.22	0.23	0.24	0.25	0.26	0.27
	11.4	12.7	14.2	15.7	17.2	18.7	20.2	21.8	23.3	24.8	26.3	27.8
36	0.17	0.19	0.20	0.21	0.22	0.23	0.24	0.25	0.26	0.27	0.28	0.29
	13.2	14.5	16.1	17.5	19.0	20.5	22.1	23.6	25.1	26.6	28.1	29.6
42	0.20	0.22	0.23	0.24	0.25	0.26	0.27	0.28	0.29	0.30	0.31	0.32
	14.7	16.1	17.5	19.1	20.6	22.1	23.6	25.1	26.6	28.1	29.6	31.1
48	0.22	0.24	0.25	0.26	0.27	0.28	0.29	0.30	0.31	0.32	0.33	0.34
	16.4	17.8	19.3	20.8	22.3	23.8	25.3	26.8	28.3	29.8	31.3	32.9

Notes:
1 Pressure drops in feet H_2O/row.
2 Top row is based on water velocity of 1.0 FPS.
3 Bottom row is based on water velocity of 8.0 FPS.
4 Water velocity (FPS) = (GPM × 1.66)/finned width.
5 Based on W type coil.

G. Electric Coils

1. Open coils: Use when personnel contact is not a concern. It is the most common type of electric coil used in HVAC applications.
   a. Air pressure drops:
      1) 400–900 fpm 0.01–0.10 WG.
   b. Minimum velocity:
      1) 400 fpm   6 KW/sq.ft. of duct.
      2) 500 fpm   8 KW/sq.ft. of duct.
      3) 600 fpm   10 KW/sq.ft. of duct.
      4) 700 fpm   12 KW/sq.ft. of duct.
      5) 800 fpm   14 KW/sq.ft. of duct.
      6) 900 fpm   16 KW/sq.ft. of duct.
      7) The manufacturer's literature should be consulted.
2. Finned tubular coils: Use when personnel contact is a concern.
   a. Air pressure drops:
      1) 400–900 fpm   0.02–0.20 WG.
   b. Minimum velocity:
      1) 400 fpm   6 KW/sq.ft. of duct.
      2) 500 fpm   9 KW/sq.ft. of duct.
      3) 600 fpm   12 KW/sq.ft. of duct.
      4) 700 fpm   15 KW/sq.ft. of duct.
      5) 800 fpm   17 KW/sq.ft. of duct.
      6) 900 fpm   20 KW/sq.ft. of duct.
      7) Manufacturer's literature should be consulted.

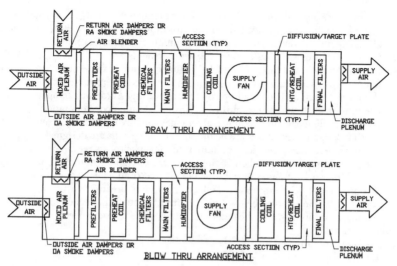

FIGURE 25.4 AIR HANDLING UNIT TERMINOLOGY.

H. Air Handling Units

1. Blow through versus draw through: The terminology of blow through and draw through air handling units is generally in reference to the cooling coil location. If the cooling coil is downstream of the fan, the unit is considered a blow through air handling unit. If the cooling coil is upstream of the fan, the unit is considered a draw through air handling unit. See Figs. 25.4 and 25.5.

2. Air handling unit terminology drawings show a number of different components. The design of air handling units may incorporate any number or combination of the components.

3. Coil arrangements:

   a. Preheat/cooling: Preheat/cooling coil arrangements are used when mixed air or outside air design temperatures are below 40°F. The preheat coil heats the air to a desired setpoint level; quite often the setpoint is the cooling coil discharge

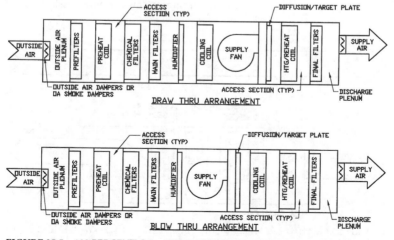

FIGURE 25.5 100 PERCENT O.A. AIR HANDLING UNIT TERMINOLOGY.

temperature plus or minus 5°F. This setpoint temperature may or may not be adequate for maintaining space temperature for human comfort; however, it is generally adequate to prevent freezing and for equipment room heating.

    b. Cooling/heating: Cooling/heating coil arrangements are used when the mixed air temperature will not fall below 40°F. Heating coils are designed to maintain space temperatures acceptable for human comfort or process requirements. Heating coils should not operate when cooling coils are operating. Automatic temperature control interlocks should be established to prevent simultaneous operation.

    c. Heating/cooling: Heating/cooling coil arrangements are used when mixed air or outside air design temperatures are below 40°F. Heating coils are designed to maintain space temperatures acceptable for human comfort or process requirements. Heating coils should not operate when cooling coils are operating. Automatic temperature control interlocks should be established to prevent simultaneous operation.

    d. Cooling/reheat: Cooling/reheat coil arrangements are used when air must be cooled to a temperature below that required to satisfy the space temperature to remove moisture and then heated to maintain space temperature.

4. Filter terminology:

    a. Prefilters (required):

      1) First stage of filtration.

      2) Filtration level guideline: 30–60 percent.

      3) Prefilters are required for air handling maintenance and operating requirements.

    b. Main filters (recommended):

      1) Second stage of filtration.

      2) Filtration level guideline: 60–90 percent.

      3) Two stages of filtration are recommended in nearly all air handling systems because of current and future indoor air quality standards and requirements.

    c. Final filters (optional):

      1) Last stage of filtration.

      2) Filtration level guideline: 90 percent to HEPA/ULPA filtration levels.

      3) Use final filters whenever clean air is required at space (hospital operating rooms, nurseries, cleanrooms, laboratories).

5. Coils and filters located immediately downstream of fans will require a target/diffusion plate to distribute air evenly over the coil or filter and to prevent damage to that device, especially filters.

6. Access sections are recommended between each and every component in the air handling unit. However, the prefilters may be adjacent to the main filters without access between them, provided both sets of filters (prefilters and main filters) can be removed without having to remove the other (side access or upstream/downstream access).

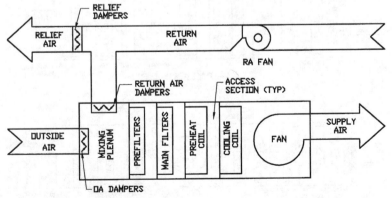

FIGURE 25.6 AHU EXAMPLE 1 AIR HANDLING UNITS—W/PREHEAT AND COOLING COILS W/RETURN AIR, RETURN FAN, AND POWERED RELIEF AIR.

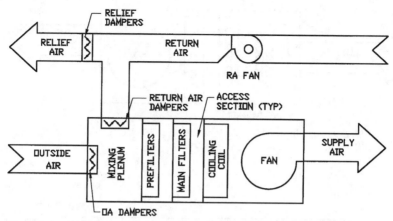

FIGURE 25.7   AHU EXAMPLE 2 AIR HANDLING UNITS—W/COOLING COIL ONLY W/RETURN AIR, RETURN FAN, AND POWERED RELIEF AIR.

7. Air blenders are used to promote proper mixing of the return air and the outside air flow streams and to prevent air stratification within the air handling unit. The use of air blenders will reduce the risk of localized freezing of water coils.
8. Smoke dampers and smoke detectors have not been shown on the air handling unit flow diagrams. Smoke dampers and smoke detectors may be required in the supply, return, or outside air ductwork depending on unit capacity, service, and code requirements. Verify smoke damper and smoke detector requirements with NFPA 90A, IBC, and local code requirements.
9. See Figs. 25.6 through 25.10 for examples of air handling units and a few of the many possible arrangements.

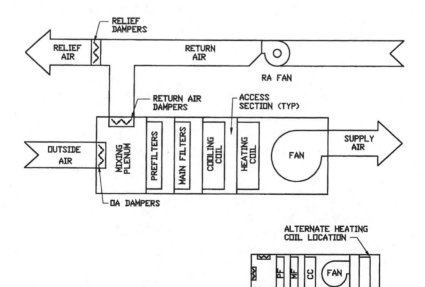

FIGURE 25.8   AHU EXAMPLE 3 AIR HANDLING UNITS—W/COOLING AND HEATING COILS W/RETURN AIR, RETURN FAN, AND POWERED RELIEF AIR.

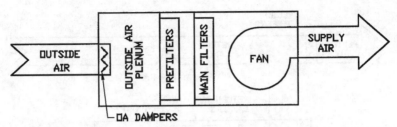

FIGURE 25.9   AHU EXAMPLE 4 AIR HANDLING UNITS—VENTILATING 100 PERCENT OUTSIDE AIR.

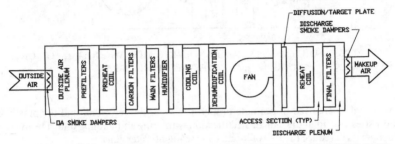

FIGURE 25.10   AHU EXAMPLE 5 CLEANROOM MAKEUP AIR HANDLING UNIT W/CARBON FILTERS AND 100 PERCENT OUTSIDE AIR.

Fans

26.01 Fan Types and Size Ranges

A. Fan types and size ranges are shown in the following table.

FAN COMPARISON TABLE					
Fan Type	Wheel\Drive Type	Sp in. W.G.	Wheel Dia. in.	CFM	hp
Utility Sets	FC/B	0–3	8–36	200–27,500	1/6–30
	BI/B	0–4	10–36	250–27,500	1/6–30
	FC/D	0–2.5	6–12	100–3,500	1/6–3
Centrifugal	SWSI-BI/B	0–12	10–73	600–123,000	1/3–200
	DWDI-BI/B	0–12	12–73	1,300–225,000	1/3–400
	SWSI-AF/B	0–14	18–120	1,400–447,000	1/3–1500
	DWDI-AF/B	0–14	18–120	2,400–804,000	3/4–2500
Tubular Centrifugal	BI/B BI/D	0–9	10–108	450–332,000	1/3–750
Vane Axial	–/B	0–5	18–72	1,400–115,000	1/3–100
	–/D	0–4	18–60	1,200–148,000	1/3–150
Tubeaxial	–/B	0–1.5	12–60	900–76,000	1/3–25
	–/D	0–1	18–48	2,600–48,000	1/4–15
Mixed Flow	–/B	0–8.5	15–54	2,000–95,000	1/4–100
	–/D	0–9.0	15–54	1,000–95,000	1/4–100
Propeller	–/B	0–1	20–72	400–80,000	1/4–15
	–/D	0–1	8–48	50–49,000	1/6–10
Roof Ventilator	BI/B	0–1.25	7–54	100–34,000	1/4–7.5
	BI/D	0–1	6–18	75–3,200	1/8–3/4
Roof Upblast	BI/B	0–1.25	9–48	200–26,000	1/4–5
	BI/D	0–1.25	9–14	300–3,100	1/8–1
Sidewall	BI/B	0–1.25	14–24	850–8,200	1/4–2
	BI/D	0–1	6–18	80–4,000	1/8–3/4
Inline Centrifugal	BI/B	0–2.25	7–36	60–22,600	1/4–10
	BI/D	0–1.75	6–16	60–5,100	1/8–2

Notes:
FC—Forward Curved   B—Belt Drive
BI—Backward Inclined   D—Direct Drive
AF—Backward Inclined Airfoil   DWDI—Double Width, Double Inlet
    SWSI—Single Width, Single Inlet

B. Refer to Figure 26.1 for a photograph of a roof ventilator in its installed condition.

26.02 Fan Construction Classes

A. Fan construction classes are shown in the following table:

Fan Class	Maximum Total Pressure
I	3–3/4" W.G.
II	6–3/4" W.G.
III	12–3/4" W.G.
IV	Over 12–3/4" W.G.

26.03 Fan Selection Criteria

A. Fan to be catalog rated for 15 percent greater static pressure (SP) than specified SP at specified volume.

B. Select the fan so that the specified volume is greater than at the apex of the fan curve.

C. Select the fan to provide a stable operation down to 85 percent of the design volume operating at a required speed for the specified conditions.

FIGURE 26.1   PHOTOGRAPH OF A ROOF VENTILATOR.

D. Specify SP at specified airflow.

E. Consider system effects. Fans are tested with open inlets and a length of straight duct on discharge. When field conditions differ from the test configuration, performance is reduced. Therefore, the fan must be selected at a slightly higher pressure to obtain the desired results.

F. Fan Design Arrangements (See Figure 26.2)

1. Series Fan Operation: At equal CFM, static pressure is additive.
2. Parallel Fan Operation: At equal static pressure, CFM is additive.
3. Standby Fans: Standby fan arrangements are often used for reliability purposes in the event of fan failure. Standby fans may be provided with coupled or headered systems (see Figs. 26.3 and 26.4).

G. Every attempt should be made to have 1.0–1.5 diameters of straight duct on the discharge of the fan as a minimum.

H. There should be a minimum of 1.0 diameter of straight duct between fan inlet and an elbow. In plenum installations, there should be a minimum of 0.75 of the wheel diameter between the fan inlet and the plenum wall.

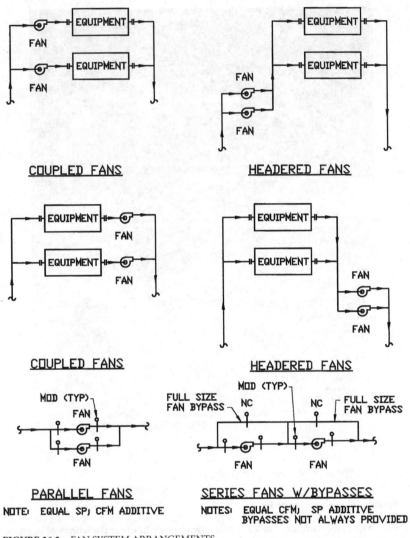

FIGURE 26.2   FAN SYSTEM ARRANGEMENTS.

26.04 Fan Terms

A. *Centrifugal*. Flow within the fan is substantially radial to the shaft.

B. *Axial*. Flow within the fan is substantially parallel to the shaft.

C. *Static Pressure*. Static pressure is the compressive pressure that exists in a confined air-stream. Static pressure is a measure of potential energy available to produce flow and to maintain flow against resistance. Static pressure is exerted in all directions and can be positive or negative (vacuum).

D. *Velocity Pressure*. Velocity pressure is the measure of the kinetic energy resulting from the fluid flow. Velocity pressure is exerted in the direction of fluid flow. Velocity pressure is always positive.

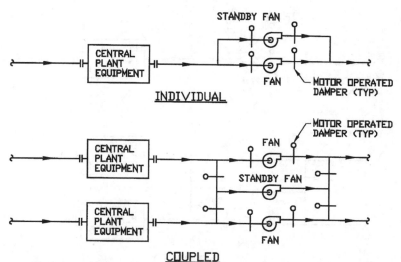

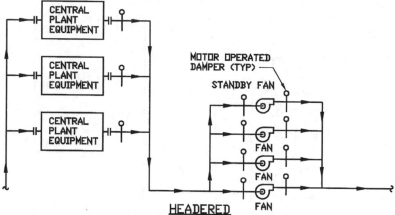

FIGURE 26.3   STANDBY FANS.

E. *Total Pressure.* Total pressure is the measure of the total energy of the airstream. Total pressure is equal to static pressure plus velocity pressure. Total pressure can be either positive or negative.

F. *Quantity* of *Airflow.* Volume measurement expressed in Cubic Feet per Minute (CFM).

G. *Fan Outlet Velocity.* Fan airflow divided by the fan outlet area.

H. *Fan Velocity Pressure.* Fan velocity pressure is derived by converting fan velocity to velocity pressure.

I. *Fan Total Pressure.* Fan total pressure is equal to the fan's outlet total pressure minus the fan's inlet total pressure.

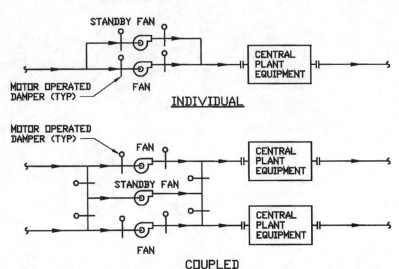

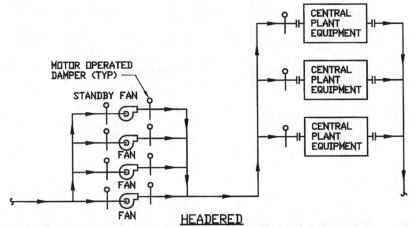

FIGURE 26.4   STANDBY FANS.

J. *Fan Static Pressure.* Fan static pressure is equal to the fan's total pressure minus the fan's velocity pressure. Numerically, it is equal to the fan's outlet static pressure minus the fan's inlet total pressure.

K. *Fan Horsepower.* Theoretical calculation of horsepower assuming there are no losses.

L. *Brake Horsepower (BHP).* Brake horsepower is the actual power required to drive the fan.

M. *System Effect.* System effect is the reduced fan performance of the manufacturer's fan catalog data due to the difference between field installed conditions and laboratory test conditions (precisely defined inlet and outlet ductwork geometry assuring uniform entrance and exit velocities).

1. Maintain a minimum of three duct diameters of straight duct upstream and downstream of the fan inlet and outlet at 2,500 feet per minute (fpm) duct velocity or less. One additional duct diameter should be added for each 1,000 fpm above 2,500 fpm.
2. Recommend maintaining a minimum of five duct diameters of straight duct upstream and downstream of the fan inlet and outlet at 2,500 feet per minute (fpm) duct velocity or less. One additional duct diameter should be added for each 1,000 fpm above 2,500 fpm.
3. The system effect may require a range of 3–20 duct diameters of straight duct upstream and downstream of the fan inlet and outlet.

26.05 AMCA Spark Resistant Construction

A. **Type A. All parts of the fan in contact with the airstream must be made of nonferrous material.**

B. **Type B. The fan shall have a nonferrous impeller and nonferrous ring about the opening through which the shaft passes. Ferrous hubs, shafts, and hardware are allowed if construction is such that a shift of the impeller or shaft will not permit two ferrous parts of the fan to rub or strike.**

C. **Type C. The fan must be so constructed that a shift of the wheel will not permit two ferrous parts of the fan to rub or strike.**

26.06 Centrifugal Fans

A. Forward Curved (FC) Fan

1. FC fans have a peak static pressure curve corresponding to the region of maximum efficiency, slightly to the right. Best efficiency at low or medium pressure (0–5 in. W.G.).
2. BHP is minimum at no delivery and increases continuously with increasing flow, with maximum BHP occurring at free delivery.
3. They have a steep pressure volume performance curve; therefore, a slight change in pressure will not greatly affect CFM.
4. Fan blades curve toward the direction of rotation.
5. Advantages:
   a. Low cost. Less expensive than BC, BI, or AF fans.
   b. Low speed (400–1,200 RPM) minimizes the shaft and bearing sizes.
   c. Large operating range: 30–80 percent wide open CFM.
   d. Highest efficiency occurs: 40–50 percent wide open CFM.
6. Disadvantages:
   a. Possibility of paralleling in multiple fan applications.
   b. Possibility of overloading.
   c. Weak structurally: Not capable of high speeds necessary for developing high static pressures.
7. Used primarily in low- to medium-pressure HVAC applications: central station air handling units, rooftop units, packaged units, residential furnaces.
8. High CFM, low static pressure.

B. Backward Inclined (BI) and Backward Curved (BC) Fans

1. BC fans have a peak static pressure curve that occurs to the left of the maximum static efficiency. Best efficiency at medium pressure (3.5–5.0 in. W.G.).
2. BHP increases to a maximum, and then decreases. They are nonoverloading fans.
3. They have a steep pressure volume performance curve; therefore, a slight change in pressure will not greatly affect CFM.

4. Fan operates at high speeds—1,200–2,400 RPM—about double that of FC fans for similar air quantity.
5. Blades curve away from, or incline from, the direction of rotation.
6. BI fans are less expensive than BC fans but do not have as great a range of high efficiency operation.
7. Advantages:
   a. Higher efficiencies than FC fans.
   b. Highest efficiency occurs: 50–60 percent wide open CFM.
   c. Good pressure characteristics.
   d. Stronger structural design makes it suitable for higher static pressures.
   e. Nonoverloading power characteristics.
8. Disadvantages:
   a. Higher speeds require larger shaft and bearings.
   b. Has a larger surge area than a forward curved fan.
   c. Operating range 40–80 percent of wide open CFM.
   d. Can be noisier than FC fans.
   e. More expensive than FC fans.
9. Used primarily in large HVAC applications where power savings are significant. Can be used in low-, medium-, and high-pressure systems.

C. Airfoil Fans (AF)

1. AF fans have a peak static pressure curve that occurs to the left of the maximum static efficiency. Best efficiency at medium pressure (4.0–8.0 in. W.G.).
2. BHP increases to a maximum, and then decreases. They are nonoverloading fans.
3. They have a steep pressure volume performance curve; therefore, a slight change in pressure will not greatly affect CFM.
4. Fan operates at high speeds—1,200–2,800 RPM—about double that of FC fans for similar air quantity.
5. Blades have an aerodynamic shape similar to an airplane wing and are backwardly curved (away from direction of rotation).
6. Advantages:
   a. Higher efficiencies than FC fans.
   b. Highest efficiency occurs: 50–60 percent wide open CFM.
   c. Good pressure characteristics.
   d. Stronger structural design makes it suitable for higher static pressures.
   e. Nonoverloading power characteristics.
7. Disadvantages:
   a. Higher speeds require a larger shaft and bearings.
   b. Has a larger surge area than a forward curved fan.
   c. Operating range 40–80 percent of wide open CFM.
   d. Can be noisier than FC fans.
   e. Most expensive centrifugal fan.
8. Used primarily in large HVAC applications where power savings are significant. Can be used in low-, medium-, and high-pressure systems.
9. Airfoil blade fans have a slightly higher efficiency and the surge area is slightly larger than backward inclined or backward curved fans.

D. Radial (RA) Fans

1. Radial fans have self-cleaning blades.
2. Fan horsepower increases with an increase in air quantity (overloads), while static pressure decreases.
3. RA fans operate at high speed and pressure—2,000–3,000 RPM.
4. Blades radiate from the center along the radius of fan.
5. Used in industrial applications to transport dust, particles, or materials handling. Not commonly used in HVAC applications.

26.07 Axial Fans

A. Propeller Fans

1. Low pressure, high CFM fans.
2. Horsepower is lowest at maximum flow.
3. Maximum efficiency is approximately 50 percent and is reached near free delivery.
4. No ductwork.
5. Blade rotation is perpendicular to the direction of airflow.
6. Advantages:
    a. High volumes, low pressures.
    b. BHP is lowest at free delivery.
    c. Inexpensive.
    d. Operates at relatively low speeds—900–1,800 RPM.
7. Disadvantages:
    a. Cannot handle static pressure.
    b. BHP increases with static pressure; could overload and shut off.
    c. Air delivery decreases with increases in air resistance.

B. Tubeaxial Fans

1. Heavy duty propeller fans arranged for duct connection. Fan blades have aerodynamic configuration.
2. Slightly higher efficiency than propeller fans.
3. Discharge air pattern is circular in shape and swirls, producing higher static losses in the discharge duct.
4. Used primarily in low- and medium-pressure, high-volume, ducted HVAC applications where the discharge side is not critical. Also used in industrial applications: fume hoods, spray booths, drying ovens.
5. Fans operate at high speeds—2,000–3,000 RPM.
6. Fans are noisy.
7. Fans may be constructed to be overloading or nonoverloading. Nonoverloading type fans are more common.
8. Advantages:
    a. Straight through design.
    b. Space savings.
    c. Capable of higher static pressures than propeller fans.
9. Disadvantages:
    a. The discharge swirl creates higher pressure drops.
    b. High noise level.

C. Vaneaxial Fans

1. Vaneaxial fans are tubeaxial fans with additional vanes to increase efficiency by straightening out airflow.
2. Vaneaxial fans are more costly than tubeaxial fans.
3. High-pressure characteristics with medium flow rate capabilities.
4. Fans operate at high speeds—2,000–3,000 RPM.
5. Fans are noisy.
6. Fans may be constructed to be overloading or nonoverloading. Nonoverloading type fans are more common.
7. Typical selection: 65–95 percent wide open CFM.
8. Used in general HVAC applications—low-, medium-, and high-pressure—where straight through flow and compact installation are required. Also used in industrial applications: usually more compact than comparable centrifugal type fans for the same duty.
9. Advantages:
    a. Discharge vanes increase efficiency and reduce discharge losses.
    b. Reduced size and straight through design.

    c. Space savings.

    d. Capable of higher static pressures than propeller fans.

10. Disadvantages:

    a. Maximum efficiency only 65 percent.

    b. Selection range: 65–90 percent wide open CFM.

    c. High noise level.

D. Tubular Centrifugal Fans

1. Tubular centrifugal fans are similar to backward inclined centrifugal fans except that the fan capacity and pressure capabilities are lower.

2. Tubular centrifugal fans have a lower efficiency than backward inclined centrifugal fans.

3. Tubular centrifugal fans have a peak static pressure curve that occurs to the left of the maximum static efficiency.

4. BHP increases to a maximum, and then decreases. They are nonoverloading fans.

5. They have a steep pressure volume performance curve; therefore, a slight change in pressure will not greatly affect CFM.

6. The fan operates at high speeds—1,200–2,400.

7. Blades curve away from, or incline from, the direction of rotation.

8. Advantages:

    a. Good pressure characteristics.

    b. Nonoverloading power characteristics.

    c. The fan has straight through flow for inline duct applications.

9. Disadvantages:

    a. Higher speeds require a larger shaft and bearings.

    b. An operating range 40–80 percent of wide open CFM.

    c. Can be noisy.

10. Primarily used for low-pressure, return air HVAC systems.

E. Mixed Flow Fans

1. Mixed flow fans combine the best properties of tubeaxial, vaneaxial, and tubular centrifugal fans.

2. Mixed flow fans operate at a lower RPM than tubeaxial, vaneaxial, or centrifugal fans, resulting in less noise.

3. Used in general HVAC applications, low-, medium-, and high-pressure, where straight through flow and compact installation are required. Also used in industrial applications: usually more compact than comparable centrifugal type fans for the same duty.

4. Advantages:

    a. Less noisy than either the tubeaxial or vaneaxial fans.

    b. More efficient and therefore reduced horsepower requirements over tubeaxial, vaneaxial, or tubular centrifugal fans.

    c. Smaller physical size for equal airflow and static pressure requirements than tubeaxial, vaneaxial, or tubular centrifugal fans.

    d. Generally less expensive than comparable tubeaxial, vaneaxial, or tubular centrifugal fans.

26.08 Installation and Clearance Requirements

A. **The minimum recommended clearance around fans is 24 inches. Maintain minimum clearance as required to open access and control doors on fans for service, maintenance, and inspection.**

B. **Mechanical room locations and placement must take into account how fans can be moved into and out of the building during initial installation and after construction for maintenance and repair and/or replacement.**

26.09 Fan Rotation and Discharge Positions

See Fig. 26.5.

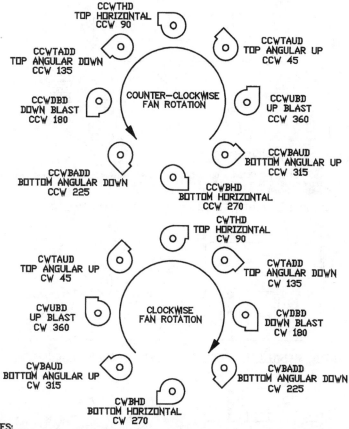

NOTES:
1. DIRECTION OF ROTATION IS DETERMINED FROM DRIVE SIDE OF FAN. ON SINGLE INLET FANS, THE DRIVE SIDE OF THE FAN IS ALWAYS CONSIDERED THE SIDE OPPOSITE THE FAN INLET.
2. ON DOUBLE INLET FANS, WHEN THE DRIVES ARE ON BOTH SIDES OF THE FAN, THE DRIVE SIDE OF THE FAN IS THE SIDE HAVING THE HIGHER HORSEPOWER DRIVING UNIT.
3. DIRECTION OF DISCHARGE IS DETERMINED IN ACCORDANCE WITH THE DIAGRAMS.
4. ANGULAR DISCHARGE IS REFERENCED TO THE HORIZONTAL AXIS OF THE FAN AND DESIGNATED IN DEGREES ABOVE OR BELOW THIS REFERENCE.
5. FANS INVERTED FOR CEILING SUSPENSION, OR SIDE WALL MOUNTING, DIRECTION OF ROTATION AND DISCHARGE IS DETERMINED WHEN FAN IS RESTING ON THE FLOOR.

FIGURE 26.5   FAN ROTATION AND DISCHARGE POSITIONS.

26.10 Fan Motor Positions

See Figs. 26.6 and 26.7.

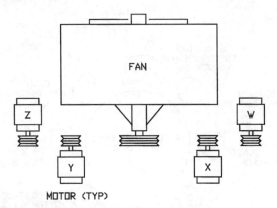

NOTES:
1. LOCATION OF THE MOTOR IS DETERMINED BY FACING THE DRIVE SIDE OF THE FAN OR BLOWER AND DESIGNATING THE MOTOR POSITION BY LETTERS W, X, Y, OR Z AS SHOWN ABOVE. FIGURE IS BASED ON AMCA STANDARD 2407.

FIGURE 26.6   CENTRIFUGAL FAN MOTOR ARRANGEMENTS.

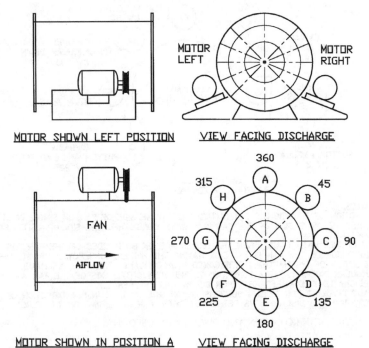

NOTES:
1. LOCATION OF THE MOTOR IS DETERMINED BY FACING THE DRIVE SIDE OF THE FAN OR BLOWER AND DESIGNATING THE MOTOR POSITION BY LETTERS A THROUGH H AS SHOWN ABOVE. FIGURE IS BASED ON AMCA STANDARD 2407.

FIGURE 26.7   INLINE FAN MOTOR ARRANGEMENTS.

Refer to the online resource for Sections 26.11 Fan Drive Arrangements, 26.12 Centrifugal Fan Inlet Box Positions, and 26.13 Centrifugal Fan Damper Arrangements for Reversible Flow. www.mheducation.com/HVACequations

Pumps

27.01 Pump Types and Size Ranges

A. Available RPM

1. 1,150 (1,200).
2. 1,750 (1,800).
3. 3,500 (3,600).

B. Pump types are shown in the following table.

Pump Type	GPM	Head Ft. H_2O	Horsepower
Circulators	0–150	0–60	1/4–5
Close coupled, end suction	0–2,000	0–400	1/4–150
Frame mounted, end suction	0–2,000	0–500	1/4–150
Horizontal split case	0–12,000	0–500	1–500
Vertical inline	0–2,000	0–400	1/4–75

C. Refer to Fig. 27.1 for a photograph of frame mounted, end suction pumps in their installed condition.

FIGURE 27.1 PHOTOGRAPH OF FRAME MOUNTED, END SUCTION PUMPS.

D. Pump Location

1. Heating water systems: Boilers to be on the suction side of pumps; pumps to draw through boilers.
2. Chilled water systems: Chillers to be on the discharge side of pumps; pumps to pump through chillers.

27.02 Pump Layout and Design Criteria

A. Pump suction piping should be kept as short and direct as possible with a minimum length of straight pipe upstream of the pump suction as recommended by the pump manufacturer. Manufacturers recommend 5–12 pipe diameters.

B. Pump suction pipe size should be at least one pipe size larger than the pump inlet connection.

C. Use flat on top, eccentric reducer to reduce pump suction piping to the pump inlet connection size.

D. Pump suction should be kept free from air pockets.

E. Horizontal elbows should not be installed at the pump suction. If a horizontal elbow must be installed at the pump suction, the elbow should be installed at a lower elevation than the pump suction. A vertical elbow at the pump suction with the flow upward toward the pump is desirable.

F. Maintain a minimum of 5 pipe diameters of straight pipe immediately upstream of pump suction unless using suction diffuser.

G. Variable speed pumping cannot be used for pure lift applications, because reduced speeds will fail to provide the required lift.

H. Variable speed pumping is well suited for secondary and tertiary distribution loops of primary/secondary and secondary/tertiary hydronic distribution systems (chilled water and heating water systems).

I. Pump Design Arrangements (see Fig. 27.2)

1. Series pumps: equal flow, head additive.
2. Parallel pumps: equal head, flow additive.
3. Standby pumps: standby pumping arrangements are often used for reliability purposes in the event of pump failure. Standby pumps may be provided with coupled or headered systems (see Figs. 27.3 and 27.4).

J. Pump Discharge Check Valves

1. Pump discharge check valves should be center-guided, spring-loaded, disc-type check valves.
2. Pump discharge check valves should be sized so the check valve is full open at the design flow rate. Generally, this will require the check valve to be one pipe size smaller than the connecting piping.
3. Condenser water system and other open piping system check valves should have globe-style bodies to prevent flow reversal and slamming.
4. Installing check valves 4–5 pipe diameters upstream of flow disturbances is recommended by most manufacturers.

K. Differential pressure control of the system pumps should never be accomplished at the pump. The pressure bypass should be provided at the end of the system or at the end each of the subsystems, regardless of whether the system is a bypass flow system or a variable speed pumping system. Bypass flow need not exceed 20 percent of the pump design flow.

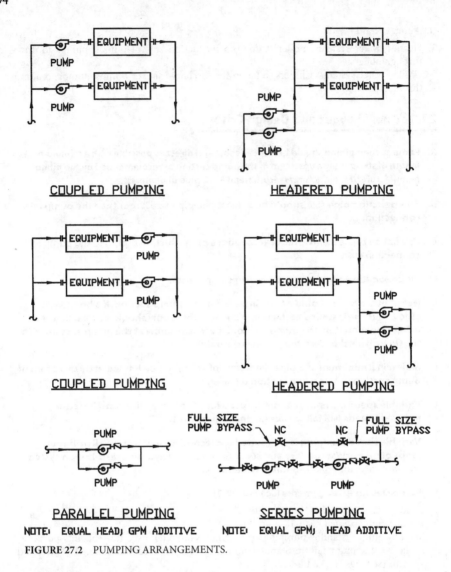

FIGURE 27.2   PUMPING ARRANGEMENTS.

27.03 Pump Selection Criteria

A. The impeller size for specified duty should not exceed 85 percent of the volute cutwater diameter.

B. The maximum cataloged impeller size should be rated to produce not less than 110 percent of the specified head at the specified flow.

C. Parallel Pump Operation: At equal head, the GPM is additive.

D. Series Pump Operation: At equal GPM, the head is additive.

E. Selection Regions:

1. Preferred selection—85–105 percent design flow.
2. Satisfactory selection—66–115 percent design flow.

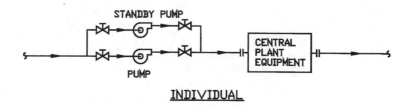

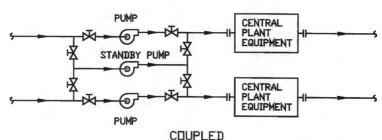

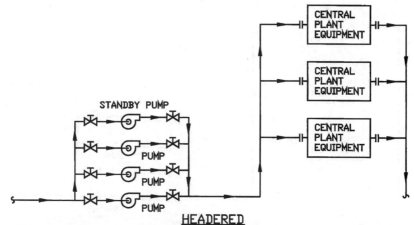

FIGURE 27.3   STANDBY PUMPS.

F. Pumps Curves

1. Flat. A 12 percent rise from design point to the shutoff head (zero flow). Flat curves should be used for variable flow systems with single pumps. A flat pump curve is a pump curve where the head at shutoff is approximately 25 percent higher than the head at the best efficiency point.

2. Steep. A 40 percent rise from design point to shutoff head (zero flow). Steep curves should be used for variable speed and constant flow systems where two or more pumps are used.

3. Hump. The developed head rises to a maximum as flow decreases and then drops to a lower value at the point of shutoff. Hump curves should be used for constant flow systems with single pumps due to increased efficiency.

G. Select pumps so the design point is as close as possible or to the left of the maximum efficiency point.

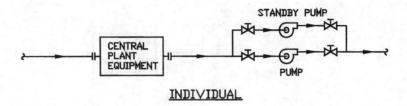

INDIVIDUAL

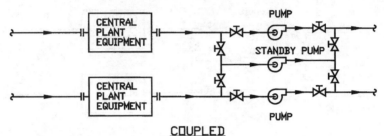

COUPLED

NOTE: STANDBY PUMP MAY SERVE EITHER PIECE OF CENTRAL PLANT EQUIPMENT.

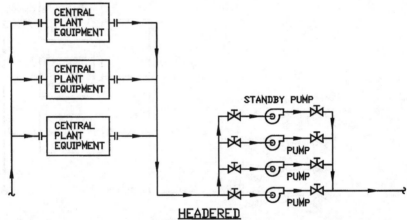

HEADERED

NOTE: STANDBY PUMP MAY SERVE ANY PIECE OF CENTRAL PLANT EQUIPMENT, PROVIDED ALL EQUIPMENT IS THE SAME CAPACITY.

FIGURE 27.4   STANDBY PUMPS.

H. Boiler warming pumps should be selected for a flow rate of 0.1 GPM/BHP (range 0.05–0.1 GPM/BHP). See Part 31 for a more detailed description of boiler warming pumps and their operation.

I. Pump Seals

1. Mechanical seal: closed systems.
2. Stuffing box seals: open systems.

J. Cavitation. Net Positive Suction Head (NPSH)

1. Cavitation: "If the pressure at any point inside the pump falls below the operating vapor pressure of the fluid, the fluid flashes into a vapor and forms bubbles. These bubbles are carried along in the fluid stream until they reach a region of higher pressure. Within this

region, the bubbles collapse or implode with tremendous shock on the adjacent surfaces. Cavitation is accompanied by a low rumbling and/or a sharp rattling noise and even vibration causing mechanical destruction in the form of pitting and erosion."[1]

2. Causes:
   a. Discharge head is far below the pump's calibrated head at peak efficiency.
   b. The suction lift or suction head is lower than the pump rating.
   c. Speeds (RPM) are higher than the pump rating.
   d. Liquid temperatures are higher than that for which the system was designed.
3. Remedies:
   a. Increase the source fluid level height.
   b. Reduce the distance and/or friction losses (larger pipe) between the source and pump.
   c. Reduce the temperature of the fluid.
   d. Pressurize the source.
   e. Use a different pump.
   f. Place the balancing valve in the pump discharge or trim the pump impeller.
4. Systems most susceptible to NPSH problems include:
   a. Boiler feedwater systems (steam systems).
   b. Cooling tower and other open systems.
   c. Medium- and high-temperature water systems.
5. Potential problems increase as:
   a. Elevation above sea level increases.
   b. Height of source above the pump decreases.
   c. Friction losses increase.
   d. Fluid temperature increases.

27.04 Pump Terms

A. *Friction Head*. Friction head is the pressure expressed in psi or in the feet of liquid needed to overcome the resistance to the flow in the pipe and fittings.

B. *Suction Lift*. Suction lift exists when the source of the supply is below the centerline of the pump.

C. *Suction Head*. Suction head exists when the source of the supply is above the centerline of the pump.

D. *Static Suction Lift*. Static suction lift is the vertical distance from the centerline of the pump down to the free level of the liquid source.

E. *Static Suction Head*. Static suction head is the vertical distance from the centerline of the pump up to the free level of the liquid source.

F. *Static Discharge Head*. Static discharge head is the vertical elevation from the centerline of the pump to the point of free discharge.

G. *Dynamic Suction Lift*. Dynamic suction lift includes the sum of static suction lift, friction head loss, and velocity head.

H. *Dynamic Suction Head*. Dynamic suction head includes static suction head minus the sum of friction head loss and velocity head.

I. *Dynamic Discharge Head*. Dynamic discharge head includes the sum of static discharge head, friction head, and velocity head.

[1]Carrier Corporation, *Carrier System Design Manuals, Part 8—Auxiliary Equipment* (Syracuse: Carrier Corporation, 1971), pp. 8–11.

J. *Total Dynamic Head*. Total dynamic head includes the sum of the dynamic discharge head plus the dynamic suction lift or discharge head minus dynamic suction head.

K. *Velocity Head*. Velocity head is the head needed to accelerate the liquid. See the following table.

Velocity (ft./sec.)	Velocity Head (feet)	Velocity (ft./sec.)	Velocity Head (feet)	Velocity (ft./sec.)	Velocity Head (feet)
0.5	0.004	7.5	0.875	14.5	3.269
1.0	0.016	8.0	0.995	15.0	3.498
1.5	0.035	8.5	1.123	15.5	3.735
2.0	0.062	9.0	1.259	16.0	3.980
2.5	0.097	9.5	1.403	16.5	4.232
3.0	0.140	10.0	1.555	17.0	4.493
3.5	0.190	10.5	1.714	17.5	4.761
4.0	0.248	11.0	1.881	18.0	5.037
4.5	0.314	11.5	2.056	18.5	5.321
5.0	0.389	12.0	2.239	19.0	5.613
5.5	0.470	12.5	2.429	19.5	5.912
6.0	0.560	13.0	2.627	20.0	6.219
6.5	0.657	13.5	2.833	21.0	6.856
7.0	0.762	14.0	3.047	22.0	7.525

L. *Specific Gravity*. Specific gravity is the direct ratio of any liquid's weight to the weight of water at 62°F (62.4 lbs./cu.ft. or 8.33 lbs./gal.).

M. *Viscosity*. Viscosity is a property of a liquid that resists any force tending to produce flow. It is the evidence of cohesion between the particles of a fluid that causes a liquid to offer resistance analogous to friction. A change in the temperature may change the viscosity depending upon the liquid. Pipe friction loss increases as viscosity increases.

N. *Static Pressure*. Static pressure is the water pressure required to fill the system.

O. *Static System Pressure*. Static system pressure is the water pressure required to fill the system plus 5 psi.

P. *Flow Pressure*. Flow pressure is the pressure the pump must develop to overcome the resistance created by the flow through the system.

27.05 Installation and Clearance Requirements

A. The minimum recommended clearance around pumps is 24 inches. Maintain minimum clearance as required to open access and control doors on pumps for service, maintenance, and inspection.

B. Mechanical room locations and placement must take into account how pumps can be moved into and out of the building during initial installation and after construction for maintenance and repair and/or replacement.

Chillers

28.01 Chiller Types and Manufacturer Offerings

Chiller Type	Capacity Range tons	kW/ton Range (1)	COP Range (1)	Turndown % Capacity	Refrigerant	Comments
Centrifugal—Water Cooled						
Carrier	200–3000	0.50–0.60	5.86–7.03	10	134a	2
Daikin	200–1250	0.60–0.62	5.86–5.67	10	134a	2, 6
	400–2500	0.61–0.64	5.49–5.76	10	134a	3, 6
Trane	200–2000	0.45–0.55	6.39–7.81	10	123	2
	1500–4000	0.45–0.55	6.39–7.81	10	123	4
York	200–3000	0.50–0.60	5.86–7.03	15	134a	2
	1800–6000	0.50–0.60	5.86–7.03	15 single, 10 dual compressor	134a	3
Centrifugal—Water Cooled with Unit-Mounted VFD						
Carrier	200–3000	0.53–0.62	5.67–6.63	10	134a	2
Daikin	125–200	0.59–0.64	5.49–5.96	25	134a	2, 7
	145–400	0.62–0.67	5.25–5.67	12.5	134a	3, 7
	400–1500	0.55–0.58	6.06–6.39	10	134a	4, 7
Trane	200–4000	0.45–0.55	6.39–7.81	10	123	2, 4
York	200–1475	0.50–0.60	5.86–7.03	15 single, 10 dual compressor	134a	2
Reciprocating—Air Cooled						
Carrier	NA	NA	NA	NA	NA	
Daikin	NA	NA	NA	NA	NA	
Trane	NA	NA	NA	NA	NA	
York	NA	NA	NA	NA	NA	
Reciprocating—Water Cooled						
Carrier	NA	NA	NA	NA	NA	
Daikin	NA	NA	NA	NA	NA	
Trane	NA	NA	NA	NA	NA	
York	NA	NA	NA	NA	NA	
Rotary Screw—Air Cooled (see Fig. 28.1)						
Carrier	80–500	1.01–1.21	2.91–3.48	6–15	134a	
Daikin	140–200	1.25–1.30	2.70–2.81	25	134a	
	170–550	1.20–1.30	2.70–2.93	25	134a	
Trane	140–500	1.05–1.16	3.03–3.35	15	134a	6
York	150–500	1.15–1.20	2.93–3.06	10	134a	
Rotary Screw—Water Cooled (see Fig. 28.2)						
Carrier	175–550	0.53	6.63	10	134a	5
	75–265	0.69–0.72	4.88–5.10	10	134a	
	150–400	0.47–0.56	6.28–7.48	10	134a	
Daikin	130–190	0.72–0.74	4.75–4.88	10	134a	
Trane	140–450	0.58–0.70	5.02–6.06	15	134a	6
York	125–300	0.58–0.70	5.02–6.05	15	134a	
Scroll—Air Cooled (see Fig. 28.3)						
Carrier	10–390	1.16–1.23	2.86–3.03	4–22	410a	
	10–150	0.76–0.91	3.86–4.63	5–20	410a	
Daikin	10–34	1.30–1.40	2.51–2.70	25	407c	
	30–70	1.10–1.20	2.93–3.20	25	410a	
Trane	20–130	1.05–1.20	2.93–3.34	25, 50	410a	
York	15–150	1.15–1.25	2.81–3.05	20	410a	

(Continued)

Chiller Type	Capacity Range tons	kW/ton Range (1)	COP Range (1)	Turndown % Capacity	Refrigerant	Comments
Scroll—Water Cooled						
Carrier	15–71	0.56–0.58	6.06–6.28	10	410a	
Daikin	30–200	0.76–0.80	4.39–4.63	10	410a	
York	60–200	0.77–0.85	4.13–4.56	12, 25 depending upon no. of compressors	410a	

Notes:

1 KW/ton and COPs are based on full load operating characteristics and are "ball park" figures. KW/ton and COPs above and below the values listed in the table are possible for all manufacturers depending on desired operating characteristics. KW/ton, COP, and capacities are driven by, and will vary from, the values listed in the table based on chilled water supply/return temperatures, condenser water supply/return temperatures (water-cooled machines), outside air temperatures (air-cooled machines), and type of refrigerant.

2 Centrifugal chillers with single compressor.

3 Centrifugal chillers with dual compressors.

4 Centrifugal chillers with dual compressors and dual refrigerant circuits (see Fig. 28.4).

5 Variable frequency drive screw chiller.

6 Variable frequency drive available.

7 Magnetic bearings.

8 COP, EER, and kW/ton relationships:
  $EER = COP \times 3.413$
  $COP = 12,000 / (kW/ton \times 3,413)$
  $kW/ton = 12,000 / (COP \times 3,413)$

28.02 Chiller Motor Types

A. Hermetic Chillers/Motors

1. Motors are refrigerant cooled.
2. Motor heat absorbed by the refrigerant must be removed by the condenser cooling medium (air or water).
3. $TONS_{COND} = TONS_{EVAP} \times 1.25$
   $= 12,000$ Btu/h ton $\times 1.25 = 15,000$ Btu/h ton.
   Therefore, motor heat gain is approximately 3,000 Btu/h ton.

FIGURE 28.1   ROTARY SCREW AIR-COOLED CHILLER WITH UNIT-MOUNTED VARIABLE FREQUENCY DRIVE. (*Material Courtesy of Trane.*)

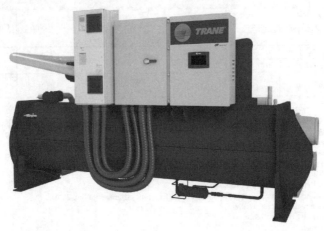

FIGURE 28.2   ROTARY SCREW WATER-COOLED CHILLER WITH
UNIT-MOUNTED VARIABLE FREQUENCY DRIVE. (*Material Courtesy
of Trane.*)

FIGURE 28.3   SCROLL AIR-COOLED CHILLER. (*Material Courtesy
of Trane.*)

FIGURE 28.4   CENTRIFUGAL WATER-COOLED CHILLER WITH
DUAL COMPRESSORS. (*Material Courtesy of Trane.*)

B. Open Chillers/Motors

1. Motors are air cooled.
2. Motor heat is rejected directly to the space. Therefore, the space HVAC system must remove approximately 3,000 Btu/h ton of motor heat gain.

C. In either case, the chillers must remove the 3,000 Btu/h ton of heat generated by the motors; the only difference is the method by which it is accomplished.

28.03 Code Required Chiller Efficiencies

Equipment Type	Equipment Capacity Size Range	2015 IECC and ASHRAE Std. 90.1-2013	
		FL kW/ton	IPLV kW/ton
Air-Cooled Chillers with Condenser—Electric	< 150 tons	10.100 EER	13.700 EER
	≥ 150 tons	10.100 EER	14.000 EER
Air-Cooled Chillers without Condenser—Electric	All Capacities	Same as air-cooled chillers	
Water-Cooled Positive Displacement—Electric	< 75 tons	0.750	0.600
	≥ 75 tons and < 150 tons	0.720	0.560
	≥ 150 tons and < 300 tons	0.660	0.540
	≥ 300 tons and < 600 tons	0.610	0.520
	≥ 600 tons	0.560	0.500
Water-Cooled Centrifugal Chillers—Electric	< 150 tons	0.610	0.550
	≥ 150 tons and < 300 tons	0.610	0.550
	≥ 300 tons and < 400	0.560	0.520
	≥ 400 tons and < 600	0.560	0.500
	≥ 600 tons	0.560	0.500
Air-Cooled Absorption Chillers—Single Effect	All Capacities	0.600 COP	–
Water Cooled Absorption Chillers—Single Effect	All Capacities	0.700 COP	–
Absorption Chillers—Double Effect, Indirect Fired	All Capacities	1.000 COP	1.050 COP
Absorption Chillers—Double Effect, Direct Fired	All Capacities	1.000 COP	1.000 COP

Notes:
1 Efficiency values apply to chillers with water temperatures above 40°F.
2 1 ton = 3.516 kW.
3 For centrifugal chillers operating at temperatures other than 44°F chilled water, 85°F condenser water, and 3.0 GPM/ton condenser water flow rate, maximum full-load kW/ton and part-load ratings shall be adjusted according to the equations given in ASHRAE Standard 90.1-2013, Section 6.4.1.2.1.

28.04 Chiller Terms

A. Refrigeration Effect. The refrigeration effect is the amount of heat absorbed by the refrigerant in the evaporator.

B. Heat of Rejection. The heat of rejection is the amount of heat rejected by the refrigerant in the condenser, which includes compressor heat.

C. Subcooling. Subcooling is the cooling of the refrigerant below the temperature at which it condenses. Subcooling the liquid refrigerant will increase the refrigeration effect of the system.

D. Superheating. Superheating is the heating of the refrigerant above the temperature at which it evaporates. Superheating the refrigerant by the evaporator is part of the system design to prevent a slug of liquid refrigerant from entering the compressor and causing damage.

E. *Coefficient of Performance (COP)*. The coefficient of performance is defined as the refrigeration effect (Btu/h) divided by the work of the compressor (Btu/h). Another way to define COP is Btu output divided by Btu input. COP is equal to EER divided by 3.413.

F. *Energy Efficiency Ratio (EER)*. The energy efficiency ratio is defined as the refrigeration effect (Btu/h) divided by the work of the compressor (watts). Another way to define EER is the Btu output divided by the watts input. The EER is equal to 3.413 times the COP.

G. *Pressure/Enthalpy Chart*. Pressure/Enthalpy chart is a graphic representation of the properties of a specific refrigerant with the pressure on the vertical axis and the enthalpy on the horizontal axis. The graph is used and is helpful in visualizing the changes that occur in a refrigeration cycle.

H. *Integrated Part Load Value (IPLV)*. ARI Specified Conditions. Acceptable tolerances for specified conditions are 6.5 percent.

I. *Application Part Load Value (APLV)*. Engineer Specified Conditions (Real World Conditions). Acceptable tolerances for specified conditions are 6.5 percent.

J. *Rupture Disc*. A relief device on low-pressure machines.

K. *Relief Valve*. A relief device on high-pressure machines.

L. *Pumpdown*. Refrigerant pumped to the condenser for storage.

M. *Pumpout*. Refrigerant pumped to a separate storage vessel. Use pumpout type storage when a reasonable size and number of portable storage containers cannot be moved into the building.

N. *Purge Unit*. Removes air from the refrigeration machine; required on low-pressure machines only.

O. *Prevac*. Device that prevents air from entering the refrigeration machine. It is used to leak test the refrigeration machine. Required on low-pressure machines only.

P. *Factory Run Tests*. 1,500 tons and smaller; most manufacturers can provide them.
   1. *Certified Test*. Certifies performance—full load and/or part load—IPLV, and/or APLV.
   2. *Witnessed Tests:*
      a. *Generic*. Any chiller the manufacturer produces of the same size and characteristics.
      b. *Specific*. The specific chiller required by the customer.

Q. *Hot Gas Bypass*. Low limit to suction pressure of the compressor. Hot gas bypass is beneficial on DX systems and generally not beneficial on chilled-water systems, except when tight temperature tolerances are required for a manufacturing process. Chillers specified with both hot gas bypass and low ambient temperature control will result in the hot gas bypass increasing the low ambient temperature operating point of the chiller (decreases the ability for the chiller to operate at low ambient conditions).

28.05 Basic Refrigeration Cycle Terminology

A. *Compressor*. Mechanical device where the refrigerant is compressed from a lower pressure and lower temperature to a higher pressure and higher temperature.

B. *Hot Gas Piping*. Refrigerant piping from the compressor discharge to the compressor suction, to the evaporator outlet, or to the evaporator inlet, or from the compressor discharge and the condenser inlet to the compressor suction.

C. *Condenser*. Heat exchanger where the system heat is rejected and the refrigerant condenses into a liquid.

D. *Liquid Piping*. Refrigerant piping from the condenser outlet to the evaporator inlet.

E. *Evaporator*. Heat exchanger where the system heat is absorbed and the refrigerant evaporates into a gas.

F. *Suction Piping*. Refrigerant piping from the evaporator outlet to the compressor suction.

G. *Thermal Expansion Valve*. Pressure and temperature regulation valve, located in the liquid line, which is responsive to the superheat of the vapor leaving the evaporator coil.

28.06 Chiller Energy Saving Techniques

A. *Constant Speed Chillers*. For each 1°F increase in chilled-water temperature, the chiller efficiency increases 1.0–2.0 percent.

B. *Variable Speed Chillers*. For each 1°F increase in chilled-water temperature, the chiller efficiency increases 2.0–4.0 percent.

C. For each 1°F decrease in condenser water temperature, the chiller efficiency increases 1.0–2.0 percent.

28.07 Cooler (Evaporator)/Chilled-Water System

A. Leaving Water Temperature (LWT): 42–46°F

B. ΔT Range: 10–20°F

C. 2.4 GPM/ton@10°F ΔT

D. 2.0 GPM/ton@12°F ΔT

E. 1.5 GPM/ton@16°F ΔT

F. 1.2 GPM/ton@20°F ΔT

G. 5,000 Btuh/GPM@10°F ΔT

H. 6,000 Btuh/GPM@12°F ΔT

I. 8,000 Btuh/GPM@16°F ΔT

J. 10,000 Btuh/GPM@20°F ΔT

K. AHRI Evaporator Fouling Factor: 0.00010 h ft.^{2}°F/Btu

L. Chilled Water Flow Range: Chiller Design Flow ±10 percent

M. Chiller Tube Velocity for Variable Flow Chilled Water

1. Minimum flow: 3.0 FPS.
2. Maximum flow: 12.0 FPS.

28.08 Condenser/Condenser Water Systems

A. **Entering Water Temperature (EWT): 85°F**

B. **ΔT Range: 10–20°F**

C. **Normal ΔT: 10°F**

D. **3.0 GPM/ton@10°F ΔT**

E. **2.5 GPM/ton@12°F ΔT**

F. **2.0 GPM/ton@15°F ΔT**

G. **1.5 GPM/ton@20°F DT**

H. **5,000 Btuh/GPM@10°F ΔT**

I. **6,000 Btuh/GPM@12°F ΔT**

J. **7,500 Btuh/GPM@15°F ΔT**

K. **10,000 Btuh/GPM@20°F ΔT**

L. **AHRI Condenser Fouling Factor: 0.00025 h ft.² °F/Btu**

28.09 Chilled Water Storage Systems

A. **10°F ΔT**
1. 19.3 cu.ft./ton h
2. 623.1 Btu/cu.ft.; 83.3 Btu/gal.

B. **12°F ΔT**
1. 16.1 cu.ft./ton h
2. 747.7 Btu/cu.ft.; 100.0 Btu/gal.

C. **16°F ΔT**
1. 12.4 cu.ft./ton h
2. 996.9 Btu/cu.ft.; 133.3 Btu/gal.

D. **20°F ΔT**
1. 9.6 cu.ft./ton h
2. 1246.2 Btu/cu.ft.; 166.7 Btu/gal.

28.10 Ice Storage Systems

A. **144 Btu/lb.@32°F + 0.48 Btu/lb. for each 1°F below 32°F.**

B. **3.2 cu.ft./ton h**

C. **Only the latent heat capacity of ice should be used when designing ice storage systems.**

28.11 Water-Cooled Condensers

A. Entering Water Temperature (EWT): 85°F

B. Leaving Water Temperature (LWT): 95°F

C. 3.0 GPM/ton@10°F ΔT

D. For each 1°F decrease in condenser water temperature, chiller efficiency increases 1.0–2.0 percent.

28.12 Refrigerant Estimate—Split Systems

A. Total 3.0 lbs./ton

B. Equipment 2.0 lbs./ton

C. Piping 1.0 lbs./ton

28.13 Chilled Water System Makeup Connection

Minimum connection size shall be 10 percent of the largest system pipe size or 1", whichever is greater. (A 20" system pipe size results in a 2" makeup water connection.)

28.14 Chemical Feed Systems for Chillers. Chemical Feed Systems are Designed to Control the Following

A. System pH, normally between 8 and 9.

B. Corrosion.

C. Scale.

28.15 Chiller Operating Sequence

A. Start chilled water and condenser water pumps. Verify chilled water and condenser water flow.

B. Start chiller and cooling tower.

C. Runtime.

D. Stop chiller and cooling tower.

E. Stop chilled water and condenser water pumps after 0- to 30-second delay because some chiller manufacturers use chilled water or condenser water to cool the solid state starter circuitry.

F. Chiller Startup Piping (see Fig. 28.5)

1. Because it takes 5–15 minutes from the time the chiller start sequence is initiated until the time the chiller starts to provide chilled water at the design temperature, the chilled water

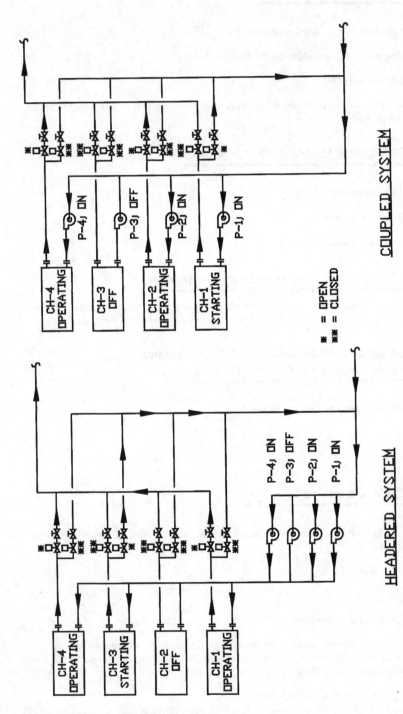

FIGURE 28.5 CHILLER STARTUP PIPING DIAGRAM.

supply temperature often rises above the desired control setpoint. If the chilled water supply temperature is critical, the method to correct this problem is to provide the chillers with startup piping which runs from the chiller discharge to the pump return main.

2. The designer should size startup piping for the flow of the largest chiller in the system. The common pipe size only needs to be sized for the flow of one chiller because it is unlikely that more than one chiller will be started at the same time.

3. Chilled-water system operation with startup piping should be as follows:

   a. On initiation of the chiller start sequence, the primary chilled water pump is started, the bypass valve is opened, and the supply header valve is closed. When the chilled water supply setpoint temperature is reached, as sensed in the bypass, the supply header valve is slowly opened, maintaining the setpoint temperature at all times. When the supply header valve is fully opened, the bypass valve is slowly closed.

   b. On initiation of the chiller stop sequence, the bypass valve is slowly opened. When the bypass valve is fully opened, the supply header valve is slowly closed. The chiller is stopped, and after a delay, the primary chilled water pump is stopped. When the primary chilled water pump stops, the bypass valve is left open to permit water to expand into, or contract from, the system. On headered systems, the chilled water return valve must be closed as well.

4. The chilled water diagram shows the chiller startup piping with motorized shutoff valves. Motorized valves are required for automatic or remote manual control. If the chilled-water system will be manually operated, these valves may be deleted. A separate manual shutoff valve has also been provided to allow for manual isolation of the system and to permit repair of the motorized valve without having to shut down the system. This manual shutoff valve may be deleted, provided the motorized shutoff valve has a manual means by which it can be opened and closed. Most motorized control valves do not have a manual means to open and close them.

28.16 Chiller Design, Layout, and Clearance Requirements/Considerations

A. Design Conditions

1. Chiller load. Tons, Btu/h, or MBH.
2. Chilled water temperatures. Entering and leaving or entering and ΔT.
3. Condenser water temperatures. Entering and leaving or entering and ΔT.
4. Chilled water flows and fluid type (correct all data for fluid type).
5. Condenser water flows and fluid type (correct all data for fluid type).
6. Evaporator and condenser pressure drops.
7. Fouling factor.
8. IPLV, desirable.
9. APLV, optional.
10. Chilled water or condenser water reset if applicable.
11. Ambient operating temperature, dry bulb and wet bulb.
12. Electrical data:
    a. Compressor or unit KW.
    b. Full load, running load, and locked rotor amps.
    c. Power factor.
    d. Energy Efficiency Ratio (EER).
    e. Voltage-phase-hertz.

B. Multiple chillers should be used to prevent complete system or building shutdown upon failure of one chiller in all chilled-water systems over 200 tons (i.e., 2@50 percent, 2@67 percent, 2@70 percent, 3@34 percent, 3@40 percent).

1. Series chiller design: Piping chillers in series can accomplish large temperature differentials without penalizing the chiller performance (see Figs. 28.6 and 28.7).

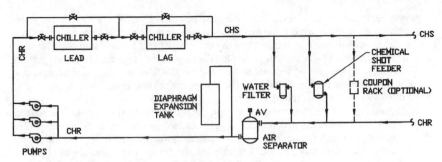

FIGURE 28.6   SERIES CHILLED-WATER SYSTEM.

2. Parallel chiller design: Piping chillers in parallel provides a simpler installation and provides for multiple chiller arrangements with standby opportunities. Standby opportunities are also available with series chiller arrangements, but they become more complex and cumbersome (see Figs. 28.8 and 28.9).
3. When designing chilled-water systems for computer centers, data centers, Internet host sites, and other mission-critical facilities where down time is not acceptable, consider utilizing a dual primary/secondary chilled-water system with primary/secondary chilled-water cross-connections and looped secondary system (see Figs. 28.10 and 28.11). This chilled-water system design permits isolating the piping segments as well as the equipment to permit service and repairs to both piping and equipment without shutdown of the system. The dual primary/secondary chilled-water system can be designed and sized to meet the Uptime Institute's Tier III classification (N+1 redundancy requirements; the arrangement actually provides N+2) and Tier IV classification (2[N+1] redundancy requirements). Chilled-water systems serving mission-critical facilities should always be designed for future expansion and growth. All future equipment and systems must provide for this growth. Space must be provided for future equipment, valved and capped connections must be provided for connections to piping mains so shutdowns are not required, piping mains must be sized for the ultimate growth of the facility, and electrical power systems must be designed and sized for the ultimate power utilized by the facility.

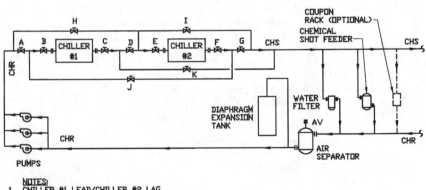

NOTES:
1 CHILLER #1 LEAD/CHILLER #2 LAG
        OPEN VALVES: A, B, C, D, E, F, AND G.
        CLOSE VALVES: H, I, J, AND K.
2 CHILLER #2 LEAD/CHILLER #1 LAG
        OPEN VALVES: B, C, E, F, H, J, AND K.
        CLOSE VALVES: A, D, G, AND I.
3 VALVES H AND I ARE BYPASS VALVES FOR CHILLER #1 AND #2, RESPECTIVELY.

FIGURE 28.7   SERIES CHILLED-WATER SYSTEM WITH LEAD-LAG CHILLER CONTROL.

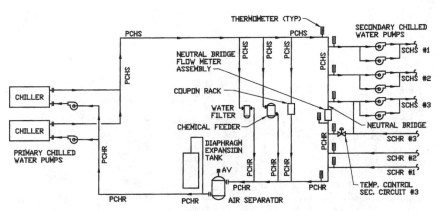

FIGURE 28.8   PARALLEL CHILLED-WATER SYSTEM—COUPLED PUMPS.

C. Water Boxes/Piping Connections

1. Marine type. Marine water boxes enable piping to be connected to the side of the chiller so piping does not need to be disconnected in order to service machine. Recommend on large chillers, 500 tons and larger.
2. Nonmarine or standard type. Recommend on small chillers, less than 500 tons.
3. Provide victaulic or flanged connections for first three fittings at chiller with nonmarine or standard type connections.
4. Locate piping connections against the wall.
5. Locate all piping connections opposite the tube clean/pull side of the chiller.
6. Locate oil cooler connections.

D. Show tube clean/pull clearances and location.

E. The minimum recommended clearance around chillers is 36 inches. Maintain minimum clearances for tube pull and cleaning of tubes as recommended by the equipment manufacturer. This is generally equal to the length of the chiller. Maintain minimum clearance as required to open access and control doors on chillers for service, maintenance, and inspection.

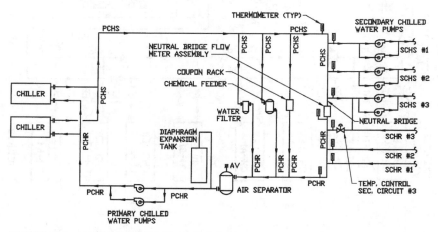

FIGURE 28.9   PARALLEL CHILLED-WATER SYSTEM—HEADERED PUMPS.

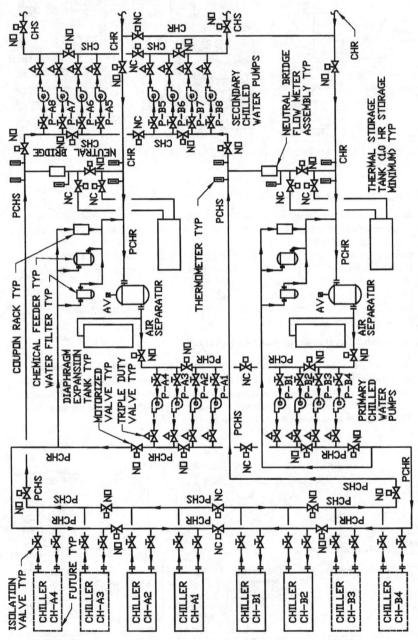

FIGURE 28.10 DUAL PRIMARY/SECONDARY CHILLED-WATER SYSTEM FLOW DIAGRAM.

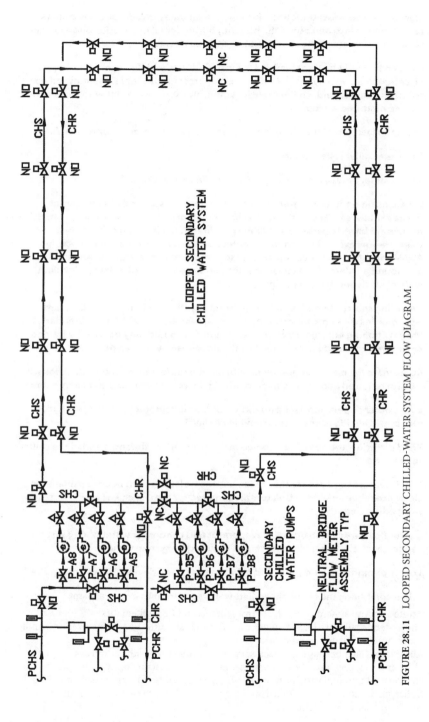

FIGURE 28.11 LOOPED SECONDARY CHILLED-WATER SYSTEM FLOW DIAGRAM.

413

F. Maintain minimum electrical clearances as required by NEC.

G. Mechanical room locations and placement must take into account how chillers can be moved into and out of the building during initial installation and after construction for maintenance and repair and/or replacement.

H. If the chiller must be disassembled for installation (the chiller cannot be shipped disassembled), specify the manufacturer's representative for reassembly; do not specify insulation with chiller (field insulate), and specify the chiller to come with remote mounted starter.

I. Show the location of the chiller starter, disconnect switch, and control panel.

J. Show the chiller relief piping.

K. Show sanitary drain locations and chiller drain connections.

L. Locate refrigerant monitoring system refrigerant sensors and the refrigerant purge exhaust fan. The refrigerant exhaust system should be designed to remove refrigerant based on its specific gravity (lighter than air—high exhaust, heavier than air—low exhaust). Refrigerant detection devices are required by code, *ASHRAE Standard 15.* Detection devices sound an alarm at certain levels (low limit) and sound an alarm and activate ventilation system at a higher level (high limit), with levels dependent on refrigerant type.

M. Providing self-contained breathing apparatus within buildings for refrigerant emergencies is not recommended as in previous versions of ASHRAE Standard 15. Pre-positioning emergency response equipment should only be used by trained emergency responders and must be labeled for use by trained personnel only.

N. Coordinate the height of the chiller with overhead clearances and obstructions. Is a beam required above the chiller for lifting the compressor or other components?

O. Low ambient operation. Is the operation of the chiller required below 40°F, 0°F, etc., or will airside economizers provide cooling?

P. Wind direction and speed (air-cooled machines). Orient the short end of the chiller to the wind.

Q. If isolators are required for the chiller, has the isolator height been considered in clearance requirements? If isolators are required for the chiller, has piping isolation been addressed?

R. Locate flow switches in both the evaporator and condenser water piping systems serving each chiller and flow meters as required by system design.

S. Locate pumpdown, pumpout, and refrigerant storage devices if they are required.

T. When combining independent chilled-water systems into a central plant
1. Create a central system concept, control scheme, and flow schematics.
2. The system shall only have a single expansion tank connection point sized to handle entire system expansion and contraction.
3. All systems must be altered, if necessary, to be compatible with central system concept (temperatures, pressures, flow concepts—variable or constant control concepts).
4. For constant flow and variable flow systems, the secondary chillers are tied into the main chiller plant return main. Chilled water is pumped from the return main through the chiller and back to the return main.

5. District chilled-water systems, due to their size, extensiveness, or both, may require that independent plants feed into the supply main at different points. If this is required, design and layout must enable isolating the plant; provide start-up and shutdown bypasses; and provide adequate flow, temperature, pressure, and other control parameter readings and indicators for proper plant operation and other design issues that affect plant operation and optimization.

U. In large systems, it may be beneficial to install a steam-to-water or water-to-water heat exchanger to place an artificial load on the chilled-water system to test individual chillers or groups of chillers during plant startup, after repairs, or for troubleshooting chiller or system problems.

V. Large and campus chilled-water systems should be designed for large delta Ts and for variable flow secondary and tertiary systems.

W. Chilled-water pump energy must be accounted for in the chiller capacity because it adds heat load to the system.

X. It is best to design chilled-water and condenser-water systems to pump through the chiller.

PART 29

Cooling Towers and Condensers

29.01 Cooling Tower Types (CTs)

A. Induced Draft—Cross Flow

1. 200–900 tons single cell.
2. 400–1,800 tons double cell.

B. Forced Draft, Counter Flow

1. 200–1,300 tons centrifugal fans.
2. 250–1,150 tons axial fans.
3. Figure 29.1 is a photograph of a forced draft, counter flow cooling tower in its installed condition.

C. Ejector Parallel Flow

1. 5–750 tons.

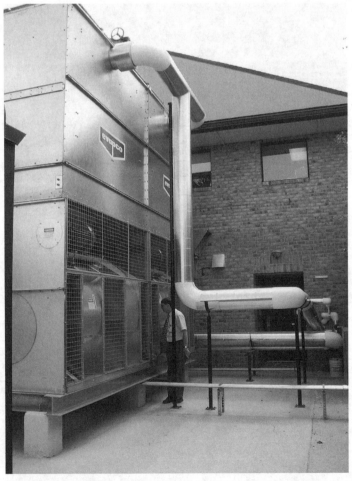

FIGURE 29.1   PHOTOGRAPH OF A FORCED DRAFT, COUNTER FLOW COOLING TOWER.

29.02 Definitions

A. *Range*. Difference between entering and leaving water, system ΔT.

B. *Approach*. Difference between leaving water temperature and entering air wet bulb.

C. *Evaporation*. Method by which cooling towers cool the water.

D. *Drift*. Entrained water droplets carried off by the cooling tower. An undesirable side effect.

E. *Blowdown* or *Bleed*. Water intentionally discharged from the cooling tower to maintain water quality.

F. *Plume*. Hot moist air discharged from the cooling tower forming a dense fog.

29.03 Condenser Water

A. Most Common Entering Water Temperature (EWT): 95°F

B. Most Common Leaving Water Temperature (LWT): 85°F

C. Range: 10–40°F ΔT

D. 3.0 GPM/ton@10°F ΔT

E. 2.5 GPM/ton@12°F ΔT

F. 2.0 GPM/ton@15°F ΔT

G. 1.5 GPM/ton@20°F ΔT

H. 0.75 GPM/ton@40°F ΔT

29.04 Power

0.035–0.040 kW/ton

29.05 TONS$_{COND}$

$= \text{TONS}_{EVAP} \times 1.25$
$= 12{,}000 \text{ Btu/h ton} \times 1.25$
$= 15{,}000 \text{ Btu/h ton}$

29.06 Condenser Water Makeup to Cooling Tower

A. Range: 0.0306–0.0432 GPM/ton

B. Range: 0.0102–0.0144 GPM/Cond. GPM
                 (1.0–1.4 percent Condenser GPM)

C. Centrifugal: 40 GPM/1,000 tons

D. Reciprocating: 40 GPM/1,000 tons

E. Screw: 40 GPM/1,000 tons

F. Scroll: 40 GPM/1,000 tons

G. Absorption: 80 GPM/1,000 tons

29.07 Cooling Tower Drains

Use two times the makeup water rate for sizing cooling tower drains.

29.08 Cycles of Concentration

A. Range: 2–10

B. Recommend: 3–5

29.09 Evaporation

A. Range: 0.024–0.03 GPM/ton

B. Range: 0.008–0.01 GPM/Cond. GPM

                 (0.8–1.0 percent Condenser GPM)

C. Recommend: 0.01 GPM/Cond. GPM

29.10 Drift

A. Range: 0.0006–0.0012 GPM/ton

B. Range: 0.0002–0.0004 GPM/Cond. GPM

                 (0.02–0.04 percent Condenser GPM)

C. Recommend: 0.0002 GPM/Cond. GPM

29.11 Blowdown or Bleed (Based on 108F Range)

A. Range: 0.006–0.012 GPM/ton

B. Range: 0.002–0.004 GPM/Cond. GPM (0.2–0.4 percent Condenser GPM)

C. Recommend: 0.002 GPM/Cond. GPM

D. Centrifugal: 10 GPM/1,000 tons

E. Reciprocating: **10 GPM/1,000 tons**

F. Screw: **10 GPM/1,000 tons**

G. Scroll: **10 GPM/1,000 tons**

H. Absorption: **20 GPM/1,000 tons**

Cooling Tower Range	Blowdown GPM—% of Cond. GPM								
	Cycles of Concentration								
	2	3	4	5	6	7	8	9	10
10	0.80	0.40	0.30	0.20	0.10	0.10	0.10	0.10	0.10
15	1.20	0.60	0.40	0.30	0.20	0.20	0.15	0.15	0.15
20	1.60	0.80	0.50	0.40	0.30	0.30	0.20	0.20	0.20
25	2.00	1.00	0.65	0.50	0.40	0.35	0.25	0.25	0.23
30	2.40	1.20	0.80	0.60	0.50	0.40	0.30	0.30	0.25
35	2.75	1.40	0.95	0.70	0.55	0.45	0.35	0.35	0.30
40	3.10	1.60	1.10	0.80	0.60	0.50	0.40	0.40	0.35

29.12 Installation Location

Cooling towers should be located at least 100 feet from the building, when located on the ground, to reduce noise and prevent moisture from condensing on the building during the intermediate seasons (spring and fall). Cooling towers should also be located 100 feet from parking structures or parking lots to prevent staining of automobile finishes due to water treatment.

29.13 Air-Cooled Condensers and Condensing Units (ACCs and ACCUs)

A. Size Range: **0.5–500 tons**

B. Air Flow: **600–1,200 CFM/ton**

C. Power:
1. Condenser Fans: 0.1–0.2 HP/ton.
2. Compressors: 1.0–1.3 KW/ton.

29.14 Evaporative Condensers and Condensing Units (ECs and ECUs)

A. Types and Sizes:
1. 10–1,600 tons centrifugal fans.
2. 10–1,500 tons axial fans.

B. Drift: **0.002 GPM/cond. GPM**

C. Evaporation: **1.6–2.0 GPM/ton**

D. Bleed: **0.8–1.0 GPM/ton**

E. Total: **2.4–3.0 GPM/ton**

29.15 Installation of CTs, ACCs, ACCUs, ECs, and ECUs

A. Allow ample space to provide the proper airflow to fans and units in accordance with the manufacturer's recommendations.

B. The top discharge of the unit should be at the same height or higher level than the adjoining building or wall to minimize recirculation caused by down drafts between the unit and wall. Raise the unit or provide a discharge hood to obtain the proper discharge height.

C. Elevating units may decrease the space required between units and between units and walls. Only decrease space in accordance with the manufacturer's recommendations.

D. Decking or metal plates over units between walls and other units may decrease the space required between units and between units and walls. Only decrease space in accordance with the manufacturer's recommendations.

E. Providing discharge hoods with units may decrease the space required between units and between units and walls. Only decrease space in accordance with the manufacturer's recommendations.

F. Chemical Feed Systems for CTs, ECs, and ECUs. Chemical feed systems are designed to control the following.
1. System pH; normally between 8 and 9.
2. Corrosion.
3. Scale.
4. Biological and microbial growth.

G. Clearance Requirements
1. The minimum recommended clearance around CTs, ACCs, ACCUs, ECs, and ECUs is 36 inches. Maintain minimum clearances as recommended by the equipment manufacturer. Maintain the minimum clearance as required to open access and control doors on equipment for service, maintenance, and inspection.
2. Mechanical room locations and placement must take into account how CTs, ACCs, ACCUs, ECs, and ECUs can be moved into and out of the building during initial installation and after construction for maintenance and repair and/or replacement.

PART 30

Heat Exchangers

30.01 Shell and Tube Heat Exchangers

A. **Used Where the Approach of the System Is Greater than 15±°F**

B. **Straight Tube or U-Tube Design**

C. **Generally Used in Heating Systems**

D. **Water to Water**

1. Maximum tube velocity: 6 ft./sec.
2. Maximum shell velocity: 5 ft./sec.

E. **Steam to Water**

1. Maximum water velocity: 6 ft./sec.
2. If system steam capacity exceeds 2" control valve size, provide 2 control valves with 1/3 and 2/3 capacity split.

30.02 Plate and Frame Heat Exchangers

A. **Used Where the Approach of the System Is Less than 15±°F**

B. **Generally Used in Cooling Systems**

C. **Refer to Fig. 30.1 for a photograph of a plate and frame heat exchanger in its installed condition.**

FIGURE 30.1   PHOTOGRAPH OF A PLATE AND FRAME HEAT EXCHANGER.

30.03 Definitions

A. *Range*: Difference between entering and leaving water, system ΔT.

B. *Approach:* Difference between hot side entering water temperature and cold side leaving water temperature.

30.04 Clearance and Design Requirements

A. The minimum recommended clearance around heat exchangers is 36 inches. Maintain minimum clearances for tube pull and the cleaning of tubes as recommended by the equipment manufacturer. This is generally equal to the length of the heat exchanger.

B. Mechanical room locations and placement must take into account how heat exchangers can be moved into and out of the building during initial installation and after construction for maintenance and repair and/or replacement.

C. Multiple heat exchangers should be used to prevent complete system or building shutdown upon failure of one heat exchanger in all water systems over 200 tons or 2,400,000 Btu/h (e.g., 2@50 percent, 2@67 percent, 2@70 percent, 3@34 percent, 3@40 percent).

D. Heat Transfer Factors

1. Change in enthalpy on the primary side (hydronic side).
2. Change in enthalpy on the secondary side.
3. Heat transfer through the heat exchanger is dependent on film coefficients and the heat transfer surface area.

E. Methods of Heat Transfer

1. Parallel flow. Both mediums flow in the same direction. The least effective method of heat transfer.
2. Counter-flow. Mediums flow in opposite directions. The most effective method of heat transfer.
3. Cross-flow. Mediums flow at right angles to each other. Heat transfer effectiveness between parallel and counter flow methods.
4. Combination. Cross-Flow/Counter-Flow or Cross-Flow/Parallel Flow. Typical in shell and tube heat exchangers.

Boilers

31.01 Boilers, General

A. Class I Boilers. ASME Boiler and Pressure Vessel Code, Section I

1. Steam boilers, greater than 15 psig
2. Hot water boilers:
   a. Greater than 160 psig.
   b. Greater than 250°F.
3. Common terminology:
   a. Process boilers.
   b. Power boilers.
   c. High-pressure boilers.

B. Class IV Boilers. ASME Boiler and Pressure Vessel Code, Section IV

1. Steam boilers, 15 psig and less
2. Hot water boilers:
   a. 160 psi and less.
   b. 250°F and less.
3. Common terminology:
   a. Commercial boilers.
   b. Industrial boilers.
   c. Heating boilers.
   d. Low-pressure boilers.

C. Common Boiler Design Pressures

1. 15 psig.
2. 30 psig.
3. 60 psig.
4. 125 psig.
5. 150 psig.
6. 200 psig.
7. 250 psig.
8. 300 psig.
9. 350 psig.

D. Boiler Sequence of Operation

1. Prepurge.
2. Pilot ignition and verification.
3. Main flame ignition and verification.
4. Run time.
5. Post purge.
6. Boiler operational considerations:
   a. Hot water and steam boilers:
      1) Prevent hot or cold shock.
      2) Prevent frequent cycling.
      3) Provide proper water treatment.
   b. Hot water boilers only:
      1) Provide continuous circulation.
      2) Balance flow through boilers.
      3) Provide proper overpressure.
   c. Causes of increased stack temperature:
      1) Soot buildup.
      2) Scale buildup.
      3) Combustion chamber and pass sealing problems.

E. Boiler Types

1. Fire tube boilers (Scotch Marine—see Fig. 31.1).
2. Water tube boilers (see Fig. 31.2).

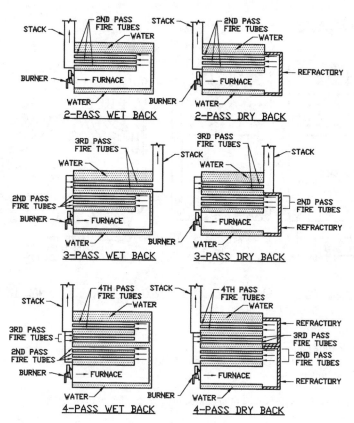

FIGURE 31.1 FIRE TUBE BOILER TYPES.

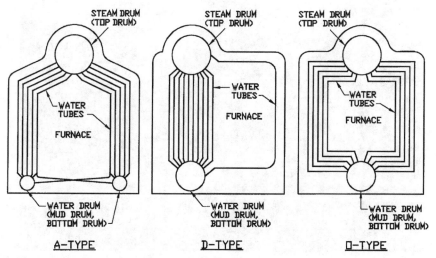

FIGURE 31.2 WATER TUBE BOILER TYPES.

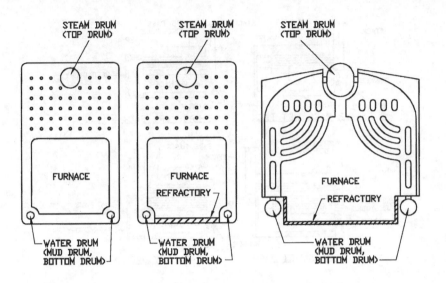

WET BASE TYPE   WET LEG TYPE   WATER TUBE/EXTERNAL HEADERS

FIGURE 31.3   CAST IRON BOILER TYPES.

3. Flexible tube boilers.
4. Cast iron boilers (see Figs. 31.3 through 31.5).
5. Modular boilers.
6. Electric boilers.
7. Fire tube versus water tube boiler characteristics are shown in the following table:

FIGURE 31.4   PHOTOGRAPH OF WET BASE CAST IRON BOILER.

FIGURE 31.5   PHOTOGRAPH OF WATER TUBE BOILER CAST IRON BOILER WITH EXTERNAL HEADERS.

FIRE TUBE VS. WATER TUBE BOILERS

Compared Item	Fire Tube Boilers	Water Tube Boilers
Steam Quality	98.5%	99.5%
Steam Purity	52.5 ppm	17.5 ppm can be modified to obtain 1 ppm
Efficiency	85% average	80% average
Design Pressure	300 psig	900 psig
Design Temperature	350°F	455°F
Super Heaters	None	Available to 750°F
Load Swings	Long recovery time	Short recovery time
Water Weight	Factor of 2.5	Factor of 1.0
Length	Longer	Shorter
Height	Shorter	Higher
Overfire	No	10–15% for short periods
Space	Door swing and tube pull	3'0" minimum all around
Electrical Load	Greater hp required	Lower hp required
Water Quality	Same	Same
Turn Down	10:1 gas; 8:1 fuel oil #2	10:1 gas; 8:1 fuel oil #2
U.L. Label	Standard entire package	Not available for entire package—components only
Soot Blowers	None	Standard option
Ultimate Decision	Customer preference	Customer preference

F.   Boiler Efficiency

1. Combustion efficiency: Indication of the burner's ability to burn fuel measured by the unburned fuel and excess air in the exhaust.
2. Thermal efficiency: Indication of the heat exchanger's effectiveness to transfer heat from the combustion process to the water or steam in a boiler. Does not account for radiation and convection losses, however.

3. Fuel-to-steam efficiency: Indication of the overall efficiency of the boiler including effectiveness of the heat exchanger, radiation losses, and convection losses (output divided by input). The test to determine fuel-to-steam efficiency is defined by *ASME Power Test Code, PTC 4.1:*
   a. Input-output method.
   b. Heat loss method.
4. Boiler efficiency: Indication of either thermal efficiency or fuel-to-steam efficiency depending on context.

G. Boiler Plant Efficiency Factors

1. Boiler, 80–85% efficient:
   a. Radiation losses.
   b. Convection losses.
   c. Stack losses.
2. Boiler room, steam:
   a. Heating of combustion air.
   b. Heating of makeup water.
   c. Steam condensate not returned.
   d. Boiler blowdown.
   e. Radiation losses:
      1) Condensate tank.
      2) Condensate pump.
      3) Feedwater pump.
      4) Deaerator or feedwater tank.
3. Boiler room, hot water:
   a. Heating of combustion air.
   b. Radiation losses:
      1) Expansion tank.
      2) Air separator.
      3) Pumps.
4. Plant, system:
   a. Steam leaks and bad steam traps.
   b. Piping, valves, and equipment radiation losses.
   c. Control valve operational problems.
   d. Flash steam losses.
   e. Water or condensate leaks/losses.

H. Steam System Energy Saving Tips

1. Insulate all hot surfaces to prevent heat loss.
2. Isolate all steam supply piping not being used.
3. Repair all steam piping leaks.
4. Repair all steam traps not operating properly which are bypassing steam.
5. Stop all internal steam leaks including venting of flash steam and open bypass valves around steam traps and control valves.
6. Produce clean, dry steam with the use of a steam separator and proper water treatment.
7. Properly control steam flow at equipment.
8. Use and properly select steam traps.
9. Use flash steam for preheating and other uses whenever possible.

I. Packaged Boiler Fuel Types

1. Natural gas.
2. Propane.
3. Light fuel oil #1 and #2.
4. Heavy fuel oil #4, #5, and #6.
5. Digester or landfill gas.

J. Gas Trains

1. Underwriter Laboratories, Standard (UL).
2. Industrial Risk Insurers (IRI).
3. Factory Mutual (FM).
4. Kemper.
5. ASME CSD-1 *Controls and Safety Devices for Automatically Fired Boilers.*
6. NFPA 8501 *Standard for Single Burner Boiler Operation.*

K. Boiler Capacity Terminology

1. Startup load. Capacity required to bring the boiler system up to temperature, pressure, or both.
2. Running load. Design capacity.
3. Maximum Instantaneous Demand (MID). A sudden peak load requirement of unusually short duration:
   a. MID loads are often hidden in process equipment loads.
   b. Cold startup or pickup loads that far exceed their normal operating demands.
   c. A full understanding of MID loads is required to properly select boiler system capacity.
   d. MID shortfall corrective actions:
      1) Change load reaction time to reduce impact; slow down valve operation, reduce number of items with simultaneous startup (staged startup).
      2) Add boiler capacity.
      3) Add back pressure regulator downstream of deaerator or feedwater tank steam supply connection.
      4) Add an accumulator.

L. Combustion

1. Improper combustion:
   a. Oxygen rich-fuel lean: Wastes energy.
   b. Oxygen lean-fuel rich: Produces CO, soot, and potentially hazardous conditions.
2. What affects combustion?
   a. Changes in barometric pressure.
   b. Changes in ambient air temperature:
      1) Oxygen trim systems compensate for ambient air temperature changes.
   c. Ventilation air:
      1) Total: 10 CFM/BHP
      2) Combustion air: 8 CFM/BHP
      3) Ventilation: 2 CFM/BHP
   d. Keep boiler room positive with respect to the stack and breeching (+0.10 in. W.G. maximum) to prevent the entrance of flue gases into the boiler room.
   e. Never exhaust boiler rooms; use supply air with relief air.

M. Stacks and Breeching. Provide a manual damper (lock damper in the open position) or a motorized damper (two-position damper) at the boiler outlet. A motorized damper interlocked with boiler operation is preferred.

1. Multiple boilers with common stack and breeching. Damper will prevent products of combustion from entering the boiler room when repairing or inspecting boilers while system is still in operation.
2. Multiple boilers with individual or common stack and breeching. Damper will prevent the natural draft through the boiler when not firing, thus reducing the energy lost up the stack.

N. 1990 Clean Air Act—Focused on the reduction of the following pollutants

1. Ozone (O_3).
2. Carbon monoxide (CO).
3. Nitrogen oxides (NO_x-NO/NO_2).

4. Sulfur oxides (SO_x-SO_2/SO_3).
5. Particulate matter, 10 ppm.
6. Lead.

O. Standard Controls

1. Steam boiler control and safeties:
   a. High limit pressure control. Provides a margin of safety.
   b. Operating limit pressure control. Starts/stops burner.
   c. Modulation pressure control. Varies burner firing rate.
   d. Low limit pressure control.
   e. Low water cutoff.
   f. Auxiliary low water cutoff.
   g. High water cutoff.
2. Hot water boiler controls and safeties:
   a. High limit pressure control. Provides a margin of safety.
   b. High limit temperature control. Provides a margin of safety.
   c. Operating limit temperature control. Starts/stops burner.
   d. Modulation temperature control. Varies burner firing rate.
   e. Low limit pressure control.
   f. Low limit temperature control.
   g. Low water cutoff.
   h. High water cutoff.
3. Fuel system controls and safeties:
   a. Low gas pressure switch.
   b. High gas pressure switch.
   c. Low oil pressure switch.
   d. High oil pressure switch.
   e. Low oil temperature.
4. Combustion controls and safeties:
   a. Pilot failure switch.
   b. Flame failure switch.
   c. Combustion air proving switch.
   d. Oil atomization proving switch.
   e. Low fire hold control.
   f. Low fire switch.
   g. High fire switch.

P. Safety, relief, and safety relief valve testing is dictated by the Insurance Underwriter.

31.02 Hot Water Boilers

A. Boiler Types

1. Fire tube boilers:
   a. 15–800 BHP.
   b. 500–26,780 MBH.
   c. 30–300 psig.
2. Water tube boilers:
   a. 350–2,400 BHP.
   b. 13,000–82,800 MBH.
   c. 30–525 psig.
3. Flexible water tube boilers:
   a. 30–250 BHP.
   b. 1,000–8,370 MBH.
   c. 0–150 psig.

4. Cast-iron boilers:
   a. 10–400 BHP.
   b. 345–13,800 MBH.
   c. 0–40 psig.
5. Modular boilers:
   a. 4–115 BHP.
   b. 136–4,000 MBH.
   c. 0–150 psig.
6. Electric boilers:
   a. 15–5,000 KW.
   b. 51–17,065 MBH.
   c. 0–300 psig.

B. Hot Water Boiler Plant Equipment

1. Boilers (see Fig. 31.6).
2. Pumps.
3. Air separators.
4. Expansion tanks.

C. Heating Water

1. Leaving water temperature (LWT): 180–200°F.
2. 20–40°F ΔT most common.
3. Boiler system design limits:
   a. Minimum flow through a boiler: 0.5–1.0 GPM/BHP.
   b. Maximum flow through a boiler: Boiler capacity divided by the temperature difference divided by 500.
   c. Pressure drop through a boiler: 3–5 feet H_2O.
   d. Minimum supply water temperature: 170°F. This temperature may vary with boiler design and with manufacturer; verify the exact temperature with the manufacturer.
   e. Minimum return water temperature: 150°F. This temperature may vary with boiler design and with manufacturer; verify the exact temperature with the manufacturer.
   f. Maximum supply water temperature: Based on the ASME Design Rating of the boiler.
4. Heating capacities:
   a. 3.45 GPM/BHP @20°F ΔT.
   b. 2.30 GPM/BHP @30°F ΔT.
   c. 1.73 GPM/BHP @40°F ΔT.
   d. 10.0 GPM/therm @20°F ΔT.
   e. 6.7 GPM/therm @30°F ΔT.
   f. 5.0 GPM/therm @40°F ΔT.
   g. 10,000 Btuh/GPM @20°F ΔT.
   h. 15,000 Btuh/GPM @30°F ΔT.
   i. 20,000 Btuh/GPM @40°F ΔT.

D. System Types

1. Low-temperature heating water systems:
   a. 250°F and less.
   b. 160 psig maximum.
2. Medium-temperature heating water systems:
   a. 251–350°F.
   b. 160 psig maximum.
3. High-temperature heating water systems:
   a. 351–450°F.
   b. 300 psig maximum.

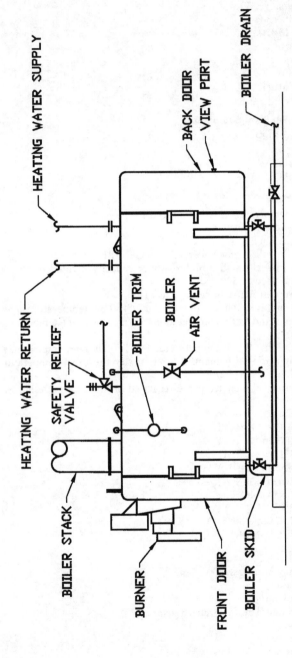

FIGURE 31.6 HEATING WATER SYSTEM AND BOILER TERMINOLOGY.

E. Heating Water Storage Systems

1. 20°F ΔT:
   a. 0.80 cu.ft./MBtu
   b. 1246.2 Btu/cu.ft.
   c. 166.6 Btu/gal.
2. 30°F ΔT:
   a. 0.54 cu.ft./MBtu
   b. 1869.3 Btu/cu.ft.
   c. 249.9 Btu/gal.
3. 40°F ΔT:
   a. 0.40 cu.ft./MBtu
   b. 2492.3 Btu/cu.ft.
   c. 333.2 Btu/gal.

F. Hot Water System Makeup Connection: Minimum connection size shall be 10 percent of largest system pipe size or 1", whichever is greater (20" system pipe size results in a 2" makeup water connection).

G. Chemical Feed Systems for Water Boilers. Chemical feed systems are designed to control the following:

1. System pH, normally between 8 and 9.
2. Corrosion.
3. Scale.

H. Design, Layout, and Clearance Requirements/Considerations

1. Design conditions:
   a. Boiler load, Btu/h, or MBH.
   b. Heating water temperatures, entering and leaving, or entering and ΔT.
   c. Heating water flows and fluid type (correct all data for fluid type).
   d. Fuel input, gas, fuel oil, electric, etc.
   e. Overall boiler efficiency.
   f. Water pressure drops.
   g. Fouling factor.
   h. Heating water reset, if applicable. Verify with boiler manufacturer that temperature limits are not exceeded.
   i. Electrical data:
      1) Unit kW, blower hp, compressor hp, and fuel oil pump hp.
      2) Full load, running load, and locked rotor amps.
      3) Voltage-phase-hertz.
2. Multiple hot water boilers should be used to prevent complete system or building shutdown upon failure of one hot water boiler in all heating water systems over 70 boiler horsepower or 2,400,000 Btu/h (i.e., 2 @ 50 percent, 2 @ 67 percent, 2 @ 70 percent, 3 @ 34 percent, 3 @ 40 percent).
3. Show tube clean/pull clearances and location.
4. The minimum recommended clearance around boilers is 36 in. Maintain minimum clearances for tube pull and the cleaning of tubes as recommended by the equipment manufacturer. This is generally equal to the length of the boiler. Maintain minimum clearance as required to open access and control doors on boilers for service, maintenance, and inspection.
5. Mechanical room locations and placement must take into account how boilers can be moved into and out of the building during initial installation and after construction for maintenance and repair and/or replacement.
6. Maintain the minimum electrical clearances as required by NEC.
7. Show the location of the boiler starter, disconnect switch, and control panel.
8. Show gas train and/or fuel oil train location.

9. Show boiler relief piping.

10. Show sanitary drain locations and boiler drain connections.

11. Design and locate combustion air louvers and motorized dampers, or engineered combustion air system. What happens if the engineered combustion air system malfunctions? Is a standby available? Verify that items that might freeze are not located in front of a combustion air intake.

12. Coordinate the height of the boiler with overhead clearances and obstructions. Is a beam required above the boiler for lifting components? Is a catwalk required to service the boiler?

13. Boiler stack and breeching. Coordinate routing in boiler room, through building, and discharge height above the building with the architect and structural engineer.

14. If isolators are required for the boiler, has the isolator height been considered in clearance requirements? If isolators are required for the boiler, has piping isolation been addressed?

15. Provide stop check valves (located closest to the boiler) and isolation valves with a drain between these valves on both the supply and return connections to all heating water boilers.

16. Boiler systems pumps should be located so the pump draws water out of the boiler, because it decreases the potential for entry of air into the system, and it does not impose the pump pressure on the boiler.

17. Interlock the boiler and the pump so the burner cannot operate without the pump operating.

18. Boiler warming pumps should be piped to both the system header and the boiler supply piping, allowing the boiler to be kept warm (in standby mode) from the system water flow or to warm the boiler when it has been out of service for repairs without the risk of shocking the boiler with the system water temperature (see Figs. 31.7 and 31.8).

19. Boiler warming pumps should be selected for 0.1 GPM/BHP (range 0.05–0.1 GPM/BHP). At 0.1 GPM/BHP, it takes 45–75 minutes to completely exchange the water in the boiler. This flow rate is sufficient to offset the heat loss by radiation and stack losses on boilers when in standby mode of operation. In addition, this flow rate allows the system to keep the boiler warm without firing the boiler, thus allowing for more efficient system operation. For example, it takes 8–16 hours to bring a boiler online from a cold start. Therefore, the standby boiler must be kept warm to enable immediate startup of the boiler upon failure of an operating boiler.

20. Circulating hot water through a boiler which is not operating, to keep it hot for standby purposes, creates a natural draft of heated air through the boiler and up the stack, especially in natural draft boilers. Forced draft or induced draft boilers have combustion dampers that close when not firing and therefore reduce, but don't eliminate, this heat loss. Although this heat loss is undesirable, for standby boilers, circulating hot water through the boiler is more energy efficient than firing the boiler. Operating (firing) a standby boiler may be in violation of air permit regulations in many jurisdictions today.

21. To provide fully automatic heating water system controls, the controls must look at and evaluate the boiler metal temperature (water temperature) and the refractory temperature prior to starting the primary pumps or enabling the boiler to fire. First, the boiler system design must circulate system water through the boilers to keep the boiler water temperature at system temperature when the boiler is in standby mode, as discussed for boiler warming pump arrangements. Second, the design must look at the water temperature prior to starting the primary pumps to verify the boiler is ready for service. And third, the design must look at refractory temperature to prevent boiler from going to high fire if the refractory is not at the appropriate temperature. However, the refractory temperature is usually handled by the boiler control package.

22. Outside air temperature reset of low temperature heating water systems is recommended for energy savings and controllability of terminal units at low load conditions.

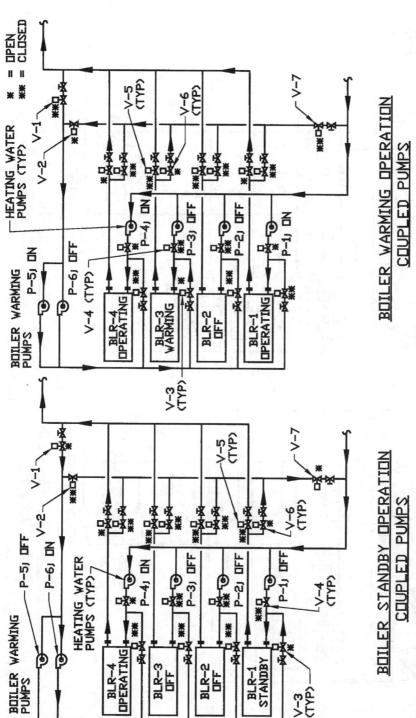

= OPEN
= CLOSED

BOILER WARMING OPERATION
COUPLED PUMPS

BOILER STANDBY OPERATION
COUPLED PUMPS

FIGURE 31.7   BOILER STANDBY AND WARMING DIAGRAM—COUPLED PUMPS.

439

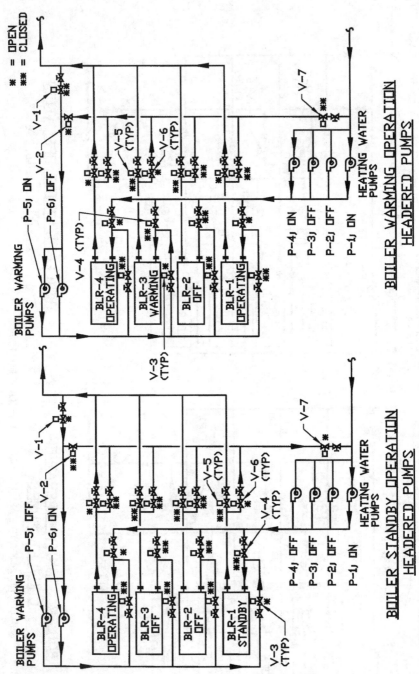

FIGURE 31.8 BOILER STANDBY AND WARMING DIAGRAM—HEADERED PUMPS.

440

However, care must be taken with the boiler design to prevent thermal shock by low return water temperatures, or to prevent condensation in the boiler due to low supply water temperatures and, therefore, a lower combustion stack discharge temperature.

23. Combustion air dampers must be extra heavy duty and should be low leakage (10 CFM/sq.ft. @ 4" WC differential) or ultralow leakage (6 CFM/sq.ft. @ 4" WC differential) type.

24. When the system design requires the use of dual fuel boilers (natural gas, fuel oil), provide a building automation control system I/O point to determine whether the boiler is on natural gas or fuel oil. Boiler control panels generally have a fuel type switch (Gas/Off/Fuel Oil Switch) which can be connected to create this I/O point. Switching from natural gas to fuel oil (or vice versa) cannot be a fully automatic operation because the boiler operator must first turn the boiler burner to the "Off" position, then turn the fuel type switch to fuel oil, then put combustion air linkage into the fuel oil position, then slide the fuel oil nozzle into position, then put the fuel oil pump into "Hand" or "Auto" position, and then turn the boiler burner to the "On" position. Remember to interlock the fuel oil pumps with operation of the boiler on fuel oil. Do not forget to include diesel generator interlocks with fuel oil pumps when the generators are fed from the same fuel oil system.

25. Heating water system warm-up procedure:
    a. Heating water system startup should not exceed a 100°F temperature rise per hour, but boiler or heat exchanger manufacture limitations should be consulted.
    b. It is recommended that no more than a 25°F temperature rise per hour be used when warming heating water systems. Slow warming of the heating water system allows for system piping, supports, hangers, and anchors to keep up with system expansion.
    c. Low temperature heating water systems (250°F and less) should be warmed slowly at a 25°F temperature rise per hour until the system design temperature is reached.
    d. Medium- and high-temperature heating water systems (above 250°F) should be warmed slowly at a 25°F temperature rise per hour until a 250°F system temperature is reached. At this temperature, the system should be permitted to settle for at least 8 hours or more (preferably overnight). The temperature and pressure maintenance time gives the system piping, hangers, supports, and anchors a chance to catch up with the system expansion. After allowing the system to settle, the system can be warmed up to 350°F or the system design temperature in 25°F temperature increments and 25 psig pressure increments, semi-alternating between temperature and pressure increases, and allowing the system to settle for an hour before increasing the temperature or pressure to the next increment. When the system reaches 350°F and the design temperature is above 350°F, the system should be allowed to settle for at least 8 hours or more (preferably overnight). The temperature and pressure maintenance time gives the system piping, hangers, supports, and anchors a chance to catch up with the system expansion. After allowing the system to settle, the system can be warmed up to 455°F or system design temperature in 25°F temperature increments and 25 psig pressure increments, semi-alternating between temperature and pressure increases, and allow the system to settle for an hour before increasing the temperature or pressure to the next increment.

26. Provide heating water systems with warm-up valves for in-service startup as follows (see Fig. 31.9). This will allow operators to warm these systems slowly and to prevent a sudden shock or catastrophic system failure when large system valves are opened. Providing warming valves also reduces wear on large system valves when only opened a small amount in an attempt to control the system warm-up speed.

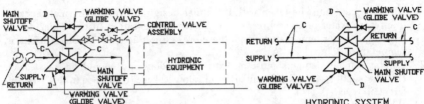

HYDRONIC EQUIPMENT APPLICATION
SERIES A

HYDRONIC SYSTEM
ISOLATION VALVE APPLICATION
SERIES B

NOTES:

1. SERIES A WARMING VALVES COVER STEAM OR MEDIUM/HIGH-TEMPERATURE HEATING WATER SERVICE FOR WARMING UP EQUIPMENT BEFORE THE MAIN SHUTOFF VALVES ARE OPENED, AND FOR BALANCING PRESSURES WHERE LINES ARE OF LIMITED VOLUME.
2. SERIES B WARMING VALVES COVER LINES CONVEYING GASES OR LIQUIDS WHERE BYPASSING MAY FACILITATE THE OPERATION OF THE MAIN VALVE BY BALANCING THE PRESSURES ON BOTH SIDES OF THE MAIN VALVE.

MAIN VALVE SIZE (C)	WARMING VALVE SIZE (D)	
	SERIES A WARMING VALVES	SERIES B WARMING VALVES
4"	1/2"	1"
5", 6"	3/4"	1-1/4"
8"	3/4"	1-1/2"
10"	1"	1-1/2"
12", 14"	1"	2"
16", 18", 20"	1"	3"
24", 30"	1"	4"
36", 42"	1"	6"
48", 54"	1"	8"
60", 72"	1"	10"
84", 96"	1"	12"

FIGURE 31.9 HYDRONIC SYSTEM WARMING VALVES.

BYPASS AND WARMING VALVES

Main Valve Nominal Pipe Size	Nominal Pipe Size	
	Series A Warming Valves	Series B Bypass Valves
4	1/2	1
5	3/4	1-1/4
6	3/4	1-1/4
8	3/4	1-1/2
10	1	1-1/2
12	1	2
14	1	2
16	1	3
18	1	3
20	1	3
24	1	4
30	1	4
36	1	6
42	1	6
48	1	8
54	1	8
60	1	10
72	1	10
84	1	12
96	1	12

Notes:

1 Series A comprehends steam service for warming up before the main line is opened, and for balancing pressures where lines are of limited volume.

2 Series B comprehends lines conveying gases or liquids where bypassing may facilitate the operation of the main valve through balancing the pressures on both sides of the disc or discs thereof. The valves in the larger sizes may be of the bolted on type.

27. Heating water system warming valve procedure (see Fig. 31.9):

   a. Open the warming return valve slowly to pressurize the equipment without flow.

   b. Once the system pressure has stabilized, slowly open the warming supply valve to establish flow and to warm the system.

   c. Once the system pressure and temperature have stabilized, perform the following
     steps, one at a time:
     1) Slowly open the main return valve.
     2) Close the warming return valve.
     3) Slowly open the main supply valve.
     4) Close the warming supply valve.

31.03 Steam Boilers

A. Boiler Types

1. Fire tube boilers:
   a. 15–800 BHP.
   b. 518–27,600 lbs./h
   c. 15–300 psig.
2. Water tube boilers:
   a. 350–2,400 BHP.
   b. 12,075–82,800 lbs./h
   c. 15–525 psig.
3. Flexible water tube boilers:
   a. 30–250 BHP.
   b. 10,000–82,000 lbs./h
   c. 15–525 psig.
4. Cast-iron boilers:
   a. 10–400 BHP.
   b. 1,035–8,625 lbs./h
   c. 0–150 psig.
5. Electric boilers:
   a. 15–5,000 KW.
   b. 51–17,065 MBH.
   c. 0–300 psig.

B. Steam Boiler Plant Equipment

1. Pretreatment systems:
   a. Filters.
   b. Softeners.
   c. Dealkalizers.
   d. RO units.
2. Feedwater systems:
   a. Deaerator:
     1) Spray type.
     2) Packed column type.
   b. Feedwater tank.
   c. Feedwater pumps.
3. Chemical feed systems:
   a. Chemical pumps.
   b. Chemical tanks.
   c. Agitators.
4. Sample coolers.
5. Blowdown coolers.
6. Surface blowdown/feedwater preheater.
7. Flue gas economizers.
8. Boilers (see Fig. 31.10).
9. Condensate return units and pumps.

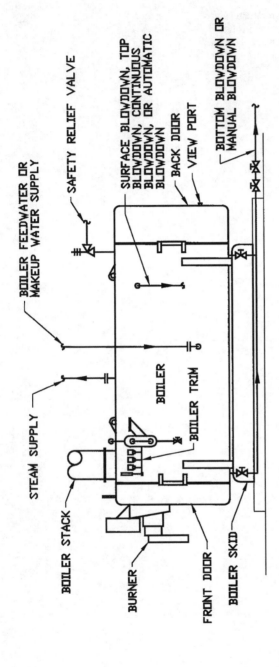

FIGURE 31.10 STEAM SYSTEM AND BOILER TERMINOLOGY.

NOTES:

1. BOILER BURNER MAY BE FORCED DRAFT, INDUCED DRAFT, OR NATURAL DRAFT TYPE DEPENDING ON BOILER TYPE AND CONSTRUCTION. FORCED DRAFT TYPE SHOWN.

2. BOILER TRIM IS COMPRISED OF THE LOW WATER LEVEL LIMIT AND ALARM, HIGH WATER LEVEL LIMIT AND ALARM, FEEDWATER CONTROLLER INCLUDING LEVEL CONTROLLER, SAFETIES, FUEL CUTOUTS, SAFETY RELIEF VALVES, PRESSURE GAUGES, THERMOMETERS, HIGH AND LOW LIMIT BURNER CONTROLS, AND OTHER APPURTENANCES.

10. Condensate receiver tank.
11. Condensate pumps.
12. Accumulators:
    a. Type:
       1) Dry.
       2) Wet.
    b. Service:
       1) Total system.
       2) Dedicated lines to specific equipment.
13. Super heaters:
    a. Internal.
    b. External.

C. Steam Capacities

1. Approx. 1,000 Btuh/1 lb. steam.
2. lbs. steam/h = lb. water/h

STEAM CAPACITY PER BOILER HORSEPOWER

Feed Water Temp.	Pounds of Dry Saturated Steam per Boiler hp @ System Pressure (psig) vs. Feedwater Temperature (°F)																	
	0	2	10	15	20	40	50	60	80	100	120	140	150	150	180	200	220	240
30	29.0	29.0	28.8	28.7	28.6	28.4	28.3	28.2	28.2	28.1	28.0	28.0	27.9	27.9	27.9	27.9	27.9	27.8
40	29.3	29.2	29.1	29.0	28.9	28.7	28.6	28.5	28.4	28.3	28.2	28.2	28.2	28.2	28.2	28.1	28.1	28.1
50	29.6	29.5	29.3	29.2	29.1	28.9	28.8	28.8	28.7	28.6	28.5	28.5	28.4	28.4	28.4	28.3	28.3	28.3
60	29.8	29.8	29.6	29.5	29.4	29.2	29.1	29.0	28.9	28.8	28.8	28.7	28.7	28.7	28.6	28.6	28.6	28.5
70	30.1	30.0	29.9	29.8	29.7	29.5	29.4	29.3	29.2	29.1	29.0	29.0	28.9	28.9	28.9	28.8	28.8	28.8
80	30.4	30.3	30.1	30.0	30.0	29.8	29.6	29.6	29.5	29.3	29.2	29.2	29.2	29.2	29.1	29.1	29.1	29.0
90	30.6	30.6	30.4	30.3	30.2	30.0	29.9	29.8	29.7	29.6	29.5	29.5	29.4	29.4	29.4	29.3	29.3	29.3
100	30.9	30.8	30.6	30.6	30.5	30.3	30.2	30.1	30.0	29.8	29.8	29.8	29.7	29.7	29.7	29.6	29.6	29.6
110	31.2	31.2	30.9	30.8	30.8	30.6	30.4	30.3	30.2	30.0	30.0	30.0	30.0	30.0	29.9	29.9	29.9	29.8
120	31.5	31.4	31.2	31.2	31.1	30.8	30.7	30.6	30.5	30.4	30.3	30.3	30.2	30.2	30.2	30.1	30.1	30.1
130	31.8	31.7	31.5	31.4	31.4	31.1	31.0	30.9	30.8	30.7	30.6	30.6	30.5	30.5	30.4	30.4	30.4	30.4
140	32.1	32.0	31.8	31.7	31.6	31.4	31.3	31.2	31.1	31.0	30.9	30.8	30.8	30.8	30.8	30.7	30.7	30.6
150	32.4	32.4	32.1	32.0	31.9	31.7	31.6	31.5	31.4	31.2	31.2	31.2	31.1	31.1	31.0	31.0	30.9	30.9
160	32.7	32.7	32.4	32.4	32.3	32.0	31.9	31.8	31.7	31.5	31.4	31.4	31.4	31.4	31.3	31.3	31.2	31.2
170	33.0	33.0	32.7	32.6	32.6	32.3	32.2	32.1	32.0	31.8	31.7	31.7	31.7	31.7	31.6	31.6	31.5	31.5
180	33.4	33.3	33.0	33.0	32.9	32.6	32.5	32.4	32.3	32.2	32.1	32.0	32.0	32.0	31.9	31.9	31.8	31.8
190	33.8	33.7	33.4	33.3	33.2	32.9	32.8	32.7	32.6	32.5	32.4	32.4	32.3	32.3	32.2	32.2	32.1	32.1
200	34.1	34.0	33.7	33.6	33.5	33.2	33.1	33.0	32.9	32.8	32.7	32.6	32.6	32.6	32.6	32.5	32.4	32.4
212	34.5	34.4	34.2	34.1	33.9	33.6	33.5	33.4	33.3	33.2	33.1	33.0	33.0	33.0	32.9	32.9	32.8	32.8
220	34.8	34.7	34.4	34.3	34.2	33.9	33.8	33.7	33.5	33.4	33.3	33.3	33.2	33.2	33.1	33.1	33.1	33.0
227	35.0	34.9	34.7	34.5	34.4	34.1	34.0	33.9	33.8	33.7	33.6	33.5	33.5	33.4	33.4	33.3	33.3	33.3
230	35.2	35.0	34.8	34.7	34.5	34.2	34.1	34.0	33.9	33.8	33.7	33.6	33.6	33.5	33.5	33.4	33.4	33.4

D. Steam Boiler Drums

1. Top drum: steam drum.
2. Bottom drum: mud or blowdown drum.

E. System Types

1. Low-pressure steam: 0–15 psig.
2. Medium-pressure steam: 16–100 psig.
3. High-pressure steam: 101 psig and greater.

F. Steam Carryover

1. Steam carryover is the entrainment of boiler water with the steam.
2. Causes of carryover:
   a. Mechanical:
      1) Poor boiler design.
      2) Burner misalignment.
      3) High water level.
   b. Chemical:
      1) High total dissolved solids (TDS).
      2) High total suspended solids (TSS).
      3) High alkalinity.
      4) High amine levels.
      5) Presence of oils or other organic materials.
3. Problems caused by carryover:
   a. Deposits minerals on valves, piping, heat transfer surfaces, and other steam-operated equipment.
   b. Causes thermal shock to the system.
   c. Contaminates process or products that have direct steam contact.
   d. If steam is used for humidification, a white dust is often left on the air handling unit components, ductwork surfaces, and furniture and other equipment within the space.
4. Carryover control:
   a. Install steam separation devices.
   b. Maintain the proper steam space in the steam drum and boiler.
   c. Maintain proper water chemistry—TDS, TSS, alkalinity, etc.

G. Design, Layout, and Clearance Requirements/Considerations

1. Design conditions:
   a. Boiler load: Btu/h, or MBH.
   b. Steam pressure and flow rate.
   c. Fuel input: gas, fuel oil, electric, etc.
   d. Overall boiler efficiency.
   e. Fouling factor.
   f. Electrical data:
      1) Unit kW, blower hp, compressor hp, and fuel oil pump hp.
      2) Full load, running load, and locked rotor amps.
      3) Voltage-phase-hertz.
2. Multiple steam boilers should be used to prevent complete system or building shutdown upon failure of 1 steam boiler in all steam systems over 70 boiler horsepower or 2,400,000 Btu/h (i.e., 2 @ 50 percent, 2 @ 67 percent, 2 @ 70 percent, 3 @ 34 percent, 3 @ 40 percent).
3. Show tube clean/pull clearances and location.
4. The minimum recommended clearance around boilers is 36 in. Maintain minimum clearances for tube pull and the cleaning of the tubes as recommended by the equipment manufacturer. This is generally equal to the length of the boiler. Maintain minimum clearance as required to open access and control doors on boilers for service, maintenance, and inspection.
5. Mechanical room locations and placement must take into account how boilers can be moved into and out of the building during initial installation and after construction for maintenance and repair and/or replacement.
6. Maintain minimum electrical clearances as required by the NEC.
7. Show the location of the boiler starter, disconnect switch, and the control panel.
8. Show gas train and/or fuel oil train location.
9. Show the boiler relief piping.
10. Show sanitary drain locations and boiler drain connections.

11. Design and locate combustion air louvers and motorized dampers or an engineered combustion air system. What happens if the engineered combustion air system malfunctions? Is a standby available? Verify that items that might freeze are not located in front of the combustion air intake.

12. Coordinate the height of the boiler with overhead clearances and obstructions. Is a beam required above the boiler for lifting components? Is a catwalk required to service the boiler.

13. Boiler stack and breeching. Coordinate routing in the boiler room, through the building, and the discharge height above building with architect and structural engineer.

14. Provide a stop check valve (located closest to the boiler) and an isolation valve with a drain between these valves on the steam supply connections to all steam boilers.

15. Combustion air dampers must be extra heavy duty and should be low leakage (10 CFM/sq.ft. @ 4" WC differential) or ultralow leakage (6 CFM/sq.ft. @ 4" WC differential) type.

16. When the system design requires the use of dual fuel boilers (natural gas, fuel oil), provide a building automation control system I/O point to determine whether the boiler is on natural gas or fuel oil. Boiler control panels generally have a fuel type switch (gas/off/fuel oil switch) that can be connected to create this I/O point. Switching from natural gas to fuel oil (or vice versa) cannot be a fully automatic operation because the boiler operator must first turn the boiler burner to the "Off" position, then turn the fuel type switch to fuel oil, then put the combustion air linkage into the fuel oil position, then slide the fuel oil nozzle into position, then put the fuel oil pump into the "Hand" or "Auto" position, and then turn the boiler burner to the "On" position. Remember to interlock the fuel oil pumps with operation of the boiler on fuel oil. Do not forget to include diesel generator interlocks with fuel oil pumps when the generators are fed from the same fuel oil system.

17. Steam system warm-up procedure:

    a. Steam system startup should not exceed a 100°F temperature rise per hour (50 psig per hour); boiler or heat exchanger manufacture limitations should be consulted.

    b. It is recommended that no more than a 25°F temperature rise per hour (15 psig per hour) be used when warming steam systems. Slow warming of the steam system allows for system piping, supports, hangers, and anchors to keep up with system expansion.

    c. Low-pressure steam systems (15 psig and less) should be warmed slowly at a 25°F temperature rise per hour (15 psig per hour) until the system design pressure is reached.

    d. Medium- and high-pressure steam systems (above 15 psig) should be warmed slowly at a 25°F temperature rise per hour (15 psig per hour) until a 250°F-15 psig system temperature-pressure is reached. At this temperature-pressure, the system should be permitted to settle for at least 8 hours or more (preferably overnight). The temperature-pressure maintenance time gives the system piping, hangers, supports, and anchors a chance to catch up with the system expansion. After allowing the system to settle, the system can be warmed up to 120 psig or the system design pressure in 25 psig pressure increments, and allow the system to settle for an hour before increasing the pressure to the next increment. When the system reaches 120 psig and the design pressure is above 120 psig, the system should be allowed to settle for at least 8 hours or more (preferably overnight). The pressure maintenance time gives the system piping, hangers, supports, and anchors a chance to catch up with the system expansion. After allowing the system to settle, the system can be warmed up to 300 psig or the system design pressure in 25 psig pressure increments, and then must be permitted to settle for an hour before increasing the pressure to the next increment.

18. Provide steam systems with warm-up valves for in-service startup, as shown in the following table. This will allow operators to warm these systems slowly and prevent

a sudden shock or catastrophic system failure when large system valves are opened. Providing warming valves also reduces wear on large system valves when only opened a small amount in an attempt to control the system warm-up speed.

BYPASS AND WARMING VALVES

Main Valve Nominal Pipe Size	Nominal Pipe Size	
	Series A Warming Valves	Series B Bypass Valves
4	1/2	1
5	3/4	1-1/4
6	3/4	1-1/4
8	3/4	1-1/2
10	1	1-1/2
12	1	2
14	1	2
16	1	3
18	1	3
20	1	3
24	1	4
30	1	4
36	1	6
42	1	6
48	1	8
54	1	8
60	1	10
72	1	10
84	1	12
96	1	12

Notes:
1. Series A comprehends steam service for warming up before the main line is opened, and for balancing pressures where lines are of limited volume.
2. Series B comprehends lines conveying gases or liquids where bypassing may facilitate the operation of the main valve through balancing the pressures on both sides of the disc or discs thereof. The valves in the larger sizes may be of the bolted on type.

19. The steam system warming valve procedure (see Fig. 31.11):

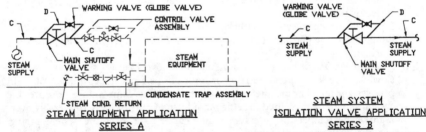

FIGURE 31.11   STEAM SYSTEM WARMING VALVES.

    a. Slowly open the warming supply valve to establish flow and to warm the system.

    b. Once the system pressure and temperature have stabilized, perform the following items one at a time:

        1) Slowly open the main supply valve.

        2) Close the warming supply valve.

20. If isolators are required for the boiler, has isolator height been considered in clearance requirements? If isolators are required for the boiler, has piping isolation been addressed?

H. Low Water Cutoffs

1. Primary: Float type.
2. Auxiliary: Probe type.
3. Low water cutoffs should be tested by using an evaporation test:
    a. Take the boiler to low fire.
    b. Shut off the feedwater to the boiler.
    c. Operate the boiler until the low water cutoff shuts down the boiler or the water level in the gauge glass falls below the low water cutoff activation point but still remains visible in glass.
    d. Conduct an evaporation test at least every 30 days; once a week is recommended.
4. Class I boilers. Low water cutoff is 3" above the top row of tubes in fire tube boilers.
5. Class IV boilers. Low water cutoff is 0"–1/4" above the top row of tubes in fire tube boilers.
6. Water should always be visible in gauge glass. If water is not visible in gauge glass, immediately perform the following two steps one after another in any order:
    a. Shut off the boiler burner.
    b. Shut off the boiler feedwater.
    c. Then allow the boiler to cool and inspect it for damage.

I. Deaerator or Feedwater Tank

1. The deaerator or the feedwater tank purpose is to remove oxygen, carbon dioxide, hydrogen sulfide, and other noncondensable gases and to heat boiler feedwater.
2. They also preheat the feedwater prior to being pumped to the boiler. Cold feedwater temperatures may cause:
    a. Thermal shock.
    b. Oxygen-rich feedwater, which causes corrosion.
3. This equipment should remove oxygen in the water to levels measured in parts per billion (ppb).
4. Steam vent on the deaerator or feedwater tank. Steam should appear 12"–18" above the top of vent. If steam appears below 12", the deaerator or feedwater tank is not removing all the oxygen, carbon dioxide, hydrogen sulfide, and other noncondensable gases.
5. Deaerators should be used when:
    a. The system pressure is 75 psig and higher.
    b. Steam systems are employed with little or no standby capacity.
    c. The system depends on continuous operation.
    d. The system requires 25 percent or more of makeup water.

J. Sizing Boiler Feed Pumps, Condensate Return Pumps, and Condensate Receivers

1. If the boiler is under 50 psi, the designer should size the boiler feed pumps or condensate return pumps so that they discharge at 5 psi above the working pressure of the boiler.
2. If the boiler is over 50 psi, the designer should size boiler feed pumps or condensate return pumps so that they discharge at 10 psi above the working pressure of the boiler.
3. The designer should size condensate receivers for 1 minute of net capacity based on the condensate return rate.
4. Size boiler feedwater system receivers for system capacity (normally estimated at 10 minutes):
    a. Deaerator systems: 10-minute supply.
    b. Feedwater tank systems: 15-minute supply.

5. Size condensate pumps at three times the condensate return rate.
6. Size boiler feedwater pumps and transfer pumps at:
   a. Turbine pumps, intermittent operation: Two times the boiler maximum evaporation rate or 0.14 GPM per boiler hp.
   b. Centrifugal pumps, continuous operation: 1.5 times the boiler maximum evaporation rate or 0.104 GPM per boiler hp.
   c. Boiler feedwater and transfer pump selection criteria:
      1) Continuous or intermittent operation.
      2) Temperature of feedwater or condensate.
      3) Flow capacity (GPM).
      4) Discharge pressure required: Boiler pressure plus piping friction loss.
      5) NPSH requirement.
7. Boiler feedwater control types:
   a. On/Off feedwater control is generally used with single boiler systems or in multiple boiler systems when one feedwater pump is dedicated to each boiler and is typically accomplished with a turbine pump.
   b. Level control is generally used with multiple boiler systems where feedwater pumps serve more than one boiler and is typically accomplished with a centrifugal pump.
8. Vacuum type steam condensate return units: 0.1 GPM/1,000 lbs./h of connected load.
9. Pumped steam condensate return units: 2.4 GPM/1,000 lbs./h.

K. Boiler Blowdown Systems

1. Bottom blowdown. Bottom blowdown, sometimes referred to as manual blowdown, functions to remove suspended solids and sediment that have settled out of the water and deposited on the bottom of the boiler. Bottom blowdown is most effective with several short discharges in lieu of one long discharge because the solids settle out between discharges; this results in the greatest removal of suspended solids with the least amount of water.
2. Surface blowdown. Surface blowdown, sometimes referred to as automatic blowdown, continuous blowdown, or periodic blowdown, depending on how the blowdown is controlled, functions to remove dissolved solids, surface water scum, and foam to maintain proper conductivity levels.
   a. Automatic:
      1) Conductivity probe.
      2) Timer.
   b. Continuous.
   c. Periodic (manual) by time.

L. Boiler Blowdown Separator Makeup

1. Noncontinuous blowdown (bottom blowdown):   5.0 GPM/1,000 lbs./h.
2. Continuous blowdown (surface blowdown):   0.5 GPM/1,000 lbs./h.

M. Blowdown Separator Drains: 10 GPM/1000 lbs./h Boiler Output

N. Steam Boiler Water Makeup

1. Boilers: 4.0 GPM/1,000 lbs./h each.
2. Deaerator/feedwater unit: 4.0 GPM/1,000 lbs./h each.
3. Makeup water for the steam system is only required at one of the boilers or one of the feedwater units at any given time for system sizing.

O. Chemical Feed Systems for Steam Boilers. Chemical feed systems are designed to control the following.

1. System pH, normally between 8 and 9.
2. Oxygen level, less than 0.007 PPM (7 ppb).
3. Water conditioning level.
4. Carbon dioxide level.
5. Scale.
6. Corrosion.

31.04 Fuel Systems and Types

A. Fuel System Design Guidelines

1. Natural gas pressure reducing valves (NGPRV):
   a. Use multiple NGPRVs when system natural gas capacity exceeds 2 NGPRV size, when normal operation calls for 10 percent of design load for sustained periods, or when there are two distinct load requirements (e.g., summer/winter) that are substantially different. Provide the number of NGPRVs to suit the project.
   b. If system capacity for a single NGPRV exceeds the 2 NGPRV size but is not larger than the 4 NGPRV size, use 2 NGPRVs with 33 percent and 67 percent or 50 percent and 50 percent capacity split.
   c. If system capacity for a single NGPRV exceeds 4 NGPRV size, use 3 NGPRVs with 25 percent, 25 percent, and 50 percent or 15 percent, 35 percent, and 50 percent capacity split to suit the project.
   d. Provide natural gas pressure regulating valves with positive shutoff ability to prevent the natural gas system from becoming equal to the gas utility system pressure when the building natural gas system is not using gas.
2. Natural gas meters should be provided as follows:
   a. Coordinate equipment, building, or site meter requirements with the local utility company. If the project budget permits, these meter readings should be logged and recorded at the building facilities management and control system.
   b. Meter for a campus or site of buildings. A site meter is generally provided by the utility company.
   c. Meter for individual buildings on a campus. If fed from a site meter, design documents should provide a meter for each building. This meter will assist in tracking energy use at each building and for troubleshooting system problems.
   d. Meter for individual buildings. A building meter will generally be provided by the utility company.
   e. Meters for individual boilers. A meter should be provided by the design documents for each and every boiler; environmental air permit requirements insist natural gas be monitored at each boiler.
   f. Meters for other major users. A meter should be provided by the design documents for each major user within the building (emergency generators, gas-fired AHUs, domestic water heaters, unit heaters, kitchens).
3. Boiler fuel oil pump flow rates and generator day tank pump flow rates are generally 2.5–3.0 times the boiler and generator consumption rates. Confirm with the manufacturer or the electrical engineer that the information received is the consumption rate of the boiler/generator or fuel oil pumping rate of the boiler/generator. When boilers are located above the fuel oil tanks, a method of preventing back siphoning through the return line must be provided. This may be accomplished by providing the return line with a pressure regulator or with an operated valve interlocked with the fuel oil pump. Also, the fuel oil pumps must be provided with a check valve in the discharge, or if large height differentials are required, a motorized discharge isolation valve interlocked with the pump may be required because check valves will leak.
4. Fuel oil meters should be provided as follows:
   a. If the fuel oil system is a circulating system with a fuel oil return line, meters must be provided in both the supply and return to determine the fuel oil consumed. Most manufacturers provide fuel oil meters with this capability with controls and software to automatically calculate the fuel oil consumed. If the project budget permits, these meter readings should be logged and recorded at the building facilities management and control system. All fuel oil meters must be shown on the design documents. Environmental regulations require the fuel oil purchased versus fuel oil consumed be recorded and tracked for determining when leaks may be occurring in the system.
   b. Meters for each group of site distribution pumps are located at the pumps.

c. Meters for individual buildings on a campus are located at each building. This meter will assist in tracking energy use at each building and for troubleshooting system problems.
d. Meters for individual boilers. A meter should be provided for each and every boiler; environmental air permit requirements require fuel oil to be monitored at each boiler.
e. Meters for other major users. A meter should be provided for each major user within the building (emergency generators, oil-fired AHUs, domestic water heaters, and unit heaters).

B. **Natural Gas**

1. 900–1200 Btu/cu.ft.
2. 1,000 Btu/cu.ft. average.

C. **Fuel Oil**

1. #2: 138,000 Btu/gal.
2. #4: 141,000 Btu/gal.
3. #5: 148,000 Btu/gal.
4. #6: 152,000 Btu/gal.

D. **LP Gas**

1. Butane:
   a. 21,180 Btu/lbs.
   b. 3,200 Btu/cu.ft.
2. Propane:
   a. 21,560 Btu/lbs.
   b. 2,500 Btu/cu.ft.

E. **Electric**

1. 3,413 Btuh/KW.
2. 3,413 Btuh/watt.

F. **Coal**

1. Anthracite: 14,600–14,800 Btu/lb.
2. Bituminous: 13,500–15,300 Btu/lb.

G. **Wood**

1. 8,000–10,000 Btu/lb.

H. **Kerosene**

1. 135,000 Btu/gal.

Motors and Motor Controllers

32.01 Motors

A. Motor Types. Items 1, 2, and 3 are the most common HVAC motor types.

1. Open drip proof (ODP): Ventilation openings arranged to prevent liquid drops falling within an angle of 15 degrees from the vertical from affecting motor performance. Use indoors and in moderately clean environments.
2. Totally enclosed fan cooled (TEFC): A fan on the motor shaft, outside the stator housing and within the protective shroud, blows air over the motor. Use in damp, dirty, corrosive, or contaminated environments.
3. Explosion proof (EXPRF): Totally enclosed with enclosure designed to withstand internal explosion of a specific gas-air or dust-air mixture to prevent escape of ignition products. Motors are approved for a specific Hazard Classification as covered by the NEC. Class I Explosion Proof and Class II Dust Ignition Resistant are the two most common types of hazardous location motors.
4. Open drip proof air over (ADAO): Ventilation openings arranged to prevent liquid drops falling within an angle of 15 degrees from the vertical from affecting motor performance. Use indoors and in moderately clean environments. Rated for motor cooling by airflow from a driven device.
5. Totally enclosed non-ventilated (TENV): No ventilation openings in housing. Motor rated for cooling by airflow from a driven device. TENV motors are usually under 5 horsepower.
6. Totally enclosed air over (TEAO): No ventilation openings in housing. Motor rated for cooling by airflow from a driven device. TEAO motors frequently have dual horsepower ratings depending on speed and cooling air temperature.

B. Motor Horsepowers, Voltage, Phase, and Operating Guidelines:

1. Suggested horsepower and phase:
    a. Motors 1/2 horsepower and larger: 3 Phase.
    b. Motors less than 1/2 horsepower: Single Phase.
    c. Considering first cost economics only, it is less costly, on average, to have motors smaller than 1 hp to be single-phase. At 3/4 hp, single-phase and three-phase motors cost about the same, but branch circuits and control equipment for three-phase motors are usually more expensive.
    d. When life cycle owning and operating costs are considered, it is often more economical to provide motors as specified in lines a. and b. earlier.
2. Do not start and stop motors more than six times per hour.
3. Motors of 5 horsepower and larger should not be cycled; they should run continuously.
4. Specify energy-efficient motors—EPAct motors as a minimum; preferred premium efficiency motors. Premium efficiency motors are a higher efficiency motor than the EPAct motors.
5. Do not use energy-efficient motors with variable speed/frequency drives.
6. For best motor life and reliability, do not select motors to run within the service factors. Specify motors with a minimum 1.15 service factor.
7. For every 50°F (28°C) increase in motor operating temperature, the life of the motor is cut in half. Conversely, for every 50°F (28°C) decrease in motor operating temperature, the life of the motor is doubled.
8. Energy-efficient motors have a higher starting current than their standard efficiency counterparts.
9. The best sign of motor trouble is smoke and/or paint discoloration.
10. In general, motors can operate with voltages plus or minus 10 percent of their rated voltage.
11. Motors in storage should be turned by hand every 6 months to keep the bearings from drying out.
12. Available motor voltages are given in the following table:

Phase	Nominal Voltage	Nameplate Voltage
Single-Phase	120	115
	240	230
	277	265
Three-Phase	208	200
	240	230
	480	460
	600	575

C. Standard motor sizes are given in the following table:

Motor Sizes (hp)	Recommended Starter Type	Standard Service Factors
1/8; 1/10; 1/12; 1/15; 1/20; 1/25; 1/30; 1/60; 1/100	SPC or PSC	1.40
1/6	SPC or PSC	*
1/4; 1/3	CS	*
1/2; 3/4; 1	MS	*
1-1/2; 2	MS	*
3; 5; 7-1/2; 10; 15; 20; 25; 30; 40; 50; 60; 75; 100; 125; 150; 200; 250	MS	*
300; 350; 400; 450; 500; 600; 700; 750; 800; 900; 1000; 1250; 1500; 1750; 2000; 2250; 2500; 3000; 3500; 4000; 4500; 5000; 5500; 6000**	MS	*

Notes:
SPC: Split phase capacitor start.
PSC: Permanent split capacitor start.
CS: Capacitor start.
MS: Magnetic start; polyphase induction motors (squirrel cage).
1/2 hp through 50 hp across-the-line starter.
60 hp and larger reduced-voltage starter.
*See paragraph E below for motor service factors for these motors.
**Motors generally not used in HVAC applications.

D. Standard Motor RPM: 3600, 1800, 1200, 900, 720, 600, and 514.

E. NEMA motor service factors are given in the following table:

hp	3600 RPM	1800 RPM	1200 RPM	900 RPM
1/6–1/3	1.35	1.35	1.35	1.35
1/2	1.25	1.25	1.25	1.15
3/4	1.25	1.25	1.15	1.15
1	1.25	1.15	1.15	1.15
1-1/2–250	1.15	1.15	1.15	1.15
300–2500	1.15	1.15	1.15	1.15

F. NEMA locked rotor indicating code letters are given in the following table:

NEMA Locked Rotor Indicating Code Letters			
Code Letter	KVA/hp	Code Letter	KVA/hp
A	0–3.14	L	9.00–9.99
B	3.15–3.54	M	10.00–11.19
C	3.55–3.99	N	11.20–12.49
D	4.00–4.49	O	Not used
E	4.50–4.99	P	12.50–13.99
F	5.00–5.59	Q	Not used
G	5.60–6.29	R	14.00–15.99
H	6.30–7.09	S	16.00–17.99
I	Not used	T	18.00–19.99
J	7.10–7.99	U	20.00–22.39
K	8.00–8.99	V	22.40 and up

1. Standard three-phase motors often have these NEMA starting locked rotor codes:
   a. 1 horsepower and smaller: Locked Rotor Code L.
   b. 1-1/2–2 horsepower: Locked Rotor Code K.
   c. 3 horsepower: Locked Rotor Code J.
   d. 5 horsepower: Locked Rotor Code H.
   e. 7-1/2–10 horsepower: Locked Rotor Code G.
   f. 15 horsepower and larger: Locked Rotor Code F.
2. Standard single-phase motors often have these locked rotor codes:
   a. 1/2 horsepower and smaller: Locked Rotor Code L.
   b. 3/4–1 horsepower: Locked Rotor Code K.
   c. 1-1/2–2 horsepower: Locked Rotor Code J.
   d. 3 horsepower: Locked Rotor Code H.
   e. 5 horsepower: Locked Rotor Code G.
3. Specify 15 horsepower and larger motors with NEMA Starting Code F or G.
4. Specify motors smaller than 15 horsepower with the manufacturer's standard starting characteristics.

G. Motor Insulation Classes are given in the following table.

1. Specify all motors with class F insulation and class B motor temperature rise.
2. Specify all motors with a minimum 1.15 service factor or NEMA standard service factor, whichever is higher.

Motor Type	Motor Insulation Class Temperature Rise							
	A		B		F		H	
	°C	°F	°C	°F	°C	°F	°C	°F
1. Motors with 1.0 Service Factor (except 3 and 4 below)	60	140	80	176	105	221	125	257
2. All Motors with 1.15 Service Factor or Higher	70	158	90	194	115	239	–	–
3. Totally Enclosed Nonventilated Motor with 1.0 Service Factor	65	149	85	185	110	230	135	275
4. Motors with Encapsulated Windings and with 1.0 Service, All Enclosures	65	149	85	185	110	230	–	–

Notes:
  1 Abnormal deterioration of insulation may be expected if the ambient temperature of 40°C/104°F is exceeded in regular operation.
  2 Temperature rise based on 40°C/104°F ambient. Temperature rises are based on operation at altitudes of 3300 feet or less.
  3 Class A Motors: Fractional hp motors, small appliances; maximum operating temperature 105°C/221°F.
  4 Class B Motors: Motors for HVAC applications, high-quality fractional hp motors; maximum operating temperature 130°C/266°F.
  5 Class F Motors: Inverter duty motors, industrial motors; maximum operating temperature 155°C/311°F.
  6 Class H Motors: High temperature, high reliability, high ambient; maximum operating temperature 180°C/356°F.

H. NEMA Motor Design Designations

1. Design A motors are built with high pullout torque and are used on injection molding machines.
2. Design B motors are built with high starting torque with reasonable starting current and are used with fans, pumps, air handling units, and other HVAC equipment. They are the most common HVAC motor.
3. Design C motors are built with high starting torque and used with hard-to-start loads and with conveyors.
4. Design D motors are built with high starting torque, low starting current, and high slip and are used with cranes, hoists, and low-speed presses.

I. Clearance Requirements

1. The minimum recommended clearance around the motors is 24 inches.
2. Mechanical room locations and placement must take into account how motors can be moved into and out of the building during the initial installation and after construction for maintenance, repair, and/or replacement.

J.   Motor Efficiencies

1. *ASHRAE Standard 90.1-2013:* NEMA Design A and B; Single Speed; 3600, 1800, or 1200 RPM; Open Drip Proof (ODP) or Totally Enclosed Fan-Cooled (TEFC) motors 1 hp and larger shall meet the following minimum nominal efficiencies:

Motor Horsepower	Minimum Nominal Efficiency (%)					
	Open Motors			Enclosed Motors		
	Number of Poles			Number of Poles		
	2	4	6	2	4	6
	Synchronous Speed (RPM)			Synchronous Speed (RPM)		
	3600	1800	1200	3600	1800	1200
1	–	82.5	80.0	75.5	82.5	80.0
1.5	82.5	84.0	84.0	82.5	84.0	85.5
2	84.0	84.0	85.5	84.0	84.0	86.5
3	84.0	86.5	86.5	85.5	87.5	87.5
5	85.5	87.5	87.5	87.5	87.5	87.5
7.5	87.5	88.5	88.5	88.5	89.5	89.5
10	88.5	89.5	90.2	89.5	89.5	89.5
15	89.5	91.0	90.2	90.2	91.0	90.2
20	90.2	91.0	91.0	90.2	91.0	90.2
25	91.0	91.7	91.7	91.0	92.4	91.7
30	91.0	92.4	92.4	91.0	92.4	91.7
40	91.7	93.0	93.0	91.7	93.0	93.0
50	92.4	93.0	93.0	92.4	93.0	93.0
60	93.0	93.6	93.6	93.0	93.6	93.6
75	93.0	94.1	93.6	93.0	94.1	93.6
100	93.0	94.1	94.1	93.6	94.5	94.1
125	93.6	94.5	94.1	94.5	94.5	94.1
150	93.6	95.0	94.5	94.5	95.0	95.0
200	94.5	95.0	94.5	95.0	95.0	95.0

Note:
1 Nominal efficiencies shall be established in accordance with NEMA MG1.

MOTOR DIMENSIONS AND WEIGHTS—EPAct MOTORS

Motor Horsepower	Open Drip Proof— EPAct			Totally Enclosed Fan-Cooled—EPAct		
	Dia. Inches	Length Inches	Weight lbs.	Dia. Inches	Length Inches	Weight lbs.
1	9	11	67	9	13	70
1.5	11	13	88	11	15	90
2	11	14	97	11	16	101
3	12	16	132	12	18	139
5	12	18	158	12	20	165
7.5	14	21	211	14	24	257
10	14	23	260	14	25	295
15	15	23	343	15	26	414
20	15	25	392	15	28	473
25	16	27	529	16	33	626
30	16	28	573	16	34	763
40	18	29	726	18	35	932
50	18	30	803	18	35	933
60	20	33	970	20	34	1212
75	20	35	1105	20	35	1481
100	22	39	1166	22	47	1671
125	22	41	1276	22	48	1775
150	22	41	1364	22	48	1897
200	22	44	1810	22	54	2730
250	22	49	2160	22	59	3240

Notes:
1 The motor dimensions and weights are based on 1,200-rpm motors. Motors above the 1,200 rpm rating are lighter and smaller in size.
2 Motor dimensions are rounded to the nearest inch.

MOTOR DIMENSIONS AND WEIGHTS—PREMIUM EFFICIENCY MOTORS

Motor Horsepower	Open Drip Proof—Premium			Totally Enclosed Fan-Cooled—Premium		
	Dia. Inches	Length Inches	Weight lbs.	Dia. Inches	Length Inches	Weight lbs.
1	9	11	67	9	14	70
1.5	11	13	88	11	16	90
2	11	14	97	11	16	101
3	12	16	132	12	20	200
5	12	18	158	12	20	220
7.5	14	21	260	14	26	315
10	14	23	310	14	26	350
15	15	24	394	15	29	460
20	15	26	436	15	29	510
25	16	27	580	18	33	700
30	16	28	639	18	34	763
40	18	29	770	19	35	1030
50	18	30	838	20	35	1070
60	20	33	1090	22	40	1480
75	20	35	1150	22	40	1540
100	22	39	1494	24	47	2060
125	22	41	1715	24	48	2130
150	22	44	2100	24	52	2860
200	22	50	2150	24	54	3070
250	22	54	2632	24	59	3440

Notes:
1 The motor dimensions and weights are based on 1,200-rpm motors. Motors above the 1,200 rpm rating are lighter and smaller in size.
2 Motor dimensions are rounded to the nearest inch.

32.02 Starters, Disconnect Switches, and Motor Control Centers

A. Starter Types

1. Manual starters (manual control):
   a. Reversing/nonreversing.
   b. Push button/toggle switch.
   c. Available for single-phase or three-phase electrical power.
2. Magnetic starters (automatic control):
   a. Full voltage/across the line.
   b. Reversing/nonreversing.
   c. Reduced voltage:
      1) Reactor.
      2) Resistance.
      3) Auto transformer.
      4) Wye-delta/star delta.
      5) Full voltage part winding.
      6) Reduced voltage part winding.
      7) Solid state.
   d. Two-speed starting:
      1) One winding. Full speed; half speed.
      2) Two winding. Full speed; 2/3 speed.

3) Constant torque.
4) Variable torque.
5) Constant horsepower.
e. Available for single-phase or three-phase electrical power.
3. Combination starter disconnect switch: see "magnetic starters":
a. Fused.
b. Nonfused.
c. Disconnect switches (locking/nonlocking—recommend locking switches).
d. Available for three-phase electrical power only, but a three-phase starter can be used with a single-phase motor (although expensive).

B. Starter Accessories

1. Pilot lights: green, run; red, off.
2. Switches (locking/nonlocking—recommend locking switches).
a. Hand-off-auto (HOA).
b. Push button.
c. Toggle switch.
3. Control transformer.
4. Overload protection:
a. Fused.
b. Nonfused.
c. Motor circuit protector.
d. Molded case circuit breaker.
e. Circuit fuse protection: size based on circuit ampacity and wire size.
f. Overload heaters: size based on motor overload capacity.
g. Two levels of overload protection:
1) Type 1: Considerable damage occurs to the contactor and overload relay when an overload happens but the enclosure remains externally undamaged. Parts of the starter or the entire starter may need to be replaced after an overload.
2) Type 2: No damage occurs to the contactor or overload relay except light contact burning is permitted when an overload happens.
h. The choice between circuit breakers and fuses is purely a matter of user preference.
5. Auxiliary contacts (NO-Normally Open/NC-Normally Closed).
6. Relays.

C. Disconnect switch sizes and accepted fuse sizes are given in the following table:

Safety Switch Size Amps	Acceptable Fuse Sizes Amps	Safety Switch Size Amps	Acceptable Fuse Sizes Amps
30	15, 20, 25, 30	1600	1600
60	35, 40, 45, 50, 60	2000	2000
100	70, 80, 90, 100	2500	2500
200	110, 125, 150, 175, 200	3000	3000
400	225, 250, 300, 350, 400	4000	4000
600	450, 500, 600	5000	5000
800	700, 800	6000	6000
1,200	1000, 1200	–	–

D. Standard Fuse and Circuit Breaker Sizes (Amperes): 1, 3, 6, 10, 15, 20, 25, 30, 35, 40, 45, 50, 60, 70, 80, 90, 100, 110, 125, 150, 175, 200, 225, 250, 300, 350, 400, 450, 500, 600, 700, 800, 1000, 1200, 1600, 2000, 2500, 3000, 4000, 5000, and 6000

E. Single-Phase Starter Types

SINGLE-PHASE MOTOR CHARACTERISTIC TABLE

Characteristics	Motor Type				
	Split Phase, Capacitor Start (SPC)	Permanent-Split Capacitor (PSC)	Capacitor Start, Induction Run (CSIR)	Capacitor Start, Capacitor Run (CSCR)	Shaded Pole (SP)
Starting Control	Speed switch	None	Speed switch	Speed switch	None
Ratings (Horsepower)	1/25–1/2	1/20–5	1/20–5	1/20–5	1/100–1/4
Full Load Speeds (RPM @ 60 Hz)	3450 1725	3450 1725	3450 1725	3500 1750	3100 1550 1000
Locked Rotor Torque (Percent @ Full Load)	125–150%	250%	250–350%	250%	250%
Breakdown Torque (Percent @ Full Load)	250–300%	250–300%	250–300%	250%	125%
Speed Classification	Constant	Constant or variable	Constant	Constant	Constant or variable
Full Load Power Factor	60%	95%	65%	95%	60%
Efficiency	Medium	High	Medium	High	Low

F. Three-phase starter types by starting method are given in the following table:

THREE-PHASE STARTERS

Starting Method	Inrush Current % LRA	Starting Torque % LRT
Across-the-Line	100	100
Auto-Transformer		
80% Tap	71	64
65% Tap	48	42
50% Tap	28	25
Primary Resistor or Reactor		
80% Applied Voltage	80	64
65% Applied Voltage	65	42
58% Applied Voltage	58	33
50% Applied Voltage	50	25
Star Delta	33	33
Part Winding	60	48
Part Winding w/Resistors	60–30	48–12
Wound Rotor (Approx.)	25	150
Solid State	3 × RLA	–

Notes:
1 % LRA = Percent full voltage locked rotor current (amps).
2 % LRT = Percent full voltage locked rotor torque.
3 RLA = Rated load amps or running load amps.

G. Disconnect Switches

1. Fused disconnect switches should be used whenever the equipment manufacturer requires fused disconnect switches or when more than one motor or piece of equipment is on a single electrical circuit. Fused disconnect switches are generally required with packaged air conditioning equipment, and some chillers. Fusing means it may be either a fuse or a circuit breaker. Circuit breakers are preferred; however, some equipment will require fuses because they have not been tested or rated with circuit breakers.
   a. Fuses shall be Class RK5 Time Delay, Dual Element Fuses.
   b. Circuit breakers shall be Thermal Magnetic Circuit Breakers.
2. Nonfused disconnect switches should be used whenever fused disconnects are not required by the equipment manufacturer. Most fans, pumps, and air-handling units do not require fused disconnect switches.

H. Motor Size, Starter and Disconnect Switch Size, and Fuse and Circuit Breaker Size are given in the following tables. The following notes are applicable to all schedules.

1. Starters and/or disconnect switches. Fuses shall be Class RK5 Time Delay, Dual Element Fuses. Circuit breakers shall be Thermal Magnetic Circuit Breakers.
2. Motor data, starters, disconnect switches, and fuses based on *2014 NEC* and Square D Company.

115 Volt (120 Volt) Single-Phase Motor Starter Schedule					
Motor hp	NEMA Starter Size	Full Load Amps Per Phase	Disc. Switch Size	Fuse Size Amperes	Circuit Breaker Size Amperes
1/8	1	3.0	30	4.5	15
1/6	1	4.4	30	7	15
1/4	1	5.8	30	9	15
1/3	1	7.2	30	12	15
1/2	1	9.8	30	15	20
3/4	1	13.8	30	20	25
1	1	16.0	30	25	30
1.5	1	20.0	30	30	40
2	1	24.0	30	30	50
3	2	34.0	60	50	70
5	3	56.0	100	80	90
7.5	4	80.0	100	100	110
10	–	–	–	–	–

230 Volt (240 Volt) Single-Phase Motor Starter Schedule					
Motor hp	NEMA Starter Size	Full Load Amps Per Phase	Disc. Switch Size	Fuse Size Amperes	Circuit Breaker Size Amperes
1/8	1	1.7	30	2.5	15
1/6	1	2.2	30	3.5	15
1/4	1	2.9	30	4.5	15
1/3	1	3.6	30	5.6	15
1/2	1	4.9	30	8	15
3/4	1	6.9	30	10	15
1	1	8.0	30	12	15
1.5	1	10.0	30	15	20
2	1	12.0	30	17.5	25
3	1	17.0	30	25	35
5	2	28.0	60	40	60
7.5	2	40.0	60	60	80
10	3	50.0	60	60	90

200 Volt (208 Volt) Three-Phase Motor Starter Schedule					
Motor hp	NEMA Starter Size	Full Load Amps per Phase	Disc. Switch Size	Fuse Size Amperes	Circuit Breaker Size Amperes
1/2	1	2.5	30	3.5	15
3/4	1	3.7	30	5	15
1	1	4.8	30	6.25	15
1.5	1	6.9	30	10	15
2	1	7.8	30	12	15
3	1	11.0	30	17.5	20
5	1	17.5	30	25	35
7.5	1	25.3	60	40	50
10	2	32.2	60	50	60
15	3	48.3	60	60	90
20	3	62.1	100	90	100
25	3	78.2	100	100	110
30	4	92.0	200	125	125
40	4	120.0	200	175	175
50	5	150.0	200	200	200
60	5	177.0	400	250	250
75	5	221.0	400	300	300
100	6	285.0	400	400	400
125	6	359.0	600	500	600
150	6	414.0	600	600	600
200	7	552.0	–	–	800

230 Volt (240 Volt) Three-Phase Motor Starter Schedule					
Motor hp	NEMA Starter Size	Full Load Amps per Phase	Disc. Switch Size	Fuse Size Amperes	Circuit Breaker Size Amperes
1/2	1	2.2	30	3.2	15
3/4	1	3.2	30	4.5	15
1	1	4.2	30	5.6	15
1.5	1	6.0	30	8	15
2	1	6.8	30	10	15
3	1	9.6	30	15	20
5	1	15.2	30	25	30
7.5	1	22.0	30	30	45
10	2	28.0	60	40	60
15	2	42.0	60	60	80
20	3	54.0	100	80	90
25	3	68.0	100	100	100
30	3	80.0	100	100	110
40	4	104.0	200	150	150
50	4	130.0	200	200	200
60	5	154.0	200	200	225
75	5	192.0	400	300	250
100	5	248.0	400	350	350
125	6	312.0	400	400	450
150	6	360.0	600	500	600
200	6	480.0	600	600	800
250	7	600.0	800	800	800
300	7	720.0	1200	1000	1000
400	–	–	–	–	–

460 Volt (480 Volt) Three-Phase Motor Starter Schedule					
Motor hp	NEMA Starter Size	Full Load Amps per Phase	Disc. Switch Size	Fuse Size Amperes	Circuit Breaker Size Amperes
1/2	1	1.1	30	1.6	15
3/4	1	1.6	30	2.25	15
1	1	2.1	30	2.8	15
1.5	1	3.0	30	4	15
2	1	3.4	30	5.6	15
3	1	4.8	30	8	15
5	1	7.6	30	12	15
7.5	1	11.0	30	17.5	20
10	1	14.0	30	20	25
15	1	21.0	30	30	40
20	1	27.0	60	40	60
25	2	34.0	60	50	70
30	2	40.0	60	60	80
40	3	52.0	100	80	90
50	3	65.0	100	100	100
60	3	77.0	100	100	110
75	4	96.0	200	150	125
100	4	124.0	200	175	200
125	5	156.0	200	200	225
150	5	180.0	400	250	250
200	5	240.0	400	350	350
250	6	302.0	600	500	500
300	6	361.0	600	600	600
400	6	477.0	800	700	700
500	7	590.0	1200	800	800

575 Volt (600 Volt) Three-Phase Motor Starter Schedule					
Motor hp	NEMA Starter Size	Full Load Amps per Phase	Disc. Switch Size	Fuse Size Amperes	Circuit Breaker Size Amperes
1/2	1	0.9	30	1.25	15
3/4	1	1.3	30	1.6	15
1	1	1.7	30	2.25	15
1.5	1	2.4	30	3.5	15
2	1	2.7	30	4.5	15
3	1	3.9	30	6.25	15
5	1	6.1	30	10	15
7.5	1	9.0	30	15	15
10	1	11.0	30	17.5	20
15	2	17.0	30	25	35
20	2	22.0	30	30	45
25	2	27.0	60	40	60
30	3	32.0	60	50	60
40	3	41.0	60	60	80
50	3	52.0	100	80	90
60	4	62.0	100	90	100
75	4	77.0	100	110	110
100	4	99.0	200	150	150
125	5	125.0	200	175	200
150	5	144.0	200	200	200
200	5	192.0	400	300	250
250	6	242.0	600	350	350
300	6	289.0	600	400	400
400	6	382.0	600	500	500
500	7	472.0	800	800	700

I. Motor Control Centers (MCCs)

1. NEMA Class I, Type A:
   a. No terminal boards for load or control connections are provided.
   b. Numbered terminals for field-wired power and control connections are provided on the starter.
   c. Starter unit mounted pilot devices are internally wired to starter.
2. NEMA Class I, Type B:
   a. Terminal boards for load connections are provided for Size 3 and smaller controllers. For controllers larger than Size 3, numbered terminals for field-wired power connections are provided on starter.
   b. Unit control terminal boards for each combination motor controller are provided for field wiring.
   c. Both terminal boards are factory-wired and mounted on, or adjacent to, the unit.
   d. No load terminal boards for feeder tap units are provided.
   e. Starter unit mounted pilot devices are internally wired to the starter.
   f. NEMA Class I, Type B will be suitable for most HVAC applications.
3. NEMA Class I, Type C:
   a. Factory-wired master section terminal board, mounted on the stationary structure, is provided for each section.
   b. Terminal boards for load connections are provided for Size 3 and smaller controllers. For controllers larger than Size 3, numbered terminals for field-wired power connections are provided on the starter.
   c. Unit control terminal boards for each combination motor controller are provided for field wiring.
   d. Complete wiring between combination controllers or control assemblies and their master terminal boards is factory installed. No wiring between sections or between master terminals is provided. No interconnections between combination controllers and control assemblies.
   e. No load terminal boards for feeder tap units are provided.

4. NEMA Class II, Type B:
   a. Terminal boards for load connections are provided for Size 3 and smaller controllers. For controllers larger than Size 3, numbered terminals for field-wired power connections are provided on the starter.
   b. Unit control terminal boards for each combination motor controller are provided for field wiring.
   c. Both terminal boards are factory-wired and mounted on, or adjacent to, unit.
   d. Complete wiring between combination controllers or control assemblies in the same and other sections is factory-wired.
   e. No load terminal boards for feeder tap units are provided.
5. NEMA Class II, Type C:
   a. A factory-wired master section terminal board, mounted on the stationary structure, is provided for each section.
   b. Terminal boards for load connections are provided for Size 3 and smaller controllers. For controllers larger than Size 3, numbered terminals for field-wired power connections are provided on the starter.
   c. Unit control terminal boards for each combination motor controller are provided for field wiring.
   d. Complete wiring between combination controllers or control assemblies and their master terminal boards in the same section and other sections is factory-wired.
   e. No load terminal boards for feeder tap units are provided.
6. MCCs are available in NEMA enclosure types 1, 2, 3R, and 12.

32.03 Variable Frequency Drives

A. Variable Frequency Drives have many names and acronyms.

1. Variable frequency drives (VFDs)—Used within this text.
2. Adjustable frequency drives (AFDs).
3. Variable frequency controllers (VFCs).
4. Adjustable frequency controllers (AFCs).

B. VFD Components (from power side to load side)

1. Rectifier section: Silicon-controlled rectifiers (SCRs) or diodes change single- or three-phase AC power to DC power.
2. DC bus section: Capacitors and an inductor smooth the rippled DC power supplied by the rectifier.
3. Inverter section: An inverter converts the DC bus power to three-phase variable frequency power.
4. Controller section: The controller turns the inverter on and off to control the output frequency and voltage.

C. VFD Types

1. Variable voltage inverters (VVIs) use an SCR to convert incoming AC power to a varying DC power and then use an inverter to convert the DC power to three-phase variable voltage and variable frequency power. The disadvantages of VVIs are:
   a. Incoming line notching, which requires isolation transformers.
   b. The power factor is proportional to speed, which may require power factor correction capacitors.
   c. Torque pulsations are experienced at low speeds.
   d. Non-sinusoidal current waveforms produce additional heating in the motor.
2. Current source inverters (CSIs) use SCRs in the rectifier and inverter sections and only an inductor in the DC bus section. The disadvantages of CSIs are:
   a. Incoming line notching, which requires isolation transformers.
   b. The power factor is proportional to the speed, which may require power factor correction capacitors.

   c. Motor drive matching is critical to proper operation.
   d. Non-sinusoidal current waveforms produce additional heating in the motor.
3. Pulse width modulated (PWM) drives use a full wave diode bridge rectifier to convert the incoming AC power to DC power. Most PWM drives use a six-pulse converter, while some offer a 12-pulse converter in the rectifier section. The DC bus section consists of capacitors, and in some cases an inductor. The inverter section uses Insulated Gate Bipolar Transistors (IGBTs), Bipolar Junction Transistors (BJTs), or Gate Turn off Thyristors (GTOs) to convert the DC bus power to a three-phase variable voltage and variable frequency power. PWM drives are the most common VFD in use in the HVAC industry today despite the fact it can punish motors electrically, especially 460 and 575 volt motors.
   a. The advantages of PWM drives are:
      1) Minimal line notching.
      2) Better efficiency.
      3) Higher power factor.
      4) Larger speed ranges.
      5) Lower motor heating.
   b. The disadvantages of the PWM drives are:
      1) Higher initial cost.
      2) Regenerative braking is caused because power is allowed to flow in both directions and can act as a drive or a brake.

D. VFD Design Guidelines

1. Provide VFDs with the following:
   a. VFDs serving motors:
      1) 10 hp and smaller: six-pulse VFD with a 3 percent impedance input line reactor.
      2) 15–40 hp: six-pulse VFD with a 5 percent impedance input line reactor.
      3) 50 hp and larger: 18-pulse VFD.
   b. NEMA-rated controller enclosure.
   c. Push-button stations, pilot lights, and selector switches: NEMA-ICS-2, heavy-duty type.
   d. Stop and lockout push-button station: Momentary-break, push-button station with a factory-applied hasp arranged so the padlock can be used to lock the push button in the depressed position with the control circuit open.
   e. Lockable disconnect switch.
   f. Control relays: Auxiliary and adjustable time-delay relays.
   g. Standard displays:
      1) Output frequency (Hz).
      2) Setpoint frequency (Hz).
      3) Motor current (amperes).
      4) DC-link voltage (VDC).
      5) Motor torque (percent).
      6) Motor speed (rpm).
      7) Motor output voltage (V).
   h. Historical logging information and displays:
      1) Real-time clock with current time and date.
      2) Running log of total power versus time.
      3) Total runtime.
      4) Fault log, maintaining the last four faults with the time and date stamp for each.
   i. Current-sensing, phase-failure relays for bypass controller: A solid-state sensing circuit with isolated output contacts for hard-wired connections; arranged to operate on phase failure, phase reversal, current unbalance of from 30 to 40 percent, or loss of supply voltage; with adjustable response delay.
2. For best motor life and reliability, do not operate motors run by VFDs into their service factor and do not select motors to run within the service factors.
3. Do not run motors below 25 percent of their rated speed or capacity.

4. Use inverter duty motors whenever possible. Inverter duty motors are built with winding thermostats that shut down the motor when elevated temperatures are sensed inside it. In addition, these motors are built with oversized frames and external blowers to cool the motor through the full range of speeds.
5. Motors that are operated with VFDs should be specified with phase insulation, should operate at a relatively low temperature rise (most high efficiency motors fit this category), and should use a high class of insulation (either insulation class F or H).
6. Generally, VFDs do not include disconnect switches; therefore, the engineer must include a disconnect switch in the project design. The disconnect switch should be fused with the fuse rated for the drive input current rating.
7. Multiple motors can be driven with one VFD.
8. All control wiring should be run separately from VFD wiring.
9. Most VFDs include the following features as standard:
   a. Overload protection devices.
   b. Short circuit protection.
   c. Ground fault protection.
10. Provide VFDs with a manual bypass in the event the drive fails.
    a. Manual bypasses may not be required when standby equipment is provided.
    b. Manual bypasses may not be required when multiple pieces of equipment are headered together, especially if three or more pieces of equipment are headered together.
11. Coordinate harmonic mitigation requirements with an electrical engineer.
    a. Line reactors.
    b. Active harmonic filters.

E. VFDs produce nonlinear loads, which cause the following unwanted effects.

1. AC system circuits containing excessive currents and unexpectedly higher or lower voltages.
2. Conductor, connector, and component heating, which is unsafe.
3. Loss of torque on motors.
4. Weaker contactor, relay, and solenoid action.
5. High heat production in transformers and motors can be destructive.
6. Poor power factor.

32.04 NEMA Enclosures

A. NEMA Type 1:   Indoor General Purpose, Standard

B. NEMA Type 2:   Indoor Drip-Proof

C. NEMA Type 3R: Outdoor, Rain Tight, Water Tight, Dust Tight

D. NEMA Types 4, 4X, 5: Outdoor Rain Tight, Water Tight, Dust Tight, Corrosion Resistant

E. NEMA Type 7X: Explosion-Proof

F. NEMA Type 12:   Indoor Oil and Dust Tight

PART 33

Humidifiers

33.01 Humidifiers

A. **The number of humidifier manifolds required is given in the following table:**

Duct Height	Number of Manifolds
Less than 37"	1
37"–58"	2
59"–80"	3
81"–100"	4
101" and Over	5

A. Humidifier Installation Requirements

1. Humidifiers shall be installed a minimum of 3'0" from any duct transformation, elbow, fitting, or outlet.
2. Consideration must be given to the length of the vapor trail and air handling unit, and ductwork design must provide sufficient length to prevent the vapor trail from coming in contact with items downstream of the humidifier before the vapor has had time to completely evaporate.

B. Humidifier Makeup Requirements

1. Steam humidifiers: 5.6 GPM/1,000 kW input or 5.6 GPM/3413 MBH.
2. Electric humidifiers: 5.6 GPM/1,000 kW input or 5.6 GPM/3413 MBH.
3. Evaporative humidifiers: 5.0 GPM/1,000 lbs./h.
4. Spray coil humidifiers: 5.0 GPM/1,000 lbs./h.

C. Humidifier Makeup Water Types

1. Potable (untreated) water.
2. Softened water.
3. Deionized water (DI).
4. Reverse osmosis water.

D. Residential Humidifier Types

1. Pan humidifiers:
   a. Basic pan.
   b. Electrically heat pan.
   c. Pan with wicking plates.
2. Wetted element humidifiers:
   a. Fan type.
   b. Bypass type.
   c. Duct mounted type.
3. Atomizing humidifiers:
   a. Spinning disk.
   b. Spray nozzles—water pressure.
   c. Spray nozzles—compressed air.
   d. Ultrasonic.
4. Portable or non-ducted humidifiers.

E. Industrial Humidifier Types

1. Heated pan humidifiers:
   a. Steam.
   b. Hot water.
2. Direct steam injection humidifiers:
   a. Single or multiple steam jacketed humidifiers.
   b. Nonjacketed manifold or panel-type distribution humidifiers.

3. Electrically heated, self-contained steam humidifiers:
   a. Electrode type humidifier.
   b. Resistance type humidifiers.
4. Atomizing humidifiers:
   a. Ultrasonic humidifiers.
   b. Centrifugal humidifiers.
   c. Compressed air nozzle humidifiers.
5. Wetted media humidifiers:
   a. Rigid media humidifiers.
6. Evaporative cooling.

Filters

34.01 Minimum Efficiency Reporting Value (MERV)

A. MERV reports a filter's ability to capture particles between 0.3 and 10 microns.

B. MERV values are used in comparing the performance of different filters.

C. MERV ratings are derived from an ASHRAE test method.

D. The higher the MERV rating, the better the filter is at removing particulates from the air.

E. MERV Values.

MERV RATINGS

MERV Rating	Average Particle Size in Microns	Efficiency	Filter Types
1–4	3.0–10.0	Less than 20%	Roll filters Flat or panel filters Electronic air cleaners Carbon filters (not designed to remove particulates)
5	3.0–10.0	20–34.9%	Flat or panel filters
6	3.0–10.0	35–49.9%	Pleated media filters
7	3.0–10.0	50–69.9%	Cartridge filters
8	3.0–10.0	70% or greater	
9	3.0–10.0 1.0–3.0	85% or greater Less than 50%	
10	3.0–10.0 1.0–3.0	85% or greater 50–64.9%	Bag filters
11	3.0–10.0 1.0–3.0	85% or greater 65–79.9%	Box filters
12	3.0–10.0 1.0–3.0	90% or greater 80% or greater	
13	3.0–10.0 1.0–3.0 0.30–1.0	90% or greater 90% or greater Less than 75%	
14	3.0–10.0 1.0–3.0 0.30–1.0	90% or greater 90% or greater 75–84.9%	Bag filters Box filters
15	3.0–10.0 1.0–3.0 0.30–1.0	90% or greater 90% or greater 85–94.9%	HEPA filters ULPA filters
16	3.0–10.0 1.0–3.0 0.30–1.0	95% or greater 95% or greater 95% or greater	

34.02 Flat or Panel Filters

A. Efficiency: 20–35% (dust spot)

B. Face Velocity: 500 FPM

C. Initial Pressure Drop: 0.25" W.G.

D. Final Pressure Drop: 0.50" W.G.

E. Nominal Sizes
 1. 1" Thick: 24×24; 20×25; 20×24; 20×20; 16×25; 16×20
 2. 2" Thick: 24×24; 20×25; 20×24; 20×20; 18×24; 16×25; 16×20; 12×24

F. Test Method: *ASHRAE 52.2-2012*, Atmospheric

34.03 Pleated Media Filters

A. Efficiency (dust spot)

1. 25–35%
2. 60–65%
3. 80–85%
4. 90–95%

B. Face Velocity: 500 FPM

C. Initial Pressure Drop

1. 25–35%:   0.25–0.45" W.G.
2. 60–65%:   0.50" W.G.
3. 80–85%:   0.60" W.G.
4. 90–95%:   0.70" W.G.

D. Final Pressure Drop

1. 25–35%:   1.20" W.G.
2. 60–65%:   1.20" W.G.
3. 80–85%:   1.20" W.G.
4. 90–95%:   1.20" W.G.

E. Nominal Sizes

1. Thicknesses (inches):
   a. 25–35%:   1; 2; 4.
   b. 60–65%:   4; 6; 12.
   c. 80–85%:   4; 6; 12.
   d. 90–95%:   4; 6; 12.
2. Face sizes:
   a. 25–35%: 24×24; 20×25; 20×24; 20×20; 18×24; 16×25; 16×20; 12×24.
   b. 60–65%: 24×24; 20×25; 20×24; 20×20; 18×24; 16×25; 16×20; 12×24.
   c. 80–85%: 24×24; 20×25; 20×24; 20×20; 18×24; 16×25; 16×20; 12×24.
   d. 90–95%: 24×24; 20×25; 20×24; 20×20; 18×24; 16×25; 16×20; 12×24.

F. Test Method: *ASHRAE 52.2-2012*, Atmospheric

34.04 Bag Filters

A. Efficiency (dust spot)

1. 40–45%
2. 50–55%
3. 60–65%
4. 80–85%
5. 90–95%

B. Face Velocity: 500 FPM

C. Initial Pressure Drop

1. 40–45%:   0.25" W.G.
2. 50–55%:   0.35" W.G.
3. 60–65%:   0.40" W.G.
4. 80–85%:   0.50" W.G.
5. 90–95%:   0.60" W.G.

D. Final Pressure Drop

1.	40–45%:	1.00" W.G.
2.	50–55%:	1.00" W.G.
3.	60–65%:	1.00" W.G.
4.	80–85%:	1.00" W.G.
5.	90–95%:	1.00" W.G.

E. Nominal Sizes

1. Thicknesses (inches):
   a. 40–45%:   12; 15.
   b. 50–55%:   21; 22; 30; 37.
   c. 60–65%:   21; 22; 30; 37.
   d. 80–85%:   21; 22; 30; 37.
   e. 90–95%:   21; 22; 30; 37.
2. Face sizes:
   a. 40–45%: 24×24; 24×20; 20×25; 20×24; 20×20; 16×25; 16×20; 12×24.
   b. 50–55%: 24×24; 24×20; 20×24; 20×20; 12×24.
   c. 60–65%: 24×24; 24×20; 20×24; 20×20; 12×24.
   d. 80–85%: 24×24; 24×20; 20×24; 20×20; 12×24.
   e. 90–95%: 24×24; 24×20; 20×24; 20×20; 12×24.

F. Test Method: *ASHRAE 52.2-2012*, Atmospheric

34.05 HEPA (High Efficiency Particulate Air) Filters

A. Efficiency: **99.97% for 0.3 micron particles and larger**

B. Face Velocity: **250 FPM maximum**

C. Initial Pressure Drop

1.	95%:	0.50" W.G.
2.	99.97–99.995%:	1.00" W.G.

D. Final Pressure Drop

1.	95%:	2.00" W.G.
2.	99.97–99.995%:	3.00" W.G.

E. Nominal Sizes

1. Thicknesses (inches): 3; 5; 6; 12.
2. Face sizes: 8×8; 12×12; 12×24; 16×20; 20×20; 24×12; 24×24; 24×30; 24×36; 24×48; 24×60; 24×72; 30×24; 30×30; 30×36; 30×48; 30×60; 30×72; 36×24; 36×30; 36×36; 36×48; 36×60; 36×72.

F. Test Method: **D.O.P. or Polystyrene Latex (PSL) Spheres (PSL preferred)**

34.06 ULPA (Ultra Low Penetrating Air) Filters

A. Efficiency: **99.9997% for 0.12 micron particles and larger**

B. Face Velocity: **250 FPM maximum**

C. Initial Pressure Drop

1. 99.997–99.9999%: 1.00" W.G.

D. Final Pressure Drop

1. 99.997–99.9999%:	3.00" W.G.

E. Nominal Sizes

1. Thicknesses (inches): 3; 5; 6; 12.
2. Face sizes: 8×8; 12×12; 12×24; 16×20; 20×20; 24×12; 24×24; 24×30; 24×36; 24×48; 24×60; 24×72; 30×24; 30×30; 30×36; 30×48; 30×60; 30×72; 36×24; 36×30; 36×36; 36×48; 36×60; 36×72.

F. Test Method: **D.O.P. or Polystyrene Latex (PSL) Spheres (PSL preferred)**

34.07 Roll Filters

A. Efficiency: **20–25% (dust spot)**

B. Face Velocity: **500 FPM**

C. Initial Pressure Drop

1. 20%:	0.20" W.G.

D. Final Pressure Drop

1. 20%:	0.45" W.G.

E. Nominal Sizes

1. Thicknesses: 2.
2. Face sizes:
   a. Height: 5'0–15'0" by increments of 4".
   b. Width: 3'0–30'0" by increments of 1'0".

F. Test Method: *ASHRAE 52.2-2012*, Atmospheric

34.08 Carbon Filters

A. Front/Back Access

1. Face velocity: 500 FPM
   a. Pressure drop: 0.35–0.45" W.G.
   b. Nominal sizes: $24 \times 24 \times 24$: 90 lbs. of carbon per 2,000 CFM. $24 \times 12 \times 24$: 45 lbs. of carbon per 1,000 CFM.
   c. Tray size: 24×24.
2. Face velocity: 250 FPM
   a. Pressure drop: 0.30–0.40" W.G.
   b. Nominal sizes: $24 \times 24 \times 8$: 30 lbs. of carbon per 1,000 CFM. $24 \times 24 \times 8$: 15 lbs. of carbon per 500 CFM.
   c. Tray size: 24×8.

B. Side Access

1. Face velocity: 500 FPM
   a. Pressure drop: 0.35–0.45" W.G.
   b. Nominal sizes: $24 \times 24 \times 24$: 108 lbs. of carbon per 2,000 CFM.
   c. Tray size: 12×24.

C. Test Method: *ASHRAE 52.2-2012*, Atmospheric

34.09 Electronic Air Cleaners

A. **Efficiency:**   30–40% (dust spot)

B. **Face Velocity:**   625 FPM

C. **Initial Pressure Drop**
1. 90%:   0.26" W.G.

D. **Final Pressure Drop**
1. 90%:   0.50" W.G.

E. **Nominal Sizes**
1. Thicknesses: 2'0–4'0".
2. Face sizes:
   a. Height: 2'4–15'8" by increments of 4".
   b. Width: 2'8–18'8" by increments of 1'0".

F. **Test Method:**   *ASHRAE 52.2-2012*, **Atmospheric**

34.10 Filter Characteristics

A. **Filter Removal Capabilities**
1. Fine mode   < 2.5 microns.
2. Coarse mode   2.5 microns.
3. Respirable   < 10.0 microns.
4. Nonrespirable   10.0 microns.

B. **Filter Design Factors**
1. Degree of air cleanliness required.
2. Particulate/contaminate size and form (solid or aerosols).
3. Concentration.
4. Cost (initial and maintenance).
5. Space requirements.
6. Pressure loss/energy use.

C. **Filter Characteristics**
1. *Efficiency.* Ability of the filter to remove particulates/contaminates.
2. *Airflow Resistance.* Static pressure drop of the filters.
3. *Dust Holding Capacity.* Amount of particulates/contaminates the filter will hold before efficiency drops drastically.

D. **Filter Classes**
1. Class 1 Filters: Filters that, when clean, do not contribute fuel when attacked by flame and emit only negligible amounts of smoke.
2. Class 2 Filters: Filters that, when clean, burn moderately when attacked by flame or emit moderate amounts of smoke, or both.
3. However, dust, trapped by filters, will support combustion and will produce smoke more than the filter itself.
4. 2015 IMC:
   a. Media-type air filters shall comply with UL-900.
   b. High-efficiency particulate air filters shall comply with UL-586.
   c. Electrostatic-type air filters shall comply with UL-867.

d. Ducts and systems shall be designed to allow even distribution of air over the entire filter.

e. Filters shall be either Class 1 or Class 2.

5. NFPA 90A-2015: Filters shall comply with UL-900.

E. Filter Test Methods

1. ASHRAE "test dust." ASHRAE test dust is composed of 72 percent standardized air cleaner test dust, fine; 23 percent powdered carbon; and 5 percent cotton linters.

2. Arrestance test:
   a. Uses ASHRAE test dust.
   b. Tests the ability of the filter to remove the larger atmospheric dust particles.
   c. Measures the concentration of the dust leaving the filter.

3. Atmospheric dust spot efficiency test:
   a. Measures the change in light transmitted by HEPA filter media targets.
   b. Intermittent flow method. Airflow upstream and downstream of the tested filter is drawn through separate target filters. Upstream airflow is intermittently drawn and the downstream airflow is continuously drawn. The test takes more time for higher efficiency filters.
   c. Constant flow method. Airflow upstream and downstream of the tested filter is drawn through separate target filters at a constant flow. Test takes the same time for high- and low-efficiency filters.

4. Dust holding capacity test. The amount of dust held by the filter when the filter pressure drop reaches its maximum or final pressure drop, or when arrestance tests drop below 85 percent for two consecutive readings, or below 75 percent for one reading.

5. DOP (dioctyl phthalate) test:
   a. High-efficiency filter tests (HEPA and ULPA).
   b. DOP or BEP (Bis-[2-Ethylhexyl] Phthalate). Test aerosols are used.
   c. A cloud of DOP or BEP is passed through the test filter, and the amount passing through the filter is measured by a light-scattering photometer.

6. Polystyrene latex (PSL) spheres test:
   a. High-efficiency filter tests (HEPA and ULPA).
   b. Filter media thickness 20 mL.
   c. Media is tested at 10.5 feet per minute with PSL.
   d. Filters are tested at 70–100 feet per minute.
   e. PSL test material is selected to allow 90 percent of the mean size to be between 0.1 and 0.3 microns.
   f. The minimum number of PSL particles in the filter test challenge will be a minimum of 10 million particles per cubic foot.
   g. The particle test challenge is monitored in accordance with the Institute of Environmental Sciences (IES) standards *IES-RP-C001* for HEPA filters and *IES-RP-C007* for ULPA filters.

7. Leak scan tests:
   a. Used with HEPA and ULPA filters.
   b. The DOP Test is used while scanning the face of the filter for air leakage through or around the filters.

8. Particle size tests. No standard exists; depends heavily on the type of aerosol used.

PART 35

Insulation

35.01 Insulation Materials and Properties

A. General

1. Insulation, adhesives, mastics, sealants, and coverings shall have a flame spread rating of 25 or less and a smoke developed rating of 50 or less as determined by an independent testing laboratory in accordance with *NFPA 255* and *UL 728* as required by *ASHRAE 90A* and *90B*. Coatings and adhesives applied in the field shall be nonflammable in the wet state.
2. Hangers on chilled water and other cold piping systems should be installed on the outside of the insulation to prevent hangers from sweating.
3. Cold surfaces: Normal operating temperatures less than 75°F.
4. Hot surfaces: Normal operating temperatures of 100°F or higher.
5. Dual-temperature surfaces: Normal operating temperatures that vary from hot to cold.
6. Thermal conductivity:
   a. K-values.
   b. Thermal conductivity values express the rate of heat loss of a homogenous substance in Btu-in./h sq.ft.°F.
7. Thermal conductance:
   a. C-values.
   b. Thermal conductance values express the rate of heat loss of a homogenous substance in Btu-in./h sq.ft.°F.
8. Thermal resistance:
   a. R-values.
   b. Thermal resistance values express the resistance of heat loss of a homogenous substance in °F sq.ft. h/Btu.
9. Overall heat transfer coefficients:
   a. U-values.
   b. Overall heat transfer coefficient values express the rate of heat loss of a nonhomogenous substance in Btu/h sq.ft.°F.

$$R = \frac{1}{C} = \frac{1}{K} \times Thickness$$

$$U = \frac{1}{\sum R}$$

B. Materials

1. Calcium silicate temperature range: 0 to +1200°F.
2. Fiberglass temperature range: −20 to +1000°F.
3. Mineral wool temperature range: +200 to +1900°F.
4. Urethane, styrene, beadboard temperature range: −350 to +250°F.
5. Cellular glass temperature range: −450 to +850°F.
6. Ceramic fiber temperature range: 0 to +3000°F.
7. Flexible tubing and sheets temperature range: −40 to +250°F.

35.02 Pipe Insulation

A. Insulation shall be sectional molded glass fiber, minimum 3.0 lbs. per cubic foot density, with a thermal conductivity not greater than 0.24 Btu-in./sq.ft./°F/h at a mean temperature difference of 75°F and a white factory-applied flame-retardant vapor barrier jacket of 0.001" aluminum foil laminated to Kraft paper reinforced with glass fibers, or all service jacket.

B. Insulation shall be flexible foamed plastic, minimum 5.0 lbs. per cubic foot density, with a thermal conductivity not greater than 0.28 Btu-in./sq.ft./°F/h at a mean temperature difference of 75°F.

C. Insulation shall be cellular glass, with a thermal conductivity not greater than 0.40 Btu-in./sq.ft./°F/h at a mean temperature difference of 75°F.

D. Insulation shall be foamglass, minimum 8.5 lbs. per cubic foot density, with a thermal conductivity not greater than 0.35 Btu-in./sq.ft./°F/h at a mean temperature difference of 75°F.

E. Code Required Pipe Insulation Thickness.

ASHRAE STANDARD 90.1-2013 AND 2015 IECC

Fluid Design Operating Temperature	Conductivity Btu-in./h ft.^{2} °F	Nominal Pipe or Tube Diameter				
		<1"	1–1-1/2"	>1-1/2–4"	>4–8"	≥8"
Heating Systems—Hot Water and Steam Condensate						
>350°F	0.32–0.34	4.5	5.0	5.0	5.0	5.0
251–350°F	0.29–0.32	3.0	4.0	4.5	4.5	4.5
201–250°F	0.27–0.30	2.5	2.5	2.5	3.0	3.0
141–200°F	0.25–0.29	1.5	1.5	2.0	2.0	2.0
105–140°F	0.22–0.28	1.0	1.0	1.5	1.5	1.5
Heating Systems—Steam						
>350°F >120 psig	0.32–0.34	4.5	5.0	5.0	5.0	5.0
251–350°F 16–120 psig	0.29–0.32	3.0	4.0	4.5	4.5	4.5
212–250°F 0–15 psig	0.27–0.30	2.5	2.5	2.5	3.0	3.0
Cooling Systems—Chilled Water, Glycol, Brine, and Refrigerant						
40–60°F	0.22–0.27	0.5	0.5	1.0	1.0	1.0
<40°F	0.20–0.26	0.5	1.0	1.0	1.0	1.5

F. Recommended pipe insulation thicknesses are provided in the following table:

Piping System (7)	Pipe Sizes	Insulation Thickness vs. Type (1, 8)			
		A	B	C	D
Chilled Water 40–60°F (3)	1-1/2" and smaller	1.0	1.5	2.0	1.5
	2" and larger	1.5	2.0	2.5	2.5
Chilled Water 32–40°F (3)	1" and smaller	1.0	1.5	2.0	1.5
	1-1/4–6"	1.5	2.0	2.5	2.5
	8" and larger	2.0	2.5	3.5	3.0
Chilled Water Below 32°F (3)	2" and smaller	1.5	2.0	2.5	2.5
	2-1/2–6"	2.0	2.5	3.5	3.0
	8" and larger	2.5	3.0	4.5	4.0
Condenser Water	All sizes	(2)	(2)	(2)	(2)
Condenser Water—Waterside Economizer	1-1/2" and smaller	1.0	1.5	2.0	1.5
	2" and larger	1.5	2.0	2.5	2.5
Heating Water—Low Temperature 100–140°F (4)	1-1/2" and smaller	1.0	1.5	2.0	1.5
	2" and larger	2.0	2.5	3.5	3.0
Heating Water—Low Temperature 141–200°F (4)	1-1/2" and smaller	1.5	1.5	2.0	1.5
	2" and larger	2.0	2.5	3.5	3.0
Heating Water—Low Temperature 201–250°F (4)	1-1/4" and smaller	2.5	2.0	2.5	2.5
	1-1/2" and larger	3.0	2.5	3.5	3.0
Heating Water—Medium Temperature 251–350°F (4)	3/4" and smaller	3.0		2.5	2.5
	1-1-1/4"	4.0	(10)	4.5	4.0
	1-1/2" and larger	4.5		5.0	4.5
Heating Water—High Temperature 351–450°F (4)	3/4" and smaller	4.5		4.5	4.0
	1–3"	5.0	(10)	5.0	4.5
	4" and larger	5.0		6.5	6.0

(Continued)

Piping System (7)	Pipe Sizes	Insulation Thickness vs. Type (1, 8)			
		A	B	C	D
Dual Temperature	All sizes	(9)	(9)	(9)	(9)
Heat Pump Loop	All sizes	(2)	(2)	(2)	(2)
Steam and Steam Condensate—Low Pressure (5) 15 psig and Lower 201–250°F	3" and smaller 4" and larger	2.5 3.0	2.0 4.0	2.5 5.0	2.5 4.5
Steam and Steam Condensate—Medium Pressure (5) 16–100 psig 251–350°F	3/4" and smaller 1–1-1/4" 1-1/2" and larger	3.0 4.0 4.5	(10)	2.5 4.5 5.0	2.5 4.0 4.5
Steam and Steam Condensate—High Pressure (5) 101–300 psig >350°F	3/4" and smaller 1" and larger	4.5 5.0	(10)	4.5 6.5	4.0 6.0
Refrigerant Suction and Liquid Lines (6)	1" and smaller 1-1/4–6" 8" and larger	1.0 1.5 2.0	1.5 2.0 2.5	2.0 2.5 3.5	1.5 2.5 3.0
Refrigerant Hot Gas (6)	All sizes	0.75	1.0	1.5	1.0
Air Conditioning Condensate	All sizes	0.5	0.5	1.0	0.75

Notes:
1 Type A: Fiberglass insulation.
   Type B: Flexible foamed plastic insulation.
   Type C: Cellular glass insulation.
   Type D: Foamglass insulation.
2 Insulation is not required on systems with temperatures between 60°F and 105°F, unless insulating the pipe for freeze protection—in which case, use chilled water (40°F and above) thicknesses. Remember to include insulation on condenser water systems used for waterside economizer operation.
3 Chilled water system piping is often insulated with fiberglass insulation; although, cellular glass and flexible foamed plastic may be more appropriate for moisture condensation protection. Other types of insulation may be used.
4 Heating water system piping is generally insulated with fiberglass pipe insulation. Other types of insulation may be used.
5 Steam system piping and steam condensate system piping are generally insulated with fiberglass pipe insulation. Other types of insulation may be used.
6 Refrigerant system piping is generally insulated with flexible foamed plastic. Other types of insulation may be used. Normally, only refrigerant suction lines are insulated, but liquid lines should be insulated where condensation will become a problem, and hot gas lines should be insulated where personal injury from contact may pose a problem.
7 Table meets or exceeds ASHRAE Standard 90.1-2013 and the 2015 IECC.
8 For piping exposed to ambient temperatures, increase the insulation thickness by 1 in.
9 For dual temperature systems, use insulation thickness for a more stringent system, usually the heating system.
10 The system temperature exceeds the temperature rating of the insulation.

35.03 Duct Insulation

A. Internal Duct Liner

1. 1-1/2 pounds per cubic foot density amber color glass fiber blanket with smooth coated matte facing to conform to *TIMA Standard AHC-101, NFPA 90A, NFPA 90B, NFPA 255, UL 181,* and *UL 723.* Duct lining shall have a thermal conductivity (k) not greater than 0.24 Btu/sq.ft./°F/h at a mean temperature difference of 75°F. Vinyl spray face shall not be permitted.
2. Thicknesses: 1", 1-1/2", 2".

B. External Duct Insulation

1. Duct wrap: Insulation shall be a flexible glass fiber blanket, minimum 3/4 lb. per cubic foot density, with a thermal conductivity (k) not greater than 0.29 Btu-in./sq.ft./°F/h at a mean temperature difference of 75°F and a factory-applied jacket of minimum 0.001"

aluminum foil reinforced with glass fiber bonded to flame-resistant Kraft paper vapor barrier. Thicknesses: 1", 1-1/2", 2".

2. Duct board: Insulation shall be glass fiber, minimum 3.0 lbs. per cubic foot density, with a thermal conductivity (k) not greater than 0.23 Btu-in./sq.ft./°F/h at a mean temperature difference of 75°F and a white factory-applied flame-retardant vapor barrier jacket of 0.001" aluminum foil reinforced with glass fibers bonded to flame-resistant Kraft paper. Thicknesses: 1", 1-1/2", 2", 3", 4".

3. Duct board: Insulation shall be rigid glass fiber board, minimum 6.0 lbs. per cubic foot density, with a thermal conductivity (k) not greater than 0.22 Btu-in./sq.ft./°F/h at a mean temperature difference of 75°F and a white factory-applied flame-retardant vapor barrier jacket of 0.001" aluminum foil reinforced with glass fibers bonded to flame-resistant Kraft paper. Thicknesses: 1", 1-1/2", 2".

C. Code Required Duct Insulation Thickness

1. 2015 IECC:
   a. Supply and return air ducts and plenums located in unconditioned spaces: R-6 insulation minimum.
   b. Supply and return air ducts and plenums located outside: R-8 insulation minimum in Climate Zones 1 through 4, and R-12 insulation minimum in Climate Zones 5 through 8.
   c. Ducts or plenums shall be separated from the building exterior or unconditioned or exempt spaces by R-8 insulation minimum in Climate Zones 1 through 4, and R-12 insulation minimum in Climate Zones 5 through 8.
   d. Duct insulation is not required where located within equipment.
   e. Duct insulation is not required when the design temperature difference between the interior and exterior of the duct or plenum does not exceed 15°F. This exception will apply to most return air ducts except when located outside.

2. ASHRAE Standard 90.1-2013

Climate Zone	Duct Location						
	Exterior	Ventilated Attic	Unvented Attic above Insul. Ceiling	Unvented Attic w/Roof Insulation	Unconditioned Space	Indirectly Conditioned Space	Buried
Heating-Only Ducts							
1, 2	None	None	None	None	None	None	None
3	R-3.5	None	None	None	None	None	None
4	R-3.5	None	None	None	None	None	None
5	R-6	R-3.5	None	None	None	None	R-3.5
6	R-6	R-6	R-3.5	None	None	None	R-3.5
7	R-8	R-6	R-6	None	R-3.5	None	R-3.5
8	R-8	R-8	R-6	None	R-6	None	R-6
Cooling-Only Ducts							
1	R-6	R-6	R-8	R-3.5	R-3.5	None	R-3.5
2	R-6	R-6	R-6	R-3.5	R-3.5	None	R-3.5
3	R-6	R-6	R-6	R-3.5	R-1.9	None	None
4	R-3.5	R-3.5	R-6	R-1.9	R-1.9	None	None
5, 6	R-3.5	R-1.9	R-3.5	R-1.9	R-1.9	None	None
7, 8	R-1.9	R-1.9	R-1.9	R-1.9	R-1.9	None	None
Combined Heating and Cooling Ducts							
1	R-6	R-6	R-8	R-3.5	R-3.5	None	R-3.5
2	R-6	R-6	R-6	R-3.5	R-3.5	None	R-3.5
3	R-6	R-6	R-6	R-3.5	R-3.5	None	R-3.5
4	R-6	R-6	R-6	R-3.5	R-3.5	None	R-3.5
5	R-6	R-6	R-6	R-1.9	R-3.5	None	R-3.5
6	R-8	R-6	R-6	R-1.9	R-3.5	None	R-3.5
7	R-8	R-6	R-6	R-1.9	R-3.5	None	R-3.5
8	R-8	R-8	R-8	R-1.9	R-6	None	R-6
Return Ducts							
1–8	R-3.5	R-3.5	R-3.5	None	None	None	None

D. **Recommended duct insulation R-values and insulation thicknesses are provided in the following table:**

Climate Zone	Exterior	Ventilated Attic	Unvented Attic above Insul. Ceiling	Unvented Attic w/ Roof Insulation	Unconditioned Space	Indirectly Conditioned Space	Buried
			Duct Location				
Heating Ducts Only							
All Climate Zones	R-8	R-8	R-8	R-5	R-6	R-5	R-6
Duct Liner	2"	2"	2"	1.5"	1.5"	1.5"	1.5"
Duct Wrap	3"	3"	3"	2"	2"	2"	2"
Duct Board	2"	2"	2"	1.5"	1.5"	1.5"	1.5"
Cooling Ducts Only							
All Climate Zones	R-8	R-8	R-8	R-5	R-6	R-5	R-5
Duct Liner	2"	2"	2"	1.5"	1.5"	1.5"	1.5"
Duct Wrap	3"	3"	3"	2"	2"	2"	2"
Duct Board	2"	2"	2"	1.5"	1.5"	1.5"	1.5"
Cooling and Heating Ducts							
All Climate Zones	R-8	R-8	R-8	R-5	R-6	R-5	R-6
Duct Liner	2"	2"	2"	1.5"	1.5"	1.5"	1.5"
Duct Wrap	3"	3"	3"	2"	2"	2"	2"
Duct Board	2"	2"	2"	1.5"	1.5"	1.5"	1.5"
Return Ducts							
All Climate Zones	R-8	R-8	R-8	None	None	None	None
Duct Liner	2"	2"	2"	None	None	None	None
Duct Wrap	3"	3"	3"	None	None	None	None
Duct Board	2"	2"	2"	None	None	None	None

Notes:
1 The duct liner represented in the table has a K-value of 0.24 and a density of 1.5 lbs./ft^3.
2 The duct wrap represented in the table has a K-value of 0.29 and a density of 0.75 lb./ft^3.
3 The duct board represented in the table has a K-value of 0.22 and a density of 3.0 lbs./ft^3.

35.04 Insulation Protection

A. **Aluminum roll jacketing and fitting covers produced from ASTM-B-209, 3003 Alloy, 0.016 in. thickness, H-14 temper with a smooth finish. Install in accordance with the manufacturer's recommendations.**

B. **Stainless steel roll jacketing and fitting covers produced from ASTM-A-167, Type 304 or 316, 0.10-in. thick, No. 2B finish, and factory cut and rolled to indicated sizes. Install in accordance with the manufacturer's recommendations.**

C. **Prefabricated PVC fitting covers and jacketing produced from 20-mil-thick, high-impact, ultra-violet-resistant PVC with the same insulation and thickness as specified. Install in accordance with the manufacturer's recommendations.**

D. **Bands: 3/4-in. wide, in one of the following materials compatible with jacket.**
1. Aluminum: 0.007-in. thick.
2. Stainless steel: Type 304, 0.020-in. thick.

PART 36

Fire-Stopping and Through-Penetration Systems

36.01 Fire-Stopping and Through-Penetration Protection Systems

A. **All openings in fire-rated and smoke-rated building construction must be protected from fire and smoke by systems that seal these openings to resist the passage of fire, heat, smoke, flames, and gases. These openings include passages for mechanical and electrical systems, expansion joints, seismic joints, construction joints, control joints, curtain wall gaps, the space between the edge of the floor slab and the exterior curtain wall and columns, and other openings or cracks.**

B. **Terms**

1. *Firestopping.* Firestopping is noncombustible building materials or a system of lumber pieces installed to prevent the movement of fire, heat, smoke, flames, and gases to other areas of the building through small concealed spaces. The term *firestopping* is used with all types of building construction, except for noncombustible and fire-resistive construction.

2. *Through-Penetration Protection Systems (TPPS).* TPPS are building materials or assemblies of materials specifically designed and manufactured to form a system developed to prevent the movement of fire, heat, smoke, flames, and gases to other areas of the building through openings made in fire-rated floors and walls to accommodate the passage of combustible and noncombustible items. The term *TPPS* is used with noncombustible and fire-resistive building construction.

3. *Combustible Penetrating Items.* Combustible penetrating items are materials such as plastic pipe and conduit, electrical cables, and combustible pipe insulation.

4. *Noncombustible Penetrating Items.* Noncombustible penetrating items are materials such as copper, iron, or steel pipe; steel conduit; EMT; electrical cable with steel jackets; and other noncombustible items.

5. *Annular Space Protection.* Annular space protection is the building materials or assembly of materials that protect the space between noncombustible penetrating items and the rated assembly. In concrete or masonry assemblies, the materials generally used for annular space protection are concrete, grout, or mortar. In all other assemblies, the materials must be tested and meet *ASTM E119* standard under positive pressure.

6. *Single-Membrane Protection.* Single-membrane protection is the building materials or assembly of materials that protect the opening through one side, or a single membrane, of a fire-resistive wall, roof/ceiling, or floor/ceiling to accommodate passage of combustible or noncombustible items. Materials protecting single membranes are annular space protection systems or TPPS.

7. *Shaft Alternatives.* A fire-rated shaft or enclosure is not required if a TPPS system with a flame rating (F-Rating) and a thermal rating (T-Rating) equal to the rating of the assembly is used to protect openings made in fire-rated floors and walls to accommodate the passage of combustible and noncombustible items.

C. **System Ratings**

1. F-Ratings define the period of time for which the fire-stopping or TPPS system prevents the passage of flames and hot gases to the unexposed side of the assembly, in accordance with *ASTM E814*. To receive an F-Rating, the system must also pass the hose stream test. F-Ratings are needed for all applications, and must be equal to the rating of the assembly.

2. T-Ratings define the period of time for which the fire-stopping or TPPS system prevents the passage of flames and hot gases to the unexposed side of the assembly (F-Rating), and must also restrict the temperature rise on the unexposed surface to 325°F in accordance with *ASTM E814*. T-Ratings must be equal to the rating of the assembly and at least 1 hour. T-Ratings are rarely applied because most penetrations in commercial structures tend to be in noncombustible concealed spaces and are generally only applied where codes require open protectives.

D. TPPS Materials

1. Intumescent materials expand to form an insulating char.
2. Subliming materials pass from solid to vapor when heated without passing through the liquid phase.
3. Ablative materials char, melt, or vaporize when heated.
4. Endothermic materials, such as concrete and gypsum, absorb heat using chemically bounded water of the material.
5. Ceramic fibers are high-temperature refractory materials.

E. Material Forms

1. Caulks.
2. Putties.
3. Mixes.
4. Sheets, strips, or collars.
5. Kits.
6. Devices.

PART 37

Makeup Water

37.01 Makeup Water Requirements

A. **Hot Water System Makeup Connection:** Minimum connection size shall be 10 percent of largest system pipe size or 1", whichever is greater (20" system pipe size results in a 2" makeup water connection).

B. **Chilled Water System Makeup Connection:** Minimum connection size shall be 10 percent of largest system pipe size or 1", whichever is greater (20" system pipe size results in a 2" makeup water connection).

C. **Condenser Water Makeup to Cooling Tower**

1. Centrifugal: 40 GPM/1,000 tons.
2. Reciprocating: 40 GPM/1,000 tons.
3. Screw chillers: 40 GPM/1,000 tons.
4. Scroll chillers: 40 GPM/1,000 tons.
5. Absorption chillers: 80 GPM/1,000 tons.

D. **Cooling Tower Blowdown and Drains**

1. Drains: Use two times the makeup water rate for sizing cooling tower drains.
2. Blowdown:
   a. Centrifugal: 10 GPM/1,000 tons.
   b. Reciprocating: 10 GPM/1,000 tons.
   c. Screw: 10 GPM/1,000 tons.
   d. Scroll: 10 GPM/1,000 tons.
   e. Absorption: 20 GPM/1,000 tons.

E. **Steam Boiler Water Makeup**

1. Boilers: 4.0 GPM/1,000 lbs./h each
2. Deaerator/feedwater unit: 4.0 GPM/1,000 lbs./h each
3. Makeup water for the steam system is only required at one of the boilers or one of the feedwater units at any given time, for system sizing.

F. **Boiler Blowdown Separator Makeup**

1. Noncontinuous blowdown (bottom blowdown): 5.0 GPM/1,000 lbs./h.
2. Continuous blowdown (surface blowdown):   0.5 GPM/1,000 lbs./h.

G. **Blowdown Separator Drains: 10 GPM/1,000 lbs./h Boiler Output**

H. **Vacuum Type Steam Condensate Return Units: 0.1 GPM/1,000 lbs./h of Connected Load**

I. **Pumped Steam Condensate Return Units: 2.4 GPM/1,000 lbs./h**

J. **Humidifiers**

1. Steam humidifiers: 5.6 GPM/1,000 kW input or 5.6 GPM/3413 MBH.
2. Electric humidifiers: 5.6 GPM/1,000 kW input or 5.6 GPM/3413 MBH.
3. Evaporative humidifiers: 5.0 GPM/1,000 lbs./h.
4. Spray coil humidifiers: 5.0 GPM/1,000 lbs./h.

K. **Air Conditioning Condensate**

1. Unitary packaged AC equipment: 0.006 GPM/ton.
2. Air handling units (100% outdoor air): 0.100 GPM/1,000 CFM.
3. Air handling units (50% outdoor air):   0.065 GPM/1,000 CFM.
4. Air handling units (25% outdoor air):   0.048 GPM/1,000 CFM.
5. Air handling units (15% outdoor air):   0.041 GPM/1,000 CFM.
6. Air handling units (0% outdoor air):   0.030 GPM/1,000 CFM.

Water Treatment and Chemical Feed Systems

38.01 Water Treatment and Chemical Feed Systems

A. General

1. Water treatment objectives:
   a. Prevent hard scale and soft sludge deposits.
   b. Prevent corrosion and pitting.
   c. Protect boiler, piping, and equipment metal chemistry.
   d. Prevent steam carryover.
2. Corrosion and scale/deposit control factors:
   a. pH Level: As the pH of the system water increases (moves toward the alkaline side of the scale), the corrosiveness of the water decreases. However, as the pH of the system water increases, the formation of scale increases. Normal pH range is 6.5 to 9. A typical pH range is 7.8 to 8.8 (Acid pH = 1; Neutral pH = 7; Alkaline pH = 14).
   b. Hardness: As the hardness of the system water increases, the corrosiveness of the water decreases. However, as the hardness of the system water increases, the formation of scale increases.
   c. Temperature: As the temperature of the system water increases, the corrosiveness of the water increases. In addition, as the temperature of the system water increases, the formation of scale increases. Corrosion rates double for every 20°F increase in water temperature.
   d. Foulants: The more scale-forming material and foulants in the system water, the greater the chances of scale and deposit formation. Foulants include calcium, magnesium, biological growth (algae, fungi, and bacteria), dirt, silt, clays, organic contaminants (oils), silica, iron, and corrosion by-products.
3. Water treatment limits:
   a. Oxygen: Less than 0.007 ppm (7 ppb).
   b. Hardness: Less than 5.0 ppm.
   c. Suspended matter: Less than 0.15 ppm.
   d. pH: 8 to 9.
   e. Silicas: Less than 150 ppm.
   f. Total alkalinity: Less than 700 ppm.
   g. Dissolved solids: Less than 7,000 mmho/cm.
4. Water source comparison:
   a. Surface water:
      1) High in suspended solids.
      2) High in dissolved gases.
      3) Low in dissolved solids.
   b. Well water:
      1) High in dissolved solids.
      2) Low in suspended solids.
      3) Low in dissolved gases.
5. Suspended solids:
   a. Dirt.
   b. Silt.
   c. Biological growth.
   d. Vegetation.
   e. Insoluble organic matter.
   f. Undissolved matter.
   g. Iron.
6. Hardness measures the amount of calcium and magnesium in the water.
7. Alkalinity measures the water's ability to neutralize strong acid.
8. Scale is the result of precipitation of hardness salts on heat exchange surfaces.
9. Corrosion is the dissolving or wearing away of metals:
   a. *General Corrosion.* General corrosion is caused by acidic conditions.
   b. *Under-Deposit Corrosion.* Under-deposit corrosion is caused by foreign matter resting on a metal surface.

    c. *Erosion.* Erosion is caused by turbulent water flow.

    d. *Pitting Corrosion.* Pitting corrosion is caused by the presence of oxygen.

    e. *Galvanic Corrosion.* Galvanic corrosion is an electrochemical reaction between dissimilar metals.

10. Problems caused by poor water quality:
    a. Scale and deposits.
    b. Decreased efficiency/heat transfer.
    c. Equipment failure/unscheduled shutdowns.
    d. Corrosion.
    e. Tube burnout or fouling.
    f. Carryover in steam systems.

11. Chemical Types:
    a. Scale inhibitors. Scale inhibitors prevent scale formation:
      1) Phosphonate.
      2) Polyacrylate.
      3) Polymethacrylate.
      4) Polyphosphate.
      5) Polymaleic acid.
      6) Sulfuric acid.
    b. Biocides. Biocides prevent biological growth:
      1) Oxidizing:
        a) Chlorine. Most common.
        b) Chlorine dioxide.
        c) Bromine. Most common.
        d) Ozone.
      2) Non-Oxidizing:
        a) Carbamate. Most common.
        b) Organo-bromide.
        c) Methylenebis-thiocyanate.
        d) Isothiazoline.
        e) Quaternary ammonium salts.
        f) Organo-tin/quaternary ammonium salts.
        g) Glutaraldehyde.
        h) Dodecylguanidine.
        i) Triazine.
        j) Thiocyanates.
        k) Quaternary ammonium metallics.
      3) Biocide treatment program should include alternate use of oxidizing and non-oxidizing biocides for maximum effectiveness (see the following table):

Biocide	Effectiveness Against			Comments
	Bacteria	Fungi	Algae	
Oxidizing Biocides				
Chlorine (Cl_2)	E	G	G	Usable pH range 5 to 8. Effective at neutral pH (pH = 7). Less effective at high pH. Reacts with–NH_2 groups.
Chlorine Dioxide (ClO_2)	E	G	G	Insensitive to pH levels. Insensitive to presence of–NH_2 groups.
Bromine	E	G	P	Usable pH range 5 to 10. Effective over broad pH range. Substitute for chlorine.
Ozone	E	G	G	pH range 7 to 9.
Non-Oxidizing Biocides				
Carbamate	E	E	G	pH range of 5 to 9. Good in high suspended solids systems. Incompatible with chromate treatment programs.

(Continued)

Biocide	Effectiveness Against			Comments
	Bacteria	Fungi	Algae	
Organo-Bromide (DBNPA)	E	P	P	pH range 6 to 8.5.
Methylenebis-Thiocyanate (MBT)	E	P	P	Decomposes above a pH of 8.
Isothiazoline	E	G	G	Insensitive to pH levels. Deactivated by HS and–NH$_2$ groups.
Quaternary Ammonium Salts	E	G	G	Tendency to foam. Surface active. Ineffective in organic-fouled systems.
Organo-Tin/Quaternary Ammonia Salts	E	G	E	Tendency to foam. Functions best in alkaline pH.
Glutaraldehyde	E	E	G	Effective over a broad pH range. Deactivated by–NH$_2$ groups.
Dodecylguanidine (DGH)	E	E	G	pH range of 6 to 9.
Triazine	N	N	E	pH range of 6 to 9. Specific for algae control. Must be used with other biocides.

Notes:
1 Table Abbreviations:
  E = Excellent Biocide Control
  G = Good Biocide Control
  P = Poor Biocide Control
  N = No Biocide Control

    c. Corrosion inhibitors. Corrosion inhibitors prevent corrosion:
        1) Molybdate. Most common and most effective.
        2) Nitrite. Most common.
        3) Aromatic azoles.
        4) Chromate.
        5) Polyphosphate.
        6) Zinc.
        7) Orthophosphate.
        8) Benzotriazole. Copper corrosion inhibitor.
        9) Tolyltriazole. Copper corrosion inhibitor.
        10) Silicate. Copper and steel corrosion inhibitor.
    d. Dispersants. Dispersants prevent suspended and dissolved solids from settling out or forming scale in the system, remove existing deposits, and enhance biocide effectiveness:
        1) Polyacrylate.
        2) Polymethacrylate.
        3) Polymaleic acid.
        4) Surfactants.
12. Corrosion monitoring is recommended with the use of corrosion coupons for closed and open hydronic systems.
13. Side stream filtration is recommended to maintain system cleanliness. Filters should be sized to filter the entire volume of the system three to five times per day.

38.02 Closed System Chemical Treatment (Chilled-Water Systems, Heating Water Systems)

A. The chemical treatment objective is to prevent and control the following:

1. Scale formation.
2. Corrosion. Major concern.
3. System pH (between 8 and 9).

B. Chemical Types Used in Closed Systems:

1. Scale inhibitors.
2. Corrosion inhibitors.
3. Dispersants.

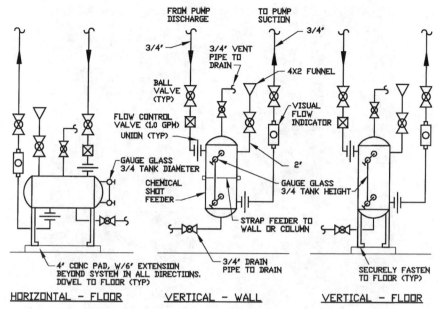

FROM PUMP DISCHARGE

TO PUMP SUCTION

3/4"

3/4"

3/4" VENT PIPE TO DRAIN

4X2 FUNNEL

BALL VALVE (TYP)

FLOW CONTROL VALVE (1.0 GPM)

UNION (TYP)

VISUAL FLOW INDICATOR

GAUGE GLASS 3/4 TANK DIAMETER

CHEMICAL SHOT FEEDER

2'

GAUGE GLASS 3/4 TANK HEIGHT

STRAP FEEDER TO WALL OR COLUMN

4' CONC PAD, W/6' EXTENSION BEYOND SYSTEM IN ALL DIRECTIONS. DOWEL TO FLOOR (TYP)

3/4" DRAIN PIPE TO DRAIN

SECURELY FASTEN TO FLOOR (TYP)

HORIZONTAL – FLOOR   VERTICAL – WALL   VERTICAL – FLOOR

FIGURE 38.1 CLOSED SYSTEM CHEMICAL SHOT FEEDER.

C. Most Common Chemicals Used:

1. Molybdate.
2. Nitrite-based inhibitors.

D. Water analysis should be conducted at least once a year, preferably semiannually or quarterly, depending on system water losses.

E. See Figs. 38.1 and 38.2 regarding the chemical treatment components used in a closed piping system.

38.03 Open System Chemical Treatment (Condenser Water Systems)

A. The chemical treatment objective is to prevent and control the following:

1. Scale formation.
2. Fouling:
   a. Particulate matter.
   b. Biological growth.
3. Corrosion.
4. System pH. Between 8 and 9.

B. Chemical Types Used in Open Systems:

1. Scale inhibitors.
2. Biocides.
3. Corrosion inhibitors.
4. Dispersants.

FIGURE 38.2   PHOTOGRAPH OF A CLOSED SYSTEM CHEMICAL SHOT FEEDER.

C. Makeup water analysis should be conducted at least twice a year, preferably quarterly.

D. System water analysis should be conducted at least once a week.

E. See Figs. 38.3 and 38.4 for chemical treatment components used in an open piping system.

38.04 Steam Systems

A. The chemical treatment objective is to prevent and control the following:

1. Scale formation.
2. Corrosion. Major concern.
3. System pH. Between 8 and 9.

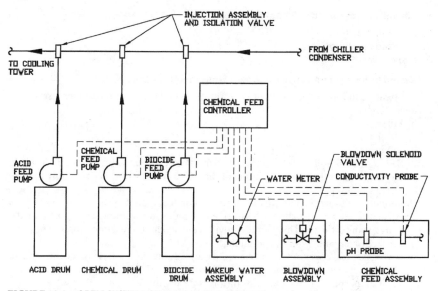

FIGURE 38.3 OPEN SYSTEM CHEMICAL TREATMENT.

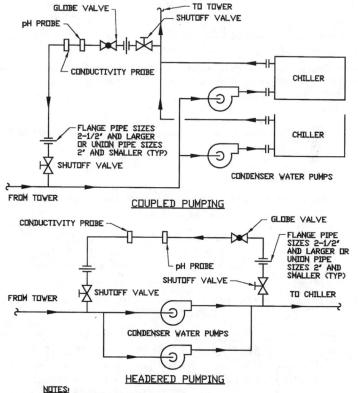

FIGURE 38.4 OPEN SYSTEM CHEMICAL FEED CONTROL ASSEMBLY.

B. Chemical Types Used in Steam Systems:

1. Scale inhibitors.
2. Corrosion inhibitors.
3. Dispersants.

C. Steam Boiler System Water Treatment Equipment:

1. Pre-treatment: Most effective way to control steam boiler chemical treatment issues:
   a. Softeners.
   b. Filters.
   c. Dealkalizers.
   d. RO units.
   e. See Figs. 38.5 and 38.6 for steam system chemical treatment. Figure 38.5 shows all the potential treatment equipment; however, many steam systems only require water softening.
2. Pre-boiler: Feedwater system treatment (deaerator, feedwater tank):
   a. An oxygen scavenger should be injected into the storage tank. Injection into the storage tank is the ideal location. It provides the maximum reaction time and protects the feedwater tank, pumps, and piping.
   b. An oxygen scavenger can be injected into the feedwater line, but is not recommended.
   c. Oxygen scavenger chemicals (see the following table):
      1) Sodium sulfite. Low- and medium-pressure systems.
      2) Hydrazine. Medium- and high-pressure systems.

Oxygen Scavenger	Feedwater Levels	Boiler Levels
Sodium Sulfite	10 to 15 ppm	30 to 60 ppm
Hydrazine	0.05 to 0.1 ppm	0.1 to 0.2 ppm

3. Boiler: Organic treatment program:
   a. Scale control chemicals should be injected directly into the boiler; however, they may be injected into the feedwater tank or feed water line as well.
   b. Polymers. Most common.
   c. Phosphonate.

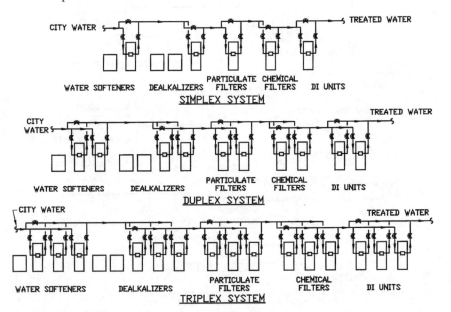

FIGURE 38.5 STEAM SYSTEM WATER TREATMENT.

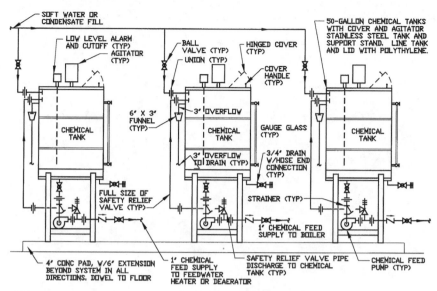

FIGURE 38.6 STEAM BOILER & FEEDWATER CHEMICAL TREATMENT SYSTEM.

4. After-boiler: Steam and Condensate Pipe Treatment:
   a. Amines:
      1) Neutralizing amines. Neutralize carbonic acid; may be injected into the boiler or steam header.
      2) Filming amines. Injected into the steam header.
   b. Injection location:
      1) Steam header. Best location.
      2) Boiler.
      3) Feedwater. Worst location; not recommended.
      4) These chemicals can be injected anywhere along the steam piping for better localized protection, especially in long piping runs.

D. Chemical Feed Methods

1. Shot feed or batch process. Not recommended.
2. Continuous:
   a. Manual control:
      1) Continuous.
      2) Clock timer.
      3) Percent timer.
   b. Automated control:
      1) Activated with feedwater pump.
      2) Activated with makeup water flow control.
      3) Activated with burner control.

E. Makeup water analysis should be conducted at least twice a year, preferably quarterly.

F. System water analysis should be conducted at least once a week.

PART 39

Automatic Controls Building Automation Systems

39.01 Automatic Controls and Building Automation Systems

A. Control Design Guidelines:

1. Today's automatic control systems and building automation systems should be designed to meet the following:
   a. Open protocol design.
   b. Web-based system design.
   c. BACnet standards preferred.
   d. Security—passwords for different levels.
      1) View only.
      2) View and modify setpoints.
      3) View, modify setpoints, and program.
   e. BAS workstations.
      1) Computers:
         a) Web-based systems—any computer connected to the network can be a workstation. However, it is good practice to provide a work station for facilities use.
         b) Speed—the faster the better.
         c) Provide laptop computers for facility maintenance staff to use on larger facilities or campus settings.
         d) Items to specify:
            Processor.
            Memory.
            Storage.
            Media Drives.
            Communication.
            Modem.
            Monitor.
            Video.
            Backup.
            Ports.
            Accessories—keyboard, mouse, UPS.
            Operating system—industry standard, professional grade.
            Data base—industry standard, professional grade, enterprise class.
      2) Report printers:
         a) LaserJet or Ink Jet
         b) Paper sizes:
            Letter—8.5" × 11".
            Legal—8.5" × 14".
            Tabloid—11" × 17" (for printing drawings).
      3) Alarm printers:
         a) LaserJet or Ink Jet.
         b) Continuous paper feed is preferred; however, hard to find.
         c) Paper sizes:
            Letter—8.5" × 11".
   f. Remote contact and alarm reporting:
      1) E-mail.
      2) Telephone.
      3) Smart phone.
      4) Other.
2. Two-way control valves should be installed upstream of equipment so that equipment is not subject to pump pressures.
3. *Proportional Band.* Throttling range over which the regulating device travels from fully closed to fully open.

4. *Drift or Offset*. Difference between the set point and the actual control point.
5. *Rangeability*. Ratio of maximum free area when fully open to the minimum free area.
6. Bypass valves should be plug valves, ball valves, or butterfly valves.
7. Control valves in HVAC systems should be the equal percentage type for output control, because equal percentage control valve flow characteristics are opposite of coil capacity characteristics.
   a. Do not oversize control valves; most control valves are at least one to two sizes smaller than the pipe size.
   b. The greater the resistance at design flow, the better the controllability.
   c. Control Valve Pressure Drop:
      1) Minimum control valve pressure drop: 5 percent of total system pressure drop.
      2) Preferred control valve pressure drop: 10 to 15 percent of total system pressure drop.
      3) Maximum control valve pressure drop: 25 percent of total system pressure drop.
   d. When specifying control valves include:
      1) Maximum design flow.
      2) Minimum design flow.
      3) Internal pressure.
      4) Pressure drop at design flow.
      5) Pressure drop at minimum flow.
8. Two-way control valves:
   a. Two-way control valves should be selected for a resistance of 20 to 25 percent of the total system resistance at the valve location. This results in selecting the control valves for the available head at each location requiring a different pressure drop for each valve in direct return systems. In reverse return systems, control valves may be selected with equal pressure drop requirements. If control valves are selected for the pressure drop at each location, balancing valves are not required for external balancing of systems unless the pressure differential at the control valve location becomes excessive. Variable volume systems will be self-balancing.
9. Three-way control valves:
   a. Three-way control valves exhibit linear control characteristics that are not suited for output control at terminal units.
   b. If three-way control action is desired to maintain minimum flow requirements, use two opposed-acting, equal percentage, two-way valves. A balancing valve must be installed in the bypass adjusted to equal the coil pressure drop. Operate valves sequentially in lieu of simultaneously, because if both valves are operated simultaneously, significant flow variations may occur.
   c. The three-way valve pressure drop should be greater than the pressure drop (up to twice the pressure drop) of the coil it serves with a balancing valve in the bypass. The bypass valve pressure drop should be adjusted to equal to the coil pressure drop. A balancing valve or flow control device should be installed in the return downstream of the three-way valve.
10. Do not use on/off type control valves, except for small line sizes (1 inch and smaller).
11. Provide a fine mesh strainer ahead of each control valve to protect the control valve.

39.02 Control Definitions

The following control definitions were taken from the *Honeywell Control Manual* listed in Part 53:

A. *Algorithm*. A calculation method that produces a control output by operating on an error signal or a time series of error signals. Operational logic affected by a control system usually resident in controlled hardware or software.

B. *Amplifiers*. Amplifiers condition the control signal, including linearization, and raise it to a level adequate for transmission and use by controllers.

C. *Analog.* Continuously variable (e.g., mercury thermometer, clock, faucet controlling water from closed to open).

D. *Authority.* The effect of the secondary transmitter versus the effect of the primary transmitter.

E. *Automatic Control System.* A system that reacts to a change or imbalance in the variable it controls by adjusting other variables to restore the system to the desired balance.

F. *Binary.* A distinct variable; a noncontinuous variable (e.g., digital clock, digital thermometer, digital radio dial); also related to computer systems and the binary numbering system (base 2).

G. *Closed Loop Control System.* Sensor is directly affected by the action of the controlled device, system feedback.

H. *Contactors.* Similar to relays, but are made with much greater current carrying capacity. Used in devices with high power requirements.

I. *Controls.* As related to HVAC, three elements are necessary to govern the operation of HVAC systems:

1. *Sensor.* A device or component that measures the value of the variable (e.g., temperature, pressure, humidity).
2. *Controller.* A device that senses changes in the controlled variable, internally or remotely, and derives the proper corrective action and output to be taken (e.g., receiver/controller, DDC panel, thermostat).
3. *Controlled Device.* That portion of the HVAC system that affects the controlled variable (e.g., actuator, damper, valve).

J. *Control Action.* Effect on a control device to create a response.

K. *Controlled Agent.* The medium in which the manipulated variable exists (e.g., steam, hot water, chilled water).

L. *Controlled Medium.* The medium in which the controlled variable exists (e.g., the air within the space).

M. *Controlled Variable.* The quantity or condition that is measured and controlled (e.g., temperature, flow, pressure, humidity, three states of matter).

N. *Control Point.* Actual value of the controlled variable (set point plus or minus set point).

O. *Corrective Action.* Control action that results in a change of the manipulated variable.

P. *Cycle.* One complete execution of a repeatable process.

Q. *Cycling.* A periodic change in the controlled variable from one value to another. Uncontrolled cycling is called "hunting."

R. *Cycling Rate.* The number of cycles completed per unit time, typically cycles per hour.

S. *Dampers.* Dampers are mechanical devices used to control airflow:

1. *Quick Opening.* Maximum flow is approached as the damper begins to open.
2. *Linear.* Opening and flow are related in direct proportion.
3. *Equal Percentage.* Each equal increment of opening increases flow by an equal percentage over the previous value.

4. *Opposed Blade.* Balancing, mixing, and modulating control applications. Half of the blades rotate in one direction, while the other half rotate in the other direction:
   a. At low pressure drops, opposed blade dampers tend to be equal percentage.
   b. At moderate pressure drops, opposed blade dampers tend to be linear.
   c. At high pressure drops, opposed blade dampers tend to be quick opening.
5. *Parallel Blade.* Two-position control applications. All the blades rotate in a parallel, or in the same, direction:
   a. At low pressure drops, parallel blade dampers tend to be linear.
   b. At high pressure drops, parallel blade dampers tend to be quick opening.

T. **Deadband.** A range of the controlled variable in which no corrective action is taken by the controlled system and no energy is used.

U. **Discriminator.** A device that accepts a large number of inputs (up to 20) and selects the appropriate output signal (averaging relay, high relay, low relay).

V. **Deviation.** The difference between the set point and the value of the controlled variable at any moment. Also called *offset.*

W. **DDC.** Direct Digital Control.

X. **Differential.** The difference between the turn-on signal and the turn-off signal.

Y. **Digital.** Series of On and Off pulses arranged to carry messages (e.g., digital radio and TV dials, digital clock, computers).

Z. **Digital Control.** A control loop in which a microprocessor-based controller directly controls equipment based on sensor inputs and set point parameters. The programmed control sequence determines the output to the equipment.

AA. **Direct Acting.** Controller is direct acting when an increase in the level of the sensor signal results in an increase in the level of the controller output.

BB. **Droop.** A sustained deviation between the control point and the set point in a two-position control system caused by a change in the heating or cooling.

CC. **Dry Bulb Control.** Control of the HVAC system based on outside air dry bulb temperature (sensible heat).

DD. **Electric Control.** A control circuit that operates on line or low voltage and uses a mechanical means, such as temperature-sensitive bimetal or bellows, to perform control functions.

EE. **Electronic Control.** A control circuit that operates on low voltage and uses solid state components to amplify input signals and perform control functions.

FF. **Enthalpy Control.** Control of the HVAC system based on outside air enthalpy (total heat).

GG. **Fail Closed.** Position device will assume when system fails (e.g., fire dampers fail closed).

HH. **Fail Open.** Position device will assume when system fails (e.g., present coil valves fail open).

II. **Fail Last Position.** Position device will assume when system fails (e.g., process coding water valve fails in last position).

JJ. **Final Control Element.** A device such as a valve or damper that acts to change the value of the manipulated variable (e.g., controlled device).

KK.   *Floating Action.* Dead spot or neutral zone in which the controller sends no signal but allows the device to float in a partly open position.

LL.   *Gain.* Proportion of control signal to throttling range.

MM.   *In Control.* Control point lies within the throttling range.

NN.   *Interlocks.* Devices that connect HVAC equipment so operation is interrelated and systems function as a whole.

OO.   *Lag.* A delay in the effect of a changed condition at one point in the system, or some other condition to which it is related. Also, the delay in response of the sensing element of a control due to the time required for the sensing element to sense a change in the sensed variable.

PP.   *Lead/Lag.* A control method in which the selection of the primary and secondary piece of equipment is obtained and alternated to limit and equalize wear on the equipment.

QQ.   *Manipulated Variable.* The quantity or condition regulated by the automatic control system to cause desired change in the controlled variable.

RR.   *Measured Variable.* A variable that is measured and may be controlled.

SS.   *Microprocessor-Based Control.* A control circuit that operates on low voltage and uses a microprocessor to perform logic and control functions. Electronic devices are primarily used as sensors. The controller often furnishes flexible DDC and energy management control routines.

TT.   *Modulating Action.* The output of the controller can vary infinitely over the range of the controller.

UU.   *Modulating Range.* Amount of change in the controlled variable required to run the actuator of the controlled device from one end of its stroke to the other.

VV.   *Motor Starters.* Electromechanical device that utilizes the principle of electromagnetism to start and stop electric motors, often containing solenoid coil actuators, relays, and overload protective devices.

WW.   *Normally Closed.* The device assumes the closed position when the control signal is removed (the device is in the closed position in the box prior to installation).

XX.   *Normally Open.* The device assumes the open position when the control signal is removed (the device is in the open position in the box prior to installation).

YY.   *Offset.* The difference between the control point and the set point.

ZZ.   *On/Off Control.* A simple two-position control system in which the device being controlled is either full On or full Off with no intermediate operating positions available.

AAA.   *Open Loop Control System.* The sensor is not directly affected by the action of the controlled device; no system feedback.

BBB.   *Out of Control.* The control point lies outside of the throttling range.

CCC.   *Pigtail.* A loop put in a sensing device to prevent the element from experiencing temperature or pressure extremes.

DDD. *Pneumatic Control*. A control circuit that operates on air pressure and uses a mechanical means, such as temperature-sensitive bimetal or bellows, to perform control functions.

EEE. *Proportional Control*. A control algorithm or method in which the final control element moves to a position proportional to the deviation of the value of the controlled variable from the set point. Cyclical control (sine/cosine).

FFF. *Proportional-Integral (PI) Control*. A control algorithm that combines the proportional (proportional response) and integral (reset response) control algorithms. Cyclical control, but automatically narrows the band between upper and lower points. Used most commonly in commercial building applications.

GGG. *Proportional-Integral-Derivative (PID) Control*. A control algorithm that enhances the PI control algorithm by adding a component that is proportional to the rate of change (derivative) of the deviation of the controlled variable. Compensates for system dynamics and allows faster control response. Cyclical control, but automatically narrows the band between upper and lower points and also calculates the time between peak high and peak low and adjusts accordingly. Used most commonly in industrial applications.

HHH. *Relays*. Electromagnetic devices for remote or automatic control actuated by variations in conditions of an electric circuit and operating, in turn, other devices (such as switches) in the same or different circuit. Carry low-level control voltages and currents.

III. *Reverse Acting*. Controller is reverse acting when an increase in the level of the sensor signal results in a decrease in the level of the controller output.

JJJ. *Sensing Element*. A device or component that measures the value of the variable.

KKK. *Sensitivity*. Proportion of the control signal to throttling range.

LLL. *Set Point*. Desired value of the controlled variable (usually in the middle of the throttling range).

MMM. *Snubber*. A component installed with a sensing device that prevents sporadic fluctuations from reaching the sensing device. These sporadic fluctuations often make the sensing device inoperative.

NNN. *Step Control*. Control method in which a multiple-switch assembly sequentially switches equipment as the controller input varies through the proportional band.

OOO. *Time Delay Relays*. Relays that provide a delay between the time the coil is energized and the time the contactors open and/or close.

PPP. *Thermistor*. A solid state device in which resistance varies with temperature.

QQQ. *Throttling Action*. Amount of change in the controlled variable required to run the actuator of the controlled device from one end of its stroke to the other.

RRR. *Throttling Range*. Amount of change in the controlled variable required to run the actuator of the controlled device from one end of its stroke to the other. Also referred to as *proportional band*.

SSS. *Transducers*. Devices that change a pneumatic signal to an electric signal and vice versa. Pneumatic-Electric (PE) or Electric-Pneumatic (EP) switches (two-position transducer or analog to analog).

TTT. *Turndown Ratio.* The minimum flow or capacity of a piece of equipment expressed as a ratio of maximum flow/capacity to minimum flow/capacity. The higher the ratio, the better the control.

UUU. *Two-Position Control.* Control system in which the device being controlled is either full On or full Off with no intermediate operating positions available (On/Off; open/closed; also called *On/Off control*).

VVV. *Valves.* Valves are mechanical devices used to control the flow of steam, water, gas, and other fluids:

1. 2-Way: Temperature control, modulate flow to controlled device, variable flow system.
2. 3-Way mixing: Temperature control, modulate flow to controlled device, constant flow system; two inlets and one outlet.
3. 3-Way diverting: Used to divert flow; generally cannot modulate flow—two positions; one inlet and two outlets.
4. Quick opening control valves: Quick opening control valves produce wide free port area with relatively small percentage of total valve stem stroke. Maximum flow is approached as the valve begins to open.
5. Linear control valves: Linear control valves produce free port areas directly related to valve stem stroke. Opening and flow are related in direct proportion.
6. Equal percentage control valves: Equal percentage control valves produce an equal percentage increase in the free port area with each equal increment of valve stem stroke. Each equal increment of opening increases flow by an equal percentage over the previous value (most common HVAC control valve).
7. Control valves are normally smaller than line size unless used in two-position applications (open/closed).
8. Control valves should normally be sized to provide 20 to 60 percent of the total system pressure drop:
   a. Water system control valves should be selected with a pressure drop equal to two to three times the pressure drop of the controlled device.
      OR
      Water system control valves should be selected with a pressure drop equal to 10 ft. or the pressure drop of the controlled device, whichever is greater.
      OR
      Water system control valves for constant flow systems should be sized to provide 25 percent of the total system pressure drop.
      OR
      Water system control valves for variable flow systems should be sized to provide 10 percent of the total system pressure drop, or 50 percent of the total available system pressure.
   b. Steam control valves should be selected with a pressure drop equal to 75 percent of the inlet steam pressure.

39.03 Types of Control Systems

A. Pneumatic:

1. Safe.
2. Reliable.
3. Proportional.
4. Inexpensive.
5. Fully modulating or two-position in nature.
6. Seasonal calibration required.
7. If there are more than a couple dozen control devices in a building, then pneumatic controls would be less expensive than electric or electronic controls.

8. Widely used in commercial, institutional, and industrial facilities.
9. Pneumatic control system pressure signals:
   a. Typical heating: 0–7 psi.
   b. Typical cooling: 8–15 psi.
   c. Max. system pressure: 30 psi.
10. Compressor runtime should be 1/3 to 1/2 the operating time.

B. Electric:

1. Simple control systems.
2. Used on small HVAC systems.
3. Mostly used for starting and stopping equipment.
4. Electric control system signals:
   a. 120 volts and less AC or DC.
   b. Typically 120 volts or 24 volts.

C. Electronic:

1. Used widely in prepackaged control systems.
2. Fully modulating in nature.
3. Reasonably inexpensive.
4. Electronic control system signals:
   a. 24 volts or less AC or DC.
   b. Typical voltage signal range of 0 to 10 volts.
   c. Typical amperage signal range of 4 to 20 milliamps.

D. Direct Digital Control (DDC):

1. Computerized control.
2. Fully modulating, start/stop and staged control.
3. Faster and more accurate than all other control systems.
4. Control systems can be adapted and changed to suit field conditions. Very flexible.
5. Able to communicate measured, control, input and output data over a network.
6. Fairly expensive.
7. Often DDC systems use DDC controllers and pneumatic actuators to operate valves, dampers, and other devices.
8. DDC system signals:
   a. Typical voltage signal range of 0 to 10 volts DC.
   b. Typical amperage signal range of 4 to 20 milliamps.
9. Most energy codes are forcing control system design to DDC.
10. Common names for DDC systems.
    a. BAS—Building Automation System.
    b. BMS—Building Management System.
    c. EMS—Energy Management System.
    d. FMS—Facility Management System.

39.04 Control System Objectives

A. Define Control Functions:

1. Start/Stop—All control systems should be provided with the following types of start/stop control.
   a. Manual—manual control at starter or variable frequency drive.
   b. Remote manual—manual control through the building automation system.
   c. Automatic control—automatic control through the building automation system.
2. Occupied/unoccupied/preparatory.

3. Fan capacity control:
   a. Variable Frequency Drives (VFDs)—energy codes are forcing the use of VFDs.
   b. Inlet Vanes.
   c. Two-speed motors.
   d. Discharge dampers—energy wasting; not typically permitted by energy codes.
   e. Scroll volume control.
   f. Supply air-, return air-, relief air-fan tracking.
4. Pump capacity control:
   a. Variable Frequency Drives (VFDs)—energy codes are forcing the use of VFDs.
   b. Two-speed motors.
   c. Variable flow pumping systems (two-way control valves).
5. Damper control (OA, RA, RFA, Inlet Vanes).
6. Valve control (two-way, three-way).
7. Temperature.
8. Humidity.
9. Pressure.
10. Flow.
11. Temperature Reset (SA, Water).
12. Terminal unit control (room, discharge, submaster).
13. Modulate, sequence, cycling.
14. Monitoring systems:
    a. HVAC systems.
    b. Plumbing systems.
    c. Medical gas, vacuum, and compressed air systems.
    d. Laboratory gas, vacuum, and compressed air systems.
    e. Fire protection systems.
    f. Electrical systems.
    g. Elevators.
    h. Other.
15. Alarms.
16. Energy/utility consumption—natural gas, fuel oil, electric, water.
17. Lighting—time of day schedule, daylighting.

B. Define Interlock Functions:

1. Fans/AHUs.
2. Pumps/boilers/chillers.
3. Smoke control system interlocks.

C. Define Safety Functions:

1. Fire.
2. Smoke.
3. Freeze protection.
4. Low/high pressure limit.
5. Low/high temperature limit.
6. Low/high water.
7. Low/high flow.
8. Over/under electrical current.
9. Vibration.

D. Alarm Functions (most often safety alarms).

E. Typical Control Algorithms:

1. Occupied/unoccupied/preparatory (time of day scheduling).
2. Night/weekend/holiday (time of day/week/year scheduling).
3. AHU dry-bulb economizer.
4. AHU enthalpy economizer.

5. Boiler/heat exchanger OA reset.
6. AHU discharge air control.
7. AHU discharge air control with room reset.
8. AHU VAV pressure independent.
9. AHU VAV pressure dependent.
10. Chiller discharge water reset.
11. Daylight savings time adjustments.
12. Electrical demand limiting.
13. Start/stop optimization.
14. Energy-performance optimization.
15. Duty cycle.
16. Enthalpy optimization.
17. Smoke control.
18. Trending.
19. Alarm instructions.
20. Maintenance work order.
21. Runtime totalizing.

F. Types of Controls:

1. *Operating Controls.* Operating controls are used to control a device, system, or entire facility in accordance with the needs of the device, system, or facility.
2. *Safety Controls.* Safety controls are used to protect the device, system, or facility from damage should some operating characteristic get out of control; to prevent catastrophic failure of the device or system; and to prevent harm to the occupants of the facility. Most safety controls come in the form of high or low limits:
   a. Automatic reset.
   b. Manual reset.
3. *Operator Interaction Controls.* Controls the building occupant would normally be provided with to activate various HVAC equipment devices or systems.

39.05 Building Automation and Control Networks (BACnet)

A. BACnet is a communication protocol. A communication protocol is a set of rules governing the exchange of data between two computers. A protocol encompasses both hardware and software specifications, including the following:

1. Physical medium.
2. Rules for controlling access to the medium.
3. Mechanics for addressing and routing messages.
4. Procedures for error recovery.
5. The specific formats for the data being exchanged.
6. The contents of the messages.

B. The BACnet goal is to enable building automation and control devices from different manufacturers to communicate.

C. BACnet Data Structures:

1. Analog input.
2. Analog output.
3. Analog value.
4. Binary input.
5. Binary output.
6. Binary value.

7. *Calendar.* Represents a list of dates that have special meaning when scheduling the operation of mechanical equipment.
8. Command.
9. *Device.* Contains general information about a particular piece of mechanical equipment (i.e., model, location).
10. *Device Table.* Shorthand reference to a list of devices.
11. *Directory.* Provides information on how to access other objects.
12. *Event Enrollment.* Provides a way to define alarms or other types of events.
13. File.
14. *Group.* Shorthand method to access a number of values in one request.
15. *Loop.* Represents a feedback control loop (PID).
16. Mailbox.
17. Multi-state input.
18. Multi-state output.
19. Program.
20. Schedule.

D. BACnet Object Properties:

1. Object identifier.
2. Object type.
3. Present value.
4. Description.
5. Status flags.
6. Reliability.
7. Override.
8. Out-of-service.
9. Polarity.
10. Inactive text.
11. Active text.
12. Change-of-state time.
13. Elapsed active time.
14. Change-of-state count.
15. Time of reset.

E. BACnet Applications:

1. Alarm and event services.
2. File access services (read, write).
3. Object access services (add, create, delete, read, remove, write).
4. Remove device management services.
5. Virtual terminal services (open, close, data).

F. BACnet Conformance Classes:

1. *Class 1.* Class 1 devices are the lowest level in BACnet system structure and consist of smart sensors.
2. *Class 2.* Class 2 devices consist of smart actuators.
3. *Class 3.* Class 3 devices consist of unitary controllers.
4. *Class 4.* Class 4 devices consist of general purpose local controllers.
5. *Class 5.* Class 5 devices consist of operator interface controllers.
6. *Class 6.* Class 6 devices are the highest level in the BACnet system structure and consist of head-end computers.

G. BACnet Functional Groups:

1. Clock.
2. Hand-held workstation.
3. Personal computer workstation.
4. Event initiation.

5. Event response.
6. Files.
7. Reinitialize.
8. Virtual operator interface.
9. Virtual terminal.
10. Router.
11. Device communications.
12. Time master.

39.06 Control Points List

A. Inputs:

1. Analog:
   a. Measured:
      1) Temperature.
         a) Air
         b) Water
         c) Steam
      2) Relative humidity.
      3) Dewpoint.
      4) Pressure.
         a) Air
         b) Water
         c) Steam
      5) Differential pressure.
      6) Airflow—CFM.
      7) Water flow—GPM.
      8) Steam flow—lbs./h.
      9) Btu/hr.
      10) Tons.
      11) Gas flow—CFH.
      12) Oil flow—GPH.
      13) kW.
      14) Current.
      15) Voltage.
      16) VFD speed.
      17) CO_2 concentration.
      18) CO sensor.
      19) Refrigerant sensor.
      20) Filter static.
      21) Water or liquid level.
   b. Calculated:
      1) Relative humidity.
      2) Dewpoint.
      3) Specific humidity.
      4) Wet bulb.
      5) Enthalpy.
      6) Steam consumption—lb.
      7) Gas consumption—CF.
      8) Oil consumption—gal.
      9) Water consumption—gal.
      10) Btu/h.

  11) Tons.
  12) kWh.
  13) Runtime.
  14) Efficiency.
  15) Volume—gal.
2. Binary:
 a. Run status—Flow Switch.
 b. Run status—Differential Pressure Switch.
 c. Run status—Current Switch.
 d. Filter.
 e. Smoke.
 f. Freeze.
 g. Airflow.
 h. Water Flow.
 i. Steam Flow.
 j. Meter.
 k. Interlocks.
 l. Status.
 m. Extinguishing agent flow.

B. Outputs:

1. Digital:
 a. Off-on.
 b. Off-auto-on.
 c. Off-high-low.
 d. Off-auto-on (VFD).
 e. Damper open-closed.
 f. Valve open-closed.
 g. Heating stages.
2. Analog:
 a. Damper position.
 b. Damper control.
 c. Valve position.
 d. Valve control.
 e. Setpoint adjustment.
 f. Load reset.
 g. Temperature reset.
 h. Electric heat—SCR.

C. System Features:

1. Alarms:
 a. High analog.
 b. Low analog.
 c. High-high digital.
 d. High digital.
 e. Low digital.
 f. Low-low digital.
 g. Run status.
 h. Filter.
 i. Smoke.
 j. Freeze.
 k. Pressure.
 l. Fire.
 m. Vibration.
2. Programs:
 a. Time scheduling.
 b. Demand limiting.

    c. Duty cycle.
    d. Start/stop optimization.
    e. Energy/performance optimization.
    f. Enthalpy optimization.
    g. Smoke control.
    h. Trends.
    i. Alarm instruction.
    j. Maintenance work order.

D. General:

1. Color graphics.
2. Summary report.
3. Alarm reports.
4. Trends reports.
5. X-Y graphic plots.
6. Statistical reports.
7. Historical reports.
8. Custom reports.
9. Expansion capacity: 10 to 25 percent spare capacity in controller, panels, and computer systems.
10. PC, monitor, keyboard.
11. Alarm printer.
12. Report printer.
13. Laptop computers.

39.07 DDC Control System Specification Outline

A. Part 1—General

1. Scope of work.
2. System description.
3. System performance.
    a. Graphic display.
    b. Graphic refresh.
    c. Object command.
    d. Object scan.
    e. Alarm response time.
    f. Program execution frequency.
    g. Performance.
    h. Multiple alarm annunciation.
    i. Reporting accuracy.
    j. Stability of control.
4. Codes and standards.
5. Products furnished and installed under this contract.
    a. Thermostats.
    b. Humidistats.
    c. Air distribution system temperature sensors.
    d. Air distribution system pressure and differential pressure sensors.
    e. Air distribution system flow switches.
    f. Refrigerant, CO_2, CO, and other gas detection systems.
    g. Current switches.
    h. All other devices not specifically mentioned in the following.
6. Products furnished but not installed under this contract.
    a. Piping temperature sensors, Wells, and Sockets.
    b. Piping pressure and differential Pressure Sensors.
    c. Piping flow switches.

      d. Control valves—may be furnished and installed under another division.
      e. Control dampers—may be furnished and installed under another division.
      f. Water flow meters—may be furnished and installed under another division.
      g. Airflow meters—may be furnished and installed under another division.
      h. Energy meters—may be furnished and installed under another division.
      i. Terminal unit controls—may be furnished and installed under another division.
      j. Air terminal unit controls—often furnished by the control contractor and manufacturer; installed under another division.
  7. Products installed but not furnished under this contract.
  8. Products not furnished or installed but integrated with under this contract.
      a. Chiller control package.
      b. Boiler control package.
      c. Cooling tower basin heater and water level control package.
      d. Packaged air handling system controllers.
      e. Variable frequency controllers.
      f. Motor controllers and disconnect switches.
      g. Fire, smoke, and fire/smoke dampers.
      h. Duct mounted smoke detectors.
      i. Control valves.
      j. Control dampers.
      k. Water flow meters.
      l. Airflow meters.
     m. Energy meters.
      n. Terminal unit controls.
      o. Air terminal unit controls.
      p. Electrical distribution systems—normal power, emergency power, and UPS systems.
      q. Emergency generator control package.
      r. Lighting control systems.
  9. Quality assurance.
 10. Submittals.
 11. Warranty.
 12. Ownership of proprietary material.

B. Part 2—Products

  1. Approved control system contractors/manufacturers.
  2. Materials.
  3. Communication.
      a. Network arrangement.
      b. Workstation communication.
      c. Controller communication.
      d. Secondary bus communication.
      e. System architecture.
      f. Communication performance.
      g. Communication protocols.
         1) Interoperability.
         2) Network communications.
  4. Integrating a proprietary system with an open protocol system.
  5. Operator interface.
      a. Number of work stations.
      b. System connection.
      c. System hardware.
         1) Fixed operator work stations:
           a) Computer terminal—computer, keyboard, monitor, mouse, modem, backup method.
           b) Report printer.
           c) Alarm printer.

      2) Portable operator work stations—laptop computers.
- d. System software:
  - 1) Operating system.
  - 2) System graphics.
  - 3) System applications:
    - a) General.
    - b) Automatic and manual system database save and restore.
    - c) System configuration.
    - d) Online help.
    - e) Security.
    - f) System diagnostics.
  - 4) Alarm processing.
  - 5) Trend, alarm, and event logs.
  - 6) Object and property status and control.
  - 7) Clock synchronization.
  - 8) Reports and logs.
  - 9) Custom reports.
    - a) Tenant override reports.
    - b) Electrical, gas, water, utility, and weather reports.
    - c) ASHRAE Guideline 3 Report—large chillers.
  - 10) Workstation application editors.
  - 11) Controller.
  - 12) Scheduling.
  - 13) Custom application programming.
6. Controller software.
   - a. System security.
   - b. Scheduling.
   - c. Grouping.
   - d. Alarm processing.
   - e. Remote communications.
   - f. Standard application programs.
   - g. Demand limiting.
   - h. Maintenance management.
   - i. Sequencing.
   - j. PID control.
   - k. Staggered start.
   - l. Energy calculations.
   - m. Anti-short cycling.
   - n. On/off control with differential.
   - o. Runtime totalization.
7. Building controllers.
   - a. General.
   - b. Background.
   - c. Internal software.
   - d. Modularity.
   - e. Operator software.
   - f. Inputs and outputs.
   - g. Power supplies, including UPS.
   - h. Listings.
   - i. Distribution of controllers—limits the number of systems on any one controller.
8. Custom application controllers.
   - a. General.
   - b. Internal software.
   - c. Modularity.
   - d. Operator software.
   - e. Inputs and outputs.

f. Power supplies, including UPS.

g. Listings.

h. Distribution of controllers—limits the number of systems on any one controller.

9. Application specific controllers.

a. General.

b. Background.

10. Input/output interface.

11. Power supplies and line filtering.

12. Auxiliary control devices.

a. Motorized control dampers.

b. Damper/valve actuators.

1) Electric.

2) Pneumatic.

c. Control valves.

d. Temperature devices.

e. Humidity devices.

f. Flow switches.

g. Relays.

h. Override timers.

i. Power monitoring.

1) Current.

2) Voltage.

3) Power sensing.

j. Equipment status sensing.

k. Pressure devices.

l. Electro-pneumatic transducers.

m. Local control panels.

13. Wiring and raceways.

14. Fiber optic cable system.

15. Compressed air supply—pneumatic actuation.

a. Air compressor.

b. Air dryer.

c. Air filters.

d. Pressure reducing valves.

e. Relief valves.

f. Condensate drains.

g. Pneumatic tubing.

C. Part 3—Execution

1. General installation.

2. Examination.

3. Protection.

4. Coordination.

5. General workmanship.

6. Field quality control.

7. Existing equipment.

8. Wiring.

9. Communication wiring.

10. Fiber optic cable.

11. Pneumatic systems installation.

12. Control air tubing installation.

13. Installation of sensors.

14. Flow switch installation.

a. Airflow.

b. Water flow.

15. Flow meter installation.

16. Control valve installation.
17. Control damper installation.
18. Valve and damper actuators.
19. Warning labels.
20. Identification of hardware and wiring.
21. Controllers—controller loading—spare capacity.
22. Programming.
    a. Project specific programming.
        1) Text-based programming.
        2) Graphic-based programming.
        3) Menu-driven programming.
    b. Point naming.
    c. Other programming and database setup.
23. Control system checkout and testing.
24. Control system demonstration and acceptance.
25. Commissioning.
26. Cleaning.
27. Training.
28. Sequences of operation—sequences of operation may be contained in a separate specification section.
29. I/O points lists—I/O points lists may be contained in a separate specification section or in a graphical matrix form.

PART 40

Sustainability Guidelines Relating to HVAC Systems

40.01 Introduction

A. Sustainability is a term used in the building industry to describe the focus on the design, construction, and operation of buildings in a manner that reduces their impact on the natural environment.

B. There are many aspects of sustainability in the building industry including site selection; water efficiency; use of recycled, renewable, and regional materials; energy efficiency; and indoor environmental quality. These aspects have been defined by the U.S. Green Building Council (USGBC) through its Leadership in Energy and Environmental Design (LEED) rating system.

C. The LEED rating system was started in 1993 to meet the need in the building industry for a system to define and measure "green buildings." Since that time, the LEED rating system has been expanded to cover all sectors of the building industry including new construction, existing buildings, commercial interiors, and others. The USGBC periodically updates the LEED rating system to refine its guidelines and incorporate emerging sustainability concepts.

D. Energy Star is an international standard for energy efficient consumer products that was created in 1992 by the U.S. Environmental Protection Agency. Products carrying the Energy Star service mark are generally 20 to 30 percent more energy efficient than what is required by federal standards. The program also includes labeling for residential HVAC equipment, such as air conditioning and heat pump units, furnaces, and boilers; lighting products; new homes; and commercial and industrial buildings.

40.02 Sustainable HVAC System Design

A. This chapter presents sustainability guidelines relating to HVAC systems. These guidelines focus mainly on energy efficiency, commissioning, refrigerant management, and indoor environmental quality.

B. Energy Efficiency

1. The minimum energy efficiency requirements for buildings and their associated energy systems are established by ANSI/ASHRAE/IESNA Standard 90.1-2013 *Energy Standard for Buildings Except Low-Rise Residential Buildings*. This standard prescribes the minimum energy efficiency requirements for all building envelope, electrical, and HVAC systems.

2. Increases in the energy efficiency of buildings can be achieved through improvements to the building envelope, reduction in lighting power densities, increases in motor efficiencies, and incorporation of energy-efficient HVAC systems.

3. Improvements in the energy efficiency of HVAC systems can be made in a number of ways, depending upon the system:

a. Energy-efficient equipment.

   1) High-efficiency boilers and chillers.

      a) Select higher-efficient central heating and cooling equipment, such as boilers and chillers, in lieu of central HVAC equipment that meets the minimum energy efficiency requirements of ASHRAE Standard 90.1-2013. For example, condensing boilers which extract the latent heat of vaporization from hot flue gases are more energy-efficient than noncondensing boilers which must keep stack temperatures high in order to avoid condensation of moisture in the flue gases. Recent chiller technologies, such as variable speed compressors and magnetic bearings, have greatly increased the energy efficiency of this equipment.

2) High-efficiency air conditioning equipment.
   a) Select air conditioning and heat pump equipment that exceeds the minimum energy efficiency requirements of ASHRAE Standard 90.1-2013.

3) NEMA premium efficiency motors.
   a) For motors 1 horsepower and larger, specify NEMA premium efficiency motors as defined by NEMA Standards Publication MG 1-2014 Table 52 in lieu of motors classified as energy efficient by NEMA Standards Publication MG 1-2014 Table 51.

b. HVAC system configuration.
   1) Variable flow systems.
      a) Variable flow water and air systems are inherently more energy-efficient than constant flow water and air systems. Variable frequency drives for pumps and fans can reduce energy use considerably when there are fluctuations in the loads this equipment serves.

   2) Heat reclaim systems.
      a) Systems which reclaim heat that would otherwise be rejected to the outdoors improve the energy efficiency of HVAC systems. Heat is most often reclaimed from building exhaust airstreams and mechanical refrigeration equipment. Heat reclaimed from exhaust airstreams can be used to preheat/precool outdoor air, and heat reclaimed from mechanical refrigeration equipment is often used to preheat domestic hot water. Heat reclaim can also be used to provide dehumidification in some cases where humidity control is required by the space usage.

   3) Thermal storage systems.
      a) Thermal storage systems can reduce energy costs by operating central cooling systems during off-peak hours when energy is less expensive. (The downside to thermal storage systems is that the central cooling equipment actually uses more, albeit less expensive, energy than the same central cooling equipment in nonthermal storage systems.)

c. HVAC system control strategies.
   1) Water and air temperature reset.
      a) Reset of water and air temperatures based on outdoor air temperature improves energy efficiency by more closely matching the HVAC systems' capacities to the building HVAC loads.

   2) Demand-control ventilation.
      a) Adjusting the outdoor air ventilation delivered by the HVAC systems to meet the ventilation requirements of the building occupants (referred to as demand-control ventilation, or DCV) is an energy-efficient strategy that only delivers the outdoor air ventilation that is necessary for acceptable indoor air quality. DCV is usually applied to HVAC systems which serve densely occupied spaces (those with a design occupant density greater than or equal to 25 people per 1,000 sq.ft.), such as conference rooms and auditoriums. Space CO_2 sensors located within the breathing zone (between 3 in. and 72 in. above the floor) of these spaces are used to control the amount of outdoor air ventilation delivered by the HVAC system. A CO_2 concentration that is about 700 ppm above outdoor air levels will satisfy a substantial majority of visitors entering a space with respect to human bioeffluents (body odor) according to ASHRAE Standard 62.1-2013, Appendix C. CO_2 concentrations in acceptable outdoor air range from 300 to 500 ppm. Therefore, the CO_2 setpoint for the space CO_2 sensors for DCV systems should be between 1,000 and 1,200 ppm.

   3) Night setback.
      a) Night setback control of the space temperature setpoint reduces energy use during unoccupied periods.

4) Economizers.
   a)   Airside and waterside economizers reduce energy use by utilizing outdoor air for cooling instead of mechanical refrigeration when the outdoor air conditions are suitable for this use. Airside economizers compare the enthalpy or dry bulb temperature of the outdoor air and return air and increase the outdoor airflow beyond the minimum value when the outdoor air has a lower enthalpy or dry bulb temperature and cooling is required by the HVAC system. Dry bulb economizers are often disabled above an outdoor air temperature of approximately 60°F in order to prevent the possibility of increasing the indoor space relative humidity above an acceptable limit. Waterside economizers can only be incorporated into central chilled water systems which utilizes water-cooled chillers. When the outdoor air wet bulb temperature is low enough for the cooling tower(s) to cool the returning chilled water, the returning chilled water is diverted to a plate and frame heat exchanger where it is cooled by the cooling tower water. Thus, the outdoor air is used to cool (or precool) the chilled water loop through the cooling tower(s) instead of (or in addition to) using mechanical refrigeration to cool the chilled water loop through the chiller(s).
d. Architectural components.
   1)   Close coordination with the architectural design can improve energy efficiency by increasing the insulating quality of walls (above and below grade), roofs, partitions, and the edges of slabs on grade or below grade. Other areas of consideration include building orientation, window-to-wall ratio, glazing properties, and both internal and external shading of windows.

C. Commissioning

1. The commissioning process is defined by ASHRAE Guideline 0-2013 *The Commissioning Process* as "a quality-oriented process for achieving, verifying, and documenting that the performance of facilities, systems, and assemblies meets the defined objectives and criteria."
2. Commissioning of HVAC systems ensures that the energy-efficient features of the project that are desired by the owner and intended by the design are actually implemented by the completed facilities, systems, and assemblies.
3. Commissioning begins in the predesign phase with stated goals called the Owner's Project Requirements (OPRs), continues during the design phase by meeting the OPRs with the Basis of Design (BOD) and the construction documents, moves into the construction phase with submittal review, installation checklists, issues log, startup checklists, prefunctional checklists, functional testing, and training in the operation and maintenance of the commissioned systems. Commissioning should continue at least through the warranty period with ongoing monitoring of the commissioned systems. Ideally, commissioning should continue throughout the life of the building.
4. When implemented properly throughout the course of design, construction, and operation, commissioning plays a key role in ensuring that the intended sustainability initiatives are fully implemented.
5. The commissioning process is defined in detail by ASHRAE Guideline 0-2013 *The Commissioning Process* and ASHRAE Guideline 1.1-2007 *HVAC&R Technical Requirements for the Commissioning Process.*

D. Refrigerant Management

1. From a sustainability standpoint, the two characteristics of a refrigerant that are of the greatest concern are the refrigerant's ozone-depletion potential (ODP) and its global warming potential (GWP). The types of refrigerants most commonly used in HVAC equipment are hydrochlorofluorocarbon (HCFC)-based refrigerants and hydrofluorocarbon (HFC)-based refrigerants. Chlorofluorocarbon (CFC)-based refrigerants, such as R-11 and R-12, are no longer produced because of their ODP.

2. The LEED rating system requires zero use of chlorofluorocarbon (CFC)-based refrigerants, such as R-11 and R-12, in new base building heating, ventilating, air conditioning, and refrigeration (HVAC&R) systems.

3. The LEED rating system discourages the use of hydrochlorofluorocarbon (HCFC)-based refrigerants, such as R-22 and R-123, because of their ODP. Although these refrigerants have a lesser ODP than CFC-based refrigerants, they are scheduled to cease production in the year 2030 due to their ODP.

4. Hydrofluorocarbon (HFC)-based refrigerants, such as R-134a and R-410A, do not have any ODP and are therefore not scheduled to cease production. However, both HCFC- and HFC-based refrigerants have GWPs that need to be considered in sustainable HVAC system design. The LEED rating system is concerned with the lifecycle direct global warming potential (LCGWP) of refrigerants used in HVAC&R systems. The LCGWP of a piece of HVAC&R equipment is a function of the refrigerant's GWP, the equipment's life and refrigerant leakage rate, the end-of-life refrigerant loss, and the equipment's refrigerant charge.

5. Small HVAC units (defined as containing less than 0.5 pounds of refrigerant) are not considered part of the base building system and are not subject to the requirements of the LEED rating system.

E. Indoor Environmental Quality

1. Many factors contribute to the quality of the indoor environment. From an HVAC standpoint, indoor air quality, thermal comfort, and acoustic quality need to be considered. Other factors, such as interior lighting, daylight, and quality views of the outdoors also contribute to the quality of the indoor environment but are not HVAC-related.

2. Indoor air quality.

  a. Some indoor air contaminants which must be controlled in order to ensure acceptable indoor air quality are airborne particulates, formaldehyde and volatile organic compounds (VOCs) from building materials, and ozone which is generated by copying and printing equipment. Airborne particulates are suspended in the air and can be removed through air filtration. Formaldehyde, VOCs, and ozone, on the other hand, cannot be removed through air filtration, but must be controlled through dilution ventilation and exhaust at their source.

    1) Air filtration.

      a) The minimum level of air filtration efficiency recommended for commercial HVAC systems is a minimum efficiency reporting value (MERV) of 8, as determined by ASHRAE Standard 55.2-2012 *Method of Testing General Ventilation Air-Cleaning Devices for Removal Efficiency by Particle Size*.

      b) For superior air filtration, MERV 13 air filters may be used, but at the expense of higher installed and maintenance costs and increased energy use of the air handling equipment.

      c) All inlets for air handling equipment used during construction should be protected with air filters having a minimum efficiency of MERV 8 to keep the air handling equipment and ductwork free from airborne particulates generated by the construction activities.

      d) Ductwork should also be kept clean during construction in accordance with the recommended control measures of the Sheet Metal and Air Conditioning Contractors' National Association (SMACNA) *IAQ Guidelines for Occupied Buildings under Construction*, 2nd edition, 2007, ANSI/SMACNA 008-2008, Chap. 3. These measures include covering all openings of air handling equipment and ductwork with plastic until the systems are operational.

    2) Dilution ventilation.

      a) The concentration of indoor air contaminants can be reduced to acceptable levels for human occupancy by introducing outdoor air to the building through the air handling equipment. The Ventilation Rate Procedure in Section 6.2 of ASHRAE Standard 62.1-2013 *Ventilation for Acceptable Indoor Air Quality* describes the procedure for determining the minimum rates of outdoor ventilation for various types of occupancy.

3) Source control of contaminants through exhaust.
   a) Where indoor air contaminants are generated within a building, these contaminants should be captured at their source and exhausted to the outdoors. Areas such as toilet rooms, copying and printing rooms, janitor's closets, garages, and storage closets where hazardous chemicals are stored should be exhausted at the rate required by the applicable mechanical code (minimum of 0.50 cfm per sq.ft.). These rooms should also be kept at under negative air pressurization with respect to adjacent spaces when the doors are closed.
4) Testing airborne pollutant levels.
   a) The best way to demonstrate that indoor pollutants have been reduced to acceptable levels after construction has been completed is to test for pollutants in accordance with the California Department of Public Health (CDPH) Standard Method v1.1. Air testing includes measurements of the concentrations of formaldehyde; particulate matter (PM) up to 10 microns in diameter (PM10); PM up to 2.5 microns in diameter (PM2.5); ozone; total VOCs, target chemicals listed in CDPH Standard Method v1.1, Table 4-1, except formaldehyde; and carbon monoxide (CO).
5) Building flush-out.
   a) Another acceptable means of reducing indoor pollutants after construction has been completed is to flush the building (or project area) with a minimum of 14,000 cu.ft. of outdoor air per sq.ft. of gross floor area. This amounts to operating a typical HVAC system at 100 percent outdoor air for approximately 2 weeks. For optimum results, the space temperature should be maintained at a minimum of 60°F and a maximum of 80°F. Space relative humidity during the flush-out process should be no higher than 60 percent.
6) Thermal comfort.
   a) Comfort conditions for humans are subject to six primary variables: surface temperature, air temperature, relative humidity, air movement, metabolic rate, and clothing. Comfort conditions are also a matter of personal preference and a certain percentage of dissatisfied building occupants can be expected for any indoor environmental condition. Minimizing this percentage of building occupants is the goal of sustainable HVAC design as it relates to thermal comfort. Providing occupants with the ability to control air temperature, relative humidity, air speed, and radiant temperature for their individual spaces or shared group spaces can reduce the percentage of dissatisfied building occupants. Since it is not possible to achieve an indoor environmental condition that is acceptable for all building occupants, *ASHRAE Standard 55-2013 Thermal Comfort Conditions for Human Occupancy* has defined an acceptable indoor environment as one in which 80 percent of occupants are satisfied. Achieving acceptable thermal comfort conditions in buildings is a part of sustainable HVAC system design because workers' satisfaction and productivity increase with improved thermal comfort in buildings.
7) Acoustic performance.
   a) Background noise of HVAC systems is the only acoustic performance consideration of the LEED rating system that is related to building HVAC systems.
   b) Building HVAC systems are to be designed to meet the background noise levels specified in 2011 *ASHRAE Handbook—HVAC Applications*, Chapter 48, Table 1; *AHRI Standard 885-2008*, Table 15; or a local equivalent. Noise levels are to be calculated or measured.

New Technologies for HVAC Systems

41.01 Variable Refrigerant Flow Systems

A. Introduction

1. Variable refrigerant flow (or VRF) systems utilize refrigerant as the working fluid to provide heating and cooling in the same way that conventional refrigeration systems do, with the exception that the refrigerant flow within the system can be modulated through a variable speed drive on the compressor motor. This modulation of refrigerant flow enables VRF systems to closely match the refrigeration system capacity to the heating or cooling load on the system. Variable speed condenser fan motors also improve the load matching capabilities of these systems. VRF systems also utilize circuiting manifolds which enable multiple indoor fan-coil units to be connected to a single outdoor unit. For some manufacturers, one outdoor unit can serve up to 50 indoor fan-coil units.

B. VRF Heat Pump and Energy Recovery Systems

1. VRF heat pump systems.
   a. VRF heat pump systems can provide either heating or cooling to all of the indoor fan-coil units served by a single outdoor unit.
2. VRF energy recovery systems.
   a. VRF energy recovery systems can simultaneously provide heating to some indoor fan-coil units and cooling to other indoor fan-coil units, all of which are served by the same outdoor unit. This exchange of energy from spaces requiring heating to spaces requiring cooling, and vice versa, reduces the use of purchased energy for the system.

C. Equipment Configurations

1. Indoor fan-coil units.
   a. Indoor fan-coil unit configurations include: ceiling-recessed cassettes, wall-mounted, ceiling-suspended, ceiling-concealed ducted, and vertical ducted units. Figures 41.1 through 41.6 illustrate these fan-coil unit configurations.
2. Outdoor units.
   a. The capacity of a single outdoor module is limited to about 25 tons; however, two or three modules can be combined through the use of a twinning kit in order to increase the capacity of the outdoor unit. Figure 41.7 illustrates a typical outdoor unit.
3. Refrigerant piping.
   a. Some manufacturers utilize three pipes between the refrigerant piping manifold and the outdoor unit, while other manufacturers utilize two pipes. Figure 41.8 illustrates the two-pipe and three-pipe configurations.

FIGURE 41.1   CEILING-RECESSED CASSETTE, FOUR-WAY BLOW.
((c) 2015 Mitsubishi Electric US. All Rights Reserved.)

FIGURE 41.2 CEILING-RECESSED CASSETTE, ONE-WAY BLOW. (*(c) 2015 Mitsubishi Electric US. All Rights Reserved.*)

FIGURE 41.3 WALL-MOUNTED FAN-COIL UNIT. (*(c) 2015 Mitsubishi Electric US. All Rights Reserved.*)

FIGURE 41.4 CEILING-SUSPENDED FAN-COIL UNIT. (*(c) 2015 Mitsubishi Electric US. All Rights Reserved.*)

FIGURE 41.5 CEILING-CONCEALED DUCTED FAN-COIL UNIT. (*(c) 2015 Mitsubishi Electric US. All Rights Reserved.*)

FIGURE 41.6 VERTICAL DUCTED FAN-COIL UNIT. (*(c) 2015 Mitsubishi Electric US. All Rights Reserved.*)

4. Air-source and water-source heat rejection/absorption
   a. Heat rejection/absorption can be accomplished through air-source equipment located outdoors, or through water-source equipment located indoors that receives cooling/heating water from an external heat rejection/absorption system. Figure 41.9 illustrates a water-source heat exchanger.
   b. Manufacturers have improved the heating capabilities of the outdoor air-source heat pump units so that approximately 85 percent of the heating capacity at 47°F is available at −13°F. High heating capacity heat pump units are also available which provide 100 percent of the heating capacity at −4°F.
5. Air conditioning condensate.
   a. Condensate that is formed on the cooling coil of the indoor fan-coil units and in the circuiting manifolds must be drained to the outdoors or to the building storm water system. Indoor fan-coil units are equipped with condensate pumps which are capable of between 20 and 33 inches of lift.
6. Controls.
   a. Wireless or hardwired room temperature sensors are available to control the indoor fan-coil units. (One temperature sensor is required for each fan-coil unit.) These controls can be networked to a central, stand-alone, proprietary, web-interface

FIGURE 41.7 TYPICAL VRF OUTDOOR UNIT. (*(c) 2015 Mitsubishi Electric US. All Rights Reserved.*)

control system. The VRF control system can also be connected to a building automation system through a LonWorks or BACnet gateway.

7. System efficiency.

a. VRF energy recovery systems are capable of obtaining an integrated energy efficiency ratio (IEER) rating as high as 22.1 (based on AHRI 1230 test method). Actual system efficiency will depend upon the level of energy recovery that can be

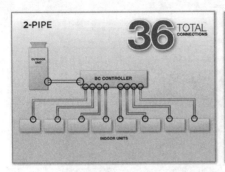
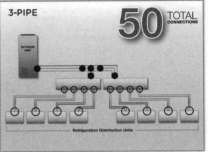

FIGURE 41.8 TWO-PIPE AND THREE-PIPE CONFIGURATIONS. (*(c) 2015 Mitsubishi Electric US. All Rights Reserved.*)

FIGURE 41.9 WATER-SOURCE HEAT EXCHANGER. (*(c) 2015 Mitsubishi Electric US. All Rights Reserved.*)

achieved. Care must be taken during the design of VRF energy recovery systems to ensure that simultaneous heating and cooling is required by the zones served by the VRF system in order to maximize the system efficiency.

41.02 Variable Frequency Drives for Chillers

A. **Variable frequency drives (VFDs), also called variable speed drives (VSDs), adjustable frequency drives (AFDs), or adjustable speed drives (ASDs) have been used for many years in the HVAC industry to control the speed of fans and pumps because of the energy savings that can be achieved during part-load operation.**

B. **VFDs are now being used by HVAC equipment manufacturers to control the speed of compressors in both water- and air-cooled chillers to achieve part-load energy savings as well.**

C. **VFDs can be used on scroll, helical rotary (or screw), and centrifugal compressors.**

D. **Application**

1. For both scroll and screw compressors, refrigeration capacity is proportional to compressor speed. Therefore, VFDs can be the sole means of capacity control for these types of compressors. Other means of capacity control are available for scroll and screw compressor, such as compressor cycling, digital on/off, and hot gas bypass. Slide valve control is available for capacity control of screw compressors. However, VFDs provide the greatest energy savings at part-load operation for both scroll and screw compressors.

2. For centrifugal compressors, refrigeration capacity can be varied not only by compressor speed, but also by guide vanes on the inlet of the impeller. The inlet guide vanes control the flow rate of refrigerant through the compressor, while the compressor speed determines the differential pressure across the compressor. Since both refrigerant flow and differential pressure affect a chiller's capacity, VFDs are commonly used in conjunction with inlet guide vanes on centrifugal compressors. Compressor cycling and hot gas bypass can also be used for capacity control of centrifugal compressors, but the use of VFDs coupled with inlet guide vanes provides the greatest energy savings at part-load operation.

3. Because only refrigerant lift (temperature difference between the condenser and evaporator conditions) is affected by the speed of a centrifugal compressor, variable-speed centrifugal chillers should be used in conjunction with a reduction in the temperature of entering condenser water in water-cooled chillers in order to maximize the energy-efficiency associated with their capacity control capabilities.

4. Using a fixed entering condenser water temperature, such as 85°F, can negate the savings that can be achieved by using VFDs on the compressors for water-cooled chillers.

5. For multiple-chiller plants, it is common to utilize one (or more) constant-speed chiller(s) and one variable-speed chiller. In this design, the constant-speed chiller(s) are staged while the capacity of the variable-speed chiller is controlled to meet the cooling load.

E. **Other Considerations**

1. It is important to note that the full-load efficiency of a variable-speed chiller is actually lower than the full-load efficiency of the same constant-speed chiller because the VFD introduces an additional electric efficiency loss. Care must be taken during the chiller plant design to ensure that the variable-speed chiller operates predominantly at part-load in order to maximize its energy efficiency.

2. The full-load energy efficiency rating of a chiller (either energy efficiency ratio [EER] or coefficient of performance [COP]) should not be the sole criterion for comparing the energy-efficiency of variable-speed and constant-speed chillers. Integrated part load value (IPLV) or nonstandard part load value (NPLV) should be used instead because

these performance characteristics also take into consideration the energy-efficiency of the chillers at part-load operation. IPLV and NPLV will provide a better relative measure of a chiller's annual energy use. For example, a chiller with an IPLV of 20 Btuh/W will use approximately 5 percent less energy on an annual basis than a chiller with an IPLV of 19 Btuh/W.

3. Chillers utilizing VFDs on the compressors are able to achieve lower sound levels at part-load conditions than constant-speed chillers.
4. VFDs are capable of handling voltage dips, surges, and other imbalances in the electrical distribution system.
5. As with all VFDs, harmonic distortion must be addressed by the VFD manufacturer to meet the IEEE 519 requirement for less than 5 percent total demand distortion.

41.03 Magnetic Bearings for Centrifugal Compressors

A. Description

1. Magnetic bearings levitate the centrifugal compressor shaft in a magnetic electric field instead of utilizing oil-lubricated bearings.
2. When coupled with a direct drive motor, the use of magnetic bearings results in a totally oil-free compressor.
3. Sensors are required at each magnetic bearing to provide real-time feedback to the bearing control system.
4. In the event of a power failure, the compressor motor acts as a generator and provides power to the bearing control system during coast down.
5. A system is required to gently de-levitate the compressor shaft.
6. Figures 41.10 through 41.12 illustrate the magnetic bearing technology.

B. Benefits

1. Reduced frictional losses within the bearing system.
2. Improved reliability and reduced maintenance cost due to oil-free operation.
   a. No need for oil-handling equipment, such as oil pumps, oil reservoirs, oil coolers, oil filters, water regulating valves, oil relief valves, oil system controls, starter, piping, heaters, etc.
   b. No possibility of oil loss at light loads.
   c. No need for oil system maintenance such as oil sampling, oil and filter changes, and oil leak repairs.
   d. No need for oil storage and disposal.

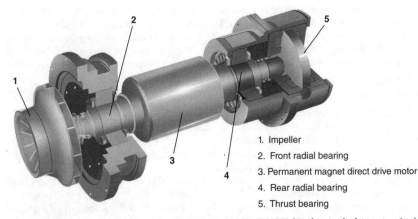

1. Impeller
2. Front radial bearing
3. Permanent magnet direct drive motor
4. Rear radial bearing
5. Thrust bearing

FIGURE 41.10   CUT-AWAY VIEW OF COMPRESSOR SHAFT. (*Daikin Applied Americas Inc.*)

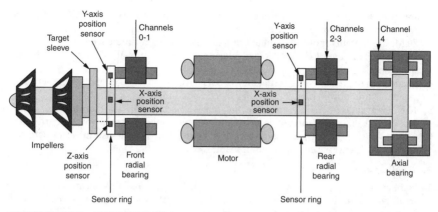

FIGURE 41.11   SCHEMATIC DIAGRAM OF COMPRESSOR SHAFT. (*Daikin Applied Americas Inc.*)

3. Consistent efficiency over the life of the equipment since there is no oil to coat heat transfer surfaces.
4. Lower compressor vibration that could be transmitted to the structure.
5. Lower compressor sound level compared to traditional centrifugal chillers.

41.04 Electronically-Commutated Motors (ECMs) for Fans and Pumps

A. Description

1. Electronically-commutated, or brushless DC permanent-magnet, motors use the same principle as AC motors—interaction of rotating magnetic fields in the rotor (rotating member in the motor) and stator (stationary members in the motor).
2. Electronically-commutated motors (ECMs) are more energy efficient than the traditional shaded pole and permanent-split capacitor (PSC) AC motors commonly used in the fractional horsepower (less than 1 hp) size range. Fractional horsepower ECMs

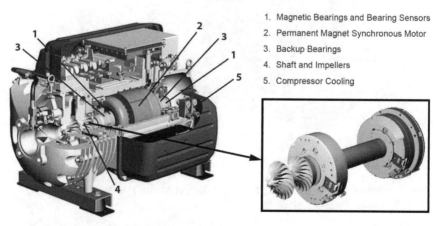

1. Magnetic Bearings and Bearing Sensors
2. Permanent Magnet Synchronous Motor
3. Backup Bearings
4. Shaft and Impellers
5. Compressor Cooling

FIGURE 41.12   CUT-AWAY VIEW OF COMPRESSOR. (*Daikin Applied Americas Inc.*)

are approximately 70 percent efficient; whereas, fractional horsepower PSC motors are approximately 50 percent efficient.

3. ECMs are not only more efficient at full-load, the efficiency difference is even greater at part-load. This gives ECMs a decided advantage over traditional AC motors in variable speed applications.

4. In addition to greater energy efficiency, ECMs have the advantages of quieter operation, greater output power, higher operating speeds, and longer service life.

5. The disadvantages of ECMs are that they have a higher first cost than traditional AC motors and also create disruptive harmonic currents in the electrical power distribution system which can overheat conductors and connectors and can also interfere with the operation of sensitive electronic equipment.

B. Applications

1. ECMs are now a commonly available option for fan-powered VAV terminal units and small fan-coil units (up to about 800 CFM). Thermostats which enable variable speed operation of the fan motors take advantage of the higher efficiencies of ECMs at part load than traditional AC motors.

2. ECMs are also available for fractional horsepower pumps for heating, cooling, and domestic hot water applications. One pump manufacturer incorporates a variable speed drive into the pump motor to provide constant speed, constant differential pressure, or proportional differential pressure operation. Pump control is accomplished without the use of external sensors, but rather by the pump control system which continuously monitors the electric current draw of the motor and adjusts the motor speed accordingly based on the pump's hydraulic performance which is mapped into memory. Figures 41.13 and 41.14 illustrate this type of pump.

FIGURE 41.13   INTEGRAL VARIABLE SPEED DRIVE FOR FRACTIONAL HORSEPOWER PUMP. (*Xylem AWS*)

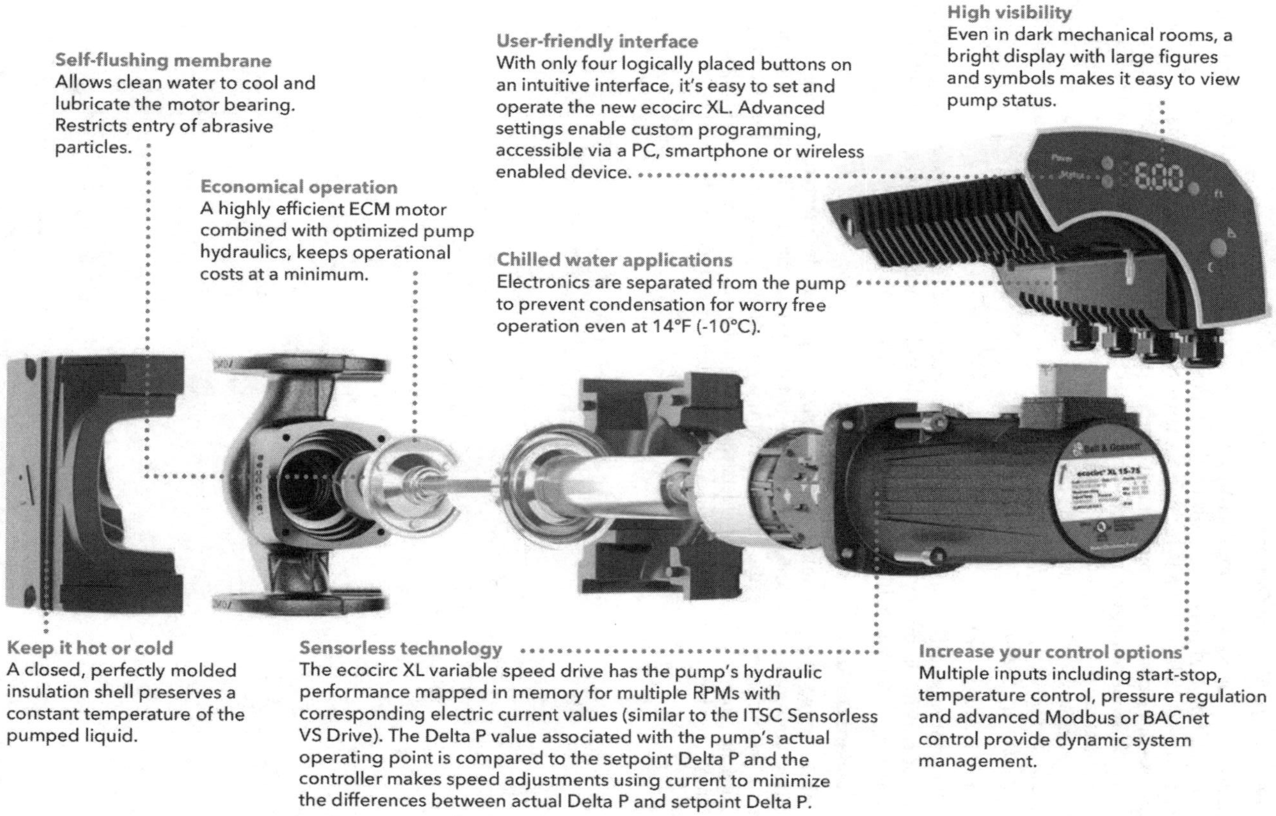

High visibility
Even in dark mechanical rooms, a bright display with large figures and symbols makes it easy to view pump status.

User-friendly interface
With only four logically placed buttons on an intuitive interface, it's easy to set and operate the new ecocirc XL. Advanced settings enable custom programming, accessible via a PC, smartphone or wireless enabled device.

Increase your control options
Multiple inputs including start-stop, temperature control, pressure regulation and advanced Modbus or BACnet control provide dynamic system management.

Chilled water applications
Electronics are separated from the pump to prevent condensation for worry free operation even at 14°F (-10°C).

Self-flushing membrane
Allows clean water to cool and lubricate the motor bearing. Restricts entry of abrasive particles.

Economical operation
A highly efficient ECM motor combined with optimized pump hydraulics, keeps operational costs at a minimum.

Sensorless technology
The ecocirc XL variable speed drive has the pump's hydraulic performance mapped in memory for multiple RPMs with corresponding electric current values (similar to the ITSC Sensorless VS Drive). The Delta P value associated with the pump's actual operating point is compared to the setpoint Delta P and the controller makes speed adjustments using current to minimize the differences between actual Delta P and setpoint Delta P.

Keep it hot or cold
A closed, perfectly molded insulation shell preserves a constant temperature of the pumped liquid.

FIGURE 41.14 EXPLODED VIEW OF PUMP, MOTOR, AND DRIVE. (*Xylem AWS*)

537

41.05 New Technologies for Small Packaged Rooftop Units

A. New Energy-Efficient Options

1. Many of the options previously available only for large packaged, and custom rooftop units (20 tons and larger) are now available for small packaged rooftop units in the 3 to 15 ton range. These options include: variable speed supply and exhaust fans (through the use of electronically-commutated motors), single-zone and multiple-zone variable air volume operation, variable speed condenser fans (through the use of electronically-commutated motors), electronic expansion valve, outdoor airflow monitoring and control, variable speed drives for the compressors, and an energy recovery wheel. These options significantly increase the energy efficiency of these units, particularly at part-load operation.

B. Other New Options

1. Other options that are now available in this size range include a modulating hot gas reheat coil for humidity control, and 2-inch and 4-inch filter racks for a filtration efficiency as high as MERV 14.

C. Standard Options

1. Other options that are not new for this size range include integrated airside economizer; communication to a building automation system through BACnet, LonWorks, or a manufacturer-specific communications protocol; natural gas, electric, or hot water heating coil; single-point electrical connection; unit-powered ground fault circuit interrupter-type receptacle; nonfused safety switch; and a field-erected roof curb.

PART 42

Plastic Piping Systems

42.01 Cross-Linked Polyethylene (PEX or XLPE)

A. Description

1. High density polyethylene plastic pipe cross-linked during or after extrusion of the piping. Cross-linking improves the temperature limitations of the material.

B. Uses

1. Domestic hot and cold water systems.
2. Radiant heating systems.
   a. PEX-AL-PEX—Aluminum tube laminated between interior and exterior layers of plastic.
3. Available in sizes from 1/4 in. to 4 in.
4. Red and blue piping used for domestic hot and cold water, respectively.

C. Advantages

1. Flexible.
2. Scale, chlorine, and corrosion resistant.
3. Fewer joints. PEX piping can be run continuously from source to outlet.
4. Less costly material than copper tubing.
5. Less costly installation than copper tubing.
6. Won't develop "pinhole" leaks.
7. Approved for use in domestic water piping systems in all 50 states of the United States as well as Canada.

D. Disadvantages

1. Costly crimping tools required for fittings.
2. Degradation from ultraviolet radiation.
3. Dezincification in yellow brass fittings having 30 percent zinc content. Red brass fittings are recommended which have 5 to 10 percent zinc content.

E. Temperature Limitations

1. Maximum operating temperature of 180°F.

F. Pressure Limitations

1. 100 psi maximum nonshock operating pressure at 180°F.

42.02 Polypropylene (PP)

A. Description

1. Chemically inert thermoplastic polymer.

B. Uses

1. Domestic hot and cold water systems.
2. High purity water systems.
3. Clean chemical processes.
4. Pharmaceutical operations.
5. Food processing.
6. Available in sizes from 1/2 in. to 6 in.

C. Advantages

1. Corrosion and chemical resistant.
2. Impact resistant.
3. Joined by heat fusion rather than gluing.

D. Disadvantages

1. Degradation from ultraviolet radiation.

E. Temperature Limitations

1. Maximum operating temperature of 180°F.

F. Pressure Limitations

1. 4" Schedule 80 PP pipe with thermo-seal joints:
   a. 160 psi maximum nonshock operating pressure at 73°F.
   b. 104 psi maximum nonshock operating pressure at 140°F.
2. Smaller pipe sizes have higher pressure ratings and larger pipe sizes have lower pressure ratings.

42.03 Polyvinyl Chloride (PVC)

A. Description

1. Thermoplastic polymer.

B. Uses

1. Sanitary (sewage) systems.
2. Intake and exhaust piping for high efficiency condensing water heaters.
3. Schedule 80 piping—chilled water systems.
4. Available in sizes from 1/4 in. to 12 in.

C. Advantages

1. Light weight.
2. Low cost.
3. Corrosion and chemical resistant.
4. Ease of joining.

D. Disadvantages

1. Brittle.

E. Temperature Limitations

1. Maximum operating temperature of 140°F.

F. Pressure Limitations

1. 12" Schedule 40 PVC pipe with solvent cemented joints:
   a. 130 psi maximum nonshock operating pressure at 73°F.
2. Smaller pipe sizes have higher pressure ratings and larger pipe sizes have lower pressure ratings.

42.04 Chlorinated Polyvinylchloride (CPVC)

A. Description

1. Thermoplastic produced by chlorination of polyvinyl chloride.

B. Uses

1. Domestic hot and cold water systems.
2. Sprinkler systems.
3. Available in sizes from 1/4 in. to 12 in.

C. Advantages

1. Higher temperature limitation than PVC.

2. Low cost.
3. Corrosion and chemical resistant.
4. Ease of joining.
5. More ductile than PVC.

D. Disadvantages

1. Degradation from ultraviolet radiation.
2. Brittle.

E. Temperature Limitations

1. Maximum operating temperature of 210°F.

F. Pressure Limitations

1. 6" Schedule 40 CPVC pipe with solvent cemented joints:
   a. 180 psi maximum nonshock operating pressure at 73°F
   b. 99 psi maximum nonshock operating pressure at 140°F.
2. Smaller pipe sizes have higher pressure ratings and larger pipe sizes have lower pressure ratings.

42.05 High-Density Polyethylene (HDPE)

A. Description

1. Linear polymer prepared from ethylene by a catalytic process.

B. Uses

1. Geothermal systems.
2. Irrigation systems.
3. Sprinkler systems.
4. Industrial systems.
5. Available in sizes from 3/4 in. to 24 in. Available in coils through 6".

C. Advantages

1. Nontoxic.
2. Corrosion, abrasion, and chemical resistant.
3. Durable and light weight.
4. Carbon black is added to provide ultraviolet protection.
5. Joined by heat fusion rather than gluing.

D. Disadvantages

1. Non-UV stabilized pipe degrades from ultraviolet radiation.

E. Temperature Limitations

1. Maximum operating temperature of 122°F.

F. Pressure Limitations

1. 4" Type IV HDPE pipe with butt fused joints:
   a. 232 psi maximum nonshock operating pressure at 68°F.
   b. 93 psi maximum nonshock operating pressure at 122°F.
2. Larger pipe sizes have lower pressure ratings.

Noise and Vibration Control

43.01 Noise Control

A. Indoor Noise Control

1. Mechanical equipment rooms.
   a. Locate mechanical equipment rooms away from noise-sensitive areas.
   b. Design spaces such as corridors and storage rooms around mechanical equipment rooms that can be used as buffer zones for noise control.
   c. Design walls of mechanical equipment rooms to be constructed of concrete masonry units.
   d. Design mechanical chases terminating in mechanical equipment rooms to be closed at the mechanical equipment room.
   e. Specify sleeves for all duct and pipe penetrations of mechanical equipment room walls, floors, ceilings, and mechanical chase enclosures. The gap between the pipe or duct and the sleeve should be packed with an appropriate material and caulked.
   f. Specify rubber flexible pipe connectors for piping connections to pumps and chillers.
   g. Specify noise suppressors for steam pressure reducing valves where recommended by the manufacturer.
2. Fans and air handling units.
   a. Design flexible duct connectors for all duct connections to fans and air handling units.
   b. Design duct silencers (sound attenuators) for supply and return air duct connections to air handling units.
   c. Consider the use of round or oval ductwork for the first 20 feet of ductwork. This will reduce low frequency breakout noise from the equipment.
   d. Keep the aspect ratio of the ductwork near the equipment as low as possible.
   e. Select fans at their point of maximum efficiency.
   f. Minimize the system effect of ductwork connected to fans and air handling units.
   g. For VAV systems, select fans for maximum efficiency at 70 to 80 percent of maximum airflow, which is the airflow at which the system will operate most of the time.
   h. Locate equipment away from noise-sensitive areas inside the building.
   i. Where rooftop units need to be located over noise-sensitive areas, design concrete pads under the units and/or design acoustical material in the areas directly beneath the condensing sections. Only the supply and return air ducts for each rooftop unit should penetrate the acoustical material and roof deck within the curb perimeter. Openings around the supply and return air duct penetrations should be sealed once the ducts are installed.
3. Ductwork.
   a. Do not exceed airflow velocities of: 950 fpm for ducts within occupied spaces, 1,200 fpm for ducts above suspended ceilings, 1,700 fpm for ducts in shafts, and 2,000 fpm for supply air ducts upstream of VAV terminal units.
   b. Maintain at least four to five duct diameters of straight duct between duct fittings, such as elbows and branch takeoffs.
   c. Specify turning vanes for all duct elbows.
   d. Design open-end ductwork for a maximum of 500 fpm airflow velocity through the opening.
   e. Design open-end return air ductwork to have at least one elbow or tee between the opening and the air handling unit.
   f. Design transfer air ducts to be sound-lined and have a Z-configuration.
4. VAV terminal units.
   a. Locate fan-powered VAV terminal units over areas that are not noise-sensitive, such as corridors and storage rooms.
   b. For most applications, select VAV terminal units for a maximum noise criterion of NC 30 (incorporating sound attenuating effects from the room).

c. Design 5 feet of sound-lined ductwork or specify duct silencers downstream of VAV terminal units.

5. Diffusers, registers, and grilles.

   a. For most applications, select diffusers, registers, and grilles for a maximum noise criterion of NC 30 (incorporating sound attenuating effects from the room).

   b. Do not exceed airflow velocities at the neck of supply and return air diffusers, registers, and grilles of: 425 fpm for supply air outlets, and 500 fpm for return and exhaust air inlets.

   c. Design insulated flexible ducts for final connections to diffusers, registers, and grilles installed above suspended ceilings. Maximum length of flexible ducts should be 8 feet. Keep flexible ducts straight with long radius bends. Avoid abrupt bend at connection to diffusers, registers, and grilles.

   d. Design branch takeoffs from duct mains that are at least 5 feet long for final connections to diffusers, registers, and grilles. Avoid direct connections of diffusers, registers, and grilles to duct mains.

   e. Locate balancing dampers at least 5 feet away from diffusers, registers, and grilles.

   f. Do not use dampers at the neck of diffusers and registers for significant throttling of airflow.

B. Outdoor Noise Control

1. Outdoor equipment.

   a. Locate equipment away from noise-sensitive areas outside of the building.

   b. Specify manufacturer-furnished sound attenuating devices, such as compressor enclosures and oversized condenser fans, when available.

   c. Design sound barriers around equipment, maintaining all required clearances between equipment and barrier.

   d. Ensure that noise at the property lines generated by mechanical equipment (usually expressed in terms of A-weighted sound pressure [dBA]) is below the daytime and nighttime limits of the local noise ordinance.

43.02 Vibration Control

A. Mechanical Equipment

1. Specify vibration isolation for all reciprocating and rotating equipment connected to the structure.

   a. Specify vibration isolation hangers for all equipment that is suspended from the building structure.

   b. Specify vibration isolators and bases for all floor-mounted equipment. Refer to the 2011 *ASHRAE Handbook—HVAC Applications*, Chap. 48, Table 47, for complete information on recommended vibration isolation for floor-mounted equipment.

   c. For roofs and floors constructed with open web joists or any unusually light construction, ensure that the isolator deflection is at least 15 times the deflection of the structure that is attributed to the mechanical equipment.

   d. Avoid both internal and external vibration isolation of air handling units.

2. Specify vibration isolation pipe hangers for all piping in mechanical equipment rooms and for piping within 50 feet of vibration-isolated equipment.

3. Specify vibration isolation roof curbs for rooftop units located above noise-sensitive areas.

B. Piping Risers

1. Locate pipe risers away from noise sensitive areas so that pipe anchors and guides rigidly attached to the building structure to accommodate pipe expansion do not need to be vibration isolated.

2. Completely spring-isolated piping riser systems must be carefully designed to avoid overstressing the piping due to movements in the riser and branch takeoffs.

43.03 Sound Information

VELOCITY OF SOUND IN VARIOUS MEDIA

Medium	Velocity	
	Feet per Second	Miles per Hour
Rubber	310	211
Air	1,130	770
Water Vapor	1,328	905
Cork	1,640	1,118
Lead	4,026	2,745
Water	4,625	3,153
Wood	10,825	7,380
Brass	11,480	7,827
Copper	11,670	7,957
Brick	11,800	8,045
Concrete	12,100	8,250
Wood	12,500	8,523
Steel and Iron	16,000	10,909
Glass	16,400	11,181
Aluminum	19,000	12,955

VOICE LEVEL COMPARISON AT VARIOUS DISTANCES

Distance Feet	Normal Voice Level dB	Raised Voice Level dB	Very Loud Voice dB	Shouting Voice dB
1	70	76	82	88
3	60	66	72	78
6	54	60	66	72
12	48	54	60	66
24	42	48	54	60

DIRECTIONAL EFFECT ON SOUND

Direction of Sound Source with Respect to Listener	Decrease in Speech Energy
Face to Face	0 dB
30 Degree Rotation Away	1.5
60 Degree Rotation Away	3.0
90 Degree Rotation Away	4.5
120 Degree Rotation Away	6.0
150 Degree Rotation Away	7.5
180 Degree Rotation Away Source Turned Away from Listener	9.0

TYPICAL SOUND LEVELS

Pressure Level dB	Typical Sound	Subjective Impression
150	Jet plane take-off	Short exposure can cause hearing loss.
140	Military jet take-off at 100 ft.	
130	Artillery fire at 10 ft. Machine gun	Deafening (threshold of pain)
120	Siren at 100 ft. Jet plane (passenger ramp) Thunder Sonic boom	

(Continued)

TYPICAL SOUND LEVELS (*Continued*)

Pressure Level dB	Typical Sound	Subjective Impression
110	Wood working shop Accelerating motorcycle Hard rock band 75-piece orchestra	Threshold of discomfort
100	Subway (steel wheels) Propeller plane, outboard motor Loud street noise Power lawn mower	Very loud
90	Truck unmuffled Train whistle Kitchen blender Pneumatic jackhammer Shouting at 5 ft.	
80	Printing press Subway (rubber wheels) Noisy office Computer printout room Average factory	Loud Intolerable for phone use
70	Average street noise Quiet typewriter Freight train at 100 ft. Average radio Speech at 3 ft.	Loud
60	Noisy home Average office Normal conversation at 3 ft.	Loud Unusual background
50	General office Quiet office Quiet radio, window AC unit Average home Quiet street	Moderate
40	Private office Quiet home/residential area	Moderate
30	Quiet conversation Broadcast studio	Noticeably quiet
20	Empty auditorium Whisper Watch ticking Buzzing inset at 3 ft. Rural ambient	Very quiet
10	Rustling leaves Soundproof room	Very faint Threshold of good hearing
0	Human breathing	Intolerably quiet Threshold of audibility (youthful hearing)

TYPICAL NOISE LEVELS

Equipment	dBA
Saturn rocket	200
Turbo jet engine	170
Jet plane/aircraft at take-off, inside jet engine test cell	150
Turbo propeller plane at take-off, military jet take-off at 100 ft.	140
Large pipe organ, artillery fire at 10 ft., machine gun	130
Jolt squeeze hammer	122
Small aircraft engine, siren at 100 ft., jet plane (passenger ramp), thunder, sonic boom, threshold of feeling (pain)	120
Blaring radio, wood working shop, accelerating motorcycle, hard rock band, 75-piece orchestra, chain saw	110

(*Continued*)

TYPICAL NOISE LEVELS (*Continued*)

Equipment	dBA
Vacuum pump, large air compressor	108
Positive displacement blower, air hammer	107
Magnetic drill press, air chisel, high-pressure gas leak	106
Banging of steel plate, wood planer	104
Air compressor, automobile at highway speed, subway (steel wheels), propeller plane, outboard motor, loud street noise, power lawn mower, helicopter	100
Turbine condenser, welder, punch press, riveter, power saws, plastic chipper	98
Small air compressor, airplane cabin normal flight	94
Heavy duty grinder	93
Heavy diesel powered vehicle, spinning machines-looms, noisy street	92
Voice, shouting, truck unmuffled, train whistle, kitchen blender, pneumatic jackhammer, shouting at 5 ft., noisy factory, blender	90
Printing press, inside average rail road car, toilet flushing	86
Garbage disposal, printing press, subway (rubber wheels), noisy office, computer printout room, average factory, lathe, police whistle, telephone ring, clothes washer, dish washer, TV–loud	80
Voice—conversational level, average street noise, quiet typewriter, freight train at 100 ft., average radio, speech at 3 ft., inside average automobile, clothes dryer, vacuum cleaner, TV–soft	70
Electronic equipment ventilation fan, noisy home, average office, normal conversation at 3 ft., hair dryer	60
Office air diffuser, general office, quiet office, quiet radio, window AC unit, average home, quiet street	50
Small electric clock, private office, quiet home/residential area, refrigerator, bird singing, wilderness ambient, agricultural land	40
Voice, soft whisper, quiet conversation, broadcast studio	30
Rustling leaves, empty auditorium, whisper, watch ticking, buzzing inset at 3 ft., rural ambient	20
Human breath, sound proof room, rustling leaves	10
Threshold of hearing	0

SUBJECTIVE EFFECT OF CHANGES IN SOUND CHARACTERISTICS

Change in Sound Pressure Level	Change in Apparent Loudness
1 dB	Insignificant
3 dB	Just perceptible
5 dB	Clearly noticeable
10 dB	Twice or half as loud
15 dB	Significant change
20 dB	Much louder or quieter

DECIBEL ADDITION

Difference between Two Levels dB	Add to Higher Level dB
0	3
1	2.5
2	2
3	2
4	1.5
5	1
6	1
7	1
8	0.5
9	0.5
10	0.5
More than 10	0

ACCEPTABLE HVAC NOISE LEVELS

Space Type	Recommended NC Level	Recommended RC Level	Equivalent Sound Level Meter Readings (A Scale) dB
Apartments	NC 25–35	RC 25–35	35–45
Assembly Halls	NC 25–30	RC 25–30	35–40
Churches	NC 30–35	RC 30–35	40–45
Concert and Recital Halls	NC 15–20	RC 15–20	25–30
Courtrooms	NC 30–40	RC 30–40	40–50
Factories	NC 40–65	RC 40–65	50–75
Hospitals and Clinics			
Private Rooms	NC 25–30	RC 25–30	35–40
Wards	NC 30–35	RC 30–35	40–45
Operating Rooms	NC 25–30	RC 25–30	35–40
Laboratories	NC 35–40	RC 35–40	45–50
Corridors	NC 30–35	RC 30–35	40–45
Public Areas	NC 35–40	RC 35–40	45–50
Hotels/Motels			
Individual Rooms/Suites	NC 25–35	RC 25–35	35–45
Meeting/Banquet Rooms	NC 25–35	RC 25–35	35–45
Halls/Corridors/Lobbies	NC 35–40	RC 35–40	45–50
Service/Support Areas	NC 40–45	RC 40–45	50–55
Legitimate Theaters	NC 20–25	RC 20–25	30–35
Libraries	NC 30–40	RC 30–40	40–50
Music Rooms	NC 20–25	RC 20–25	30–35
Movie/Motion Picture Theaters	NC 30–35	RC 30–35	40–45
Offices			
Executive	NC 25–30	RC 25–30	35–40
Conference Rooms	NC 25–30	RC 25–30	35–40
Private	NC 30–35	RC 30–35	40–45
Open-Plan Offices/Areas	NC 35–40	RC 35–40	45–50
Business Mach/Computers	NC 40–45	RC 40–45	50–55
Public Circulation	NC 40–45	RC 40–45	50–55
Private Residences	NC 25–35	RC 25–35	35–45
Recording Studios	NC 15–20	RC 15–20	25–30
Restaurants	NC 40–45	RC 40–45	50–55
Retail Stores	NC 40–45	RC 40–45	50–55
Schools			
Lecture and Classrooms	NC 25–30	RC 25–30	35–40
Open-Plan Classrooms	NC 35–40	RC 35–40	45–50
Sports Coliseums	NC 45–55	RC 45–55	55–65
TV/Broadcast Studios	NC 15–25	RC 15–25	25–35

PART 44

Building Construction Business Fundamentals

44.01 Engineering/Construction Contracts

A. Methods of Obtaining Contracts

1. *Competitive Bidding Contracts.* Contracts in which engineers/contractors are selected on the basis of their competitive bids.
2. *Negotiated Contracts.* Contracts in which engineers/contractors are selected on the basis of ability, reputation, past experience with the owner, or type of project, etc., and fees are then negotiated.

B. Contract Types

1. *Lump Sum Contract.* A contract in which the engineer/contractor agrees to carry out the stipulated project for a fixed sum of money.
2. *Unit Price Contract.* A contract based on estimated quantities of adequately specified items of work, and the costs for these items of work are expressed in dollars per unit of work. For example, the unit of work may be dollars per foot of caisson drilled, dollars per cubic yard of rock excavated, or dollars per cubic yard of soil removed.
   a. This contract is generally only applicable to construction contracts.
   b. Unit price contracts are usually used when quantities of work cannot be accurately defined by the construction documents (driving piles, foundation excavation, rock excavation, contaminated soil removal, etc.). Unit prices may be included in part of a lump sum or other type of contract.
3. *Cost Plus Contracts.* A contract in which the owner reimburses the engineer/contractor for all costs incurred and compensates them for services rendered. Cost plus contracts are always negotiated. Compensation may be based on the following:
   a. Fixed percentage of the cost of the work (cost plus fixed percentage contract). Compensation is based on an agreed percentage of the cost.
   b. Sliding-scale percentage of the cost of the work (cost plus sliding-scale percentage contract). Compensation is based on an agreed sliding-scale percentage of the cost (federal income taxes are paid on an increasing sliding scale).
   c. Fixed fee (cost plus fixed fee contract). Compensation is based on an agreed fixed sum of money.
   d. Fixed fee with guaranteed maximum price (cost plus fixed fee with guaranteed maximum price contract). Compensation is based on an agreed fixed sum of money and the total cost will not exceed an agreed upon total project cost.
   e. Fixed fee with bonus (cost plus fixed fee with bonus contract). Compensation is based on an agreed fixed sum of money and an agreed upon bonus is established for completing the project ahead of schedule, under budget, for superior performance, etc.
   f. Fixed fee with guaranteed maximum price and bonus (cost plus fixed fee with guaranteed maximum price and bonus contract). Compensation is based on an agreed fixed sum of money, a guaranteed maximum price, and an agreed upon bonus is established for completing the project ahead of schedule, under budget, for superior performance, etc.
   g. Fixed fee with agreement for sharing any cost savings (cost plus fixed fee with agreement for sharing any cost savings contract). Compensation is based on an agreed upon fixed sum of money and an agreed upon method of sharing any cost savings.
   h. Other fixed fee contracts can be generated using variations on those listed earlier or by negotiating certain aspects particular to the project into a cost plus fixed fee contract with the owner.
4. *Incentive Contracts.* A contract in which the owner awards or penalizes the engineer/contractor for performance of work in accordance with an agreed upon target. The target is often project cost or project schedule.
5. *Liquidated Damages Contracts.* A contract in which the engineer/contractor is required to pay the owner an agreed upon sum of money in accordance with an agreed upon target. The target is often for each calendar day of delay in completion of the project.
   a. Liquidated damages, when included in the contract, must be a reasonable measure of the damages suffered by the owner due to delay in the completion of the project

to be enforceable in a court of law. The owner must also be able to demonstrate and prove the damages suffered due to delay in the completion of the project. Weather, strikes, contract changes, natural disasters, and other events beyond the control of the contractor can void the claim for liquidated damages.

6. *Percentage of Construction Fee Contracts.* A contract in which the engineer's fee is based on an agreed upon the percentage of the project's construction cost.

7. *Scope of Work.* The scope of work is part of the engineer's contract defining the engineer's responsibilities and work required to produce the contract documents required by the owner to get the project built. The engineer's scope of work can be compared to the Contract Documents defining a construction contract.

44.02 Building Construction Business Players

A. *Owner.* The individual (or individuals) who initiates the building design process (may be a business, corporation, developer, hospital, local government, municipality, state government, or federal government).

B. *Architect.* Design team member responsible for internal and external space planning, space sizes, relative location and interconnection of spaces, emergency egress, internal and external circulation, aesthetics, life safety, etc. Generally, the architect is the lead and the driving force behind the project.

C. *Civil Engineers.* The design team members responsible for site drainage, roadways, parking, site grading, site circulation, retaining walls, site utilities (sometimes done by the mechanical and electrical engineers), etc.

D. *Structural Engineers.* The design team members responsible for building structure (design of beams, columns, foundations, floors, roof). Responsible for making the building stand.

E. *Interior Designers.* The design team members responsible for building finishes (wall coverings, floor coverings, ceilings); often assist with, or are responsible for, space planning. Frequently, this is also done by the architect.

F. *Landscape Architect.* The design team member responsible for interior as well as external plantings (grass, shrubs, trees, flowers), etc.

G. *Surveyors.* Design team members responsible for establishing contours and site boundaries and locating existing benchmarks, trees, roads, water lines, sanitary and storm sewers, electric and telephone utilities, etc.

H. *Geologists/Soils Analysts.* Design team members responsible for establishing soil characteristics for foundation analysis, potential ground water problems, rock formations, etc.

I. *Transportation Engineer.* The design team member responsible for elevators, escalators, dumbwaiters, and other modes of vertical and/or horizontal transportation.

J. *Electrical Engineer.* The design team member responsible for the design of electrical distribution systems, lighting, powering mechanical and other equipment, receptacles, communication systems (telephone, intercom, paging), fire alarm and detection systems, site lighting, site electrical (or civil engineer), emergency power systems, uninterruptible power systems, security systems, etc.

K. Mechanical Engineers

1. *Plumbing Engineer.* The design team member responsible for water supply and distribution systems; sanitary, vent, and storm water systems; natural gas systems; medical and laboratory gas and drainage systems; underground storage tanks; plumbing fixtures; etc.

2. *Fire Protection Engineer*. The design team member responsible for sprinkler and other fire protection systems, standpipe and hose systems, fire pumps, site fire mains, fire extinguishers (sometimes fire extinguishers are designated by the architect), etc.
3. *HVAC Engineer*. The design team member responsible for the design of the heating, ventilating, and air conditioning systems; ductwork and piping systems; automatic temperature control systems; industrial ventilation systems; environmental control; indoor air quality; heat loss and heat gain within the building; human comfort; etc.

L. Contractors

1. *General Contractor*. Also referred to as prime contractor in single-contract construction projects. The general contractor is the construction team member responsible for construction of the building structure and foundations, building envelope, interior partitions, building finishes, roofing, site work, elevators, project schedule, project coordination, project management, etc. The general contractor may subcontract some or all of the work to other contractors. In single-contract projects, the general contractor is also responsible for mechanical and electrical work as well, but this work is most often done by subcontractors.
2. *Mechanical Contractor*. Also referred to as a subcontractor in single-contract construction projects. The mechanical contractor is the construction team member responsible for construction of the building HVAC, plumbing, and fire protection systems. The mechanical contractor may be broken into one, two, or three subcontracts for HVAC and plumbing and/or fire protection. The mechanical contractor may subcontract some or all of the work to other contractors (plumbing, sheet metal, fire protection, automatic temperature controls, etc.).
3. *Electrical Contractor*. Also referred to as a subcontractor in single-contract construction projects. The electrical contractor is the construction team member responsible for construction of the building electrical systems, fire alarm systems, communication systems, security systems, lighting systems, etc. The electrical contractor may subcontract some or all of the work to other contractors (communication, security fire alarm, etc.).
4. *Prime Contractor*. The contractor who signs a contract with the owner to perform the work.
5. *Multiple Prime Contractors*. When more than one contractor signs a contract with the owner to perform the work. Often this is accomplished with four prime contracts as follows, but may be done with any number of contracts:
   a. General contract.
   b. Mechanical (HVAC) contract.
   c. Plumbing/fire protection contract.
   d. Electrical contract.
6. *Subcontractor*. The contractor or contractors who sign a contract with the general or prime contractor to perform a particular portion of the prime contractor's work.
7. *Sub-Subcontractor*. The contractor or contractors who sign a contract with a subcontractor to perform a particular portion of the subcontractor's work.

PART **45**

Architectural, Structural, and Electrical Information

45.01 Ceiling Plenum Space Requirements

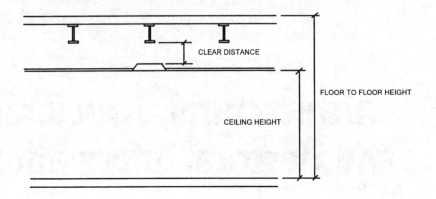

CLEAR DISTANCE

FLOOR TO FLOOR HEIGHT

CEILING HEIGHT

CEILING PLENUM SPACE

Floor to Floor	Ceiling Height	Clear Distance—Light to Beam in Inches									
		Beam Depth									
		12"	14"	16"	18"	21"	24"	27"	30"	33"	36"
9'0"	7'0"	*	*	*	*	*	*	*	*	*	*
	7'6"	*	*	*	*	*	*	*	*	*	*
	8'0"	*	*	*	*	*	*	*	*	*	*
	8'6"	*	*	*	*	*	*	*	*	*	*
	9'0"	*	*	*	*	*	*	*	*	*	*
10'0"	7'0"	10.5	8.5	6.5	4.5	1.5	*	*	*	*	*
	7'6"	4.5	2.5	*	*	*	*	*	*	*	*
	8'0"	*	*	*	*	*	*	*	*	*	*
	8'6"	*	*	*	*	*	*	*	*	*	*
	9'0"	*	*	*	*	*	*	*	*	*	*
11'0"	8'0"	10.5	8.5	6.5	4.5	1.5	*	*	*	*	*
	8'6"	4.5	2.5	*	*	*	*	*	*	*	*
	9'0"	*	*	*	*	*	*	*	*	*	*
	9'6"	*	*	*	*	*	*	*	*	*	*
	10'0"	*	*	*	*	*	*	*	*	*	*
	10'6"	*	*	*	*	*	*	*	*	*	*
12'0"	8'0"	22.5	20.5	18.5	16.5	13.5	10.5	7.5	4.5	1.5	*
	8'6"	16.5	14.5	12.5	10.5	7.5	4.5	1.5	*	*	*
	9'0"	10.5	8.5	6.5	4.5	1.5	*	*	*	*	*
	9'6"	4.5	2.5	0.5	*	*	*	*	*	*	*
	10'0"	*	*	*	*	*	*	*	*	*	*
	10'6"	*	*	*	*	*	*	*	*	*	*
13'0"	8'0"	34.5	32.5	30.5	28.5	25.5	22.5	19.5	16.5	13.5	10.5
	8'6"	28.5	26.5	24.5	22.5	19.5	16.5	13.5	10.5	7.5	4.5
	9'0"	22.5	20.5	18.5	16.5	13.5	10.5	7.5	4.5	1.5	*
	9'6"	16.5	14.5	12.5	10.5	7.5	4.5	1.5	*	*	*
	10'0"	10.5	8.5	6.5	4.5	1.5	*	*	*	*	*
	10'6"	4.5	2.5	0.5	*	*	*	*	*	*	*
14'0"	8'0"	46.5	44.5	42.5	40.5	37.5	34.5	31.5	28.5	25.5	22.5
	8'6"	40.5	38.5	36.5	34.5	31.5	28.5	25.5	22.5	19.5	16.5
	9'0"	34.5	32.5	30.5	28.5	25.5	22.5	19.5	16.5	13.5	10.5
	9'6"	28.5	26.5	24.5	22.5	19.5	16.5	13.5	10.5	7.5	4.5
	10'0"	22.5	20.5	18.5	16.5	13.5	10.5	7.5	4.5	1.5	*
	10'6"	16.5	14.5	12.5	10.5	7.5	4.5	1.5	*	*	*
	11'0"	10.5	8.5	6.5	4.5	1.5	*	*	*	*	*
	11'6"	4.5	2.5	0.5	*	*	*	*	*	*	*

(Continued)

CEILING PLENUM SPACE (*Continued*)

Floor to Floor	Ceiling Height	Clear Distance—Light to Beam in Inches									
		Beam Depth									
		12"	14"	16"	18"	21"	24"	27"	30"	33"	36"
	8'0"	58.5	56.5	54.5	52.5	49.5	46.5	43.5	40.5	37.5	34.5
	8'6"	52.5	50.5	48.5	46.5	43.5	40.5	37.5	34.5	31.5	28.5
	9'0"	46.5	44.5	42.5	40.5	37.5	34.5	31.5	28.5	25.5	22.5
	9'6"	40.5	38.5	36.5	34.5	31.5	28.5	25.5	22.5	19.5	16.5
15'0"	10'0"	34.5	32.5	30.5	28.5	25.5	22.5	19.5	16.5	13.5	10.5
	10'6"	28.5	26.5	24.5	22.5	19.5	16.5	13.5	10.5	7.5	4.5
	11'0"	22.5	20.5	18.5	16.5	13.5	10.5	7.5	4.5	1.5	*
	11'6"	16.5	14.5	12.5	10.5	7.5	4.5	1.5	*	*	*
	12'0"	10.5	8.5	6.5	4.5	1.5	*	*	*	*	*
	9'0"	106	104	102	100	97.5	94.5	91.5	88.5	85.5	82.5
	9'6"	100	98.5	96.5	94.5	91.5	88.5	85.5	82.5	79.5	76.5
	10'0"	94.5	92.5	90.5	88.5	85.5	62.5	79.5	76.5	73.5	70.5
20'0"	10'6"	88.5	86.5	84.5	82.5	79.5	76.5	73.5	70.5	67.5	64.5
	11'0"	82.5	80.5	78.5	76.5	73.5	70.5	67.5	64.5	61.5	58.5
	11'6"	76.5	74.5	72.5	70.5	67.5	64.5	61.5	58.5	55.5	52.5
	12'0"	70.5	68.5	66.5	64.5	61.5	58.5	55.5	52.5	49.5	46.5

Notes:
1 Assumptions: 2" fire proofing on beam, 6" fluorescent light depth, 5-1/2" floor slab thickness, 2" suspended ceiling thickness.
2 For depth from beam to finished ceiling, add 6" to the preceding figures.
3 For depth from underside of the slab to light, add depth of beam plus 2".
4 * Indicates a beam protruding through the ceiling.

45.02 Building Structural Systems

A. Standard Nominal Structural Steel Depths

1. W-Shapes (Wide Flange Beams): 4, 5, 6, 8, 10, 12, 14, 16, 18, 21, 24, 27, 30, 33, 36, 40, 44.
2. S-Shapes (I beams): 3, 4, 5, 6, 7, 8, 10, 12, 15, 18, 20, 24.
3. C-Shapes (Channels): 3, 4, 5, 6, 7, 8, 9, 10, 12, 15.

B. Standard Nominal Joist Depths as Manufactured by Vulcraft

1. K-Series: 8, 10, 12, 14, 16, 18, 20, 22, 24, 26, 28, 30.
2. LH-Series and DLH-Series: 18, 20, 24, 28, 32, 36, 40, 44, 48, 52, 56, 60, 64, 68, 72, 84.

C. Building mechanical equipment support points should not deflect more than 0.33 in. for cooling towers and no more than 0.25 in. for all other mechanical equipment.

D. Maximum duct and pipe sizes that may pass through steel joists are given in the following table:

Joist Depth	Round Duct or Pipe Size	Square Duct Size	Rectangle Duct Size
8"	5"	4×4	3×8
10"	6"	5×5	3×8
12"	7"	6×6	4×9
14"	8"	6×6	5×9
16"	9"	7×7	6×10
18"	11"	8×8	7×11
20"	11"	9×9	7×12
22"	12"	9×9	8×12
24"	13"	10×10	8×13
26"	15"	12×12	9×18
28"	16"	13×13	9×18
30"	17"	14×14	10×18

Notes:
1 Table based on Vulcraft K Series joists. For LH or DLH Series joists, consult with Vulcraft.
2 The preceding values are maximum sizes. The designer must consider duct insulation or duct liner thickness.
3 Do not recommend running ductwork through joists or between joists because it generally becomes a problem in the field. If you must run ductwork through joists or between joists, notify the structural engineer and verify the locations of joist bridging.

E. Floor Span vs. Structural Member Depths is given in the following table:

Floor—Structural Member Depth (1)							
	Structural Steel Shapes				Structural Steel Joists		
Structural Member Span	Beams		Girders		Joists (9)		Joists Girders	
	Min. (2,4)	Max. (3,4,8)	Min. (2,5,7)	Max. (3,5,8)	Min. (2,4,6)	Max. (3,6)	Min. (2,5)	Max. (3,5)
20 ft.	10"	14"	16"	24"	12"	14"	18"	28"
30 ft.	16"	18"	21"	33"	16"	24"	20"	40"
40 ft.	21"	24"	24"	36"	20"	24"	24"	52"
50 ft.	N/A	N/A	N/A	N/A	N/A	N/A	N/A	N/A
60 ft.	N/A	N/A	N/A	N/A	N/A	N/A	N/A	N/A

Notes:
1 Floor spans generally do not exceed 40 ft.
2 Assumed Floor Dead Load (DL) = 50 psf; Live Load (LL) = 50 psf.
3 Assumed Floor Dead Load (DL) = 50 psf; Live Load (LL) = 150 psf.
4 Assumed Spacing = ± 5'0".
5 Assumed Spacing = ± 30'0".
6 Assumed Spacing = ± 2'0".
7 Assumed Steel Grade 50 ksi.
8 Assumed Steel Grade 36 ksi.
9 K Series Joists for 20' and 30' spans; LH Series for 40' spans.
10 Rule of Thumb: Beam and joist depths (in inches) are approximately 1/2 the length of the span (in feet).
11 Rule of Thumb: Girder and joist girder depths (in inches) are approximately 3/4 the length of the span (in feet).

F. Roof Span vs. Structural Member Depths is given in the following table:

Roof—Structural Member Depth							
	Structural Steel Shapes				Structural Steel Joists		
Structural Member Span	Beams		Girders		Joists (7)		Joists Girders	
	Min. (1,3)	Max. (2,3)	Min. (1,4,5)	Max. (2,4,6)	Min. (1,3)	Max. (2,3)	Min. (1,4)	Max. (2,4)
20 ft.	8"	10"	10"	18"	12"	14"	18"	28"
30 ft.	14"	16"	16"	24"	16"	20"	20"	40"
40 ft.	18"	21"	21"	30"	20"	24"	24"	52"
50 ft.	N/A	N/A	27"	36"	28"	32"	32"	64"
60 ft.	N/A	N/A	30"	36"	32"	36"	44"	84"

Notes:
1 Assumed Roof Dead Load (DL) = 20 psf; Live Load (LL) = 20 psf.
2 Assumed Roof Dead Load (DL) = 35 psf; Live Load (LL) = 50 psf.
3 Assumed Spacing = ± 5'0".
4 Assumed Spacing = ± 30'0".
5 Assumed Steel Grade 50 ksi.
6 Assumed Steel Grade 36 ksi.
7 K Series Joists for 20' and 30' spans; LH Series for 40', 50', and 60' spans.

45.03 Architectural and Structural Information

A. Equipment Weights. Provide equipment weights, sizes, and locations to the architect and structural engineer. The architect does not normally need the weights of equipment, but this information is needed by the structural engineer.

Obtain weights and sizes from the manufacturers' catalogs or the manufacturers' representatives. Equipment weights should include the following information at minimum.

1. Item designation.
2. Location.
3. Size—length, width, height—include curb height if required.
4. Weight. Operating weight if substantially different from the installed weight.
5. Floor/roof openings. Wall openings if load bearing or shear walls are used.
6. Special remarks.

B. **Ductwork Weight. Coordinate all ductwork with the structural engineer, especially when ductwork weight is 20 lbs./lf. or more. Provide ductwork weight and drawings showing the location of ductwork and sizes. See Part 17 for ductwork weight information.**

C. **Piping Weight. Coordinate all piping with the structural engineer, especially pipe sizes 6 in. and larger. Provide piping weight, location of anchors and forces, and drawings showing the location of piping and pipe sizes.**

Structural List

PROJECT STRUCTURAL LIST

PROJECT NAME: _____ OF _____ PROJECT NO. _____

SHEET NO. _____

SUBMITTAL   BY   DATE

PRELIMINARY _____ _____

FIRST _____ _____

SECOND _____ _____

THIRD _____ _____

FINAL _____ _____

ITEM NO.	ITEM DESIGNATION	LOCATION	SIZE			WEIGHT LBS.	ROOF/FLOOR OPENINGS		SPECIAL REMARKS
			LENGTH	WIDTH	HEIGHT		LENGTH	WIDTH	

45.04 Electrical Information

A. **Provide electrical information for all mechanical equipment requiring electrical power to the electrical engineer. Electrical information should include the following information at minimum.**

1. Item designation.
2. Location.
3. Voltage-phase-hertz.
4. Horsepower, full load amps, locked rotor amps, kW, minimum circuit amps: provide 1 or more.
5. Is equipment to be on emergency power?
6. Who provides the starter? Who provides the disconnect switch?
7. Control type, hand-off-automatic (HOA), manual, two-speed, etc.
8. Special requirements?

Electrical List

PROJECT ELECTRICAL LIST

PROJECT NAME: _____

SHEET NO. _____ OF _____ PROJECT NO. _____

SUBMITTAL   BY   DATE

PRELIMINARY   _____   _____

FIRST   _____   _____

SECOND   _____   _____

THIRD   _____   _____

FINAL   _____   _____

ITEM NO.	ITEM DESIGNATION	LOCATION	VOLTAGE/ PHASE	HP OR FLA	EMERG. POWER	STARTER BY	DISC. SW. BY	CONTROL TYPE	SPECIAL REQUIREMENTS

45.05 Mechanical/Electrical Equipment Space Requirements

A. Commercial Buildings

1. 18–20% of gross floor area. Most of the mechanical equipment is located indoors (i.e., no rooftop AHUs).
2. 1/4–1/3 of total building volume. This includes the ceiling plenum as mechanical/electrical space.

B. Hospital and Laboratory Buildings

1. 15–50% of gross floor area. Most of the mechanical equipment is located indoors (i.e., no rooftop AHUs).
2. 1/3–1/2 of total building volume. This includes the ceiling plenum as mechanical/electrical space.

C. The original building design should allow from 10 to 15 percent additional shaft space for future expansion and modification of the facility. This additional shaft space will also reduce the initial installation cost.

D. Minimum recommended clearance around the boilers and chillers is 36 in. The minimum recommended clearance around all other mechanical equipment is 24 in. Maintain minimum clearances for coil pull, tube pull, and the cleaning of tubes as recommended by the equipment manufacturer. This is generally equal to the length of the tubes and width of the piece of equipment. Maintain minimum clearance as required to open access and control doors on equipment for service, maintenance, and inspection.

E. Minimum recommended clearance between the top of the lights and the deepest structural member is 24 in.

F. Mechanical and electrical rooms should be centrally located to minimize ductwork, pipe, and conduit runs (size and length). Centrally locating mechanical and electrical spaces will minimize construction, maintenance, and operating costs. Additional space is quite often required when mechanical and electrical equipment rooms cannot be centrally located or when space requirements are fragmented throughout the building. In addition, centrally located equipment rooms will simplify distribution systems and will in some cases decrease above ceiling space requirements.

G. Mechanical rooms with fans and air handling equipment should have at least 10–15 sq. ft. of floor area for each 1,000 CFM of equipment air flow.

H. Mechanical rooms with refrigeration equipment must have an exit door that opens directly to the outside or through a vestibule type exit equipped with self-closing, tight-fitting doors.

I. Mechanical rooms must be clear of electrical rooms, elevators, and stairs on at least two sides, preferably on three sides.

J. Electrical rooms must be clear of elevators and stairs on at least two sides, preferably on three sides.

K. In general, mechanical equipment rooms require from 12–20 ft. clearance from the floor to the underside of the structure.

L. Mechanical and electrical shafts must be clear of elevators and stairs on at least two sides. Rectangular shafts with aspect ratios of 2:1 to 4:1 are easier to work mechanical and electrical distribution systems in and out of the shafts than square shafts.

M. The main electrical switchgear room should be located as close as possible to the incoming electrical service. If an emergency generator is required, the emergency generator room should be located adjacent to the main switchgear room to minimize electrical costs and interconnection problems. The emergency generator room should be located on an outside wall, preferably a corner location to enable proper ventilation, combustion air, and venting of engine exhaust.

N. A mechanical equipment room should be located on the first floor or basement floor to accommodate the incoming domestic water service main, the fire protection service mains, and the gas service. These service mains may include meter and regulator assemblies if these assemblies are not installed in meter vaults or outside the building. Consult your local utility company for service and meter/regulator assembly requirements.

O. The locations and placement of mechanical and electrical rooms must take into account how large pieces of equipment (chillers, boilers, cooling towers, transformers, and others) can be moved into and out of the building during initial installation and after construction for maintenance and repair and/or replacement.

45.06 Americans with Disabilities Act (ADA)

A. ADA Titles

1. Title I—Equal Employment Opportunity.
2. Title II—State and Local Governments.
3. Title III—Public Accommodations and Commercial Facilities.

B. Drinking Fountains

1. Where only one drinking fountain is provided on a floor, a drinking fountain with two bowls, one high bowl and one low bowl, is required.
2. Where more than one drinking fountain is provided on a floor, 50% shall be handicapped accessible and shall be on an accessible route.
3. Spouts shall be no higher than 36 in. above the finished floor or grade.
4. Spouts shall be located at the front of the unit and shall direct the water flow parallel or nearly parallel to the front of the unit.
5. Controls shall be mounted on the front or side of the unit.
6. Clearances:
   a. Knee space below the unit should be 27 in. high, 30 in. wide, and 17–19 in. deep, with a minimum front clear floor space of 30 in. × 48 in.
   b. Units without clear space below: 30-in. × 48-in. clearance is suitable for parallel approach.

C. Water Closets

1. The height of the water closet shall be 17–19 in. to the top of the toilet seat.
2. Flush controls shall be hand-operated or automatic. Controls shall be mounted on the wide side of toilet areas, and no more than 44 in. above the floor.
3. At least one toilet shall be handicapped accessible.

D. Urinals

1. Urinals shall be stall-type or wall hung with an elongated rim at a maximum of 17 in. above the floor.
2. Flush controls shall be hand-operated or automatic. Controls shall be mounted no more than 44 in. above the floor.
3. If urinals are provided, at least one shall be handicapped accessible.

E. Lavatories

1. Lavatories shall be mounted with the rim or counter surface no higher than 34 in. above the finished floor with a clearance of at least 29 in. to the bottom of the apron.

2. Hot water and drain pipe under lavatories shall be insulated or otherwise configured to protect against contact.
3. Faucets shall be lever-operated, push-type, or electronically controlled. Self-closing valves are acceptable, provided they remain open a minimum of 10 seconds.

F. Bathtubs

1. Bathtub controls shall be located toward the front half of the bathtub.
2. Shower units shall be provided with a hose at least 60 in. long that can be used both as a fixed shower head and a handheld shower head.

G. Shower Stalls

1. The shower controls shall be opposite the seat in a 36 in. × 36 in. shower stall and adjacent to the seat in a 30 in. × 60 in. shower stall.
2. Shower units shall be provided with a hose at least 60 in. long that can be used both as a fixed shower head and a handheld shower head.

H. Forward Reach

1. Maximum high forward reach: 48 in.
2. Minimum low forward reach: 15 in.

I. Side Reach

1. Maximum high side reach: 54 in.
2. Minimum low side reach: 9 in.

J. Areas of Rescue Assistance

1. A portion of a stairway landing within a smokeproof enclosure.
2. A portion of an exterior exit balcony located immediately adjacent to an exit stairway.
3. A portion of a 1-hour fire-resistive corridor located immediately adjacent to an exit enclosure.
4. A portion of a stairway landing within an exit enclosure that is vented to the exterior and is separated from the interior of the building with not less than 1-hour fire-resistive doors.
5. A vestibule located immediately adjacent to an exit enclosure and constructed to the same fire-resistive standards as required for corridors.
6. When approved by the authorities having jurisdiction, an area or room that is separated from other portions of the building by a smoke barrier.
7. An elevator lobby when elevator shafts and adjacent lobbies are pressurized as required for smokeproof enclosures by local regulations and when complying with the requirements herein for size, communication, and signage.
8. Size:
   a. Each area of rescue assistance shall have at least two accessible areas 30 × 48 minimum.
   b. Area shall not encroach on the exit width.
   c. The total number of areas per floor shall be one for every 200 persons. If the occupancy per floor is less than 200, the authorities having jurisdiction may reduce the number of areas to one.
9. A method of two-way communication, with both visible and audible signals, is required between the primary fire entry and the areas of rescue assistance.
10. Each area must be identified.

K. Stairway Width, 48 in. Between Handrails Minimum

L. Protruding Objects

1. Objects protruding from the wall with their leading edges between 27 and 80 in. above the finished floor shall protrude no more than 4 in. into walks, halls, corridors, passageways, or aisles.
2. Objects mounted with their leading edges at or below 27 in. above the finished floor may protrude any amount.

3. Protruding objects shall not reduce the clear width of an accessible route or maneuvering space.
4. Walks, halls, corridors, passageways, aisles, or other circulation spaces shall have 80 in. minimum clear head room.

M. Controls and Operating Mechanisms

1. The highest operable part of controls, dispensers, receptacles, and other operable equipment shall be placed within at least one of the reach ranges.
2. Electrical and communication system receptacles on walls shall be mounted no less than 15 in. above the floor.
3. Controls and operating mechanisms shall be operable with one hand and shall not require tight grasping, pinching, or twisting of the wrist. The force required to activate shall be no greater than 5 lbf.

Properties of Air

46.01 Thermodynamic Properties of Air/Water Vapor Mixtures

A. Psychrometric Definitions

1. Dry bulb temperature: The temperature of air read on a standard thermometer. Units: °F.DB. Symbol: T_{DB} or DB.
2. Wet bulb temperature: The wet bulb temperature is the temperature indicated by a thermometer whose bulb is covered by a wet wick and exposed to air moving at a velocity of 1,000 ft./min. Units: °F.WB. Symbol: T_{WB} or WB.
3. Humidity ratio: The weight of water vapor in each pound of dry air; also known as specific humidity. Units: lbs.H_2O/lbs.DA or Gr.H_2O/lbs.DA. Symbol: W.
4. Enthalpy: A thermodynamic property that serves as a measure of the heat content above some datum temperature (air 0°F.DB and water 32°F). Units: Btu/lbs.DA or Btu/lbs.H_2O. Symbol: h.
5. Specific volume: The cubic feet of air/water mixture per pound of dry air. Units: cu.ft./lbs.DA. Symbol: SpV.
6. Dewpoint temperature. The temperature at which moisture will start to condense from the air. Units: °F.DP. Symbol: T_{DP} or DP.
7. Relative humidity: The ratio of water vapor in the air/water mixture to the water vapor in saturated air/water mixture. Units: %RH. Symbol: RH.
8. Sensible heat: Heat that causes a rise in temperature. Units: Btu/h. Symbol: H_S.
9. Latent heat: Heat that causes a change in state (e.g., liquid water to gaseous water). Units: Btu/h. Symbol: H_L.
10. Total heat: Sum of sensible heat and latent heat. Units: Btu/h. Symbol: H_T.
11. Sensible heat ratio: The ratio of the sensible heat to the total heat. Units: None. Symbol: SHR.
12. Vapor pressure: Pressure exerted by water vapor in the air. Units: in. Hg. Symbol: P_W.
13. Standard barometric pressure: Pressure at sea level (29.921 in. Hg. = 14.7 psi).

B. Thermodynamic properties of air/water mixtures are given in the following table:

Temperature Range °F	Specific Heat Btu/lb. °F
−80–129	0.240
130–215	0.241
216–280	0.242
281–330	0.243
331–370	0.244
371–400	0.245
401–440	0.246
441–460	0.247
461–470	0.248
471–500	0.249

Refer to the online resource for thermodynamic and barometric properties of moist air, and physical properties of gases. www.mheducation.com/HVACequations

PART 47

Properties of Water

47.01 Properties of Water—Effects on Standard HVAC Water Equations

WATER EQUATION FACTORS

System Type	System Temperature Range °F	Equation Factor
Low Temperature (Glycol) Chilled Water	0–40	See Note 2
Chilled Water	40–60	500
Condenser Water Heat Pump Loop	60–110	500
Low Temperature Heating Water	110–150	490
	151–200	485
	201–250	480
Medium Temperature Heating Water	251–300	475
	301–350	470
High Temperature Heating Water	351–400	470
	401–450	470

Notes:
1 Water equation corrections for temperature, density, and specific heat.
2 For glycol system equation factors, see Part 20.

A. Water Equation Factor Derivations

1. Standard water conditions:
   a. Temperature: 60°F.
   b. Pressure: 14.7 psia (sea level)
   c. Density: 62.4 lbs./ft.^{3}
2. Water equation examples:
   $H = m \times c_w \times \Delta T$

   Water @ 250°F
   $c_w = 1.02$ Btu/~~Lb H$_2$O~~ °F $\times$ 62.4 ~~Lbs.H$_2$O/ft.~~$^3 \times 1.0$ ~~ft.~~^{3} / 7.48052 gal. $\times$ 60 min./h $\times$ 0.94 (SG)
   $= 480$ Btu min./h °F gal.
   $H_{250F} = 480$ Btu min./h °F Gal. $\times$ GPM (gal./min.) $\times \Delta T$ (°F)
   $H_{250F} = 480 \times$ GPM $\times \Delta T$ (°F)

   Water @ 450°F
   $c_w = 1.13$ Btu/~~Lb H$_2$O~~ °F $\times$ 62.4 ~~Lbs.H$_2$O/ft.~~$^3 \times 1.0$ ~~ft.~~^{3}/7.48052 gal. $\times$ 60 min./h $\times$ 0.83 (SG)
   $= 470$ Btu min./h °F gal.
   $H_{450F} = 470$ Btu min./h °F gal. $\times$ GPM (gal./min.) $\times \Delta T$ (°F)
   $H_{450F} = 470 \times$ GPM $\times \Delta T \cdot$(°F)

Refer to the online resource for Section 47.02 Thermodynamic Properties of Water.

Cleanroom Criteria

48.01 Airborne Contaminants

A. Particle Classifications

1. Fine   <2.5 microns
2. Course   2.5 microns
3. Respirable   <10.0 microns
4. Nonrespirable   10.0 microns

B. Relative Sizes

1. Micron = 1 millionth of a meter (0.000001 meter) = 39 millionths of an in. (0.000039 in.)
2. Visible to the naked eye: 25 microns
3. Human hair: 100 microns
4. Dust: 25 microns
5. Optical microscope: 0.25 microns
6. Scanning electron microscope: 0.002 microns
7. Macro particle range   25 microns and larger
8. Micro particle range   1.0–25 microns
9. Molecular macro range   0.085–1.0 microns
10. Molecular range   0.002–0.085 microns
11. Ionic range   0.002 microns and smaller

C. Airborne particle sizes are given in the following table:

AIRBORNE PARTICLE SIZE TABLE

Particle	Particle Size Microns	Particle	Particle Size Microns
Plant			
Pollen	10–100	Tea dust	8–300
Spanish moss pollen	150–750	Grain dusts	5–1,000+
Mold	3–12	Sawdust	30–600
Spores	3–40	Corn starch	0.09–0.75
Starches	3–100	Pudding mix	3–160
Milled flour	1–100	Cayenne pepper	15–1,000
Milled corn	1–100	Snuff	3–30
Mustard	6–10	Textile fibers	8–1,000+
Ginger	25–40	Corn cob chaff	30–100
Coffee	5–400	Carbon black	0.2–10
Coffee roast soot	0.6–3.5	Channel black	0.2–100
Animal			
Bacteria	0.3–60	Human hair	60–600
Viruses	0.005–0.1	Hair	5–200
Dust mites	100–300	Red blood cells	5–10
Spider web	2.5	Liquid droplets:	–
Disintegrated feces	0.8–1.5	sneezed	0.5–5
Feces	10–45	Bone dust	3–350
Combustion			
Combustion	0.01–0.1	Smoke particles:	–
Tobacco smoke	0.01–4.5	natural materials	0.01–0.1
Burning wood	0.2–3	synthetic materials	1–50
Rosin smoke	0.01–1	Smoldering	–
Coal flue gas	0.08–0.2	cooking oil	0.3–0.9
Oil smoke	0.03–1	Flaming cooking oil	0.3–0.9
		Auto emissions	1–150
Fly ash	0.9–1000		

(Continued)

AIRBORNE PARTICLE SIZE TABLE (*Continued*)

Particle	Particle Size Microns	Particle	Particle Size Microns
Mineral			
Asbestos	0.7–90	Carbon dust	0.25–5
Cement dust	3–100	Carbon dust-graphite	0.02–2
Coal dust	1–100	Fertilizer	10–1,000
Sea salt	0.035–0.5	Ground limestone	10–1,000
Textiles	6–20	Lead	0.1–0.7
Clay	0.1–50	Bromine	0.1–0.7
Calcium, zinc	0.7–20	Glass wool	1,000
Iron	4–20	Fiberglass	8
Lead dust	2	Insulation	1–1,000
Talc	0.5–50	Metallurgical dust	0.1–1,000
NH_3Cl fumes	0.1–3	Metallurgical fumes	0.1–1,000
Other			
Atmospheric dust	0.001–40	Yeast cells	2–75
Lung damaging dust	0.6–7	Sugars	0.0008–0.005
Mist	70–350	Gelatin	5–90
Oxygen	0.00050	Beach sand	100–10,000
Carbon dioxide	0.00065	Copier toner	0.5–15
Atomic radii	0.0001–0.001	Fabric protector	2.5–5
Air freshener	0.2–2	Face powder	0.1–30
Hairspray	3–7	Lint	10–90
Spray paint	8–10	Humidifier	0.9–3
Antiperspirant	6–10	Artificial textile	–
Dusting aid	6–15	fibers	10–30
Paint pigments	0.1–5	Insecticide dusts	0.5–10

D. Cleanroom Definitions

1. A *clean zone* is a defined space in which the concentration of airborne particles is controlled to meet a specified airborne particulate cleanliness class.
2. A *cleanroom* is a room in which the concentration of airborne particles is controlled and which contains one or more clean zones.
   a. An as-built cleanroom is a cleanroom complete and ready for operation, certifiable, with all services connected and functional, but without equipment or operating personnel in the facility.
   b. An at-rest cleanroom is a cleanroom that is complete, with all services functioning and with equipment installed and operable or operating, as specified, but without operating personnel in the facility.
   c. An operational cleanroom is a cleanroom in normal operation, with all services functioning and with equipment and personnel, if applicable, present and performing their normal work functions in the facility.

48.02 Cleanroom Class Designations: FED-STD-209E

CLEANROOM CLASS DESIGNATIONS

Cleanroom Class Name		Class Limits								
		0.1 μm		0.2 μm		0.3 μm		0.5 μm		5 μm	
		Volume Units		Volume Units		Volume Units		Volume Units		Volume Units	
SI	English	M^3	Ft.^{3}	M^3	Ft.^{3}	M^3	Ft.^{3}	M^3	Ft.^{3}	M^3	Ft^3
M1		350	9.91	75.7	2.14	30.9	0.875	10.0	0.283	–	–
M1.5	1	1,240	35.0	265	7.50	106	3.00	35.3	1.00	–	–
M2		3,500	99.1	757	21.4	309	8.75	100	2.83	–	–
M2.5	10	12,400	350	2,650	75.0	1,060	30.0	353	10.0	–	–

(*Continued*)

CLEANROOM CLASS DESIGNATIONS (*Continued*)

Cleanroom Class Name		0.1 µm Volume Units		0.2 µm Volume Units		0.3 µm Volume Units		0.5 µm Volume Units		5 µm Volume Units	
						Class Limits					
M3		35,000	991	7,570	214	3,090	87.5	1,000	28.3	–	–
M3.5	100	–	–	26,500	750	10,600	300	3,530	100	–	–
M4		–	–	75,700	2,140	30,900	875	10,000	283	–	–
M4.5	1,000	–	–	–	–	–	–	35,300	1,000	247	7.00
M5		–	–	–	–	–	–	100,000	2,830	618	17.5
M5.5	10,000	–	–	–	–	–	–	353,000	10,000	2,470	70.0
M6		–	–	–	–	–	–	1,000,000	28,300	6,180	175
M6.5	100,000	–	–	–	–	–	–	3,530,000	100,000	24,700	700
M7		–	–	–	–	–	–	10,000,000	283,000	61,800	1,750

Notes:
1 Federal Standard 209E is obsolete and superseded by the International Organization Standard ISO 14644.
2 Federal Standard 209E information provided for comparison purposes only.

48.03 Cleanroom Class Designations: ISO Standard 14644-1

CLEANROOM CLASS DESIGNATIONS

ISO Class	Maximum Number of Particles in the Air (Particles in Each Cubic Meter Equal to or Greater than the Specified Particle Size)					
	Particle Size					
	>0.1 µm	>0.2 µm	>0.3 µm	>0.5 µm	>1.0 µm	>5.0 µm
ISO Class 1	10	2	0	0	0	0
ISO Class 2	100	24	10	4	0	0
ISO Class 3	1,000	237	102	35	8	0
ISO Class 4	10,000	2,370	1,020	352	83	0
ISO Class 5	100,000	23,700	10,200	3,520	832	29
ISO Class 6	1,000,000	237,000	102,000	35,200	8,320	293
ISO Class 7				352,000	83,200	2,930
ISO Class 8				3,520,000	832,000	29,300
ISO Class 9				35,200,000	8,320,000	293,000

Notes:
1 Cleanrooms are maintained virtually free of contaminants, such as dust or bacteria, are used in laboratory work, and in the production of precision parts for electronic or aerospace equipment.
2 In the cleanroom standard ISO 14644-1 *Classification of Air Cleanliness*, the classes are based on the formula:
$C_n = 10^N (0.1/D)^{2.08}$
where
C_n = the maximum permitted number of particles per cubic meter equal to or greater than the specified particle size, rounded to a whole number
N = the ISO Class number, which must be a multiple of 1 and be 9 or less
D = the particle size in micrometers
3 ISO Cleanroom Standards
ISO 14644-1 Classification of Air Cleanliness
ISO 14644-2 Cleanroom Testing for Compliance
ISO 14644-3 Methods for Evaluating and Measuring Cleanroom and Associated Controlled Environment
ISO 14644-4 Cleanroom Design and Construction
ISO 14644-5 Cleanroom Operations
ISO 14644-6 Terms, Definitions, and Units
ISO 14644-7 Enhanced Clean Devices
ISO 14644-8 Molecular Contamination
ISO 14698-1 Bio-contamination: Control General Principles
ISO 14698-2 Bio-contamination: Evaluation and Interpretation of Data
ISO 14698-3 Bio-contamination: Methodology for Measuring Efficiency of Cleaning Inert Surfaces

48.04 Cleanroom Design Criteria

CLEANROOM DESIGN CRITERIA

Cleanroom Design Criteria	Federal Standard 209e Classifications English/Metric					
	1	10	100	1,000	10,000	100,000
	M1.5	M2.5	M3.5	M4.5	M5.5	M6.5
Circulation Rate AC/h (8)	360–540	360–540	210–540	120–300	30–120	12–60
Room Air Velocity ft./min.	60–90	60–90	35–90 (1)	20–50	5–20	2–10
% Filter Coverage	100	100	50–100 (1)	25–60	10–40	5–20
Room Characteristics	Laminar	Laminar	Laminar/ non-laminar	Non-laminar	Non-laminar	Non-laminar
Unidirectional Flow	Yes	Yes	Yes/No	No	No	No
Parallelism Degrees (2)	10–35	10–35	10–35 N/A	N/A	N/A	N/A

Notes:

1 Velocity and filter coverage could be reduced possibly as low as 35 fpm and 50 percent coverage if parallelism requirements are relaxed by the client.
2 Parallelism requirements are often driven by a client's standard facility criteria.
3 Makeup air: 1–6 CFM/sq.ft.
4 Pressurization requirement: 1/4–1/2 CFM/sq.ft.
5 Temperature
  a. Range: 68–74°F
  b. Tolerance: ±0.1–±2.0°F
  c. Change rate: 0.75–2.0°F/h
  d. Example: 72°F, ±2.0°F
6 Relative humidity
  a. Range: 30–50 percent RH
  b. Tolerance: ±1.0–± 5.0 percent RH
  c. Change rate: 1.0–5.0 percent RH/h
  d. Example: 45 percent RH, ±5.0 percent RH
7 Fire protection/smoke purge exhaust: 3–5 CFM/sq.ft.
8 The air change rate is based on a 10'0" ceiling height.

Refer to the online resource for Sections 48.05 Areas and Circumferences of Circles, 48.06 Fraction/Decimal Equivalents, 48.07 Physical Properties of Fuels and Oils, and 48.08 U.S. Postal Service Abbreviations. www.mheducation.com/HVACequations

Wind Chill and Heat Index

49.01 Wind Chill Index

WIND CHILL INDEX

°F Dry Bulb	Wind Velocity (mph)										
	0 Calm	5	10	15	20	25	30	35	40	45	50
35	35	33	21	16	12	7	5	3	1	1	0
30	30	27	16	11	3	0	−2	−4	−5	−6	−7
25	25	21	9	1	−4	−7	−11	−13	−15	−17	−17
20	20	16	2	−6	−9	−15	−18	−20	−22	−24	−24
15	15	12	−2	−11	−17	−22	−26	−27	−29	−31	−31
10	10	7	−9	−18	−24	−29	−33	−35	−37	−38	−39
5	5	0	−15	−25	−32	−37	−41	−43	−45	−46	−47
0	0	−6	−22	−33	−40	−45	−49	−52	−53	−54	−56
−5	−5	−11	−27	−40	−46	−52	−56	−60	−62	−63	−63
−10	−10	−15	−31	−45	−52	−58	−63	−67	−69	−70	−70
−15	−15	−20	−38	−51	−60	−67	−70	−72	−76	−78	−79
−20	−20	−25	−45	−60	−68	−75	−78	−83	−87	−87	−88
−25	−25	−31	−52	−65	−76	−83	−87	−90	−94	−94	−96
−30	−30	−35	−58	−70	−81	−89	−94	−98	−101	−101	−103
−35	−35	−41	−64	−78	−88	−96	−101	−105	−107	−108	−110
−40	−40	−47	−70	−85	−96	−104	−109	−113	−116	−118	−120
−45	−45	−54	−77	−90	−103	−112	−117	−123	−128	−129	−130

Notes:
1 The table provides equivalent wind chill temperatures at various outside dry bulb temperatures and corresponding wind velocities.
2 Wind speeds greater than 40 mph have little additional chilling effect.
3 $WCF \cong T_{DB} - (1.5 \times W_S)$
WCF = Wind Chill Factor
T_{DB} = Dry Bulb Air Temperature
W_S = Wind Speed

49.02 Heat Index

HEAT INDEX

	Apparent Temperature, °F														
	Temperature, °F														
%RH	70	75	80	85	90	95	100	105	110	115	120	125	130	135	140
0	64	69	73	78	83	87	91	95	99	103	107	111	117	120	125
5	64	69	74	79	84	88	83	97	102	107	111	115	122	128	
10	65	70	75	80	85	90	95	100	105	111	116	123	131		
15	65	71	76	81	85	91	97	102	106	115	123	131			
20	66	72	77	82	87	93	99	105	112	120	130	141			
25	66	72	77	83	88	94	101	109	117	127	139				
30	67	73	78	84	90	96	104	113	123	135	148				
35	67	73	79	85	91	98	107	118	130	143					
40	68	74	79	86	93	101	110	123	137	151					
45	68	74	80	87	95	104	115	129	143						
50	69	75	81	88	96	107	120	135	150						
55	69	75	81	89	96	110	126	142							
60	70	76	82	90	100	114	132	149							
65	70	76	83	91	102	119	138								
70	70	77	85	93	106	124	144								
75	70	77	86	95	109	130									
80	71	77	86	97	113	136									
85	71	78	87	99	117										
90	71	79	88	102	122										
95	71	79	89	105											
100	72	80	91	106											

Notes:
1 The table provides equivalent heat index temperatures at various temperatures and corresponding relative humidities.
2 The heat index is a measure of how the average person perceives temperature and humidity and how it affects the body's ability to cool itself.
3 Sunstroke and heat exhaustion are likely when the heat index is 105 or greater.

General Notes

50.01 General

A. Provide all materials and equipment and perform all labor required to install complete and operable mechanical systems as indicated on the drawings, as specified, and as required by code.

B. Contract document drawings for mechanical work (HVAC, plumbing, and fire protection) are diagrammatic and are intended to convey scope and general arrangement only.

C. Install all mechanical equipment and appurtenances in accordance with manufacturers' recommendations, contract documents, and applicable codes and regulations.

D. Provide vibration isolation for all mechanical equipment to prevent transmission of vibration to building structure.

E. Provide vibration isolators for all piping supports connected to, and within 50 ft. of, isolated equipment (except at base elbow supports and anchor points) throughout mechanical equipment rooms. Do the same for supports of steam mains within 50 ft. of boiler or pressure-reducing valves.

F. Provide vibration isolators for all piping supports of steam mains within 50 ft. of boilers and pressure-reducing valves.

G. The location of existing underground utilities is shown in an approximate way only. The contractor shall determine the exact location of all existing utilities before commencing work. The contractor shall pay for and repair all damages caused by failure to exactly locate and preserve any and all underground utilities unless otherwise indicated.

H. Coordinate construction of all mechanical work with architectural, structural, civil, electrical work, etc., shown on other contract document drawings.

I. Maintain a minimum 6'8" clearance to the underside of pipes, ducts, conduits, suspended equipment, etc., throughout access routes in mechanical rooms.

J. All tests shall be completed before any mechanical equipment or piping insulation is applied.

K. Locate all temperature, pressure, and flow measuring devices in accessible locations with the straight section of pipe or duct up- and downstream as recommended by the manufacturer for good accuracy.

L. Testing, adjusting, and balancing agency shall be a member of the Associated Air Balance Council (AABC) or the National Environmental Balancing Bureau (NEBB). Testing, adjusting, and balancing shall be performed in accordance with the AABC standards.

M. Where two or more items of the same type of equipment are required, the product of one manufacturer shall be used.

N. Reinforcement, detailing, and placement of concrete shall conform to *ASTM 315* and *ACI 318*. Concrete shall conform to *ASTM C94*. Concrete work shall conform to *ACI 318*, part entitled "Construction Requirements." Compressive strength in 28 days shall be 3,000 psi. Total air content of exterior concrete shall be between 5 and 7 percent by volume. Slump shall be between 3 and 4 in. Concrete shall be cured for 7 days after placement.

O. Coordinate all equipment connections with manufacturers' certified drawings. Coordinate and provide all duct and piping transitions required for final equipment connections to furnished equipment. Field verify and coordinate all duct and piping dimensions before fabrication.

P. All control wire and conduit shall comply with the National Electric Code and Division 16 of the specification.

Q. Concrete housekeeping pads to suit mechanical equipment shall be sized and located by the mechanical contractor. Minimum concrete pad thickness shall be 6 in. Pad shall extend beyond the equipment a minimum of 6 in. on each side. Concrete housekeeping pads shall be provided by the general contractor. It shall be the responsibility of the mechanical contractor to coordinate the size and location of concrete housekeeping pads with the general contractor.

R. All mechanical room doors shall be a minimum of 4'0" wide.

S. Where beams are indicated to be penetrated with ductwork or piping, coordinate ductwork and piping layout with beam opening size and opening locations. Coordination shall be done prior to the fabrication of ductwork, cutting of piping, or fabrication of beams.

T. When mechanical work (HVAC, plumbing, sheet metal, fire protection, etc.) is subcontracted, it shall be the mechanical contractor's responsibility to coordinate subcontractors and the associated contracts. When discrepancies arise pertaining to which contractor provides a particular item of the mechanical contract or which contractor provides final connections for a particular item of the mechanical contract, it shall be brought to the attention of the mechanical contractor, whose decision shall be final.

U. The locations of all items shown on the drawings or called for in the specifications that are not definitely fixed by dimensions are approximate only. The exact locations necessary to secure the best conditions and results must be determined by the project site conditions and shall have the approval of the engineer before being installed. Do not scale drawings.

V. All miscellaneous steel required to ensure proper installation and as shown in details for piping, ductwork, and equipment (unless otherwise noted) shall be furnished and installed by the mechanical contractor.

W. Provide access panels for installation in walls and ceilings, where required, to service dampers, valves, smoke detectors, and other concealed mechanical equipment. Access panels shall be turned over to the general contractor for installation.

X. All equipment, piping, ductwork, etc., shall be supported as detailed, specified, and required to provide a vibration-free installation.

Y. All ductwork, piping, and equipment supported from structural steel shall be coordinated with the general contractor. All attachments to steel bar joists, trusses, or joist girders shall be at panel points. Provide beam clamps meeting MSS standards. Welding to structural members shall not be permitted. The use of C-clamps shall not be permitted.

Z. Mechanical equipment, ductwork, and piping shall not be supported from a metal deck.

AA. All roof-mounted equipment curbs for equipment provided by the mechanical contractor shall be furnished by the mechanical contractor and installed by the general contractor.

BB. Locations and sizes of all floor, wall, and roof openings shall be coordinated with all other trades involved.

CC. All openings in fire walls due to ductwork, piping, conduit, etc., shall be fire stopped with a product similar to 3M or an approved equal.

DD. All air conditioning condensate drain lines from each air handling unit and rooftop unit shall be piped full size of the unit drain outlet, with "P" trap, and piped to the nearest drain. See the details shown in the drawings or the contract specifications for the depth of the air conditioning condensate trap.

EE. Refer to typical details for ductwork, piping, and equipment installation.

50.02 Piping

A. Provide all materials and equipment and perform all labor required to install complete and operable piping systems as indicated on the drawings, as specified, and as required by code.

B. Elevations shown on the drawings are to the bottom of all pressure piping and to the invert of all gravity piping unless otherwise noted.

C. Maintain a minimum of 36" of ground cover over all underground HVAC piping (edit the depth of the ground cover to suit frost line depth and project requirements).

D. Unless otherwise noted, all chilled water and heating water piping shall be 3/4 in. size (edit system type or pipe size to suit project requirements).

E. Provide an air vent at the high point of each drop in the heating-water, chilled-water, and other closed-water piping systems (edit system types to suit the project requirements). All piping shall grade to low points. Provide hose end drain valves at the bottom of all risers and low points.

F. Unless otherwise noted, all piping is overhead, tight to the underside of the structure or slab, with space for insulation if required.

G. Install piping so all valves, strainers, unions, traps, flanges, and other appurtenances requiring access are accessible.

H. All valves shall be installed so that the valve remains in service when equipment or piping on the equipment side of the valve is removed.

I. All balancing valves and butterfly valves shall be provided with position indicators and maximum adjustable stops (memory stops).

J. Provide chainwheel operators for all valves in equipment rooms mounted greater than 7'0" above floor level; chain shall extend to 7'0" above floor level.

K. All valves (except control valves) and strainers shall be the full size of the pipe before reducing in size to make connections to equipment and controls.

L. Unions and/or flanges shall be installed at each piece of equipment, in bypasses, and in long piping runs (100 ft. or more) to permit disassembly for alteration and repairs.

M. Pitch steam piping downward in the direction of flow 1/4 in. per 10 ft. (1 in. in 40 ft.) minimum. Pitch all steam return lines downward in the direction of condensate flow 1/2 in. per 10 ft. (1 in. in 20 ft.) minimum. Where the length of branch lines is less than 8 ft., pitch branch lines toward mains 1/2 in. per ft. minimum.

N. Pitch up all steam and condensate runouts to risers and equipment 1/2 in./ft. Where this pitch cannot be obtained, runouts over 8 ft. in length shall be one size larger than noted.

O. Tap all branch lines from the top of steam mains (45 degrees preferred; 90 degrees acceptable).

P. Provide an end of main drip at each rise in the steam main. Provide condensate drips at the bottom of all steam risers, downfed runouts to equipment, radiators, etc., at the end of mains and low points, and ahead of all pressure regulators, control valves, isolation valves, and expansion joints.

Q. On straight steam piping runs with no natural drainage points, install drip legs at intervals not exceeding 200 ft. where the pipe is pitched downward in the direction of steam flow and a maximum of 100 ft. where the pipe is pitched up so that condensate flow is opposite of steam flow.

R. Steam traps shall be minimum 3/4" size.

S. Install all piping without forcing or springing.

T. All piping shall clear doors and windows.

U. All valves shall be adjusted for smooth and easy operation.

V. All piping work shall be coordinated with all trades involved. Offsets in piping around obstructions shall be provided at no additional cost to the owner.

W. Provide flexible connections in all piping systems connected to pumps, chillers, cooling towers, and other equipment which require vibration isolation except water coils. Flexible connections shall be provided as close to the equipment as possible or as indicated on the drawings.

X. Slope refrigerant piping one percent in the direction of oil return. Liquid lines may be installed level.

Y. Install horizontal refrigerant hot gas discharge piping with 1/2" per 10 ft. downward slope away from the compressor.

Z. Install horizontal refrigerant suction lines with 1/2" per 10 ft. downward slope to the compressor, with no long traps or dead ends that may cause oil to separate from the suction gas and return to the compressor in damaging slugs.

AA. Provide line size liquid indicators in the main liquid line leaving the condenser or receiver. Install moisture-liquid indicators in liquid lines between filter dryers and thermostatic expansion valves, and in liquid line to receiver.

BB. Provide a line size strainer upstream of each automatic valve. Provide a shutoff valve on each side of the strainer.

CC. Provide permanent filter dryers in low-temperature systems and systems using hermetic compressors.

DD. Provide replaceable cartridge filter dryers with a three-valve bypass assembly for solenoid valves, adjacent to receivers.

EE. Provide refrigerant charging valve connections in the liquid line between the receiver shutoff valve and the expansion valve.

50.03 Plumbing

A. Provide all materials and equipment and perform all labor required to install complete and operable plumbing systems as indicated on the drawings, as specified, and as required by code.

B. Run all soil waste and vent piping with 2 percent minimum grade unless otherwise noted (edit the slope to suit project requirements). Horizontal vent piping shall be graded to drip back to the soil or waste pipe by gravity.

C. Elevations shown on the drawings are to the bottom of all pressure piping and to the invert of all gravity piping.

D. Adjust sewer inverts to keep the tops of pipes in line where the pipe's size changes.

E. Maintain a minimum of 3'6" of ground cover over all underground water mains and a minimum of 3'0" of ground cover over all underground sewers and drains (edit the depth of the ground cover to suit frost line depth and project requirements).

F. Provide shutoff valves in all domestic water piping system branches in which branch piping serves two or more fixtures.

G. Unless otherwise noted, all domestic cold and hot water piping shall be 1/2" size (edit the system type or pipe size to suit project requirements).

H. Unless otherwise noted, all piping is overhead, tight to the underside of the slab, with space for insulation if required.

I. Install piping so all valves, strainers, unions, traps, flanges, and other appurtenances requiring access are accessible.

J. Where domestic cold and hot water piping drops into a pipe chase, the size shown for the pipe drops shall be used to the last fixture.

K. Install all piping without forcing or springing.

L. All piping shall clear doors and windows.

M. All piping shall grade to low points. Provide hose end drain valves at the bottom of all risers and low points.

N. Unions and/or flanges shall be installed at each piece of equipment, in bypasses, and in long piping runs (100 ft. or more) to permit disassembly for alteration and repairs.

O. All valves shall be adjusted for smooth and easy operation.

P. All valves (except control valves) and strainers shall be the full size of the pipe before reducing the size to make connections to the equipment and controls.

Q. Provide chainwheel operators for all valves in equipment rooms mounted greater than 7'0" above floor level; chain shall extend to 7'0" above floor level.

R. Provide all plumbing fixtures and equipment with accessible stops.

S. Unless otherwise noted, drains shall be installed at the low point of roofs, areaways, floors, etc.

T. Provide cleanouts in sanitary and storm drainage systems at ends of runs, at changes in direction, near the base of stacks, every 50 ft. in horizontal runs, and elsewhere as indicated (edit horizontal cleanout spacing to suit code and project requirements).

U. All cleanouts shall be the full size of the pipe for pipe sizes 6 in. and smaller, and shall be 6 in. for pipe sizes larger than 6 in.

V. All balancing valves and butterfly valves shall be provided with position indicators and maximum adjustable stops (memory stops).

W. All valves shall be installed so the valve remains in service when the equipment or piping on the equipment side of the valve is removed.

X. All piping work shall be coordinated with all trades involved. Offsets in piping around obstructions shall be provided at no additional cost to the owner.

Y. Provide flexible connections in all piping systems connected to pumps and other equipment that require vibration isolation. Flexible connections shall be provided as close to the equipment as possible or as indicated on the drawings.

50.04 HVAC/Sheet Metal

A. Provide all materials and equipment and perform all labor required to install complete and operable HVAC systems as indicated on the drawings, as specified, and as required by code.

B. Certain items such as rises and drops in ductwork, access doors, volume dampers, etc., are indicated on the contract document drawings for clarity for a specific location requirement and shall not be interpreted as the extent of the requirements for these items.

C. In corridors where ceiling speakers and air diffusers are indicated between the same light fixtures, install both devices at the quarter points between the same fixture.

D. Unless otherwise shown, locate all room thermostats and humidistats 4'-0" (centerline) above the finished floor. Notify the engineer of any rooms where the preceding location cannot be maintained or where there is a question on location.

E. All ductwork shall clear doors and windows.

F. All ductwork dimensions, as shown on the drawings, are internal clear dimensions. Duct size shall be increased to compensate for duct lining thickness.

G. Provide all 90-degree square elbows with double radius turning vanes unless otherwise indicated. Elbows in dishwasher, kitchen, and laundry exhausts shall be of unvaned smooth radius construction with a centerline radius equal to 1-1/2 times the width of the duct. Provide access doors upstream of all elbows with turning vanes.

H. Coordinate diffuser, register, and grille locations with architectural reflected ceiling plans, lighting, and other ceiling items and make minor duct modifications to suit.

I. Field-erected and factory-assembled air handling unit coils shall be arranged for removal from the upstream side without dismantling supports. Provide galvanized structural steel supports for all coils (except the lowest coil) in banks over two coils high to permit the independent removal of any coil.

J. All air handling units shall operate without moisture carryover.

K. Locate all mechanical equipment (single duct, dual duct, variable volume, constant volume and fan-powered boxes, fan coil units, cabinet heaters, unit heaters, unit ventilators, coils, steam humidifiers, etc.) for unobstructed access to unit access panels, controls, and valving.

L. Finned tube radiation enclosures shall be wall-to-wall unless otherwise indicated.

M. Provide flexible connections in all ductwork systems (supply, return, and exhaust) connected to air handling units, fans, and other equipment that require vibration isolation. Flexible connections shall be provided at the point of connection to the equipment unless otherwise indicated.

N. Unless otherwise noted, all ductwork is overhead, tight to the underside of the structure, with space for insulation if required.

O. Runs of flexible duct shall not exceed 5 ft. (edit the maximum length of the flexible duct to suit the project; 5 ft. maximum recommended length, 8 ft. maximum length).

P. All ductwork shall be coordinated with all trades involved. Offsets in ducts, including divided ducts and transitions around obstructions, shall be provided at no additional cost to the owner.

Q. Provide access doors in ductwork to provide access for all smoke detectors, fire dampers, smoke dampers, volume dampers, humidifiers, coils, and other items located in the ductwork that require service and/or inspection.

R. Provide access doors in ductwork for the operation, adjustment, and maintenance of all fans, valves, and mechanical equipment.

S. All ducts shall be grounded across flexible connections with flexible copper grounding straps. Grounding straps shall be bolted or soldered to both the equipment and the duct.

T. Smoke detectors shall be furnished and wired by the electrical contractor. The mechanical contractor shall be responsible for mounting the smoke detector in ductwork as shown on the drawings and in accordance with the manufacturer's printed instructions.

U. Terminate gas vents for unit heaters, water heaters, high-pressure parts washers, high-pressure cleaners, and other gas appliances a minimum of 30" above the roof with rain cap (edit any appliances and the height above the roof to meet the code and suit project requirements).

V. See specifications for ductwork gauges, bracing, hangers, and other requirements.

W. Exterior louvers are indicated for information only. Detailed descriptions are provided in the architectural specifications.

X. Exterior louvers are indicated for information only. Louver sizes, locations, and details shall be coordinated with the general contractor.

Y. Exterior louvers are indicated for information only. Louver sizes, locations, mounting, and details shall be coordinated with other trades involved.

50.05 Fire Protection

A. Provide all materials and equipment and perform all labor required to install complete and operate fire protection systems as indicated on the drawings, as specified, and in compliance with the standards of the National Fire Protection Association, Industrial Risk Insurers, Factory Mutual, and all state and local regulations.

B. The entire building sprinkler system shall be hydraulically designed unless otherwise noted on the drawings. Head spacing in general and water quantity shall be based on Light Hazard Occupancy (edit occupancy classification to suit project requirements; see NFPA 13—Light Hazard Occupancy, Ordinary Hazard Group I Occupancy, Ordinary Hazard Group II Occupancy, Extra Hazard Group I Occupancy, Extra Hazard Group II Occupancy).

C. The entire building sprinkler system shall be pipe schedule designed unless otherwise noted on the drawings. Head spacing in general and water quantity shall be based on Light Hazard Occupancy (edit the occupancy classification to suit project requirements; see NFPA 13—Light Hazard Occupancy, Ordinary Hazard Group I Occupancy, Ordinary Hazard Group II Occupancy, Extra Hazard Group I Occupancy, Extra Hazard Group II Occupancy).

D. Provide an automatic wet pipe sprinkler system throughout the entire building, complete in all respects and ready for operation including all test and drain lines, pressure gauges, hangers and supports, signs, and other standard appurtenances. Wiring shall be provided under the electrical division.

E. Provide an automatic dry pipe sprinkler system throughout the entire building, complete in all respects and ready for operation, including all test and drain lines, pressure gauges, dry pipe valves, air compressors, hangers and supports, signs, and other standard appurtenances. Wiring shall be provided under the electrical division.

F. See the architectural drawings for the exact location of fire extinguisher cabinets, fire hose cabinets, and Siamese connections.

G. All shutoff valves in the sprinkler, standpipe, and combined systems shall be approved, indicating type.

H. Coordinate sprinkler head locations with the architectural reflected ceiling plans, lighting, and other ceiling items, and make minor modifications for suitability purposes.

I. Sprinklers installed in the ceilings of finished areas shall be symmetrical in relation to ceiling system components and centered in the ceiling tile.

Designer's Checklist

51.01 Boilers, Chillers, Cooling Towers, Heat Exchangers, and Other Central Plant Equipment

A. Have owner redundancy requirements been met? Has future equipment space been clearly indicated on the drawings? Has move-in route and replacement access been determined?

B. Have multiple pieces of central plant equipment been provided to prevent system shutdown in the event of equipment failure? Has low load been evaluated and is equipment selected capable of operating at this low-load condition?

C. Has proper service access been provided? Has tube pull or cleaning space been provided?

D. Have final loads been calculated and the final equipment selection been made? Has equipment been specified and capacity scheduled?

E. Has chemical treatment of hydronic and steam systems been properly addressed? Have flushing and passivation of the hydronic and steam systems been adequately covered in particular waste treatment handling of spent flushing water and chemicals?

F. Does central plant equipment need to be on emergency power?

G. When multiple pieces of equipment are headered together, have adequate provisions for expansion and contraction been provided, especially regarding boiler systems? Recommendation: Multiple boiler connections to header, from boiler nozzles to header main, should be U-shaped (first traveling away from the header, then traveling parallel to the header, and finally traveling back toward the header) to accommodate expansion and contraction of piping to prevent excess stress on the boiler nozzles.

H. When specifying boiler control and oxygen trim systems, chillers with remote starters and remote control panels, cooling tower basin heaters, and other electrical or control systems associated with central plant equipment, has field wiring required for these systems been coordinated with the electrical and instrumentation and control (I&C) engineers? This includes panel installation, interconnecting power and control wiring, instrument air, and the mounting of devices.

I. Have starter, disconnect switch, variable frequency drive, and/or motor control center spaces been coordinated and/or located?

J. When specifying dual fuel boilers, does the owner want a dual fuel pilot (natural gas and fuel oil) or is a tee connection preferred for connection to a portable propane bottle?

51.02 Air Handling Equipment—Makeup, Recirculation, and General Air Handling Equipment

A. Have owner redundancy requirements been met? Has future equipment space been clearly indicated on the drawings? Has move-in route and replacement access been determined?

B. Have multiple pieces of air handling equipment been provided to prevent system shutdown in the event of equipment failure?

C. Has adequate coil pull space and service space been provided? Recommendation: The service access space should be a minimum of the unit width plus 2 ft. on at least one side and a minimum of 2 ft. on the other side.

D. Have unit components and capacities been properly specified, detailed, and scheduled—coils, filters, fans, motors, humidifiers, outside air and return air dampers, smoke detectors, smoke dampers, access section, service vestibules, access doors, interior lighting (incandescent, fluorescent), etc.? Have coil and filter air pressure drops been scheduled? Have coil water pressure drops been scheduled?

E. Have outside air and return air been mixed prior to entering any air handling unit filters or coils?

F. Has proper length downstream of humidifiers been provided to absorb humidification vapor trail? The first air handling unit section downstream of the humidifier should be stainless steel, including coil frames, especially with deionized (DI), reverse osmosis (RO), or ultra pure water (UPW).

G. Have cooling coils been locked out during the air handling unit preheat and humidification operation?

H. Has piping in service vestibules been checked for adequate space? Recommendation: A minimum of 6'0" wide and a minimum of 9'0" high clearance should be maintained to allow for pipe installation for the full length of the unit.

I. Are access doors of adequate size to remove fans, motors, filters, dampers, actuators, inlet guide vanes or other variable flow device, and other devices requiring service and/or replacement?

J. Do all air handling unit preheat coils with a design mixed air temperature below 40°F have preheat pumps? To reduce the risk of freezing, preheat pumps are recommended for all preheat coils with a design mixed air temperature below 40°F.

K. Have coil selections been made so that low water flows, in direct response to low loads, do not fall into laminar flow region?

L. Have air conditioning condensate drains been piped to an appropriate drainage system? Have drains been provided for storm water and sanitary?

M. Have receptacles been provided for roof-mounted equipment in accordance with the NEC?

N. Have the starter, disconnect switch, adjustable frequency drive, and/or motor control center spaces been coordinated and/or located?

O. Does air handling equipment need to be on emergency power?

51.03 Piping Systems—General

A. Expansion tank: Has size, location, adequate space, support, makeup water pressure, and makeup water location been coordinated with the plumbing engineer?

B. Are there provisions for piping expansion and contraction, anchors, guides, loops vs. joints? Have anchor locations and forces been coordinated with the structural engineer? Locate anchors at steel beams and avoid joists if possible. Is piping coordinated with building expansion joints?

C. Do the drawings clearly indicate where ASME code piping and valves are required at the boilers in accordance with ASME code requirements for high temperature (over 250°F) and high-pressure boilers (over 15 psig)?

D. Does the boiler layout and design have enough expansion and flexibility in the boiler connection piping to prevent overstressing the boiler nozzle? It is best to use a U-shaped layout to the header.

E. Have flexible connections been clearly shown on the drawings and have they been properly detailed? Have the appropriate flexible connections been specified for the application?

F. Is there structural support for large water risers?

G. Are there drains and air vents on water systems and adequate space for service?

H. Are balancing valves required on parallel piping loops?

I. Is adequate space available for the pitching of pipes?

J. Is there space for coil and tube removal or cleaning (e.g., AHUs, chillers, boilers, etc.) and is it clearly shown on the drawings where it is required?

K. Is coil piped for counterflow or parallel flow as indicated by detail (parallel flow for preheat coils only; all others counter flow)?

L. Condensate drains from room terminals with chilled, dual temperature water and packaged cooling units: Do local authorities require condensate drains to be piped to sanitary or to storm? Can condensate drains be discharged onto roof? Onto grade?

M. Are relief valve settings noted on drawings or schedules?

N. Is there adequate straight pipe up- and downstream of flow meter orifices?

O. Have all required equipment valves not covered by standard details been indicated? Avoid duplications.

P. Do not run horizontal piping in solid masonry walls or in narrow stud partitions.

Q. Has all piping been eliminated from electrical switchgear, transformer, motor control center, and emergency generator rooms? If not, have drain troughs or enclosures been provided?

R. Are shutoff valves provided at the base of all risers?

S. Are all systems compatible with flow requirements established by control diagrams?

T. Is cathodic protection required for buried piping?

U. Has required heat tracing been included, coordinated, and insulated?

V. Will large mains or risers transmit noise to occupied spaces? Are isolators required in supply and return at the pump?

W. Is the present and future duty for pumps, boilers, chillers, cooling towers, heat exchangers, terminal units, coils, AHUs, etc., specified? Scheduled?

X. Are air conditioning and steam condensate (when wasted) piped to storm water or sanitary? Is steam condensate cooled?

51.04 Steam and Condensate Piping

A.   See the "Piping Systems—General" section earlier for additional requirements.

B.   Are the ends of main drips shown, detailed, and specified?

C.   Will condensate drain? Are pipes oversized for opposing flow?

D.   Will humidifier arms add excessive sensible heat to the air stream (likely on small flat ducts and some AHUs)? Insulate where needed. Provide motor-operated shutoff valves if steam is live during the mechanical cooling season.

E.   Are riser drips shown, detailed, and specified?

F.   Flash tanks for medium- and high-pressure condensate. Vent flash tanks either to low pressure steam or outdoors.

G.   Are relief valves piped to outside? Have they been sized?

H.   Has steam consumption for humidification been considered in establishing the water makeup quantity for the boiler?

I.   Has adequate space been allowed for pressure reducing stations? Have standard details been edited?

J.   Are water sampling connections provided?

K.   Are steam injectors piped to the floor drains?

L.   Avoid cross-connections between gravity condensate returns and pumped condensate return lines.

M.   Is there adequate height between the condensate receiver and/or feedwater heater and the pump to prevent flashing at the pump, particularly with condensate above 200°F?

N.   Has bypass around the boiler feedwater heater been provided for maintenance?

O.   Are there drip runouts to equipment such as sterilizers and glassware washers?

P.   Are the ends of main drips piped?

Q.   Are condensate return systems compatible?

R.   Have noise suppressors been provided on the reduced pressure side of PRVs? Will radiated noise be a problem? Are there adequate numbers of stages of pressure reduction for quiet operation and an adequate number of valves for capacity control?

S.   Are steam and/or condensate flow meters and recorders required?

T.   Is there adequate access to components requiring service on the boilers? Is a catwalk required?

U.   Are boilers piped in accordance with the ASME code? Is there a nonreturn plus a shutoff valve on the HP boiler?

V.   Is the condensate tank vented to the outside?

W.   Are chemicals used in the treatment system suitable for humidification? Are chemical feed systems shown, detailed, and specified?

X. Is a feedwater heater or deaerator required?

Y. Are water softeners required on makeup? Are they shown, detailed, and specified?

Z. Are bottom blowdown and continuous blowdown shown, detailed, and specified?

AA. Avoid lifting steam condensate, if possible.

BB. Are proper traps being used? Have they been specified and scheduled?

CC. Are air conditioning and steam condensate (when wasted) piped to storm water or to sanitary? Is steam condensate cooled?

DD. Are large system isolation valves provided with the bypass warming valve?

51.05 Low Temperature Hot Water and Dual Temperature Systems

A. See the "Piping Systems—General" section earlier for additional requirements.

B. Are balancing valves indicated? Are flow measuring stations needed and indicated?

C. Is pressure regulation needed?

D. Is a bypass filter required? Is GPM included in pump capacity?

E. Is a standby pump needed?

F. Converter support: Are details needed? Is elevation indicated?

G. Are service valves shown?

H. Will branch piping and ducts fit in the allotted space or enclosure?

I. Are riser shutoff valves shown?

J. Are riser drains and vents shown?

K. Is there adequate space for the installation and use of riser valves?

L. Will the minimum allowable circulation be maintained through the hot water boiler?

M. Is the distribution system reverse return? If not, will balancing problems result?

51.06 Chilled Water and Condenser Water Systems

A. See the "Piping Systems—General" earlier for additional requirements.

B. Are balancing valves indicated? Are flow measuring stations needed? Have they been indicated?

C. Is pressure regulation needed?

D. Is a bypass filter required? Is the GPM included in the pump capacity?

E. Is a standby pump needed?

F.   Are service valves shown?

G.   Will branch piping and ducts fit in the allotted space or enclosure?

H.   Are riser shutoff valves shown?

I.   Are riser drains and vents shown?

J.   Is there adequate space for the installation and use of riser valves?

K.   Will the minimum allowable circulation be maintained through the chiller?

L.   Is the distribution system reverse return? If not, will balancing problems result?

M.   Condenser water piping: loop traps to avoid excessive drainage, submerged impeller. Has the available NPSH been calculated? Is the NPSH indicated in the pump schedule?

N.   For cooling tower makeup, overflow, and drain splash blocks, are there balancing valves in branch lines to tower cells? Coordinate the makeup with the plumbing engineer.

51.07 Air Systems

A.   Are adequate balancing dampers provided to prevent noise at outlets due to excessive pressure, or to avoid complicated balancing procedures on extensive low-pressure systems or exhaust systems (e.g., each zone of a multizone system; to limit flow variation due to stack effect in vertical low pressure and exhaust systems)?

B.   Are fire damper locations, type, and flow restrictions indicated? Is there adequate height for a damper recess pocket at the shaft wall? Is breakaway ductwork at the fire damper wall sleeve detailed or specified?

C.   Are smoke damper locations, type, and flow restrictions indicated? Is there adequate height for a damper at the shaft wall? Is breakaway ductwork at the smoke damper wall sleeve detailed or specified? Is the smoke damper operator located on the supported duct and not on a breakaway duct?

D.   Are access doors at fire dampers, smoke dampers, turning vanes, humidifiers, coils, etc., properly specified and included in the general notes?

E.   Are proper relief air provisions provided?

F.   Is a return air fan needed? Is an outside air fan needed?

G.   Are condensate drains provided? Are outside air intake drains provided?

H.   Are flexible connections shown and specified?

I.   Is sound lining required? Is it properly located and specified?

J.   Will the duct arrangement permit the transfer of excessive noise between offices, toilet rooms, and rooms of a different function?

K.   Is there objectionable fan noise from intakes or exhaust points to nearby buildings?

L.   Are outlets located in supply mains? Are there noisy conditions?

M.   Do trunk ducts pass above quiet rooms? Will noise be a problem?

N.   Have fan class, bearing arrangement, motor location, etc., been shown, scheduled, or specified?

O.   Are air intakes on party walls?

P.   Will outlets blow at lights, beams, sprinkler heads, or smoke detectors? Sprinkler head and smoke detector locations must meet code requirements. Locate them in accordance with code.

Q.   Have outlet and return grille elevations been coordinated with the architect and indicated?

R.   Adjust outlet air quantities for duct heat gain and duct leakage.

S.   Are isotope and chemical exhaust ducts accessible?

T.   Is there interference between sill grille discharge and drapes or blinds? Beware of the annoying movement of vertical blinds or light drapes caused by sill air discharge nearby.

U.   Are the present and future duties for air terminal units, AHUs, fans, etc., specified and scheduled?

V.   Is the exhaust or relief discharge or plumbing stack effluent near intakes? Maintain a minimum of 10 ft. of clearance.

W.   Is there an anti-stratification provision at intakes, large mixing box outlets, and downstream of steam coils or water coils? Are air blenders indicated on all AHUs?

X.   Are there aluminum grilles on the shower, sterilizer, etc., exhaust? Is stainless steel ductwork or aluminum ductwork required? Is it clearly indicated on drawings as to extent? Has it been specified?

Y.   Are there sealing and sloping of shower, cage washer, etc., exhaust ducts? When more than one type of duct material is used, is the extent and location clearly defined?

Z.   Has adequate relief from rooms been provided? Are there door louvers, undercut doors, transfer grilles, and direct exhaust? Have they been coordinated?

AA.   Will door louvers defeat the needed acoustical privacy (e.g., conference rooms, private offices, VP office)? Will door louvers defeat the needed door fire rating? Are door louvers located in accordance with code?

BB.   Are the types of branch takeoffs and duct splits shown? Are details included on drawings?

CC.   Are there intermediate drip pans on cooling coil banks? Are they piped to the floor drain? Include detail.

DD.   Are there drains for kitchen exhaust duct risers?

EE.   Is there excessive duct heat gain from nearby steam pipes and other heat sources?

FF.   Are there combustion air intakes for boilers, water heaters, etc.? Are vents, stacks, breeching, and chimneys shown, specified, and detailed? Are termination heights clearly indicated?

GG.   Locate exhaust grilles near the floor in operating rooms, flammable storage rooms, chlorine storage rooms, battery rooms (high and low), etc.

HH. Do not use corridors as return air plenums in hospitals, nursing homes, offices, and other facilities.

II. Have insulated louver blank-off panels or sheets been included where required?

JJ. Are filters provided in makeup air to elevator equipment rooms? Are filters provided for air-cooled condensers and condensing units located indoors?

KK. Are there motor-operated dampers in wall louvers? Do not use operable louvers. Use stationary louvers with motor-operated dampers behind when required.

LL. Are casings adequately described as prefabricated or field-fabricated? Is the extent of the sound paneling clear? Has an adequate pressure rating been specified?

MM. Has the architect provided adequate framing for the linear diffuser in the metal lath and plaster or dry wall bulkheads? Do not dimension diffuser lengths for wall-to-wall installations—note the dimension as "wall to wall."

NN. Have fan systems been checked for excessive sound transmission?

OO. Is there adequate space for servicing fans, motors, belts, etc.?

PP. Has sufficient space been provided between coils of AHUs to accommodate temperature sensors?

QQ. Are adequate service space or equipment size access panels noted on drawings for equipment installed above ceilings? Coordinate with the architect who furnishes, installs, provides.

RR. Are there adequate straight duct branch length or straightening vanes between the main duct and diffuser?

SS. Do ducts pierce partitions at 90-degree angles wherever possible?

TT. Are wash down systems or fire protection systems required for fume hoods or kitchen hoods?

UU. Are fume hood exhaust systems balanceable? Are orifice plates required?

VV. Are correct outside air quantities and pressurization included?

WW. Is a smoke control system required?

XX. Avoid contamination of air intake from exhaust air, contaminated vents, vehicle exhaust, etc. Are locations in accordance with code?

YY. Are static pressure sensors indicated or specified?

ZZ. Are fire and smoke dampers coordinated with fire and smoke walls? Are fire rated floor/ceiling assemblies used? Will diffusers, registers, and grilles require fire dampers? Are smoke dampers required for air handling units or fans?

AAA. Is the floor suitable for "built-up" air handling units?

BBB. Have ventilation systems been provided for equipment rooms and other non-air-conditioned spaces?

CCC. Are flow measuring devices located? Is there adequate straight run?

DDD. Is there adequate straight duct upstream of terminal units? VAV, constant volume reheat, dual duct, fan-powered, and other air terminal unit runouts should be sized based on the ductwork criteria established for sizing the ductwork upstream of the air terminal unit, and not on the terminal unit connection size. The transition from the run-out size to the air terminal unit connection size should be made at the terminal unit. A minimum of 3 ft. of straight duct should be provided upstream of all air terminal units.

EEE. Is the system compatible with architectural floor/ceiling assemblies?

FFF. Do toilet rooms have the code-required minimum exhaust?

GGG. Locate exterior wall louvers, especially intake louvers, a minimum of 2'0" above the roof, finished grade, etc.

HHH. Locate gravity roof ventilators, especially intake ventilators, a minimum of 1'0" from the finished roof to the top of the roof curb.

III. Are air-conditioning condensate drains piped to storm water or sanitary as required by the local authority?

51.08 Process Exhaust Systems

A. Branches and laterals should be connected above the duct centerline. If branches and laterals are connected below the duct centerline, drains will be required at the low point.

B. Provide blast gates or butterfly dampers at each branch, at each submain, and at each equipment or tool connection. Wind loading on blast gates needs to be considered when installed on the roof or outside the building, especially those blast gates that are normally open.

C. Blast gate blades for process exhaust systems should be specified with an EPDM wiper gasket to provide a tight seal. For blast gates installed for future use, it is recommended that the blade be removed and a gasketed blind flange be provided where the blade goes in the duct to reduce leakage.

D. Does duct pitch to low points and drains? Are drains provided at all low points?

E. Has correct duct material been specified? Is it Stainless Steel, Halar Coated Stainless Steel, FRP, or PVC? PVC is not recommended and the maximum size is 8" round.

F. Has the proper pressure class been specified upstream and downstream of scrubbers and other abatement equipment?

G. Is ductwork installed outside or in unconditioned spaces and will condensation occur on the outside or inside of this duct? Is duct insulation or heat tracing required?

H. Are adequate butterfly balancing dampers shown for system balancing?

I. Are bubble tight dampers specified and shown when and where required?

J. Are process exhaust fans on emergency power as required by code?

K. Process exhaust ductwork cannot penetrate fire-rated construction. Fire dampers are generally not desirable. If penetrating fire-rated construction cannot be avoided, process exhaust ductwork must be enclosed in a fire-rated enclosure until it exits the building, or sprinkler protection inside the duct may be used if approved by the authority having jurisdiction.

L. Are pressure ports provided at the ends of all laterals, submains, and mains?

M. Are drains required in fan scroll, scrubber, or other abatement equipment?

N. Are flexible connections provided at fans and are flexible connections specified suitable for application?

O. Are stacks properly located and is the discharge height adequate to prevent contamination of outside air intakes, CT intakes, and combustion air intakes? Are termination heights clearly indicated?

P. Have redundancy requirements been met?

Q. Are variable frequency drives required, located, and coordinated with the electrical engineer?

51.09 Refrigeration

A. See the "Piping Systems—General" section earlier for additional requirements.

B. Is future machine space indicated on the drawings?

C. Is the space for servicing indicated on the drawings?

D. Are there rigging supports for large water boxes and compressor shells?

E. Is noise transmission likely to occupied spaces?

F. Is there adequate control of chilled water temperature?

G. Are sprinklers required for wood fill towers? *NFPA 214.*

H. Is refrigerant relief piping shown on the drawings? Is it piped to the outside?

I. Is noise from the cooling towers likely to be a problem?

J. Will cooling tower discharge air pocket or recirculate?

K. Should the cooling tower be winterized?

L. Have the cooling tower support locations been cleared with the structural engineer. When determining the cooling tower enclosure height, has the height of vibration isolators been considered (8–12 in. high) and has the height of the safety rail been considered?

M. Are cooling tower discharge duct connections necessary?

N. Are flow diagrams required? Have they been coordinated?

O. Are present and ultimate duties noted where applicable and coordinated with pumps and coils, etc.?

P. Is ethylene or propylene glycol required? Has it been specified and equipment capacities derated?

Q. Has additional insulation been included for low temperature systems?

R. Has single-phase protection been included for packaged (single and/or split systems) air conditioning and heat pump compressor motors?

51.10 Controls

A. Are all panels located? Have they been coordinated with the Electrical Engineer? Are they local or central?

B. Are flow meter locations an adequate distance up- and downstream of the orifice?

C. Are thermostat and humidistat locations indicated? Do not mount stats on glass panels or door frames. Avoid middle-of-the-wall locations.

D. Are control settings, schedules, and diagrams indicated or specified?

E. Are temperature tolerances in lab areas clearly specified?

F. Are power and control wiring diagrams shown? Is interlocking wiring included?

G. Have reheat coils requiring full capacity in summer been supplied from a constant temperature hot water supply?

H. Are low-leak dampers specified on intakes and elsewhere as required?

I. Have compressor location and motor size been coordinated with the Electrical Engineer?

J. Are all AHUs and systems accounted for on control design?

K. Coordinate the purchase and installation of duct smoke detectors and duct fire stat locations with the Electrical Department for connection to the building fire detection system.

L. Are direct digital controls appropriate?

M. Are valve positions (normally open or normally closed) indicated where applicable?

N. Is the compressor sized for ultimate duty?

51.11 Sanitary and Storm Water Systems

A. See the "Piping Systems—General" section earlier for additional requirements.

B. Adjust sewer inverts to keep the tops of pipes in line where the pipe's size changes (note this on the drawings).

C. Maintain at least a minimum cover on sewers for the entire run.

D. Has the sewer authority been contacted for the following:
1. Are sewer authority mains capable of handling additional discharge?
2. The location, size, and depth of sanitary and storm sewer mains.
3. Connection requirements.
4. Requirements for grease traps, sand interceptors, oil/water separators, etc.
5. Has the DER or EPA been contacted?
6. Have storm water management requirements been determined?

E. Sewer profiles are usually required where contours vary extensively or where possible interference with other lines exists. Indicate contours where required.

F.   Indicate sewer inverts at points of connection to public sewers, at building walls, at crossover points, and at points of possible interference. Are all underground utilities coordinated with foundations and grade beams?

G.   Indicate foundation drain tile inverts. Provide back water valves (BWVs) at connections to the storm water system. Check accessibility. Is a manhole required?

H.   Is there a dry manhole for BWVs outside the building or deep BWVs inside the building?

I.   Provide headwall and rip rap for storm water discharge to a drainage ditch, storm water retention pond/tank, or stream.

J.   Size site storm sewers large enough to prevent stoppage by leaves, paper, silt, etc. Except for light duty sewers, use an 8" or 10" pipe minimum.

K.   Are all plumbing fixtures designated and scheduled?

L.   Coordinate fixture locations with final architectural plans. Check ADA requirements. Are handicapped fixtures identified?

M.   Provide BWVs for drains and groups of drains connected to the storm water below grade or where backflow is possible above grade.

N.   Vent sumps for sanitary and storm water drainage.

O.   Is the elevation of mains selected to be above the footings? Advise the Structural Engineer if mains must run below footings or through footings.

P.   Is there adequate ceiling space for AHU floor drain traps on upper floors? Are deep seal traps required? Are they indicated?

Q.   Are drains for overflows piped?

R.   Are there separate vapor vents for sterilizer and bed pan washers?

S.   Are grease traps required for commercial kitchens? Are sand interceptors and/or oil/water separators required for garages and parking areas? Is oil and/or water collected by the oil/water separator to be treated as hazardous waste?

T.   If an oil-filled transformer is located inside the building, provide the transformer room with a drain and pipe to an accessible storage tank.

U.   Provide floor drains for air handling units, boilers, chemical feed equipment, air compressors, pumps, generators, etc., especially for relief valve discharge and pump stuffing box discharge.

V.   Are disposals directly connected to heavy flow mains? Do not connect to a grease interceptor.

W.   Provide a floor drain to create an indirect waste connection for commercial dishwashers, kitchen sinks, and kitchen equipment processing food.

X.   Is the plumbing fixture connection schedule included?

Y.   Does the general piping or equipment interfere with the overhead door's travel?

Z.   Do not run horizontal piping in solid masonry walls.

AA. Is there adequate AHU pad height to allow condensate drain from the pan to be properly trapped? Are condensate drains piped to the storm or sanitary made with indirect connections? Do local authorities require condensate drains to be piped to sanitary or to storm? Can condensate drains be discharged onto the roof or grade?

BB. Are floor drains, roof drains, and trench drains coordinated with the structural system? Are drains coordinated with building expansion joints?

CC. Are air conditioning and steam condensate (when wasted) piped to storm water or sanitary? Is steam condensate cooled?

DD. Are automatic trap priming systems required?

EE. Are floor drain, roof drain, and trench drain types suitable for duty and traffic rating?

FF. Are flow or riser diagrams required by plumbing authorities? Are fixture units clearly indicated on riser diagrams when required?

GG. Is the minimum size of the vent through the roof indicated (e.g., recommend 3")? Has the minimum size pipe below floor been coordinated with local codes (e.g., Allegheny Co. 4" minimum pipe size below floor)?

HH. Are fixtures and drains trapped and vented in accordance with applicable code?

II. Will drainage to grade freeze and create a slippery condition?

JJ. Is tub overflow assembly accessible? Use a solid connection, if not.

KK. Are cooling tower and evaporative cooler overflows, bleeds, and drains piped to sanitary?

LL. Do not use cleanouts on Washington, D.C., projects. Verify requirements.

MM. Are acid waste and vent systems clearly indicated on the drawings and specified?

NN. Site drainage: Are adequate manholes, catch basins, and other items shown on the drawings and specified?

OO. Are future connections and/or expansions considered in the slope of piping, size of piping, and sewer connection sizing?

PP. Provide manways for septic and sewage holding tanks. Manholes and covers should be waterproof/watertight.

51.12 Domestic Water Systems

A. See the "Piping Systems—General" section earlier for additional requirements.

B. Has the water authority been contacted to obtain the following:
1. Water static and residual pressures and flows at the water main. Are these pressures and flows adequate?
2. The location and size of water mains.
3. Water hardness and the corrosiveness of the water.
4. Backflow prevention requirements.
5. Water meter location requirements and meter pit requirements if necessary.

C. Are pressure regulating valves required? Do pressures exceed 60 psi? If so, pressure reducing valves should be provided.

D. Are there submain section valves?

E. Are there provisions for piping and building expansion?

F. Have all wall, box, and yard hydrants been provided and specified?

G. Are water softeners for laundry and boiler makeup required?

H. Is makeup water connected to the boiler, heating, chilled, condenser, and other HVAC water systems? Is freeze protection required? Is sufficient pressure available to overcome static head?

I. Provide hose bibbs at cooling towers and in boiler rooms, mechanical rooms, large toilet rooms, dormitory toilet rooms, and kitchens.

J. In boiler and chiller rooms, provide service sink and water sampling connections.

K. Are flow or riser diagrams required by plumbing authorities? Are fixture units clearly indicated on riser diagrams when required?

L. Is a hot water recirculating pump required, located, scheduled, and specified?

M. Are all hospital, laboratory, kitchen, and other special equipment connections shown on the drawings? Are hospital, laboratory, kitchen, and other special equipment connection schedules required and included?

N. Are backflow preventers provided at the service entrance, at the fire protection service, and at the connection to the HVAC water systems fill connections? Use reduced pressure backflow preventers on all HVAC systems and double-check backflow preventers on domestic water and fire protection service.

O. Is a pressure boosting system required?

P. Is a main shutoff valve provided? Are shutoff valves shown at each toilet room and groups of two or more plumbing fixtures?

Q. Are all plumbing fixtures shown on the drawings and specified?

R. Is a water meter required? Is submetering required?

S. Are balancing valves on the hot water recirculation system shown?

T. Use a 3/4" cold water connection to eye wash units.

U. Are water heater connections shown (gas, water, vents, etc.)?

V. Is a dishwasher booster heater connected?

W. Are future connections and/or expansions considered in the size of the piping and service entrance?

X. Are all underground utilities coordinated with foundations?

51.13 Fire Protection

A. See the "Piping Systems—General" section earlier for additional requirements.

B. Are Siamese connections shown and coordinated with the architect?

C. Are check valves and shutoff valves shown on the drawings?

D. Have fire extinguishers and/or cabinets been specified by the architect or engineer? Have fire hoses and/or cabinets been specified by the architect or engineer?

E. Is fire protection for kitchen hoods required?

F. Is there adequate space for sprinkler mains?

G. Are dry systems provided for areas subject to freezing?

H. Is there a sprinkler for trash and linen chutes?

I. Are there drains for ball drips of Siamese connections?

J. Are pressures noted for hydraulically calculated systems?

K. Is the extent of the sprinklered area indicated? If more than one type of sprinkler system is required (wet, dry, pre-action, deluge, etc.), are they clearly indicated on the drawings?

L. Are fire department valves clearly indicated on the drawings?

M. Are special fire protection systems included?

N. Are standpipes and fire department valves shown?

O. Is sprinkler zoning compatible with the fire alarm zoning?

P. Are all test connections shown and locations coordinated with the Architect? Are drains for test connections provided?

Q. Has the water authority been contacted to obtain the following:
1. Water static and residual pressures and flows at the water main. Are these pressures and flows adequate or is a fire pump required?
2. The location and size of water mains.
3. The water hardness and the corrosiveness of the water.
4. Backflow prevention requirements.
5. Water meter location requirements and meter pit requirements if necessary.
6. Street or onsite fire hydrant requirements.
7. The fire hydrant and fire department connection size, thread type, etc.

R. Have electrical requirements for the fire pump, tamper switches, flow switches, etc., been coordinated with the electrical department?

S. Have fire pump requirements been coordinated between the spec and the drawings?

T. Have the fire hose and fire extinguisher locations been coordinated with the Electrical Department for the wiring of the blue indicator light?

U. Who paints fire protection piping and what color (red)?

51.14 Natural Gas Systems

A. See the "Piping Systems—General" section earlier for additional requirements.

B. Determine the minimum gas pressure required. Is the gas company pressure available at the street adequate for the equipment? Has the gas company been contacted to obtain the following:

1. Pressures and flows at the gas main. Are these pressures and flows adequate?
2. The location and size of the gas mains.
3. Gas meter location requirements and meter pit requirements if necessary.

C. Has the gas meter size been coordinated with the gas company? Has the capacity requirement and site location been given to the gas company? Is the meter required to be located inside or outside? Who provides gas meter and regulator assembly? The gas company? Who provides gas piping from the main to the curb box, from the curb box to the meter assembly, from the meter assembly to the building, and inside the building?

D. Have gas pressure regulators been evaluated for low-load conditions and during startup? It is recommended that multiple gas pressure regulators be used, especially on large central utility plant natural gas systems, not only for low-load conditions but for the replacement of regulators without a shutdown of the entire plant. For instance, the natural gas system design may use two regulators sized at 50–50, 33–67, or 40–60 percent, or it may use three regulators sized at 15–35–50 or 25–25–50 percent.

E. Is there gas meter access and room ventilation (when required)?

F. Are there drip pockets if gas lines cannot drain back to the meter, and adequate space for the pitch?

G. Are there submain section gas cocks?

H. Are gas vent valves and vents from pressure regulating valves piped to the outside?

I. Do not locate natural draft burners in the room under "negative" pressure.

J. Coordinate the gas train with gas pressure available and with the Owner's insurance carrier.

K. Are stacks, vents, and breeching shown on the drawings and are they properly sized and specified? Coordinate with the design team other equipment requiring gas vents (e.g., water heaters, shop equipment, kitchen equipment, lab equipment, hospital equipment).

L. Is combustion air for fuel-fired equipment properly designed in accordance with code? Watch for water heaters in janitor closets.

M. What pressures are permitted to be run inside the building?

N. Is piping run in plenum? If so, valves cannot be located in plenum, including walls.

O. Check with the local gas company for welded and screwed pipe requirements (concealed, exposed, etc.). Screwed pipes and fittings may only be used if gas service is less than 1 psig and vertical runs are less than four stories. Otherwise, use welded pipe.

P. Plastic pipe can only be used for underground service. Require the contractor to install #14 insulated tracer wire 4 to 6 in. above all underground plastic lines.

51.15 Fuel Oil Systems

A. See the "Piping Systems—General" section earlier for additional requirements.

B. Do not locate natural draft burners in rooms under "negative" pressure.

C. Is the suction lift within allowable limits of the fuel oil pump?

D. Is the underground fuel oil tank location coordinated with the site plan? Does it have adequate cover? Has truck traffic been considered? Are leak detection systems and double wall piping systems shown on the drawings and specified?

E. Are the tank vent and fill indicated and away from air intakes? Are vents properly sized?

F. Are fuel oil heaters required (#4, #5, #6 fuel oils)?

G. Is a tank heater required? (They are not permitted with fiberglass tanks.)

H. Is compressed air for the tank gauge provided?

I. Is a specified tank suitable for installation? Has it been coordinated with the owner? Is future conversion to heavy oil a consideration?

J. Are leak detection, double wall piping, spill containment, double wall tanks, etc., properly specified and shown on the drawings?

K. Are stacks, vents, and breeching shown on the drawings and properly sized and specified? Coordinate with the design team other equipment requiring vents (e.g., water heaters, shop equipment, lab equipment, hospital equipment).

L. Is combustion air for fuel-fired equipment properly designed in accordance with code? Watch for water heaters in janitor closets.

M. Are EPA tank requirements met? Have state police requirements been met?

N. Are emergency vents properly sized for indoor tanks?

O. Are manholes and covers for fill and access openings specified and/or detailed to be waterproof/watertight?

51.16 Laboratory and Medical Gas Systems

A. Is a separate zone valve required?

B. Are medical gas alarm panels required?

C. Is the air intake for the hospital compressor indicated? Is it outside? Does it provide clean air?

D. Vacuum pump discharge should not be at rubber membrane roofs, due to the adverse reaction of oil with membrane materials.

E. Are *NFPA 99* requirements met?

51.17 General

A.   Are all mechanical items specified and coordinated with other disciplines as to who provides, furnishes, and/or installs? Have all items on the specification coordination list been coordinated? Do all disciplines have the most current drawings showing mechanical equipment?

B.   Is there a north arrow, title block, and engineer's stamp with signature?

C.   Are scales noted on the plans? Does the project or client require graphic scales?

D.   Are there client and project numbers on all projects, and the company name, logo, address, etc., on all drawings?

E.   Check for completeness of general notes, legend, abbreviations, and title blocks.

F.   Check column numbers and grids.

G.   Check room names and numbers.

H.   Is the extent of the demolition clearly defined? Is what is to remain clearly defined? Are points of connection between the new and old clearly defined?

I.   Check the coordination and contrast of new and existing work.

J.   Coordinate the following with architectural, structural, and electrical departments:
 1. Clearances between lighting fixtures, structure, and ducts and pipes.
 2. Clearances between conduits out of electrical panels and pull boxes, structure, and ducts and pipes.
 3. Wiring of filters (roll filters and air purification systems).

K.   Does the electrical department have the final motor list and heater list?

L.   Have existing mechanical/electrical services and available space for new work been adequately field checked?

M.   Advise the electrical department of any relocated mechanical equipment having electrical components.

N.   Has the division of work between the architectural, structural, mechanical, and electrical disciplines been coordinated (as to who furnishes, installs, and/or provides) on such items as:
 1. Starters and disconnect switches.
 2. Line and low voltage control wiring and power wiring to control panels.
 3. Access panels.
 4. Fire extinguishers, fire hoses, and/or cabinets.
 5. Catwalks and ladders.
 6. Under-window unit discharge grilles on built-in cabinets.
 7. Louvers.
 8. Door grilles, undercut doors.
 9. Generators, mufflers, fuel oil piping, engine exhausts, engine cooling air ductwork, and accessories.
 10. Painting and priming.
 11. Mechanical equipment screens.
 12. Equipment supports and concrete housekeeping pads.
 13. Roof curbs (equipment, ductwork, and piping), flashing, and counter flashing.
 14. Site work/building utility design termination (5'0" outside of the foundation wall).

15. Foundation drains.
16. Excavation.
17. Kitchenette units.
18. Bus washer, vehicle lifts, hydraulic piping and accessories, and paint booths and accessories.
19. Countertop plumbing fixtures; built-in showers.
20. Kitchen hoods.
21. Laboratory fume hoods.

O. Where the ceiling height and door or window head heights provide no leeway to lower ceiling, have mechanical and electrical work space above the ceiling been closely checked?

P. Check the framing of holes in existing structures.

Q. Is the structure adequate for new mechanical equipment in existing buildings?

R. Is there adequate clearance for the removal of ceiling systems for access to equipment? A tee bar system requires 3" minimum from the underside of the ceiling to the equipment.

S. Have the heating and ventilation of bathrooms and toilet rooms been provided?

T. Is there equipment room, PRV room, electrical room, and electrical closet ventilation?

U. Has insulation or ventilation been provided to overcome radiant heat from boiler or incinerator stacks?

V. Has specified equipment been properly described by current model designation?

W. Have all items specified "As indicated on the drawings" been coordinated? Coordinate references between drawings, details, sections, risers, and specifications.

X. Is there any material or equipment for which there is no catalog data in the office library?

Y. Have details been coordinated?

Z. Has space for future ducts, pipes, fans, pumps, chillers, boilers, cooling towers, water heaters, and other equipment been clearly indicated?

AA. Are "floating floors" required for noise control? Have they been specified and detailed?

BB. Has the existing area been adequately field checked?

CC. Are elevator machine rooms free of piping, ductwork, and equipment except elevator machine equipment? Is the elevator machine room ventilated? Does the elevator machine room need to be air conditioned?

DD. Have chemical treatment systems been included?

EE. Have handwash sinks been included in mechanical equipment rooms?

FF. Have chain operators for valves more than 7'0" above the finished floor been specified?

GG. Are general notes, drawing notes, and keyed notes included?

HH. Is a key plan needed?

II. Are applicable standard details included and coordinated?

JJ. Have applicable codes been researched?

KK. Should smoke and fire walls be indicated?

LL. Are present and ultimate duties included in schedules where applicable and coordinated with the Electrical Engineer? Are future flows accounted for in duct and pipe sizing and appropriate provisions made?

MM. Have authorities having jurisdiction been consulted regarding fire detection and protection systems, applicable codes, etc.?

NN. Is the minimum head room (6'8") maintained in equipment rooms?

OO. Is verification that the building meets *ASHRAE Standard 90.1* or other Energy Conservation Codes required?

PP. Is access to equipment with electrical connections (such as ceiling-mounted heat pumps) adequate to satisfy the NEC?

QQ. Have all equipment housekeeping pads been indicated, specified, and coordinated?

RR. Is asbestos present in the existing building? Is preparation of the removal documents part of the contract?

51.18 Architect and/or Owner Coordination

A. Have all shafts/chases been coordinated? Are they large enough?

B. Do shafts/chases line up floor to floor? Are structural members located in the shaft space?

C. Have pipe or duct chases been provided where required?

D. Will partitions accommodate piping and plumbing fixtures?

E. Has a suitable type stationary louver been specified?

F. Are bird screens (not insect screens) specified? Are bird screens located on the inside or outside of louver? The outside of louver is easier to clean but its appearance is undesirable.

G. Have louver locations and sizes been coordinated? Who provides, furnishes, and/or installs louvers?

H. Have plumbing fixtures, as required, been specified under the architectural section?

I. Have all plumbing fixtures been coordinated?

J. Has all special equipment been coordinated?

K. Have not in contract (NIC) or future items requiring "stub-up" services been identified?

L. Have masonry air shafts been avoided? If not, are they specified to be airtight?

M. Has proper access to roof mounted equipment been provided?

N. Have provisions for equipment replacement been made?

O. Have supply air ceiling plenums been coordinated? Are partitions floor-to-floor where required? Is the supply air plenum area sealed where required?

P. Have return air ceiling plenums been coordinated? Are partitions floor-to-floor? If so, have provisions been provided to return air from these spaces?

Q. Have trenches, sumps, and covers been coordinated?

R. Have under-window units been coordinated?

S. Have air outlet types been coordinated?

T. Have thermostat types been selected and approved by the owner?

U. Have plumbing fixtures and types been approved? Have countertop fixtures been coordinated? Who provides, furnishes, and/or installs countertop fixtures?

V. Include vibration isolators, grillage, and cooling tower safety rails when dimensioning the height of the cooling tower for the architectural screen.

W. Have all skylights, roof hatches, bulkheads, and multiple height ceilings been coordinated with ductwork, piping, and other mechanical equipment?

X. Who provides, furnishes, and/or installs roof curbs for mechanical equipment?

Y. Who provides, furnishes, and/or installs flashing and counterflashing?

Z. Who provides cutting and patching?

51.19 Structural Engineer Coordination

A. Have equipment locations, sizes, and weights been given to the Structural Engineer? Have equipment housekeeping pad locations and sizes been coordinated? Has the final and complete structural list been given to the Structural Engineer?

B. Have all floor, roof, and wall openings been coordinated?

C. Have pipes 6 in. and larger been located and coordinated with the Structural Engineer?

D. Have all sleeved beams, grade beams, and foundations been coordinated? Have pipes and ducts been coordinated?

E. Has structural framing in the shafts been considered?

F. Has the mechanical layout been coordinated with the structural system, especially in post-tensioned concrete structural systems? (Penetrations at columns and column lines are not normally possible.)

G. Is the structural system adequate for future equipment?

H. Where equipment must be "rolled" into place, is the structure over which equipment will be rolled adequate?

I. Have catwalks been coordinated?

J.   Have pipe risers been coordinated?

K.   Do structural openings allow for insulation and ductwork reinforcing?

L.   Have anchor locations and associated forces been given to the Structural Engineer? Avoid locating anchors at joist or joist girder locations.

M.   Have louver openings, sizes, and framing been coordinated with the Structural Engineer?

51.20 Electrical Engineer Coordination

A.   Has the final and complete motor list been given to the Electrical Engineer?

B.   Have all electrical and telecommunication rooms and closets been ventilated? Do they need to be air conditioned?

C.   Have duct smoke detectors, duct fire stats, and/or smoke dampers been coordinated?

D.   Have valve position indicators/tamper switches been coordinated?

E.   Have sprinkler flow switches and alarms been coordinated?

F.   Have fuel tank level alarms and gauges been coordinated?

G.   Have cooling tower electric basin heaters and vibration switches for propeller fans been coordinated?

H.   Have medical gas alarms been coordinated?

I.   Have the automatic trap priming systems for the kitchen and other areas been coordinated?

J.   Has the automatic trap priming system for the AHUs been coordinated?

K.   Has lighting inside the AHUs been coordinated?

L.   Has power at the pneumatic tube stations been coordinated?

M.   Has power for the ATC compressors and refrigerated air dryers been coordinated?

N.   Who provides starters and disconnect switches? Who provides line voltage and low-voltage control wiring? Who provides power wiring to the control panels? Have starters, wall switches, remote starter pushbuttons, and disconnect switches been located on the mechanical drawings?

O.   Have two disconnects been provided at duplex pumps?

P.   Are there automatic fire suppression systems for fume hoods and kitchen hoods?

Q.   Are there alarms on sump pumps, condensate pumps, sewage pumps, hot water generators, and similar items?

R.   Are there chiller oil heaters and control circuits (winterize air-cooled chillers)?

S.   Are there diesel generator fuel oil pumps on emergency power? Who provides the engine exhaust, fuel-oil piping, day tank, muffler, cooling air, fuel storage tank, etc.?

T.   Steam or water flow on the BTU meter recorders?

U.   Have shower controls been coordinated?

V.   Are there automatic fire suppression systems for the computer rooms? Are AHUs interlocked with the computer room shutdown system?

W.   Are there smoke or thermal detectors for AHUs and RA fans? Who furnishes, installs, and/or provides them?

X.   Has heat tracing for piping systems been coordinated?

Y.   Are there electric fuel tank heating systems?

Z.   Is there auxiliary equipment on the water chillers?

AA.   Has the motor list been coordinated with equipment schedules?

BB.   Has the motor list been coordinated with control diagrams?

CC.   Have electric humidifiers been coordinated?

DD.   Have hot water generator or boiler circulating pumps been coordinated?

EE.   Has relocated equipment been coordinated?

FF.   Have allowances been made for lighting fixture access? Have the heights of lighting fixtures been coordinated, especially high hat fixtures?

GG.   No ductwork, piping, or other mechanical equipment should be in electrical rooms or closets.

HH.   Are motor control centers (MCCs) shown and specified? Are starters shown and specified?

II.   Is there adequate space for MCCs?

JJ.   Is there enough space for electric water level detectors?

KK.   Are electric motor-operated dampers wired?

LL.   Are there air handling light fixtures, supply, return, and heat transfer?

MM.   Has the extent of return air ceilings been coordinated with the Electrical Engineer?

NN.   Is the equipment on emergency power clearly defined and coordinated? Include the control air compressor and dryer.

OO.   Are explosion proof motors, starters, disconnect switches, etc., required?

PART 52

Professional Societies and Trade Organizations

52.01 Professional Societies and Trade Organizations

AABC	Associated Air Balance Council
AACC	American Automatic Control Council
AAHC	American Association of Health Care Consultants
ABMA	American Boiler Manufacturers' Association
ACCA	Air Conditioning Contractors of America
ACGIH	American Conference of Governmental and Industrial Hygienists
ACI	American Concrete Institute
ACS	American Ceramic Society
ACS	American Chemical Society
ACSM	American Congress on Surveying and Mapping
ADA	Americans with Disabilities Act
ADAAG	ADA Accessibility Guidelines for Buildings and Facilities
ADC	Air Diffusion Council
AEE	Association of Energy Engineers
AEI	Architectural Engineering Institute
AFBMA	American Fan and Bearing Manufacturers' Association
AFS	American Foundrymen's Society
AGA	American Gas Association
AGMA	American Gear Manufacturers Association
AHA	American Hospital Association
AHCA	American Health Care Association
AHRI	Air-Conditioning, Heating, and Refrigeration Institute
AIA	American Institute of Architects
AIA	American Insurance Association
AICE	American Institute of Consulting Engineers
AIChE	American Institute of Chemical Engineers
AIHA	American Industrial Hygiene Association
AIIE	American Institute of Industrial Engineers, Inc.
AIPE	American Institute of Plant Engineers
AISC	American Institute of Steel Construction
AISE	Association of Iron and Steel Engineers
AISI	American Iron and Steel Institute
AMCA	Air Movement and Control Association International, Inc.
ANSI	American National Standards Institute
APCA	Air Pollution Control Association
APFA	American Pipe and Fittings Association
APHA	American Public Health Association
API	American Petroleum Institute
APWA	American Public Works Association
ASA	Acoustical Society of America
ASCE	American Society of Civil Engineers
ASCET	American Society of Certified Engineering Technicians
ASEE	American Society for Engineering Education
ASHRAE	American Society of Heating, Refrigerating, and Air Conditioning Engineers
ASLE	American Society of Lubricating Engineers
ASME	American Society of Mechanical Engineers International
ASNT	American Society for Nondestructive Testing
ASPE	American Society of Plumbing Engineers
ASQC	American Society of Quality Control, Inc.
ASSE	American Society of Safety Engineers
ASSE	American Society of Sanitary Engineers
ASTM	American Society for Testing and Materials
ATBCB	Architectural and Transportation Barrier Compliance Board

AWS	American Welding Society
AWWA	American Water Works Association, Inc.
BCMC	Board for the Coordination of Model Codes (a Board of CABO)
BDC	Building Design and Construction
BEPS	Building Energy Performance Standards
BICSI	Building Industries Consulting Services International
BOCA	Building Officials and Code Administrators
BOMA	Building Owners' and Managers' Association
BRI	Building Research Institute
BSI	British Standards Institute
CABO	Council of American Building Officials
CAGI	Compressed Air and Gas Institute
CANENA	North American Electro/Technical Standards Harmonization Council
CEC	Consulting Engineers Council of the United States
CEN	European Standards Organization
CENELEC	European Committee for Electro/Technical Standardization
CGA	Compressed Gas Association, Inc.
CISPI	Cast Iron Soil Pipe Institute
CSA	Canadian Standards Association
CSI	Construction Specifications Institute
CTI	Cooling Tower Institute
DER	Department of Environmental Resources
DOE	Department of Energy
DOH	Department of Health
ECPD	Engineers' Council for Professional Development
EF	Engineering Foundation
EJC	Engineers' Joint Council
EJMA	Expansion Joint Manufacturers' Association
EPA	Environmental Protection Agency
ETL	ETL Testing Laboratories
FM	Factory Mutual System
FPS	Fluid Power Society
HAP	Hospital & Healthsystem Association of Pennsylvania
HEI	Heat Exchange Institute
HI	Hydraulic Institute
HTFMI	Heat Transfer and Fluid Mechanics Institute
HYDI	Hydronics Institute
IAHHS	International Association for Healthcare Security and Safety
IAPMO	International Association of Plumbing and Mechanical Officials
IBR	Institute of Boiler and Radiator Manufacturers
ICBO	International Conference of Building Officials
ICC	International Code Council (BOCA, CABO, ICBO, and SBCCI combined)
ICET	Institute for the Certification of Engineering Technicians
IEC	International Electro/Technical Commission
IEEE	Institute of Electrical and Electronics Engineers
IES	Illuminating Engineering Society
IESNA	Illuminating Engineering Society of North America
IFCI	International Fire Code Institute
IFI	Industrial Fasteners Institute
IFMA	International Facility Managers' Association
IHCA	Integrated Health Care Association
IIAR	International Institute of Ammonia Refrigeration
IRI	HSB Industrial Risk Insurers
IRI	Industrial Research Institute, Inc.
ISA	Instrument Society of America
ISO	International Organization for Standardization

JCAHO	Joint Commission on Accreditation of Healthcare Organizations
MCAA	Mechanical Contractors Association of America
MSS	Manufacturers' Standardization Society of the Valve and Fittings Industry
NACE	National Association of Corrosion Engineers
NAE	National Academy of Engineering
NAHC	National Association of Health Consultants
NAHSE	National Association of Health Services Executives
NAIMA	North American Insulation Manufacturers Association
NAPE	National Association of Power Engineers, Inc.
NAPHCC	National Association of Plumbing-Heating-Cooling Contractors
NAS	National Academy of Sciences
NBFU	National Board of Fire Underwriters
NBS	National Bureau of Standards
NCEE	National Council of Engineering Examiners
NCPWB	National Certified Pipe Welding Bureau
NCSBCS	National Conference of States on Building Codes and Standards
NEBB	National Environmental Balancing Bureau
NEC	National Electric Code
NEMA	National Electrical Manufacturers' Association
NEMI	National Energy Management Institute
NFPA	National Fire Protection Association
NFRC	National Fenestration Rating Council
NFSA	National Fire Sprinkler Association
NIAOP	National Association of Industrial and Office Properties
NICE	National Institute of Ceramic Engineers
NICET	National Institute of Certified Engineering Technicians
NIOSH	National Institute for Occupational Safety and Health
NIST	National Institute of Standards and Technology
NRC	National Research Council
NRCA	National Roofing Contractors' Association
NRCC	National Research Council of Canada
NSAE	National Society of Architectural Engineers
NSF	National Sanitation Foundation International
NSPE	National Society of Professional Engineers
NUSIG	National Uniform Seismic Installation Guidelines
OSHA	Occupational Safety and Health Administration
PDI	Plumbing and Drainage Institute
PFI	Pipe Fabrication Institute
RESA	Scientific Research Society of America
SAE	Society of Automotive Engineers
SAME	Society of American Military Engineers
SAVE	Society of American Value Engineers
SBCCI	Southern Building Code Congress International
SES	Solar Energy Society
SFPE	Society of Fire Protection Engineers
SMACNA	Sheet Metal and Air Conditioning Contractors' National Association
SPE	Society of Plastics Engineers, Inc.
SSPC	Structural Steel Painting Council
SSPMA	Sump and Sewage Pump Manufacturers' Association
SWE	Society of Women Engineers
TEMA	Tubular Exchanger Manufacturers Association
TIMA	Thermal Insulation Manufacturers' Association
UL	Underwriters' Laboratories, Inc.
WPCF	Water Pollution Control Federation

References and Design Manuals

53.01 References and Design Manuals

A. The references listed in the paragraphs to follow form the basis for most of the information contained in this manual. In addition, these references are excellent HVAC design manuals and will provide expanded explanations of the information contained within this text. These references are recommended for all HVAC engineers' libraries.

B. American Society of Heating, Refrigerating, and Air-Conditioning Engineers (ASHRAE) Handbooks

ASHRAE. *ASHRAE Handbook, 2015 HVAC Applications Volume, Inch-Pound Edition.* Atlanta, GA: ASHRAE, 2015.

ASHRAE. *ASHRAE Handbook, 2014 Refrigeration Volume, Inch-Pound Edition.* Atlanta, GA: ASHRAE, 2014.

ASHRAE. *ASHRAE Handbook, 2013 Fundamentals Volume, Inch-Pound Edition.* Atlanta, GA: ASHRAE, 2013.

ASHRAE. *ASHRAE Handbook, 2012 HVAC Systems and Equipment Volume, Inch-Pound Edition.* Atlanta, GA: ASHRAE, 2012.

ASHRAE. *ASHRAE Handbook, 2011 HVAC Applications Volume, Inch-Pound Edition.* Atlanta, GA: ASHRAE, 2011.

ASHRAE. *ASHRAE Handbook, 2010 Refrigeration Volume, Inch-Pound Edition.* Atlanta, GA: ASHRAE, 2010.

ASHRAE. *ASHRAE Handbook, 2009 Fundamentals Volume, Inch-Pound Edition.* Atlanta, GA: ASHRAE, 2009.

ASHRAE. *ASHRAE Handbook, 2008 HVAC Systems and Equipment Volume, Inch-Pound Edition.* Atlanta, GA: ASHRAE, 2008.

ASHRAE. *ASHRAE Handbook, 2007 HVAC Applications Volume, Inch-Pound Edition.* Atlanta, GA: ASHRAE, 2007.

ASHRAE. *ASHRAE Handbook, 2006 Refrigeration Volume, Inch-Pound Edition.* Atlanta, GA: ASHRAE, 2006.

ASHRAE. *ASHRAE Handbook, 2005 Fundamentals Volume, Inch-Pound Edition.* Atlanta, GA: ASHRAE, 2005.

ASHRAE. *ASHRAE Handbook, 2004 HVAC Systems and Equipment Volume, Inch-Pound Edition.* Atlanta, GA: ASHRAE, 2004.

ASHRAE. *ASHRAE Handbook, 2003 HVAC Applications Volume, Inch-Pound Edition.* Atlanta, GA: ASHRAE, 2003.

ASHRAE. *ASHRAE Handbook, 2002 Refrigeration Volume, Inch-Pound Edition.* Atlanta, GA: ASHRAE, 2002.

ASHRAE. *ASHRAE Handbook, 2001 Fundamentals Volume, Inch-Pound Edition.* Atlanta, GA: ASHRAE, 2001.

ASHRAE. *ASHRAE Handbook, 2000 HVAC Systems and Equipment Volume, Inch-Pound Edition.* Atlanta, GA: ASHRAE, 2000.

ASHRAE. *ASHRAE Handbook, 1999 HVAC Applications Volume, Inch-Pound Edition.* Atlanta, GA: ASHRAE, 1999.

ASHRAE. *ASHRAE Handbook, 1998 Refrigeration Volume, Inch-Pound Edition.* Atlanta, GA: ASHRAE, 1998.

ASHRAE. *ASHRAE Handbook, 1997 Fundamentals Volume, Inch-Pound Edition.* Atlanta, GA: ASHRAE, 1997.

ASHRAE. *ASHRAE Handbook, 1996 HVAC Systems and Equipment Volume, Inch-Pound Edition.* Atlanta, GA: ASHRAE, 1996.

ASHRAE. *ASHRAE Handbook, 1995 HVAC Applications Volume, Inch-Pound Edition.* Atlanta, GA: ASHRAE, 1995.

ASHRAE. *ASHRAE Handbook, 1994 Refrigeration Volume, Inch-Pound Edition.* Atlanta, GA: ASHRAE, 1994.

ASHRAE. *ASHRAE Handbook, 1993 Fundamentals Volume, Inch-Pound Edition.* Atlanta, GA: ASHRAE, 1993.

ASHRAE. *ASHRAE Handbook, 1992 HVAC Systems and Equipment Volume, Inch-Pound Edition.* Atlanta, GA: ASHRAE, 1992.

ASHRAE. *ASHRAE Handbook, 1991 HVAC Applications Volume, Inch-Pound Edition.* Atlanta, GA: ASHRAE, 1991.

ASHRAE. *ASHRAE Handbook, 1990 Refrigeration Volume, Inch-Pound Edition.* Atlanta, GA: ASHRAE, 1990.

ASHRAE. *ASHRAE Handbook, 1989 Fundamentals Volume, Inch-Pound Edition.* Atlanta, GA: ASHRAE, 1989.

ASHRAE. *ASHRAE Handbook, 1988 Equipment Volume, Inch-Pound Edition.* Atlanta, GA: ASHRAE, 1988.

ASHRAE. *ASHRAE Handbook, 1987 HVAC Systems and Applications Volume, Inch-Pound Edition.* Atlanta, GA: ASHRAE, 1987.

ASHRAE. *ASHRAE Handbook, 1986 Refrigeration Volume, Inch-Pound Edition.* Atlanta, GA: ASHRAE, 1986.

ASHRAE. *ASHRAE Handbook, 1985 Fundamentals Volume, Inch-Pound Edition.* Atlanta, GA: ASHRAE, 1985.

ASHRAE. *ASHRAE Handbook, 1984 Systems Volume.* Atlanta, GA: ASHRAE, 1984.

ASHRAE. *ASHRAE Handbook, 1983 Equipment Volume.* Atlanta, GA: ASHRAE, 1983.

ASHRAE. *ASHRAE Handbook, 1982 Applications Volume.* Atlanta, GA: ASHRAE, 1982.

ASHRAE. *ASHRAE Handbook, 1981 Fundamentals Volume.* Atlanta, GA: ASHRAE, 1981.

ASHRAE. *ASHRAE Handbook, 1980 Systems Volume.* Atlanta, GA: ASHRAE, 1980.

C. American Society of Heating, Refrigerating, and Air-Conditioning Engineers (ASHRAE) Standards, and Manuals

1. Standards:

ASHRAE. *ASHRAE Standard 15-2013, Safety Standard for Refrigeration Systems.* Atlanta, GA: ASHRAE, 2013.

ASHRAE. *ASHRAE Standard 34-2013, Design and Safety Classification of Refrigerants.* Atlanta, GA: ASHRAE, 2013.

ASHRAE. *ASHRAE Standard 52.1-2007, Gravimetric and Dust-Spot Procedures for Testing Air-Cleaning Devices Used in General Ventilation for Removing Particulate Matter.* Atlanta, GA: ASHRAE, 2007.

ASHRAE. *ASHRAE Standard 52.2-2012, Method of Testing General Ventilation Air-Cleaning Devices for Removal Efficiency by Particle Size.* Atlanta, GA: ASHRAE, 2012.

ASHRAE. *ASHRAE Standard 55-2013, Thermal Environmental Conditions for Human Occupancy.* Atlanta, GA: ASHRAE, 2013.

ASHRAE. *ASHRAE Standard 62.1-2013, Ventilation for Acceptable Indoor Air Quality.* Atlanta, GA: ASHRAE, 2013.

ASHRAE. *ASHRAE Standard 62.2-2013, Ventilation and Acceptable Indoor Air Quality in Low-Rise Residential Buildings.* Atlanta, GA: ASHRAE, 2013.

ASHRAE. *ASHRAE Standard 90.1-2013, Energy Standard for Buildings Except Low-Rise Residential Buildings.* Atlanta, GA: ASHRAE, 2013.

ASHRAE. *ASHRAE Standard 90.2-2007, Energy Efficient Design of Low-Rise Residential Buildings.* Atlanta, GA: ASHRAE, 2007.

ASHRAE. *ASHRAE Standard 100-2006, Energy Conservation in Existing Buildings.* Atlanta, GA: ASHRAE, 2006.

ASHRAE. *ASHRAE Standard 110-1995, Method of Testing Performance of Laboratory Fume Hoods.* Atlanta, GA: ASHRAE, 1995.

ASHRAE. *ASHRAE Standard 111-2008, Practices for Measurement, Testing, Adjusting, and Balancing of Building Heating, Ventilation, Air-Conditioning, and Refrigeration Systems.* Atlanta, GA: ASHRAE, 2008.

ASHRAE. *ASHRAE Standard 135-2012, BACnet: A Data Communication Protocol for Building Automation Control Networks.* Atlanta, GA: ASHRAE, 2012.

ASHRAE. *ASHRAE Standard 135.1-2013, Method of Test for Conformance to BACnet.* Atlanta, GA: ASHRAE, 2013.

ASHRAE. *ASHRAE Standard 154-2011, Ventilation for Commercial Cooking Operations.* Atlanta, GA: ASHRAE, 2011.

ASHRAE. *ASHRAE Standard 170-2013, Ventilation of Health Care Facilities.* Atlanta, GA: ASHRAE, 2013.

2. Guidelines:

ASHRAE. *ASHRAE Guideline 1.1-2007, HVAC&R Technical Requirements for the Commissioning Process.* Atlanta, GA: ASHRAE, 2007.

ASHRAE. *ASHRAE Guideline 1.5-2012, Commissioning Process for Smoke Control Systems.* Atlanta, GA: ASHRAE, 2012.

ASHRAE. *ASHRAE Guideline 4-2008, Preparation of Operating and Maintenance Documentation for Building Systems.* Atlanta, GA: ASHRAE, 2008.

ASHRAE. *ASHRAE Guideline 12-2000, Minimizing the Risk of Legionellosis Associated with Building Water Systems.* Atlanta, GA: ASHRAE, 2000.

ASHRAE. *ASHRAE Guideline 13-2014, Specifying Direct Digital Control Systems.* Atlanta, GA: ASHRAE, 2014.

3. Manuals:

ASHRAE. *Design of Smoke Control Systems for Buildings.* 1st Ed., Atlanta, GA: ASHRAE, 1983.

ASHRAE. *Pocket Handbook for Air Conditioning, Heating, Ventilation, Refrigeration.* Atlanta, GA: ASHRAE, 1987.

McIntosh, Ian B.D., Dorgan, Chad B., and Dorgan, Charles E. *ASHRAE Laboratory Design Guide.* Atlanta, GA: ASHRAE, 2001.

ASHRAE. *ASHRAE HVAC Design Manual for Hospitals and Clinics.* Atlanta, GA: ASHRAE, 2003.

Klote, John H. and Milke, James A. *Principles of Smoke Management.* Atlanta, GA: ASHRAE, 2002.

Grumman, David L., Editor. *ASHRAE Green Guide.* Atlanta, GA: ASHRAE, 2003.

Ross, Donald E. HVAC *Design Guide for Tall Commercial Buildings.* Atlanta, GA: ASHRAE, 2004.

D. American National Standards Institute (ANSI) and American Society of Mechanical Engineers (ASME)

ANSI/ASME. *ANSI/ASME A13.1 Scheme for the Identification of Piping Systems, 2007.* New York, NY: ANSI/ASME, 2007.

ANSI/ASME. *ANSI/ASME B31.1 Power Piping, 2014.* New York, NY: ANSI/ASME, 2014.

ANSI/ASME. *ANSI/ASME B31.3 Process Piping, 2014.* New York, NY: ANSI/ASME, 2014.

ANSI/ASME. *ANSI/ASME B31.5 Refrigeration Piping and Heat Transfer Components, 2013.* New York, NY: ANSI/ASME, 2013.

ANSI/ASME. *ANSI/ASME B31.9 Building Services Piping, 2014.* New York, NY: ANSI/ASME, 2014.

ANSI/ASME. *ANSI/ASME Boiler and Pressure Vessel Code, 2015.* New York, NY: ANSI/ASME, 2015.

E. Bell and Gossett Manuals

ITT Corporation. *Pump and System Curve Data for Centrifugal Pump Selection and Application.* Morton Grove, IL: ITT Corporation, Training and Education Department, Fluid Handling Division, 1967.

ITT Corporation. *Pump Data Book.* Morton Grove, IL: ITT Corporation, Training and Education Department, Fluid Handling Division, 1970.

ITT Corporation. *Parallel and Series Pump Application.* Morton Grove, IL: ITT Corporation, Training and Education Department, Fluid Handling Division, 1965.

ITT Corporation. *Principles of Centrifugal Pump Construction and Maintenance.* Morton Grove, IL: ITT Corporation, Training and Education Department, Fluid Handling Division, 1965.

ITT Corporation. *Cooling Tower Pumping and Piping.* Morton Grove, IL: ITT Corporation, Training and Education Department, Fluid Handling Division, 1968.

ITT Corporation. *Variable Speed/Variable Volume Pumping Fundamentals.* Morton Grove, IL: ITT Corporation, Training and Education Department, Fluid Handling Division, 1985.

ITT Corporation. *Heat Exchangers, Application and Installation.* Morton Grove, IL: ITT Corporation, Training and Education Department, Fluid Handling Division, 1965.

ITT Corporation. *Primary Secondary Pumping Application Manual.* Morton Grove, IL: ITT Corporation, Training and Education Department, Fluid Handling Division, 1968.

ITT Corporation. *One Pipe Primary Systems, Flow Rate and Water Temperature Determination.* Morton Grove, IL: ITT Corporation, Training and Education Department, Fluid Handling Division, 1966.

ITT Corporation. *Primary Secondary Pumping Adaptations to Existing Systems.* Morton Grove, IL: ITT Corporation, Training and Education Department, Fluid Handling Division, 1966.

ITT Corporation. *Dual Temperature Change Over Single Zone.* Morton Grove, IL: ITT Corporation, Training and Education Department, Fluid Handling Division, 1967.

ITT Corporation. *Single Coil Instantaneous Room by Room Heating-Cooling Systems.* Morton Grove, IL: ITT Corporation, Training and Education Department, Fluid Handling Division, 1965.

ITT Corporation. *Equipment Room Piping Practice.* Morton Grove, IL: ITT Corporation, Training and Education Department, Fluid Handling Division, 1965.

ITT Corporation. *Pressurized Expansion Tank Sizing/Installation Instructions for Hydronic Heating/Cooling Systems.* Morton Grove, IL: ITT Corporation, Training and Education Department, Fluid Handling Division, 1988.

ITT Corporation. *Snow Melting System Design and Problems.* Morton Grove, IL: ITT Corporation, Training and Education Department, Fluid Handling Division, 1966.

ITT Corporation. *Hydronic Systems Anti-Freeze Design.* Morton Grove, IL: ITT Corporation, Training and Education Department, Fluid Handling Division, 1965.

ITT Corporation. *Air Control for Hydronic Systems.* Morton Grove, IL: ITT Corporation, Training and Education Department, Fluid Handling Division, 1966.

ITT Corporation. *Basic System Control and Valve Sizing Procedures.* Morton Grove, IL: ITT Corporation, Training and Education Department, Fluid Handling Division, 1970.

ITT Corporation. *Hydronic Systems: Analysis and Evaluation.* Morton Grove, IL: ITT Corporation, Training and Education Department, Fluid Handling Division, 1969.

ITT Corporation. *Circuit Setter Valve Balance Procedure Manual.* Morton Grove, IL: ITT Corporation, Training and Education Department, Fluid Handling Division, 1971.

ITT Corporation. *Domestic Water Service.* Morton Grove, IL: ITT Corporation, Training and Education Department, Fluid Handling Division, 1970.

F.   Carrier Manuals

Carrier Corporation. *Carrier System Design Manuals, Part 1—Load Estimating.* Syracuse, NY: Carrier Corporation, 1972.

Carrier Corporation. *Carrier System Design Manuals, Part 2—Air Distribution.* Syracuse, NY: Carrier Corporation, 1974.

Carrier Corporation. *Carrier System Design Manuals, Part 3—Piping Design.* Syracuse, NY: Carrier Corporation, 1973.

Carrier Corporation. *Carrier System Design Manuals, Part 4—Refrigerants, Brines, Oils.* Syracuse, NY: Carrier Corporation, 1969.

Carrier Corporation. *Carrier System Design Manuals, Part 5—Water Conditioning.* Syracuse, NY: Carrier Corporation, 1972.

Carrier Corporation. *Carrier System Design Manuals, Part 6—Air Handling Equipment.* Syracuse, NY: Carrier Corporation, 1968.

Carrier Corporation. *Carrier System Design Manuals, Part 7—Refrigeration Equipment.* Syracuse, NY: Carrier Corporation, 1969.

Carrier Corporation. *Carrier System Design Manuals, Part 8—Auxiliary Equipment.* Syracuse, NY: Carrier Corporation, 1966.

Carrier Corporation. *Carrier System Design Manuals, Part 9—Systems and Applications.* Syracuse, NY: Carrier Corporation, 1971.

Carrier Corporation. *Carrier System Design Manuals, Part 10—Air-Air Systems.* Syracuse, NY: Carrier Corporation, 1975.

Carrier Corporation. *Carrier System Design Manuals, Part 11—Air-Water Systems.* Syracuse, NY: Carrier Corporation, 1966.

Carrier Corporation. *Carrier System Design Manuals, Part 12—Water and DX Systems.* Syracuse, NY: Carrier Corporation, 1975.

G. Cleaver Brooks Manuals

Cleaver Brooks. *The Boiler Book: A Complete Guide to Advanced Boiler Technology for the Specifying Engineer.* 1st Ed., Milwaukee, WI: Cleaver Brooks, 1993.

Cleaver Brooks. *Hot Water Systems, Components, Controls, and Layouts.* Milwaukee, WI: Cleaver Brooks, 1972.

Cleaver Brooks. *Application . . . and Misapplication of Hot Water Boilers.* Milwaukee, WI: Cleaver Brooks, 1976.

H. Johnson Controls Manuals

Johnson Controls. *Fundamentals of Pneumatic Control.* Milwaukee, WI: Johnson Controls.

Johnson Controls. *Johnson Field Training Handbook, Fundamentals of Electronic Control Equipment.* Milwaukee, WI: Johnson Controls.

Johnson Controls. *Johnson Field Training Handbook, Fundamentals of Systems.* Milwaukee, WI: Johnson Controls.

I. Honeywell Manual

Honeywell. Engineering Manual of Automatic Control for Commercial Buildings, Heating, Ventilating, and Air Conditioning. Inch-Pound Edition, Minneapolis, MN: Honeywell, 1991.

J. Industrial Ventilation Manual

American Conference of Governmental and Industrial Hygienists. *Industrial Ventilation, A Manual of Recommended Practice.* 28th Ed., Cincinnati, OH: American Conference of Governmental and Industrial Hygienists, 2013.

K. SMACNA (Sheet Metal and Air-Conditioning Contractors' National Association, Inc.) Manuals

SMACNA. *Fibrous Glass Duct Construction Standards.* 7th Ed., Vienna, VA: SMACNA, 2003.

SMACNA. *Fire, Smoke, and Radiation Damper Installation Guide for HVAC.* 5th Ed., Vienna, VA: SMACNA, 2002.

SMACNA. *HVAC Air Duct Leakage Test Manual.* 2nd Ed., Vienna, VA: SMACNA, 2012.

SMACNA. *HVAC Duct Construction Standards—Metal and Flexible.* 3rd Ed., Vienna, VA: SMACNA, 2005.

SMACNA. *HVAC Systems Duct Design.* 4th Ed., Vienna, VA: SMACNA, 2006.

SMACNA. *HVAC Systems Testing, Adjusting, and Balancing.* 3rd Ed., Vienna, VA: SMACNA, 2002.

SMACNA. *Rectangular Industrial Duct Construction Standards.* 2nd Ed., Vienna, VA: SMACNA, 2004.

SMACNA. *Round Industrial Duct Construction Standards.* 2nd Ed., Vienna, VA: SMACNA, 1999.

SMACNA. *Seismic Restraint Manual Guidelines for Mechanical Systems.* 3rd Ed., Vienna, VA: SMACNA, 2008.

SMACNA. *Thermoplastic Duct (PVC) Construction Manual.* 2nd Ed., Vienna, VA: SMACNA, 1995.

L. Trane Manuals

The Trane Company. *Trane Air-Conditioning Manual.* LaCross, WI: The Trane Company, 1988.

The Trane Company. *Psychrometry.* LaCross, WI: The Trane Company, 1988.

M. United McGill Corporation

United McGill Corporation. *Engineering Design Reference Manual for Supply Air Handling Systems.* Westerville, OH: United McGill Corporation, 1989.

United McGill Corporation. *Underground Duct Installation (No. 95).* Westerville, OH: United McGill Corporation, 1992.

United McGill Corporation. *Flat Oval vs. Rectangular Duct (No. 150).* Westerville, OH: United McGill Corporation, 1989.

United McGill Corporation. *Flat Oval Duct—The Alternative to Rectangular (No. 151).* Westerville, OH: United McGill Corporation, 1989.

United McGill Corporation. *Underground Duct Design (No. 155).* Westerville, OH: United McGill Corporation, 1992.

N. Manufacturers Standardization Society of the Valve and Fitting Industry

Manufacturers Standardization Society of the Valve and Fitting Industry. *Standard Marking System for Valves, Fittings, Flanges and Unions (Standard SP-25-1988).* Vienna, VA: Manufacturers Standardization Society of the Valve and Fitting Industry, 1988.

Manufacturers Standardization Society of the Valve and Fitting Industry. *Pipe Hangers and Supports—Materials, Design, and Manufacturers (Standard SP-58-1988).* Vienna, VA: Manufacturers Standardization Society of the Valve and Fitting Industry, 1988.

Manufacturers Standardization Society of the Valve and Fitting Industry. *Pipe Hangers and Supports—Selection and Application (Standard SP-69-1983).* Vienna, VA: Manufacturers Standardization Society of the Valve and Fitting Industry, 1983.

Manufacturers Standardization Society of the Valve and Fitting Industry. *Pipe Hangers and Supports—Fabrication and Installation Practices (Standard SP-89-1985).* Vienna, VA: Manufacturers Standardization Society of the Valve and Fitting Industry, 1985.

Manufacturers Standardization Society of the Valve and Fitting Industry. *Guidelines on Terminology for Pipe Hangers and Supports (Standard SP-90-1986).* Vienna, VA: Manufacturers Standardization Society of the Valve and Fitting Industry, 1986.

Manufacturers Standardization Society of the Valve and Fitting Industry. *Guidelines for Manual Operation of Valves (Standard SP-91-1984).* Vienna, VA: Manufacturers Standardization Society of the Valve and Fitting Industry, 1984.

Manufacturers Standardization Society of the Valve and Fitting Industry. *MSS Valve User Guide (Standard SP-92-1987).* Vienna, VA: Manufacturers Standardization Society of the Valve and Fitting Industry, 1987.

Manufacturers Standardization Society of the Valve and Fitting Industry. *Guidelines on Terminology for Valves and Fittings (Standard SP-96-1986).* Vienna, VA: Manufacturers Standardization Society of the Valve and Fitting Industry, 1986.

O. Miscellaneous

Angel, W. Larsen. *HVAC Design Sourcebook.* New York, NY: The McGraw-Hill Companies, Inc., 2012.

Armstrong. *Steam Conservation Guidelines for Condensate Drainage.* Three Rivers, MI: Armstrong Machine Works, 1976.

Avallone, Eugene A. and Baumeister, III Theodore. *Mark's Standard Handbook for Mechanical Engineers.* 9th Ed., New York, NY: McGraw-Hill Book Co., 1986.

Bolz, D. and Tuve, George L. *CRC Handbook of Tables for Applied Engineering Science.* 2nd. Ed., Boca Raton, FL: CRC Press, Inc., 1980.

Clough, Richard H. *Construction Contracting.* 4th Ed., New York, NY: John Wiley and Sons, Inc., 1981.

Dryomatic, Div. Airflow, Co. *Dehumidification Engineering Manual.* Frederick, MD: Dryomatic, Div. Airflow, Co., 1965.

Haines, Roger W. *Control Systems for Heating, Ventilating, and Air Conditioning.* 4th Ed., New York, NY: Van Nostrand Reinhold Company, 1987.

Hansen, Erwin G. *Hydronic System Design and Operation, A Guide to Heating and Cooling with Water.* New York, NY: The McGraw-Hill Companies, Inc., College Customs Series, 1996.

Harris, Norman C. *Modern Air Conditioning Practice.* 3rd Ed., New York, NY: Glencoe Div. of Macmillan/McGraw-Hill, 1992.

Hauf, Harold D. *Architectural Graphic Standards.* 6th Ed., New York, NY: John Wiley and Sons, Inc., 1970.

Heald, C. C. *Cameron Hydraulic Data.* 17th Ed., Woodcliff Lake, NJ: Ingersoll Rand, 1988.

Leslie Control, Inc. *Steam Pressure Control Systems.* Tampa, FL: Leslie Controls, Inc.

The Marley Cooling Tower Co. *Cooling Tower Fundamentals.* 2nd Ed., Kansas City, MO: The Marley Cooling Tower Co., 1985.

McGuinness, William J. *Mechanical and Electrical Equipment for Buildings.* 6th Ed., New York, NY: John Wiley and Sons, Inc., 1980.

Nayyar, Mohinder L. *Piping Handbook.* 6th Ed., New York, NY: McGraw Hill, Inc. 1992.

The Singer Company. *Designing the Installation of the Electro-Hydronic Energy Conservation System.* Auburn, NY: The Singer Company, Climate Control Div., 1978.

Spence Engineering Co. *Steam Pressure Reducing Station Noise Treatment.* Walden, NY: Spence Engineering Co.

Spirax/Sarco. *Design of Fluid Systems, Steam Utilization.* Allentown, PA: Spirax/Sarco, 1991.

Spirax/Sarco. *Design of Fluid Systems, Hook-ups.* Allentown, PA: Spirax/Sarco, 1992.

Strock, Clifford. *Handbook of Air Conditioning, Heating, and Ventilating.* 1st Ed., New York, NY: The Industrial Press, 1959.

Systecon, Inc. Distributed Pumping (Pressure Gradient Control) for Chilled Water and Hot Water Systems. Cincinnati, OH: Systecon, Inc., 1992.

53.02 Building Codes

A. 2015 International Code Council Series of Codes (ICC)

1. ICC. *2015 International Building Code.* 2015 Ed., Country Club Hills, IL: ICC, 2015.
2. ICC. *2015 International Mechanical Code.* 2015 Ed., Country Club Hills, IL: ICC, 2015.
3. ICC. *2015 International Energy Conservation Code.* 2015 Ed., Country Club Hills, IL: ICC, 2015.
4. ICC. *2015 International Plumbing Code.* 2015 Ed., Country Club Hills, IL: ICC, 2015.
5. ICC. *2015 International Fire Code.* 2015 Ed., Country Club Hills, IL: ICC, 2015.
6. ICC. *2015 International Fuel Gas Code.* 2015 Ed., Country Club Hills, IL: ICC, 2015.
7. ICC. *2015 International Residential Code.* 2015 Ed., Country Club Hills, IL: ICC, 2015.
8. ICC. *2015 International Existing Building Code.* 2015 Ed., Country Club Hills, IL: ICC, 2015.
9. ICC. *2015 International Performance Code for Buildings and Facilities.* 2015 Ed., Country Club Hills, IL: ICC, 2015.
10. ICC. *2015 International Private Sewage Disposal Code.* 2015 Ed., Country Club Hills, IL: ICC, 2015.
11. ICC. *2015 International Property Maintenance Code.* 2015 Ed., Country Club Hills, IL: ICC, 2015.
12. ICC. *2015 International Zoning Code.* 2015 Ed., Country Club Hills, IL: ICC, 2015.
13. ICC. *2015 International Wildland-Urban Interface Code.* 2015 Ed., Country Club Hills, IL: ICC, 2015.

B. National Fire Protection Association (NFPA)

NFPA. *NFPA 1 Fire Code.* Quincy, MA: NFPA, 2015.

NFPA. *NFPA 10 Standard for Portable Fire Extinguishers.* Quincy, MA: NFPA, 2013.

NFPA. *NFPA 13 Standard for the Installation of Sprinkler Systems.* Quincy, MA: NFPA, 2013.

NFPA. *NFPA 13 Standard for the Installation of Sprinkler Systems Handbook.* Quincy, MA: NFPA, 2013.

NFPA. *NFPA 14 Standard for the Installation of Standpipe and Hose Systems.* Quincy, MA: NFPA, 2013.

NFPA. *NFPA 16 Standard for the Deluge Foam-Water Sprinkler and Foam-Water Spray Systems.* Quincy, MA: NFPA, 2015.

NFPA. *NFPA 17 Standard for Dry Chemical Extinguishing Systems.* Quincy, MA: NFPA, 2013.

NFPA. *NFPA 17A Standard for Wet Chemical Extinguishing Systems.* Quincy, MA: NFPA, 2013.

NFPA. *NFPA 20 Standard for the Installation of Stationary Pumps for Fire Protection.* Quincy, MA: NFPA, 2013.

NFPA. *NFPA 24 Standard for the Installation of Private Fire Service Mains and Their Appurtenances.* Quincy, MA: NFPA, 2013.

NFPA. *NFPA 30 Flammable and Combustible Liquids Code.* Quincy, MA: NFPA, 2015.

NFPA. *NFPA 31 Standard for the Installation of Oil-Burning Equipment.* Quincy, MA: NFPA, 2011.

NFPA. *NFPA 45 Standard on Fire Protection for Laboratories Using Chemicals.* Quincy, MA: NFPA, 2015.

NFPA. *NFPA 52 Vehicular Gaseous Fuel Systems Code.* Quincy, MA: NFPA, 2013.

NFPA. *NFPA 54 National Fuel Gas Code.* Quincy, MA: NFPA, 2015.

NFPA. *NFPA 54 National Fuel Gas Code Handbook.* Quincy, MA: NFPA, 2015.

NFPA. *NFPA 55 Compressed Gases and Cryogenic Fluids Code.* Quincy, MA: NFPA, 2013.

NFPA. *NFPA 58 Liquefied Petroleum Gas Code.* Quincy, MA: NFPA, 2014.

NFPA. *NFPA 70 National Electrical Code.* Quincy, MA: NFPA, 2014.

NFPA. *NFPA 70 National Electrical Code Handbook.* Quincy, MA: NFPA, 2014.

NFPA. *NFPA 72 National Fire Alarm and Signaling Code.* Quincy, MA: NFPA, 2013.

NFPA. *NFPA 75 Standard for the Protection of Information Technology Equipment.* Quincy, MA: NFPA, 2013.

NFPA. *NFPA 76 Standard for the Protection of Telecommunications Facilities.* Quincy, MA: NFPA, 2012.

NFPA. *NFPA 88A Standard for Parking Structures.* Quincy, MA: NFPA, 2015.

NFPA. *NFPA 90A Standard for the Installation of Air-Conditioning and Ventilating Systems.* Quincy, MA: NFPA, 2015.

NFPA. *NFPA 90B Standard for the Installation of Warm Air Heating and Air-Conditioning Systems.* Quincy, MA: NFPA, 2015.

NFPA. *NFPA 91 Standard for Exhaust Systems for Air Conveying of Vapors, Gases, Mists, and Particulate Solids.* Quincy, MA: NFPA, 2015.

NFPA. *NFPA 92A Standard for Smoke Control Systems.* Quincy, MA: NFPA, 2015.

NFPA. *NFPA 92B Standard for Smoke Management Systems in Malls, Atria, and Large Spaces.* Quincy, MA: NFPA, 2009.

NFPA. *NFPA 96 Standard for Ventilation Control and Fire Protection of Commercial Cooking Operations.* Quincy, MA: NFPA, 2014.

NFPA. *NFPA 99 Health Care Facilities Code.* Quincy, MA: NFPA, 2015.

NFPA. *NFPA 101 Life Safety Code.* Quincy, MA: NFPA, 2015.

NFPA. *NFPA 110 Standard for Emergency and Standby Power Systems.* Quincy, MA: NFPA, 2013.

NFPA. *NFPA 204 Standard for Smoke and Heat Venting.* Quincy, MA: NFPA, 2015.

NFPA. *NFPA 211 Standard for Chimneys, Fireplaces, Vents, and Solid Fuel-Burning Appliances.* Quincy, MA: NFPA, 2013.

NFPA. *NFPA 214 Standard on Water-Cooling Towers.* Quincy, MA: NFPA, 2011.

NFPA. *NFPA 318 Standard for the Protection of Semiconductor Fabrication Facilities.* Quincy, MA: NFPA, 2015.

NFPA. *NFPA 418 Standard for Heliports.* Quincy, MA: NFPA, 2011.

NFPA. *NFPA 750 Water Mist Fire Protection Systems.* Quincy, MA: NFPA, 2015.

NFPA. *NFPA 900 Building Energy Code.* Quincy, MA: NFPA, 2013.

NFPA. *NFPA 909 Code for the Protection of Cultural Resource Properties—Museums, Libraries, and Places of Worship.* Quincy, MA: NFPA, 2013.

NFPA. *NFPA 914 Code for Fire Protection of Historic Structures.* Quincy, MA: NFPA, 2015.

NFPA. NFPA 5000 Building Construction and Safety Code. Quincy, MA: NFPA, 2015.

C. Miscellaneous

The Facility Guidelines Institute. *Guidelines for Design and Construction of Hospitals and Outpatient Facilities.* Washington, D.C.: American Hospital Association, 2014.

The Facility Guidelines Institute. *Guidelines for Design and Construction of Residential Health, Care, and Support Facilities.* Chicago, IL: American Society for Healthcare Engineering, 2014.

The American Institute of Architects Center for Advanced Technology Facilities Design. *Guidelines and Planning and Design of Biomedical Research Laboratory Facilities.* Washington, D.C: The American Institute of Architects' Press, 1999.

American Industrial Hygiene Association and The American National Standards Institute. *American National Standard—Laboratory Ventilation.* Fairfax, VA: American Industrial Hygiene Association, 2003.

Air Conditioning Contractors' Association. *Manual J Residential Load Calculations.* Version 2.10, Arlington, VA: Air Conditioning Contractors' Association, 2011.

Mower, Joe. *Updating Your Old Steam Heating System Using Modern Components.* Shippensburg, PA: Burd Street Press, 2003.

Associated Air Balance Council. AABC *Commissioning Guideline for Building Owners, Design Professionals, and Commissioning Service Providers.* Washington, D.C.: Associated Air Balance Council, 2002.

The Pennsylvania Housing Research/Resource Center (PHRC). *Pennsylvania's Alternative Residential Energy Provisions.* University Park, PA: PHRC, 2003.

Index